Thermodynamics and Statistical Mechanics

Thermodynamics and Statistical Mechanics

An Integrated Approach

ROBERT J. HARDY and CHRISTIAN BINEK

Department of Physics, University of Nebraska-Lincoln, USA

WILEY

This edition first published 2014

Registered office
John Wiley & Sons Ltd, The Atrium, Southern Gate, Chichester, West Sussex, PO19 8SQ, United Kingdom

For details of our global editorial offices, for customer services and for information about how to apply for permission to reuse the copyright material in this book please see our website at www.wiley.com.

Library of Congress Cataloging-in-Publication Data

Hardy, Robert J., 1935- author.
Thermodynamics and statistical mechanics : an integrated approach / Professor Robert J. Hardy and Professor Christian Binek.
pages cm
Includes index.
ISBN 978-1-118-50101-6 (cloth)–ISBN 978-1-118-50100-9 (pbk.) 1. Thermodynamics–Textbooks. 2. Statistical mechanics–Textbooks. I. Binek, Christian, author. II. Title.
QC311.H3273 2014
536′.7 – dc23

2013039738

A catalogue record for this book is available from the British Library.

ISBN: 9781118501009

Set in 10/12pt TimesLTStd by Laserwords Private Limited, Chennai, India
Printed and bound in Great Britain by TJ International Ltd, Padstow, Cornwall

3 2014

Contents

Supplementary material including detailed worked solutions can be downloaded online at http://booksupport.wiley.com

Preface

This is a text on the fundamentals of thermal physics and includes both the macroscopic and microscopic aspects of the subject. Thermodynamics and statistical mechanics, like classical mechanics, electricity and magnetism, and quantum mechanics, are axiomatic theories based on small numbers of postulates. The two approaches to the subject give very different insights into the nature of thermal phenomena, supply different tools of analysis, and are essential research tools of the practicing physicist. Thermodynamics establishes the basic concepts for describing observed phenomena and establishes general relationships that are exact and independent of assumptions about the structure of matter. Statistical mechanics gives the microscopic basis of thermal behavior and supplies the tools for determining the properties of systems from their atomic level characteristics. The text is ideal for three types of courses: a one-semester course for undergraduates, a two-semester course for advanced undergraduates, and a statistical mechanics course for beginning graduate students. The book is written on the principle that a good text is also a good reference. By combining the macroscopic and microscopic approaches in a single book, we give students studying statistical mechanics a readily available source for reviewing the macroscopic concepts and relationships used in atomic-level studies.

We present thermodynamics and statistical mechanics as self-contained theories. Many texts merge the macroscopic and microscopic approaches and select the approach according to the result being sought. Although practicing scientists often mix concepts from different theories when investigating new situations, it is not the best approach when teaching a subject. It leaves students with a hodgepodge of ideas instead of a coherent understanding of fundamentals. Once understood, the fundamentals are available for a wide range of applications.

Our presentation of the macroscopic aspects of the subject is tailored to the backgrounds of modern physics students. The logic of the Clausius–Kelvin formulation of thermodynamics is used, but our presentation is not traditional. We focus on the properties of equilibrium state space and utilize the student's understanding of multidimensional spaces obtained from their studies of mechanics, electricity, and magnetism. The fundamental macroscopic concepts are developed while keeping in mind the student's previous exposure to the atomic-level structure of matter.

Another unique aspect of our book is the treatment of the mathematics involved. We assume that students have completed the standard introductory sequence of calculus courses through partial differentiation and multidimensional integration. Nevertheless, we find that many students, even those who did well in those courses, have difficulties with the mathematics of thermal physics. This is especially true of thermodynamics, partially because they have not seen some of the methods used outside of their mathematics courses and partially because of the notation used. To overcome these difficulties, the essential mathematical concepts are reviewed before using them and the similarity of the mathematics to that employed in the other fields of physics is emphasized.

The book is divided into five parts. Part I introduces the concepts of thermal physics by applications to gases, liquids, and solids while making use of the first law of thermodynamics. Everyday phenomena are used to illustrate the concepts involved. The logical development of thermodynamics is the subject of Part II. The macroscopic concept of entropy is deduced and its importance for addressing concerns about energy conservation is outlined. The thermodynamic potentials are obtained and their role as succinct summaries of the equilibrium properties of systems is described.

Part III acts as a bridge between the macroscopic and microscopic approaches by presenting some intuitive descriptions of the atomic-level origins of thermodynamic behavior. Maxwell's velocity distribution is derived, and Boltzmann's microscopic formula for entropy is introduced.

Statistical mechanics is developed in Parts IV and V and begins with the derivation of the microcanonical, canonical, and grand canonical ensembles. We then develop the general methods for predicting the properties of systems from the microscopic description contained in their Hamiltonians. Starting from the fundamentals, we analyze ideal gases, the Einstein model, harmonic solids, fluctuations, paramagnetism, blackbody radiation, Fermi–Dirac statistics, and Bose–Einstein condensations.

We know from our experience as teachers that a semester is never long enough to present all of the topics we would like to discuss. To facilitate the selection of material, the less essential (but still important) topics are placed toward the ends of chapters. As studying a topic once does not automatically make it part of a student's working knowledge, we make frequent references to results obtained earlier. We are grateful to our students and colleagues for their many suggestions for ways to clarify our presentation of the subject.

We recommend teaching thermodynamics soon after the beginning sequence of courses and before the typical physics curriculum focuses on atomic and subatomic phenomena. A one-semester course for undergraduates can be based on Parts I and II with selections from Part III to introduce the microscopic origins of thermal behavior. The prerequisites for Parts I, II, and III are the standard two- or three-semester sequence of calculus-based physics courses for incoming majors. Since the thermodynamic potentials and the properties of entropy are fundamental to thermodynamics, the course should definitely include the first three sections of both Chapters 11 and 12. However, if time is running short, we recommend skipping the later sections of Part II and going to first chapter of Part III.

Parts IV and V form the core of a one-semester graduate course on statistical mechanics. The prerequisite is an understanding of quantum mechanics through familiarity with the eigenstates of the Schrödinger equation. The first two chapters of Part III should be included as an introduction to the microscopic aspects of statistical physics. Because of the diversity in the undergraduate preparations of beginning graduate students, we recommend beginning with a review of the technical meanings of terms from Chapter 1 and the properties of materials from Chapter 3. Because of their importance of in statistical mechanics, we recommend a brief review the thermodynamic potentials, as presented in the first three sections of Chapter 12. All five parts of the book make an excellent text for a two semester course for advanced undergraduates.

Part I
Elements of Thermal Physics

Thermal physics is concerned with systems that contain large numbers of particles, molecules, atoms, ions, electrons, photons, etc. Thermodynamics describes their *macroscopic* behavior, while the *microscopic* origin of thermal behavior is described by statistical mechanics. Since familiarity with the phenomena involved at the macroscopic level is essential for understanding their microscopic origins, we start with the macroscopic aspects while keeping in mind the atomic-level nature of the systems being studied. Part I introduces the concepts of thermal physics by applications to gases, liquids, and solids, and makes extensive use of the principle of energy conservation as expressed by the first law of thermodynamics. Part II focuses on the second law and the property of entropy.

Thermodynamics and Statistical Mechanics: An Integrated Approach, First Edition.
Robert J Hardy and Christian Binek.

1
Fundamentals

As many of the terms of thermal physics are also used in everyday speech, it is important to understand their more restricted technical meanings. It is especially important to understand the fundamental distinction between heat and work, which describe the interaction of systems with each other, and the intrinsic property of individual systems called internal energy. Other important concepts are equilibrium, state, state function, process, and temperature.

1.1 *PVT* Systems

Pressure P, volume V, and temperature T are essential properties of solids, liquids, and gases. In scientific usage both liquids and gases are fluids, *i.e.*, substances that flow to conform to the shape of their container. Before introducing two relatively simple mathematical models of *PVT* systems, we review the units used in the description of thermodynamic behavior.

Units

The SI units for pressure, volume, and temperature are the pascal (Pa) named after Blaise Pascal, the cubic meter (m^3), and the kelvin (K) named after Lord Kelvin, who is also known as William Thompson. Other commonly used units for volume are the cubic centimeter (cm^3), the liter (L), and the cubic foot (ft^3).

$$1\,\mathrm{m}^3 = 10^3\,\mathrm{L} = 10^6\,\mathrm{cm}^3 = 35.31\,\mathrm{ft}^3. \tag{1.1}$$

One pascal equals one newton per square meter ($1\ \mathrm{Pa} = 1\,\mathrm{N\ m}^{-2}$). In many applications the convenient SI unit for pressure is the kilopascal (kPa). An important reference pressure is the *standard atmosphere* (1 atm), which is approximately 100 kPa.

$$1\ \text{standard atmosphere} = 101.325\,\mathrm{kPa}. \tag{1.2}$$

Thermodynamics and Statistical Mechanics: An Integrated Approach, First Edition.
Robert J Hardy and Christian Binek.

Other units for pressure are the bar (1 bar = 100 kPa), the pound per square inch (lb in^{-2}), the millimeter of mercury (mmHg), and the torr (1 torr = 1 mmHg) named after Evangelista Torricelli. The unit "mmHg" is based on the height of the mercury column in a barometer. These units are related by

$$1\,\text{atm} = 101.3\,\text{kPa} = 760\,\text{mmHg} = 760\,\text{torr} = 14.70\,\text{lb in}^{-2}. \tag{1.3}$$

The unit "atm" is sometimes called an atmosphere. The unit specifies a specific force per unit area and must not be confused with the less precise "atmospheric pressure," which is the pressure of the atmosphere and is only *approximately* equal to 1 atm at elevations not too far above sea level.

The commonly used temperature scales are the Kelvin, Celsius, and Fahrenheit scales. The relationship between a Celsius temperature Θ^{C} and a Kelvin temperature T^{K} is

$$\Theta^{\text{C}} = T^{\text{K}} - 273.15. \tag{1.4}$$

The relationship between a Fahrenheit temperature Θ^{F} (in units of °F) and a Celsius temperature Θ^{C} (in units of °C) is

$$\Theta^{\text{F}} = \frac{9}{5}\Theta^{\text{C}} + 32. \tag{1.5}$$

The symbol T is reserved for temperatures measured on an absolute temperature scale, a concept made precise in Chapters 7 and 8. The symbol Θ is used with other temperature scales. The Kelvin scale is an absolute temperature scale. (The symbol for the unit is K, not °K). The Celsius and Fahrenheit scales are not absolute scales. Although the numerical values of temperatures are different in units of K and °C, the values of the *temperature differences* are the same,

$$\Delta T^{\text{K}} = \Delta\Theta^{\text{C}}. \tag{1.6}$$

As indicated in (1.5), a temperature difference of five Celsius degrees is equivalent to a difference of nine Fahrenheit degrees.

Standard temperature and pressure (STP) refers to 273.15 K (0 °C) and 1 standard atmosphere. Room temperature and atmospheric pressure are not precisely defined. This book considers "room temperature" to be 300 K and "atmospheric pressure" to be 100 kPa.

Internal energy U is the thermodynamic property of a system that represents the sum of the kinetic and potential energies of its microscopic constituents. The SI unit for energy is the joule (J) named after James Prescott Joule. In thermal physics the calorie (cal) and kilocalorie (kcal) are convenient units for energy. The food Calorie (written with a capital C) is actually a kilocalorie. The British thermal unit (Btu) is used in engineering. The units of energy are related by

$$4184\,\text{J} = 1\,\text{kcal} = 1000\,\text{cal} = 3.968\,\text{Btu}. \tag{1.7}$$

The internal energy and pressure of PVT systems can be expressed as functions of temperature and volume.[1]

$$U = U(T, V) \quad \text{and} \quad P = P(T, V). \tag{1.8}$$

[1] Equations like $V = V(T, P)$ and $U = U(T, P)$ are *constitutive relations*. The relationship $V = V(T, P)$ is sometimes called the *thermal equation of state* and $U = U(T, P)$ is called the *caloric equation of state*. In this book the phrase "equation of state" is used without modifier and refers to a relationship between P, V, and T.

It is helpful when explaining the concepts of thermodynamics to have explicit expressions for these functions. Ideal gases and simple solids are useful for this purpose.

Ideal Gases

A gas is a collection of molecules that move about in random directions with occasional collisions with each other and with the walls of their container. In many applications real gases are accurately modeled as ideal gases, which are also called *perfect gases*. In the ideal gas model the relationship between pressure, volume, and temperature is

$$PV = nRT. \tag{1.9}$$

where the function $P(T, V)$ is nRT/V. This is the *ideal gas equation of state*, and R is the *universal gas constant*.

$$R = 8.314\,\mathrm{J\,mol^{-1}\,K^{-1}} = 1.987\,\mathrm{cal\,mol^{-1}\,K^{-1}} \tag{1.10}$$

and n is the number of moles.

One *mole* is the quantity of material that has a mass in grams equal to its *molecular weight*. The mass m of one molecule of a pure substance is

$$m = (\text{molecular weight}) \cdot m_u = \frac{(\text{molecular weight in grams})}{N_A}, \tag{1.11}$$

where the *atomic mass constant* is

$$m_u = 1.661 \times 10^{-27}\,\mathrm{kg}$$

and the number of molecules per mole is given by *Avogadro's number* named after Amadeo Avogadro.

$$N_A = 6.022 \times 10^{23}\mathrm{mol^{-1}}. \tag{1.12}$$

The internal energy of an ideal gas depends on temperature only, which is a special case of the relationship indicated in (1.8). As there is no dependence on volume, it is expressed by

$$U = U(T). \tag{1.13}$$

It is shown in Chapter 10 that this special case is a consequence of $PV = nRT$. Until then, equations (1.9) and (1.13) are considered the defining characteristics of an ideal gas.

Simple Solids

More than one type of model system is needed to illustrate the generality of thermodynamics. The functions for the internal energy and pressure of another type of system are

$$U = Mc_oT + b_oV\left[\ln\left(\frac{V}{V_o}\right) - 1\right] \tag{1.14}$$

and

$$P = a_oT - b_o\ln\left(\frac{V}{V_o}\right), \tag{1.15}$$

where a_o, b_o, c_o, and V_o are constants and M is the mass of the system. Although less significant that the ideal gas model, the above equations give a useful description of the behavior of solids over the limited ranges of temperatures and pressures of interest in many applications. For this reason a system described by (1.14) and (1.15) will be referred to as a *simple solid*.

It is shown in Chapter 3 that the above equations describe the internal energy and pressure of a system whose coefficient of thermal expansion α_V, bulk modulus B_T, and specific heat c_V^M are constant. Their values for a few solids are given in Table 1.2 on page 15. The equations that relate the a_o, b_o, and c_o to α_V, B_T, and c_V^M are $a_o = \alpha_V B_T$, $b_o = B_T$, and $c_o = c_V^M$ (see (3.28)). The constant V_o is chosen so that (1.15) gives the pressure at some specific temperatures, such as $P = 100\,\text{kPa}$ and $T = 300\,\text{K}$.

The significant differences in the functions for internal energy and pressure for ideal gases and simple solids reflect differences in their microscopic structures. The behavior of solids is dominated by the forces that bind the atoms into a rigid structure. These forces can be obtained from a potential energy function. An approximate expression for the internal energy function is

$$U(T, V) = Mc_oT + \Phi(V). \tag{1.16}$$

The term Mc_oT describes the kinetic and potential energies associated with the vibrational motion of the atoms. The *static energy* $\Phi(V)$ is the energy or the system when the atoms are stationary. The volume of a solid is changed by subjecting it to high pressure. The contribution to pressure resulting from the changes to $\Phi(V)$ is $-d\Phi/dV$. An approximate expression for the pressure is

$$P = a_oT - \frac{d\Phi}{dV}. \tag{1.17}$$

The term a_oT accounts for the tendency of solids to expand as temperature increases. The static energy $\Phi(V)$ is represented in the simple solid model by the term $b_oV[\ln(V/V_o) - 1]$ in (1.14), which implies that $d\Phi/dV = b_o \ln(V/V_o)$. Combining this with (1.17) yields the expression for pressure in (1.15).

1.2 Equilibrium States

System

Many of the systems that will be studied are homogeneous (uniform composition) and isotropic (same in all directions), such as the gas or liquid in a container or a sample of dielectric or magnetic material. Nevertheless, the concepts of thermal physics are applicable to much more complex systems, sometimes referred to as *devices*, such as engines, refrigerators, and electrical generating plants.

A system may be made up of smaller systems, called *subsystems*. We may refer to something as a subsystem at one time and as a system at another time. Sometimes we concentrate on one system and refer to everything it interacts with as its *surroundings*, or the *environment*. Since the focus of a discussion often changes, it is important to clearly identify the system to which the principles of thermodynamics are being applied.

A system with a fixed quantity of matter is a *closed* system, and one that can exchange matter with its surrounding is an *open* system. An automobile engine, which takes in fuel and air and exhausts combustion products, is an open system. The mass of a closed system is constant, and in the absence of chemical reactions the numbers of moles of its different constituents are constant.

Equilibrium

A system is in equilibrium if it does not spontaneously change. A glass of warm water with a cube of ice in it is *not* in equilibrium, as the ice will spontaneously melt until either the water cools to 0.0 °C or all of the ice melts. The tendency of systems to spontaneously change until their temperatures are uniform throughout is a widely observed phenomenon. Systems in *thermal equilibrium* have a uniform temperature.

A system can be in equilibrium even when it is changing, provided the change is caused by a change in the *constraints* on it. Consider the gas confined in a cylinder with a piston at one end, as diagrammed in **Figure 1.1**. The temperature and pressure of the gas will change when the piston is moved. The gas is considered to be in equilibrium as it changes, because it is not changing spontaneously. However, this is only true if the piston is not moved too rapidly. If moved fast enough, the motion will create sound waves that move back and forth within the cylinder, so that the gas will not be in equilibrium until the sound waves die out.

A system may not be changing and yet *not* be in equilibrium. This is the case when its condition is maintained by a continuous exchange of energy with its surroundings. For example, the pane of glass that separates the warm interior of a building from the cold exterior is in a *steady state*, provided the inside and outside temperatures are steady, but the glass is *not* in equilibrium. To maintain the temperature distribution in the glass, heat must be continuously transferred to it from the room and removed by the air outside. If the pane of glass is removed from the wall and isolated, the temperature distribution in it will spontaneously change until it is uniform.

The concept of equilibrium is also important in mechanics. An object is in *mechanical equilibrium* when its translational and rotational velocities do not spontaneously change,

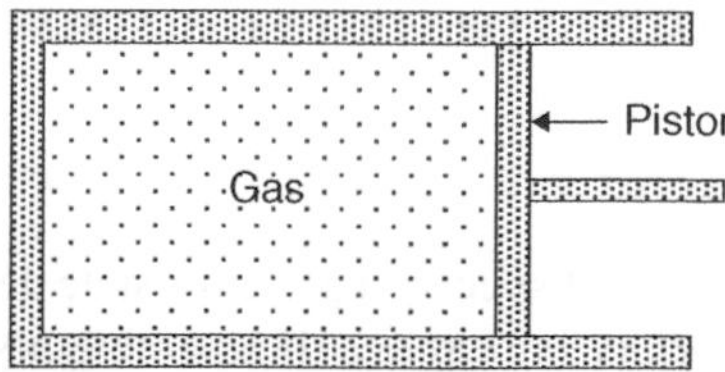

Figure 1.1 *Gas Confined in a Cylinder*

i.e., when its linear acceleration and its angular acceleration are zero. The conditions that must be satisfied for an object to be in mechanical equilibrium are

$$\sum_i \boldsymbol{F}_i = 0 \quad \text{and} \quad \sum_i \boldsymbol{\tau}_i = 0, \tag{1.18}$$

where $\boldsymbol{F}_i$ and $\boldsymbol{\tau}_i$ are the external forces and torques on the object.

State

The concept of state is fundamental but not easy to define. The different *states* of a system refer to different conditions or configurations. For example, a ball could be resting on a table or moving across the room. These mechanical states are described by giving its position, velocity, and angular momentum. The state of a system in quantum mechanics is described by its wave function, which is also called a state function. In thermodynamics we are interested in the *thermodynamic states* of systems, which are also called *equilibrium states*.

The number of properties needed to specify the equilibrium state of a system is the number of *thermodynamic degrees of freedom*. The number depends on the system involved and the type of behavior being investigated. A simple *PVT* system has two degrees of freedom, which can be specified by giving its temperature and volume (T, V) or its temperature and pressure (T, P). Other choices such as internal energy and volume are also used. A homogeneous fluid whose dielectric properties are being investigated has three thermodynamic degrees of freedom: In addition to temperature and pressure, we need to specify the strength of the electric field $\mathcal{E}$.

State Function

Properties whose values are determined by the system's state are called *state functions*. Temperature T, volume V, pressure P, internal energy U, entropy S, enthalpy H, etc. are state functions. When the state of a system is specified by T and V, the values of P, U, S, and H are given by functions $P(T, V)$, $U(T, V)$, $S(T, V)$, and $H(T, V)$. The independent variables T and V are the *thermodynamic coordinates* of the system. The number of thermodynamic coordinates, *i.e.*, the number of independent variables, equals the number of thermodynamic degrees of freedom. State functions are also called state variables. It should be noted that heat Q and work W are *not* state functions. They are process dependent quantities.

State Space

The thermodynamic coordinates that specify a system's state define a space called *thermodynamic state space* or *equilibrium state space*. If no ambiguity results, it is simply called *state space*. As the state space of a homogeneous *PVT* system has two coordinates, we can represent its states by points on a two dimensional graph. More coordinates are required for more complex systems. For example, in a system of ice and water, a third coordinate is required to specify the fraction of the system that is ice (or water).

Internal Energy

When viewed microscopically, thermodynamic systems are collections of electrons, atoms, and molecules. The energy associated with these microscopic constituents is determined by the velocities and positions of the huge number of constituent particles, which is typically of the order of Avogadro's number. In contrast, when in thermal equilibrium, the system's internal energy is determined by its thermodynamic coordinates which are few in number. The existence of the internal energy state function is fundamental to thermodynamics.

Equation of State

The relationship between pressure, volume, and temperature in a *PVT* system is called the *equation of state* and can be expressed as $P = P(T, V)$, $V = V(T, P)$, or $T = T(P, V)$. A general form that treats the properties symmetrically is

$$F(P, V, T) = 0. \tag{1.19}$$

For ideal gases the explicit expressions for the functions $P(T, V)$, $V(T, P)$, and $T(P, V)$ are nRT/V, nRT/P, PV/nR, and PV/nR, respectively. The expression for $F(P, V, T)$ is $(PV - nRT)$, and the equation of state is usually written as $(PV = nRT)$.

Example 1.1 $V(T, P)$ and $U(T, P)$ for simple solids

Find the functions that give the dependence of the volume and internal energy of a simple solid on its temperature and pressure.

Solution. According to (1.14) and (1.15), the functions for the pressure and internal energy of a simple solid are

$$P(T, V) = a_o T - b_o \ln(V/V_o)$$

and

$$U(T, V) = M c_o T + b_o V[\ln(V/V_o) - 1].$$

To find the dependence V and U on the system's state when the thermodynamic coordinates are (T, P), instead of (T, V), we rewrite the above expression for pressure as $\ln(V/V_o) = (a_o T - P)/b_o$. It then follows that the function for the dependence of volume on temperature and pressure is

$$V(T, P) = V_o e^{(a_o T - P)/b_o}. \tag{1.20}$$

The function that gives the dependence of the internal energy on temperature and pressure is $U(T, P) = U(T, V(T, P))$. Substituting the expression for $V(T, P)$ into the expression for $U(T, V)$ gives

$$U(T, P) = M c_o T + (a_o T - P - b_o) V_o e^{(a_o T - P)/b_o}. \tag{1.21}$$

The significant difference in the appearance of the expressions for $U(T, V)$ and $U(T, P)$ illustrates the extent to which the functional form of state functions depend on the choice of thermodynamic coordinates.

1.3 Processes and Heat

A process starts at an initial time t_0 and ends at some final time t_f ($t_0 < t_f$) and brings about a change in the state of the system. Thermodynamics is concerned with the direction in which processes evolve but not in its rate of evolution. Although the system is often not in equilibrium at all times throughout a process, it is usually assumed to be in thermal equilibrium at the start and end.

Equilibrium and Quasi-Static Processes

A process in which the system is in equilibrium at all times between time t_0 and time t_f is an *equilibrium process*. If at any time during the process the system is not in equilibrium, it is a *nonequilibrium process*. In equilibrium processes, a system changes in response to changes in the constraints and stops changing when the constraints stop changing. Many processes can be idealized as equilibrium processes, provided they proceed sufficiently slowly, where the meaning of slowly depends on the process. For example, the compression of the air-fuel mixture in the cylinders of an automobile engine, which occurs in a time of the order of a hundredth of a second, is often idealized as an equilibrium process. In contrast, the melting of an ice cube placed in a glass of water, which may take several minutes, is not an equilibrium process.

Equilibrium processes are sometimes called quasi-static processes. A *quasi-static* (almost-unchanging) process is one that is performed slowly enough that the system is effectively in equilibrium at all times. The term "equilibrium process" is preferred to "quasi-static process," because it suggests the useful characteristic of being representable by a line in state space.

As the state space of a *PVT* system is two dimensional, an equilibrium process can be represented by a line on a two dimensional graph. When the thermodynamic coordinates are pressure and volume, the graph is called a *PV*-diagram. Some important equilibrium processes in ideal gases are shown in **Figure 1.2**. The point "0" represents the initial state of the gas. The points labeled "*f*" represent the final states. An *isothermal process* is a

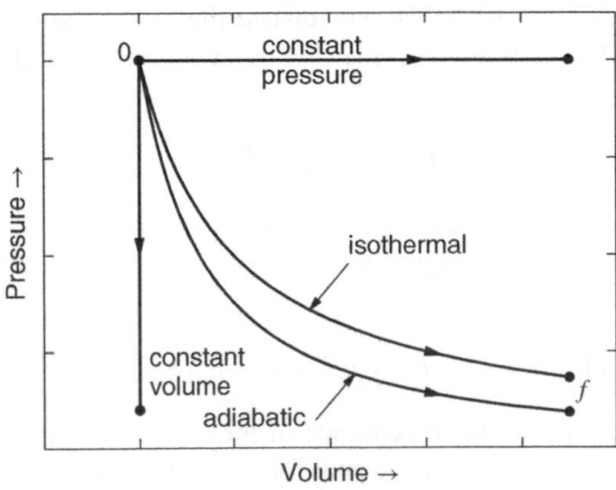

Figure 1.2 *PV-Diagram of Equilibrium Processes*

constant-temperature process, and a line at constant temperature is an *isotherm*. An *isobaric process* is a constant-pressure process, and a line at constant pressure is an *isobar*. A constant-volume process in a *PVT* system is sometimes called an *isochoric process*, which indicates that no work is done. Adiabatic processes are discussed below. As the points on lines represent equilibrium states, nonequilibrium processes are not representable by lines in state space.

Reversible Processes

A *reversible process* is one whose direction in time can be reversed. A process that starts in state "0" and proceeds along a line in state space to state "*f*" is reversed by starting in state "*f*" and traversing the same line in the opposite direction. Processes that can be idealized as reversible are especially important in thermodynamics. In practice, a reversible process is made to evolve in the reversed direction by a small change in the constraints on the system. Although equilibrium, quasi-static, and reversible processes have different defining characteristics, processes with the characteristics of one type also have the characteristics of the other types.

Heat

The word "heat" is familiar to us from everyday life. Although its technical meaning is consistent with colloquial usage, it is much more restricted. Unlike temperature and internal energy, heat does not describe the condition of a system, but instead describes what is happening during a process. The heat transferred to a system is represented by Q. By convention, Q is positive when the energy is transferred *to* the system and negative when energy is transferred *from* it. The heat transferred *to* one system *from* another is the negative of the heat transferred *to* the second system *from* the first.

We can increase the temperature of water by putting it in a container over a flame or on a hot plate. We can cool a hot object by putting it in cold water. When doing these things, something appears to pass between the water and its surroundings. That something is heat: *Heat is the energy transferred from one system to another because of a temperature difference.*

Adiabatic Processes

Adiabatic processes are important both in applications and in establishing the foundations of thermodynamics. *An adiabatic process is one in which no heat is accepted or rejected by the system at any time during the process.* An *adiabatically isolated* system is one that cannot exchange heat with its surroundings. In everyday speech we would say the system is perfectly insulated. The contents of a high quality Dewar flask, named after its inventor Sir James Dewar, is an example of an adiabatically isolated system. Dewar flasks are double-walled containers with the space between the walls evacuated to prevent the transfer of heat by conduction, and the inside surfaces of the walls are silvered to minimize heat transfer by radiation.

Adiabatic processes can be either reversible or irreversible. When a gas is compressed with no exchange of heat with its container, the process in the gas is both reversible and adiabatic. In contrast, when some ice is added to the hot water in a Dewar flask, the process in the contents of the flask is adiabatic but *not* reversible. Interfaces across which heat *cannot* be exchanged are *adiabatic boundaries*. Interfaces across which heat *can* be exchanged are sometimes called *diathermic boundaries*.

1.4 Temperature

Thermally Interacting Systems

Systems that can exchange heat with each other are in *thermal contact*, and a system with no adiabatically isolated parts is a *thermally interacting system*. When in thermal equilibrium, a thermally interacting system possesses a single temperature. In contrast, systems made up of two or more adiabatically isolated subsystems can have more than one temperature. For example, the contents of a Dewar flask and the space outside can be thought of as a single system. Although the system may be in equilibrium, *i.e.*, not spontaneously changing, it will possesses two temperatures, *i.e.* the temperature of the contents and the temperature of the space outside.

Temperature

The concepts of heat and temperature are closely related. Temperature is a state function that describes the tendency of a system (or subsystem) to exchange heat with other systems: (i) All systems in thermal contact have the same temperature when in thermal equilibrium and no heat is exchanged between them. (ii) Systems with different initial temperatures spontaneously exchange heat when brought into thermal contact. The direction of the exchange is described by saying that heat is transferred from the *hotter* system to the *colder* system. On the Kelvin, Celsius, and Fahrenheit temperature scales the hotter system has the higher temperature. The data in Table 1.1 illustrates the wide range of temperatures possessed by various systems.

Thermometers

A thermometer is a device that possesses an equilibrium property that is easily observed and depends on the tendency of the device to exchange heat. The significant property of a

Table 1.1 *Miscellaneous temperatures*

1.5×10^7 K	Estimated temperature at the center of sun
5800 K	Approximate temperature of the surface of sun
1809 K	Iron melts (1536 °C)
273.15 K	Ice point (0 °C)
234 K	Mercury freezes (−39 °C)
195 K	Solid carbon dioxide (dry ice) sublimes at 1 atm (−78 °C)
77 K	Nitrogen liquefies at 1 atm (−196 °C)
4.2 K	Helium liquefies at 1 atm
10^{-8} K	Bose–Einstein condensation of Rb gas (first observed in 1996)

liquid-in-glass thermometer is the length of the column of liquid. The significant property in many laboratory thermometers is an electrical resistance or a voltage. To measure the temperature of a system, we bring it into thermal contact with the thermometer and let the combined system come into thermal equilibrium. As the system and thermometer have the same temperature, the system is assigned the temperature of the thermometer.

Local Temperature

We sometimes need to distinguish between the fundamental concept of temperature and the reading on a thermometer. The distinction is made by calling the equilibrium state function an *equilibrium temperature* and the reading on a thermometer a *local temperature*. Systems that are not in equilibrium do not have an equilibrium temperature but may have local temperatures, which in general are different at different locations. The local temperature at a particular location is the temperature a thermometer would indicate if placed in thermal contact at that point. A system in thermal equilibrium has the same local temperature throughout, and its value is the same as the equilibrium temperature. When a system is rapidly changing, very small, or far from equilibrium, it may not be possible to assign values to the local temperature.

Zeroth Law

Two systems in thermal equilibrium with a third will be in thermal equilibrium with each other when brought into contact. This statement is often called the *zeroth law of thermodynamics*. Essentially, it is an assertion that the concept of thermal equilibrium is transitive. Transitive relationships are important in science, but are usually not called laws. For example, an equal arm balance can be used to compare the weights of objects, but it is assumed without calling it a law that two objects that individually balance a third can balance each other. As its unusual numbering suggests, the zeroth law does not have the significance of the first and second laws of thermodynamics.

Heat Reservoirs

A heat reservoir is an idealization of a system that can transfer significant amounts of heat to or from a system with no change in its temperature. It is a source or sink of heat with a fixed tendency to exchange heat. They are also called *heat baths*. For example, to maintain an object at a constant temperature of 0 °C, we can use a mixture of ice and water as a heat reservoir.

1.5 Size Dependence

Intensive Versus Extensive

Thermodynamic properties can be either intensive or extensive. *Intensive properties are independent of the size of the system. Extensive properties are proportional to size*. Temperature T and pressure P are intensive, while mass M, number of moles n, volume V, and internal energy U are extensive. *Extensive properties are additive*. For example, the values of the mass, volume, and internal energy of systems consisting of several subsystems are the sums of the masses, volumes, and internal energies of the subsystems.

When a system is made up of N identical subsystems, the values of its extensive variables (properties) are N times their values in the subsystems, while the values of its intensive variables are the same as the values in the subsystems. The sizes of homogeneous systems can be varied continuously. *When the size of a homogeneous system is scaled by a factor of* λ, *the values of its extensive properties are multiplied by* λ, *while its intensive properties remain unchanged.* For example, if a system whose mass, volume, internal energy, temperature, and pressure are M_0, V_0, U_0, T_0, and P_0 has its size increased by 20%, so that $\lambda = 1.20$, its mass, volume, and internal energy will become $M = 1.2M_0$, $V = 1.2\,V_0$, and $U = 1.2\,U_0$, while its temperature and pressure remain unchanged.

Intensive variables can be formed by dividing one extensive variable by another. For example,

$$\rho = \frac{M}{V}, \tag{1.22}$$

where the *mass density* ρ is intensive while mass M and volume V are extensive. An intensive property, obtained by dividing by the mass M or the number of moles n, is called a *specific* property. For example, depending on the divisor, the *specific volume* of a system is

$$v_M = \frac{V}{M} \quad \text{or} \quad v = \frac{V}{n}. \tag{1.23}$$

The subscript on v_M identifies it as a "per unit mass" property. A "per mole" value is also called a molar value. When expressed in terms of the specific volume V/n, the ideal gas equation of state simplifies to

$$Pv = RT. \tag{1.24}$$

When an extensive property is represented by a capital letter, it is convenient (but not always practical) to represent the associated intensive property by the same letter in lower case. The characteristics of the material that makes up a homogeneous system are specific values that can be listed in tables of material properties, like those in Table 1.2. The intensive property M/n obtained by dividing the mass of a pure substance in grams by the number of moles is the *molecular weight* of the substance.

In beginning physics courses, we learn that terms that are added or subtracted in an equation must all have the same units. Similarly, terms with thermodynamic significance that are added or subtracted in an equation must have the same size dependence, *i.e.*, the same intensive or extensive character. Also, except for functions that are powers, such as $f(x) = cx^n$, the arguments of the functions used should be dimensionless and intensive. Logarithms are a special case.

1.6 Heat Capacity and Specific Heat

Heat capacities describe the change in the temperature of a system that results from a transfer of heat. More heat is required to raise the temperature by (say) 10° when the system is a kilogram of water than when it is a kilogram of aluminum. Heat capacities are found by measuring an initial temperature T_0, transferring a small amount of heat Q to a system, and

Table 1.2 Material properties. The data are for room temperature and atmospheric pressure, except for ice. The coefficients of thermal expansion α_L and α_V and the bulk modulus B_T are defined in Section 3.2

	Aluminum	Copper	Iron	Lead	Ice (0 °C)	Water
ρ (kg m^{-3})	2700.	8960.	7880.	11340.	917.	1000.
c_P^M (kcal kg^{-1} K^{-1})	0.216	0.092	0.106	0.0306	0.502	1.000
c_P^M (kJ kg^{-1} K^{-1})	0.904	0.385	0.444	0.128	2.10	4.184
c_V^M (kJ kg^{-1} K^{-1})	0.863	0.374	0.436	0.120	2.03	
α_L (10^{-6} K^{-1})	23.2	16.6	11.9	29.0	55.	
α_V (10^{-6} K^{-1})	69.6	49.8	35.7	87.0	165.	
B_T (10^9 Pa)	75.	135.	160.	41.	7.7	

then measuring the final temperature T_f. Its heat capacity is $Q/\Delta T$, where $\Delta T = T_f - T_0$. Heat capacities can be associated with a single temperature by introducing successively smaller amounts of heat, which yield successively smaller temperature changes, and extrapolating the values of $Q/\Delta T$ to $Q = 0$. The extrapolated quantity is a state function. The value of the heat capacity depends on the process involved. The *constant-volume* and *constant-pressure heat capacities* are

$$C_V = \lim_{Q\to 0} \frac{Q}{\Delta T}\bigg|_{V=\text{const}} \quad \text{and} \quad C_P = \lim_{Q\to 0} \frac{Q}{\Delta T}\bigg|_{P=\text{const}}, \tag{1.25}$$

where "lim" refers to an extrapolation of experimental measurements.

Heat capacities are extensive. The intensive property formed by dividing them by the number of moles n or the mass M are called *specific heats*. The constant-pressure specific heats are

$$c_P = \frac{C_P}{n} \quad \text{and} \quad c_P^M = \frac{C_P}{M}. \tag{1.26}$$

Specific heats based on moles are commonly used for gases, while those based on mass are used for solids. As the units given indicate whether the specific heat was obtained by dividing by n or by M, the superscript M is often omitted. Specific heats are convenient because their values are independent of the size of the system and thus can be listed in tables of material properties. A few values are given in Table 1.2.

Constant Specific Heats

Although specific heats depend on the state of the system, they often change sufficiently slowly that they can be treated as constants. When constant, the heat transferred in a constant-pressure process is $Q = C_P \Delta T$, which in terms of the specific heats is

$$Q = n c_P \Delta T \quad \text{or} \quad Q = M c_P^M \Delta T. \tag{1.27}$$

Similarly, for a constant-volume process $Q = n c_V \Delta T = M c_V^M \Delta T$. The temperature dependence of the constant-pressure specific heat c_P^M of water at a pressure of 1 atm is shown in **Figure 1.3**. In units of J kg^{-1} K^{-1} the values range from 4178 to 4218. The range is small

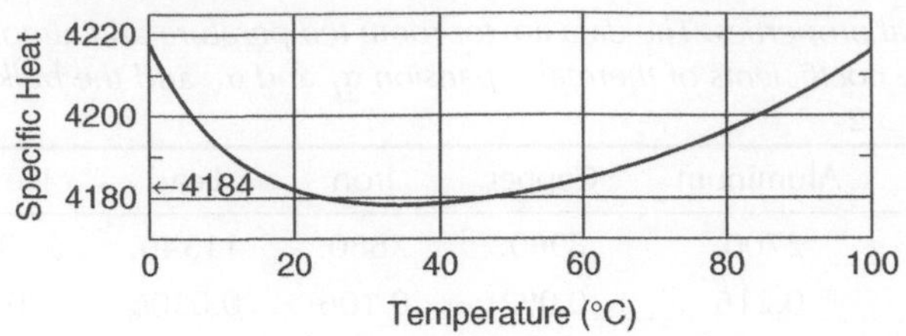

Figure 1.3 *Specific Heat of Water. c_P^M at 1.0 atm in Units of J kg^{-1}K^{-1}*

enough that the specific heat of water is often treated as a constant. The commonly used value for c_P^M is 4184 J kg^{-1} K^{-1}, which in terms of kilocalories is 1.00 kcal kg^{-1} K^{-1}.

Specific heats are measured with a *calorimeter*. Simple calorimeters use an insulated cup of known heat capacity, a thermometer, and a stirring device. A measured quantity of water is placed in the cup, the cup and water are allowed to come to equilibrium, and the temperature T_0 is measured. The mass and initial temperature of a heated sample is measured, the sample is placed in the cup of water, the combined system is allowed to come to equilibrium (with the aid of the stirrer), and the final temperature T_f is measured. The following example illustrates the procedure.

Example 1.2 Calorimetry

A 150 g piece of glass at 85 °C is placed in a calorimeter cup containing 500 g of water at 20 °C. The combined system comes to equilibrium at 22.1 °C. Find the specific heat of the glass. The heat capacity of the calorimeter cup is 0.026 kcal K^{-1}. See **Figure 1.4**.

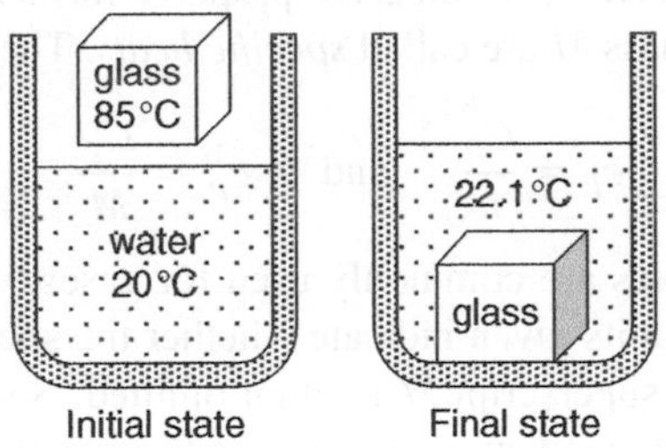

Figure 1.4 *Calorimetry Experiment*

Solution. The system consists of two subsystems: One is the glass (g), the other is the water and calorimeter cup (w + c). In principle, the states of the subsystems depend on T and P. As the experiment is presumably carried out at atmospheric pressure, P is constant, so that temperature is the important variable. In the initial state the subsystems are in equilibrium at different temperatures, while in the final state they are in equilibrium at a single temperature. Assume that no heat is exchanged between the combined system (glass, water, and cup) and the environment. As no heat is exchanged with the surroundings, the heat transferred *to* one subsystem equals the heat transferred *from* the other

(heat *to* water and cup) = (heat *from* glass).

According to the sign convention for Q, this implies that

$$Q^{\mathrm{w+c}} = -Q^{\mathrm{g}},$$

where the superscripts identify the subsystems. With $Q = C_P\Delta T$, $\Delta T = \Theta_f - \Theta_0$, and $C_P = Mc_P$ this becomes

$$(M^{\mathrm{w}}c_P^{\mathrm{w}} + C_P^{\mathrm{c}})(\Theta_f - \Theta_0^{\mathrm{c+w}}) = -M^{\mathrm{g}}c_P^{\mathrm{g}}(\Theta_f - \Theta_0^{\mathrm{g}}).$$

As both subsystems have the same final temperature, a superscript identifying the subsystem is not needed on Θ_f. Solving for the specific heat of the glass gives

$$c_P^{\mathrm{g}} = \frac{(M^{\mathrm{w}}c_P^{\mathrm{w}} + C_P^{\mathrm{c}})(\Theta_f - \Theta_0^{\mathrm{c+w}})}{M^{\mathrm{g}}(\Theta_0^{\mathrm{g}} - \Theta_f)}.$$

or

$$c_P^{\mathrm{g}} = \frac{[(0.5\,\mathrm{kg})(1.0\,\mathrm{kcal\,kg^{-1}K^{-1}}) + 0.026\,\mathrm{kcal\,K^{-1}}](22.1 - 20\,^\circ\mathrm{C})}{(0.15\,\mathrm{kg})(85 - 22.1\,^\circ\mathrm{C})}.$$

Thus, the constant-pressure specific heat of the glass is

$$c_P^{\mathrm{g}} = 0.117\,\mathrm{kcal\,kg^{-1}K^{-1}}.$$

Problems

1.1 Express common room temperature (68 °F) and normal body temperature (98.6 °F) in kelvins and degrees Celsius.

1.2 The gauge pressure in the tires of your car is 210 kPa (30.5 psi) when the temperature is 25 °C (77 °F). Several days later it is much colder. Use the ideal gas equation of state to estimate the gauge pressure in your tires, when the temperature is 0 °C. Give you answer both in kilopascal and pounds per square inch (Gauge pressure is the difference between absolute pressure and atmospheric pressure, which is approximately 100 kPa.)

1.3 When expressed as a function of temperature and volume, the entropy of an ideal gas is given by the function $S(T, V) = nR\ln(T/T_o) + nc_o\ln(V/V_o)$, where n, R, c_o, T_o, V_o, and P_o are constants. Find the entropy $S(T, P)$ of the gas expressed as a function of temperature and pressure. (Remember: $\ln(xy) = \ln x + \ln y$, $\ln(1/z) = -\ln z$, and $\ln 1 = 0$.)

1.4 300 g of aluminum at 90 °C is placed in a calorimeter cup containing 400 g of water. The mass of the copper calorimeter cup 80 g. The initial temperature of the water and cup is 22.0 °C. What is the final temperature?

1.5 A 250 g sample at 80 °C is placed in a calorimeter containing 300 g of water. The mass of the aluminum calorimeter cup is 100 g. The initial temperature of the water and cup is 20.0 °C, and the final temperature is 21.8 °C. What is the specific heat of the sample?

2

First Law of Thermodynamics

The first law tells us that energy is conserved and that there are two ways of transferring it between systems: as heat and as work. Unfortunately, the precise meanings of the terms heat and work are often obscured by their colloquial usage. We begin by reviewing the technical meaning of work which, like energy, is a concept that originates in mechanics.

2.1 Work

Work is the concept that describes how energy is transferred to an object by the forces acting on it. Specifically, the work done by a force F on an object as it moves along path L is given by the line integral

$$W = \int_L F \cdot dx. \tag{2.1}$$

The work-energy theorem equates the work done on an object to the change in its kinetic energy, and potential energy is introduced as the negative of the work done by different kinds of forces. By using Newton's third law, we find that the work done *by* an object 1 *on* object 2 is the negative of the work done *by* object 2 *on* object 1. As a result, the total amount of energy is unchanged, *i.e.*, energy is conserved. In thermal physics, objects are called systems, and the work done by a system through the *macroscopic* forces it exerts on other systems is represented by W. By convention, W is positive when work is done *by* a system and is negative when work is done *on* it.[1]

There are many ways of doing work in thermodynamics. Some devices (machines) do work by using a rotating shaft to exert a torque through an angle. PVT systems do work through the forces exerted by pressure. Electrical devices do work by transferring energy over the wires connected between them. Electrical energy is treated as work as it can be used to exert a force through a distance.

[1] Different conventions for the sign of W are used in different disciplines. In mechanical engineering W is positive when work is done *by* the system, as it is in this book. In physical chemistry W is positive when work is done *on* the system.

Thermodynamics and Statistical Mechanics: An Integrated Approach, First Edition.
Robert J Hardy and Christian Binek.

Pressure

As pressure is the force per unit area on a surface ($P = F/A$), the magnitude of the force F exerted *on* the piston *by* the fluid in **Figure 2.1** is $F = PA$, where A is the cross-sectional area of the piston. The pressure of fluids at rest is called *hydrostatic pressure* and is considered to be uniform when the change in pressure with increasing depth is negligible compared to the total pressure. The hydrostatic force on a surface in contact with a fluid is perpendicular to the surface and directed toward it. For example, the hydrostatic force exerted by the atmosphere on the front cover of this book is perpendicular to the cover and rotates as the book is rotated. The force on the back cover is in the opposite direction.

The expression for the work done by a fluid can be found by considering a process in which the piston in the figure is moved to the right a short distance Δx which causes a volume change of $\Delta V = A\,\Delta x$. As the force is parallel to the displacement, the increment of work done is

$$F \cdot \Delta x = F\Delta x = (PA)(\Delta V/A) = P\Delta V. \tag{2.2}$$

Since the pressure changes as the volume changes, the work done during a finite process is

$$W = \int F\,dx = \int_{V_0}^{V_f} P(V)\,dV, \tag{2.3}$$

where $P(V)$ specifies the dependence of pressure on volume and V_0 and V_f are the initial an final volumes. Although derived for a special case, this is the general expression for the work done by a PVT system in an equilibrium process. The work done in the special case when the pressure is constant is

$$W = P(V_f - V_0). \tag{2.4}$$

The above comments assume that the pressure in fluids in equilibrium is uniform throughout, which is not strictly true. Because of gravity, pressure decreases with increasing height h as ρgh, where ρ is mass density and g is the acceleration of gravity. However, the decrease is often negligible. For example, a typical pressure of the atmosphere is 10^5 Pa, while the decrease in pressure with each meter increase in elevation is about 13 Pa. (The pressure of the atmosphere must not be confused with the unit of pressure called an atmosphere that is abbreviated "atm.")

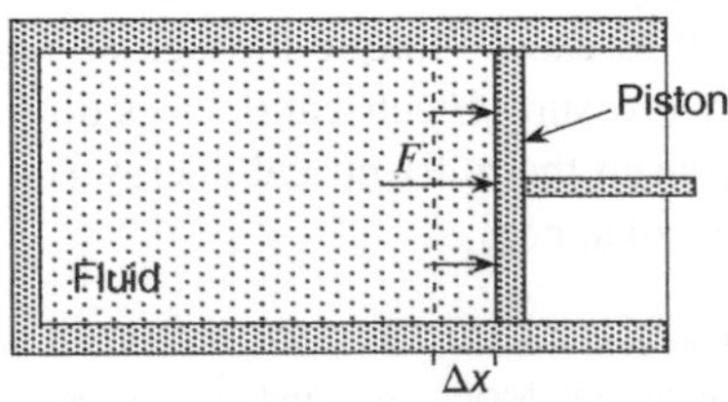

Figure 2.1 *Work done by the Force Exerted by a Fluid*

2.2 Heat

It is essential to use the terms heat, work, internal energy, and temperature correctly. In casual conversation a student might say: "The heat in the classroom was oppressive." Such a statement is technically *incorrect*. As presumably it was the air that was oppressive, the phrase "in the classroom" suggests that some property of the air – some state function – is being referred to. Since heat is *not* a state function, but temperature and internal energy are, the phrase indicates that heat is being confused with temperature, or perhaps internal energy, but it is the temperature that was oppressive. Temperature is intensive and independent of the size of the room, while internal energy is extensive and thus proportional to its size. The internal energy of the air in a cold lecture hall at 0 °C is much greater than the internal energy of the air in a small seminar room at 30 °C (86 °F). To be technically correct the student should have said: "The temperature in the classroom was oppressive."

Heat is described in Chapter 1 as the energy transferred from one system to another because of a temperature difference. That description is useful for determining whether or not heat has been transferred, but it is more fundamental to describe heat without reference to temperature: *Heat is the energy transferred to a system by non-mechanical means.*

The mechanics referred to in the above statement is the mechanics that describes the *macroscopic* behavior of systems. When systems are viewed microscopically, *i.e.*, as collections of particles, both heat Q and work W represent energy that is transferred from one system to another by the forces acting between them. What distinguishes them is the origin of the forces. Heat is the energy transferred by the forces resulting from the random uncontrolled microscopic interactions of the particles in one system with the particles in another. In contrast, work is the energy transferred by forces that are macroscopically observable. The distinction between the two ways of transferring energy is central to thermal physics.

2.3 The First Law

Finding the relationship between heat and energy was one of the crucial steps in the historical development of thermodynamics. It was known that something was being redistributed in calorimetry experiments and that the redistribution could be described quantitatively, as in Example 1.2. However, it was not obvious that the something – called heat – was a form of energy. The identification of heat with energy was made possible by the demonstration that the change in state caused by heat transfers could also be brought about by mechanical means. The changes in the state of a system caused by friction or other dissipative forces can also be brought about by transferring heat to it. These facts established the relationship between energy, work, and heat, *i.e.*, that a state function U exists with the characteristics of energy and that the change in U equals the heat Q transferred to the system minus the work W done by it. The relationship is summarized by the first law:

> ***First Law of Thermodynamics***
> 1. *All systems have an internal energy state function U.*
> 2. *Energy is conserved and is transferred between systems as either heat Q or work W.*

The first law is summarized by the equation

$$Q = \Delta U + W, \tag{2.5}$$

where Q is positive when heat is transferred *to* the system during a process and W is positive when work is done *by* the system. The change in the system's internal energy is $\Delta U = U_f - U_0$, where U_0 and U_f are the initial and final values of U. As ΔU represents a difference, an arbitrary constant can be added to U without changing its significance. This kind of arbitrariness is familiar from mechanics, where the value of the gravitational potential energy Mgh depends on the choice of the level above which height h is measured. An equation that equates the change in the property of a *single* system – a state function – to quantities that describe the interaction of two or more systems is obtained by rewriting (2.5) as

$$\Delta U = Q - W. \tag{2.6}$$

Internal energy U and its change ΔU are extensive. Since terms that are added or subtracted must have the same size dependence, both Q and W are extensive. This means that the values of Q and W for a system consisting of several subsystems are sums of the subsystem values of Q and W. Also, when the size of a homogeneous system is multiplied by λ, the values of Q and W are multiplied by λ. The macroscopic kinetic and potential energies associated with the center-of-mass motion of a system are not included in its internal energy. As a result, when the velocity of the entire system or its gravitational potential energy has changed, the first law becomes

$$Q = W + \Delta U + \Delta E_{\text{macro}}, \tag{2.7}$$

where ΔE_{macro} is the change in the macroscopic energy.

2.4 Applications

The meanings of Q, W, and ΔU are best understood by using them in the analysis of easily visualized examples. We start by considering the free expansion of a gas.

Free Expansions

The apparatus used in a *free expansion experiment* consists of two rigid containers with insulating walls, as diagrammed in **Figure 2.2**. Initially one container is filled with gas, the other evacuated, and the valve between them is closed. The valve is then opened and the gas allowed to flow between the containers. The process is complete when the gas, which now occupies both containers, has come to equilibrium.

The gas is the system we are interested in. To the extent that the walls of the containers are perfectly insulating, no heat is transferred to or from the gas, *i.e.*, the expansion is an adiabatic process. Hence,

$$Q = 0. \tag{2.8}$$

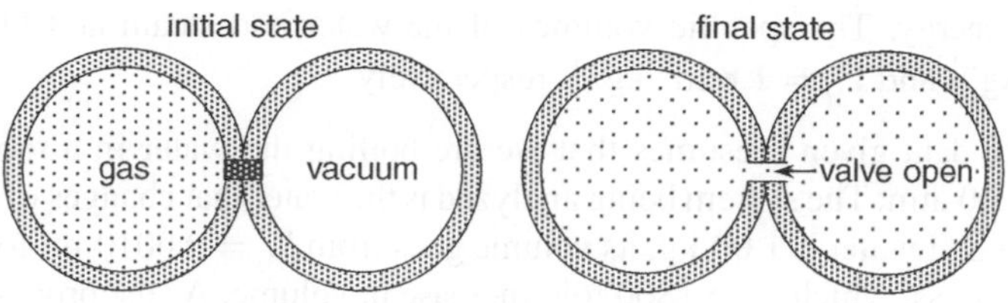

Figure 2.2 Free Expansion Experiment

For a system to do work it must exert a force on something that moves. Some of the gas (the system) exerts forces on other parts of the system, but the first law is not concerned with those forces, as they transfer energy *within* the system. The first law is concerned with forces that are exerted through a distance on *other* systems. Since the walls of the containers are rigid, there is no place where the system can exert a force through a distance on another system. Hence,

$$W = 0. \tag{2.9}$$

Since both Q and W are zero, the first law tells us that

$$\Delta U = Q - W = 0. \tag{2.10}$$

Thus, *the internal energy of a gas remains unchanged in a free expansion.*

Note that the formula $W = \int P(V)\,dV$ is *not* applicable to a free expansion. Although, the volume V of the gas has changed, the change was *not* the result of an equilibrium process. The system (the gas) was not in equilibrium at all times, but changed spontaneously after the valve was opened. Furthermore, the pressure of the gas cannot be described by a function $P(V)$, as it is not uniform while the gas is expanding.

Free expansion experiments were historically important. The temperature of the gas was measured at the start and finish of the experiment, and it was found that the initial and final temperatures were the same ($T_f = T_0$). As U did not change but the volume did, it was concluded that the internal energy was independent of volume. Since other experiments clearly indicate that internal energy depends on temperature, it was concluded that the internal energy of a gas is a function of temperature only, *i.e.*, $U = U(T)$. This is one of the defining characteristics of an ideal gas stated earlier (see 1.13) Although other experiments have shown that the result $T_f = T_0$ is not exactly true, the ideal gas model is sufficiently accurate to be very useful in many applications.

Work Done on the Atmosphere

It may not be obvious that systems open to the atmosphere are doing work when they expand, but they are. Consider what happens when we boil a liter of water.

Example 2.1 Changing Water to Steam

It requires 539 kcal of heat to change 1 kg of water at 100 °C entirely into steam. Find the percentage of the heat that is used to do work, and the percentage that goes to increasing the internal energy. The specific volumes of the water and steam at 100 °C are $v_M^{w} = 0.0010\ \text{m}^3\ \text{kg}^{-1}$ and $v_M^{st} = 1.8\ \text{m}^3\ \text{kg}^{-1}$, respectively.

Solution. The data given presumes that we are boiling the water in a room where the pressure is 1.00 atm. The system being analyzed is the water that changes to steam. As the mass of 1.00 L of water is 1.00 kg, its volume goes from $V_0 = 0.0010\ \text{m}^3$ to $V_f = 1.8\ \text{m}^3$ during the process, which is an 1800 fold increase in volume. As the process is done at a constant pressure of 1.00 atm, the work done is

$$W = P(V_f - V_0) = 1.00\ \text{atm}(1.8\ \text{m}^3 - 0.001\ \text{m}^3) = 1.80\ \text{atm m}^3.$$

Converting this to kilocalories gives

$$W = 1.80\ \text{atm m}^3 \left(\frac{1.013 \times 10^5\text{Pa}}{1\ \text{atm}}\right)\left(\frac{1\ \text{Nm}^{-2}}{1\ \text{Pa}}\right)\left(\frac{1\ \text{J}}{1\ \text{N}\cdot\text{m}}\right)\left(\frac{1\ \text{kcal}}{4184\ \text{J}}\right) = 43.6\ \text{kcal}.$$

The percentage of the heat transferred to the water that does work against the atmosphere is

$$\frac{W}{Q} = \frac{43.6\ \text{kcal}}{539\ \text{kcal}} = 0.0809 = 8.1\%.$$

The percentage of the heat that goes into increasing the internal energy of the water-steam system is

$$\frac{\Delta U}{Q} = \frac{Q - W}{Q} = 1 - \frac{W}{Q} = 1 - 0.0809 = 91.9\%.$$

Dissipative Forces

In classical mechanics, the principle of conservation of energy is limited to situations that involve conservative forces. When friction or other dissipative forces are present, the total amount of kinetic and potential energy decreases, so that mechanical energy is not conserved. The first law of thermodynamics completes the principle of energy conservation by accounting for the decrease in mechanical energy by the increase in internal energy.

A familiar type of dissipative force is the dry friction between objects sliding against each other. When the interface between the objects is flat, the frictional force is parallel to the interface and opposes their relative motion. The magnitude of the frictional force is $f = \mu N$, where N is the normal component of the force between them and μ is the coefficient of sliding friction.

Example 2.2 Dry Friction

Move a block back and forth against a stationary block through a total distance of 160 m. Find the change in the temperature of the blocks. The normal force between the blocks is 500 N, and the coefficient of friction is $\mu = 0.40$. The blocks are made of iron and their combined mass is 2.0 kg.

Solution. The forces on the moving block as it is pushed to the right are shown in **Figure 2.3**. The normal force N is the sum of the downward force F_{down} and the weight of the upper block. As the external force F_{ext} is pushing to the right and

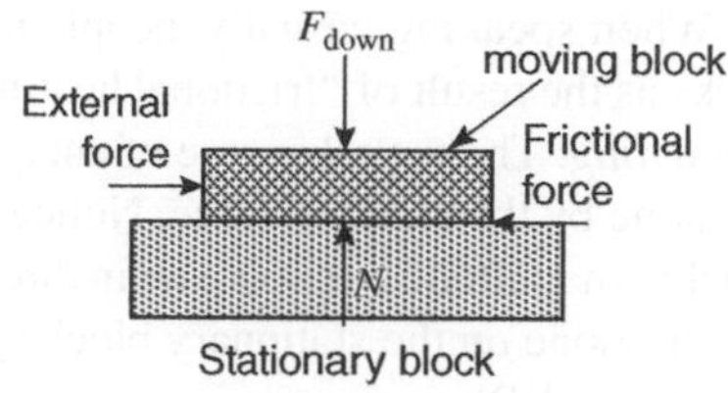

Figure 2.3 *Forces on a Moving Block*

the frictional force f is pushing to the left, applying Newton's second law indicates that $F_{ext} - f = ma$, where m is the mass of the moving block and a is its acceleration. As $f = \mu N$, the work done by the external force when the block moves a distance d is

$$w = \int F_{ext} dx = \int (f + ma)\, dx = \mu N d + \int ma\, dx.$$

As the acceleration changes sign each time the block starts and stops, the contribution of the term $\int ma\, dx$ is negligible. The work done by the external force after the block has traveled a total distance of $D = 160$ m is

$$W_{ext} = F_{ext} D = \mu N D = (0.40)(500\ \text{N})(160\ \text{m}) = 32000\ \text{J} = 32\ \text{kJ}.$$

We now consider the two blocks to be our system and apply the first law. Presumably, any heat transfers between the blocks and the environment is negligible, so that

$$Q = 0.$$

The $P\Delta V$ work done on the atmosphere by the expansion of the blocks is negligible. Then, as the work done *on* the system by the external force is the negative of the work done *by* the system, we have $W = -W_{ext} = -32$ kJ. It then follows from the first law, $Q = \Delta U + W$, that the change in the system's internal energy is

$$\Delta U = 32\ \text{kJ}.$$

This change could also be achieved by heating the blocks over a flame and transferring heat $Q_{flame} = M c_p^M \Delta T$ to them (see 1.27). When the $P\Delta V$ work is neglected, the first law asserts that $Q_{flame} = \Delta U$. Combining the two expressions for Q_{flame} indicates that $\Delta U = M c_p^M \Delta T$. Since the change in the state of the system is independent of whether energy is transferred as work or as heat, we can use $\Delta U = M c_p^M \Delta T$ to find ΔT. By using the value of c_p^M for iron in Table 1.2, the change in temperature of the blocks after they come to thermal equilibrium is

$$\Delta T = \frac{\Delta U}{M c_p^M} = \frac{32\ \text{kJ}}{(2.0\ \text{kg})(0.444\ \text{kJ kg}^{-1}\text{K} - 1)} = 36\ \text{K}.$$

The route to finding ΔT in the example may seem unduly complicated, as it "almost obvious" that $\mu N D = M c_p^M \Delta T$. However, the purpose of the example is to illustrate the meanings of Q, W, and ΔU. When speaking casually, people might describe the increase in the temperature of the blocks as the result of "frictional heating." However, since $Q = 0$, that description is technically *wrong*. The actual source of energy that caused the temperature to increase was the *work* done by the external force. Notice that, as the frictional force on the stationary block is equal in magnitude but opposite in direction to the frictional force on the stationary block, the work done *on* the stationary block (μND) is the negative of the work done *on* the moving block ($-\mu ND$).

Before leaving the subject of dissipative forces, it should be pointed out that processes that involve dissipative forces are always *irreversible*. As an illustration of the meaning of irreversible, imagine an object that is pushed across the floor and released. The force of friction will slow it down and bring it to rest. The reversed process would have the object start from rest and spontaneously move across the floor without any cause other than its interaction with the floor. Of course, that never happens, which indicates that the process cannot be reversed, *i.e.*, it is irreversible.

Problems

2.1 The vigorous stirring of 0.60 kg of a thick liquid causes its temperature to increase by 3.0 °C with negligible change in volume. The specific heat of the liquid is $c_V^M = 2.4\,\text{kJ}\,\text{kg}^{-1}\,\text{K}^{-1}$.

(a) Find the work done *on* the fluid (*i.e*, find $-W$).
(b) If the stirring is done by a 75 W (0.10 hp), how long was the liquid stirred?

2.2 An ideal gas expands as it is heated so that its temperature is constant at 300 K. Its initial pressure and volume are $P_0 = 0.99$ atm and $V_0 = 1.1\,\text{m}^3$. The pressure during the process is $P = P_0V_0/V$ and the final volume is $V_f = 2.2\,\text{m}^3$.

(a) How many moles of gas are there? (Use the equation of state.)
(b) What is the final pressure of the gas?
(c) What is W for the gas during the process? (Remember: $\int V^{-1}dV = \ln V + \text{const.}$)
(d) What is ΔU for the gas?
(e) What is Q for the gas?

2.3 Ten grams (10.0 g) of aluminum foil at 30 °C and 0.50 moles of gas at 15 °C are placed in a container whose volume changes to maintain the pressure at 200 kPa. The foil and air come to equilibrium with negligible heat exchange with the container. The specific heat of the gas is $c_P = 3.5R$, where R is the universal gas constant.

(a) Find the final temperature of the foil and gas.
(b) Find the values of Q and ΔU for the foil. (Its volume change is negligible.)
(c) Find the work done by the gas.
(d) What are the values of Q and ΔU for the gas?

2.4 What percent of the 80 kcal of heat Q needed to change one kilogram of ice at 0 °C to water at 0 °C is involved in doing work W? Is W positive or negative? The density of ice and water are $917\,\text{kg}\,\text{m}^{-3}$ and $1000\,\text{kg}\,\text{m}^{-3}$, respectively.

3

Properties and Partial Derivatives

The study of thermodynamic behavior is facilitated by the small number of variables required. Most of the mathematics employed is familiar from other branches of physics, but it is often used somewhat differently and with different notation. In mechanics we typically deal with functions of one variable: time t. For example, the trajectory of a particle is described by functions $x(t)$, $y(t)$, and $z(t)$ which are determined from their derivatives. In contrast, time derivatives (rates of change) are much less important in thermal physics. Time remains important, but it is the temporal sequence of events that is important, not the rates at which they occur. This chapter focuses on the properties of PVT systems and their representation by partial derivatives.

3.1 Conventions

As different authors sometimes use notation differently, we start by indicating some conventions. The Greek letter Δ indicates a difference, which may or may not be small. State functions are assumed to be continuous differentiable functions of several variables, except at phase transitions. The currently important quantities are usually called variables, while the supplementary quantities used to characterize systems are called parameters. As a discussion evolves, a parameter sometimes becomes a variable and vice versa.

Functions of one variable are conveniently represented by lines on a graph, and functions of two variables are represented by surfaces in three dimensions. For example, the ideal gas equation of state $P = nRT/V$ is shown in **Figure 3.1**, where the numbers designating pressure, volume, and temperature on the axes refer to P/P_o, V/V_o, and T/T_o, where P_o, V_o, and T_o are the values in some reference state.

Thermodynamics and Statistical Mechanics: An Integrated Approach, First Edition.
Robert J Hardy and Christian Binek.

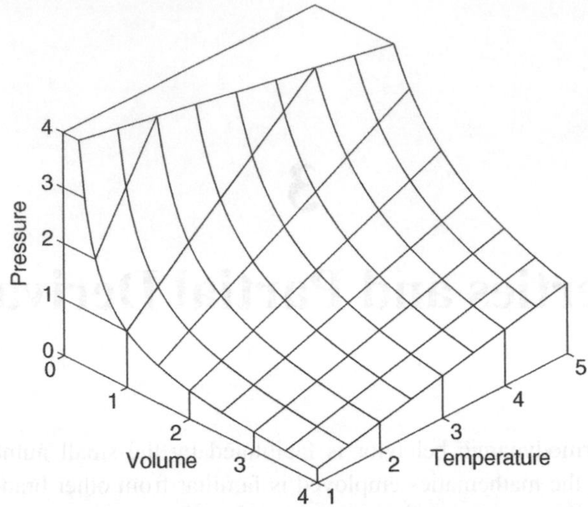

Figure 3.1 *Ideal Gas Equation of State*

Partial Derivatives

Many properties of systems are expressed as partial derivatives. The partial derivative with respect to x of $F(x, y)$ is

$$\frac{\partial F(x,y)}{\partial x} = \lim_{\Delta x \to 0} \frac{F(x+\Delta x, y) - F(x,y)}{\Delta x}, \tag{3.1}$$

where Δx is an *infinitesimal, i.e.*, a non-zero quantity that can be made arbitrarily small. Mathematics texts often use h instead of Δx. There are several different notations for partial derivatives:

$$\frac{\partial F(x,y)}{\partial x} = \frac{\partial F}{\partial x} = F_x(x,y) = \left(\frac{\partial F}{\partial x}\right)_y. \tag{3.2}$$

The first, third, and fourth of these notations indicate both the variable being varied and the fixed variable. The second notation should be avoided, as knowledge of the fixed variable is important in thermodynamics. The notation at the left, *conventional notation*, is preferred when deriving mathematical relationships. The more concise notation at the right is *thermodynamic notation*. We begin by expressing the heat capacities and specific heats as partial derivatives.

3.2 Equilibrium Properties

Constant-Volume Heat Capacity

Heat capacities are state functions that describe the change in temperature that results from a transfer of heat and depend on the process involved. Since no work is done by a *PVT* system at constant volume ($W = 0$), it follows from the first law $Q = \Delta U + W$ that

$$Q = \Delta U = U_f - U_0 = U(T + \Delta T, V) - U(T, V). \tag{3.3}$$

Substituting this into the empirical definition of C_V in (1.25) and considering limitingly small values of Q gives

$$C_V = \lim_{Q\to 0} \left.\frac{Q}{\Delta T}\right|_{V=\text{const}} = \lim_{\Delta T\to 0} \frac{U(T+\Delta T, V) - U(T, V)}{\Delta T} = \frac{\partial U(T, V)}{\partial T}. \tag{3.4}$$

Note the difference in the two limits indicated here. In the first limit, ΔT is the change in T caused by the transfer of heat Q. In the second, ΔT is the change that causes the internal energy to change. In thermodynamic notation the constant-volume heat capacity is

$$\boxed{C_V(V, T) = \left(\frac{\partial U}{\partial T}\right)_V.} \tag{3.5}$$

The constant-volume specific heat is $c_V = C_V/n$.

Ideal Gas with Constant c_V

Although the specific heats of ideal gases are in general functions of temperature, it is often a good approximation to treat c_V as a constant. In this case the internal energy function of an ideal gas is especially simple. As the internal energy of ideal gases depends on temperature only, it can be written as $U(T)$ (see (1.13)). In this case, (3.5) indicates that

$$C_V = \frac{dU(T)}{dT}. \tag{3.6}$$

When c_V is constant, it follows that

$$U(T) = nc_V T + \text{const}, \tag{3.7}$$

where $C_V = nc_V$. By representing the integration constant by nu_o, we obtain

$$U = n(u_o + c_V T). \tag{3.8}$$

Enthalpy

When heat is transferred in a constant-pressure processes, work is done by the system as its volume changes. To account for this work, we use the state function for energy called *enthalpy*, which for PVT systems is

$$H = U + PV. \tag{3.9}$$

As PV is the required to create a region of volume V in an environment held at pressure P, *the enthalpy of a system is the state function which includes both its internal energy and the energy needed to make room for it in a constant-pressure environment.*

The work done at constant pressure as a system's volume goes from V_0 to volume V is $W = P(V - V_0)$. By using this, the first law for a constant-pressure process becomes

$$Q = \Delta U + W = (U - U_0) + P(V - V_0) = (U + PV) - (U_0 + PV_0). \tag{3.10}$$

Combining this with (3.9) indicates that

$$Q = H - H_0 = \Delta H. \tag{3.11}$$

The heat transferred at constant pressure equals the change in enthalpy.

By using enthalpy, we can express the constant-pressure heat capacity C_P in a relationship similar to (3.5). Substituting (3.11) into the empirical definition of C_P in (1.25) gives

$$C_P = \lim_{Q\to 0} \left.\frac{Q}{\Delta T}\right|_{P=\text{const}} = \lim_{\Delta T\to 0} \frac{H(T+\Delta T, P) - H(T,P)}{\Delta T} = \frac{\partial H(T,P)}{\partial T}, \tag{3.12}$$

where the enthalpy is a function of temperature and pressure. Thus,

$$\boxed{C_P(T,P) = \left(\frac{\partial H}{\partial T}\right)_P.} \tag{3.13}$$

Example 3.1 $c_P - c_V$ for ideal gases

Find the difference between the constant-pressure and constant-volume specific heats of an ideal gas.

Solution. As $V = nRT/P$ in an ideal gas, its enthalpy is

$$H = U + PV = U + nRT.$$

Combining this and $U = U(T)$ with (3.13) indicates that

$$C_P(T,P) = \frac{dU(T)}{dT} + nR.$$

Since $C_V = dU(T)/dT$ in an ideal gas, this becomes

$$C_P(T) = C_V(T) + nR,$$

where the heat capacities C_P and C_V are independent of P and V. By dividing by n, rearranging terms, and omitting the explicit mention of T, we obtain

$$c_P - c_V = R. \tag{3.14}$$

Note that this result is valid whether or not c_V is a function of temperature. Also, note that c_P is constant if c_V is constant.

The internal energies of *PVT* systems are in general *not* independent of volume or pressure. To illustrate the importance of knowing the variable that is fixed, as well as the one that is varied, consider the derivatives $(\partial U/\partial T)_V$ and $(\partial U/\partial T)_P$. In the simple solid model, the internal energy expressed as a function of T and V is (see (1.14))

$$U(T,V) = M c_o T + b_o V \left[\ln\left(\frac{V}{V_o}\right) - 1\right]. \tag{3.15}$$

Thus, the temperature derivative of U with V fixed is

$$\left(\frac{\partial U}{\partial T}\right)_V = M c_o. \tag{3.16}$$

We found in (1.21) that the internal energy of a simple solid expressed as a function of T and P is

$$U(T,P) = Mc_oT + (a_oT - P - b_o)V_oe^{(a_oT-P)/b_o}. \tag{3.17}$$

Hence, the temperature derivative of U with P fixed is

$$\left(\frac{\partial U}{\partial T}\right)_P = Mc_o + \left(\frac{a_o}{b_o}\right)(a_oT - P)V_oe^{(a_oT-P)/b_o}, \tag{3.18}$$

which is very different from the derivative with V fixed.

Example 3.2 $C_P - C_V$ for simple solids

Find the difference between the constant-pressure and constant-volume heat capacities of a simple solid.

Solution. Since $C_V = (\partial U/\partial T)_V$, it follows from (3.16) that

$$C_V = Mc_o \quad \text{and} \quad c_V^M = \frac{C_V}{M} = c_o. \tag{3.19}$$

According to (1.20), the volume of a simple solid is

$$V = V_oe^{(a_oT-P)/b_o}.$$

Substituting this and the internal energy in (3.17) into $H = U + PV$ gives

$$H = Mc_oT + (a_oT - P - b_o)V_oe^{(a_oT-P)/b_o} + PV_oe^{(a_oT-P)/b_o}.$$

Thus, the enthalpy of a simple solid as a function of T and P is

$$H = Mc_oT + (a_oT - b_o)V_oe^{(a_oT-P)/b_o}.$$

Since $C_V = Mc_o$, the constant-pressure heat capacity is

$$C_P = \left(\frac{\partial H}{\partial T}\right)_P = C_V + (a_o^2/b_o)TV_oe^{(a_oT-P)/b_o}.$$

By using the above expression for V, the difference between the constant-pressure and constant-volume heat capacities becomes

$$C_P - C_V = (a_o^2/b_o)TV. \tag{3.20}$$

Thermal Expansion and Bulk Modulus

The *volume coefficient of thermal expansion* α_V describes the change in volume of a substance caused by a change in temperature at fixed pressure. The *isothermal bulk modulus* B_T describes the change in pressure caused by a change in volume at fixed temperature. Their definitions are

$$\boxed{\alpha_V = \frac{1}{V}\left(\frac{\partial V}{\partial T}\right)_P} \quad \text{and} \quad \boxed{B_T = -V\left(\frac{\partial P}{\partial V}\right)_T}. \tag{3.21}$$

Both α_V and B_T are intensive properties. When expressed in terms of the specific volume $v = V/n$, they become

$$\alpha_V = \frac{1}{v}\left(\frac{\partial v}{\partial T}\right)_P \quad \text{and} \quad B_T = -v\left(\frac{\partial P}{\partial v}\right)_T. \tag{3.22}$$

Values of α_V and B_T for a few materials are given in Table 1.2.

The coefficient of thermal expansion and the bulk modulus are in general functions of temperature and pressure. However, they can often be treated as constants. For example, the density of water between 0 and 100 °C is usually considered to be 1 kilogram per liter (1000 kg m^{-3} or 1.00 g cm^{-3}), but this is not exactly true. The specific volume $v(T,P)$ of water at $P = 1.0$ atm is plotted in **Figure 3.2**. Notice the small range of values on the vertical axis and the minimum at 3.98 °C. Since the specific volume *decreases* as the temperature is increased from 0 to 3.98 °C, the coefficient of thermal expansion is negative in that temperature range. Most liquids and solids expand with increasing temperature and thus have positive coefficients of thermal expansion. Although negative values of α_V are rare, they do exist in substances other than water.

The shape of a solid object made of homogeneous isotropic material does not change as it expands or contracts. As a result, its volume is proportional to the cube of the distance L between any two points on it, that is, $V = kL^3$ where k is a proportionality constant. The *linear coefficient of thermal expansion* α_L is defined as

$$\alpha_L = \frac{1}{L}\left(\frac{\partial L}{\partial T}\right)_P. \tag{3.23}$$

By substituting $V = kL^3$ into the definition of the volume coefficient α_V, it follows that

$$\alpha_V = \frac{1}{V}\left(\frac{\partial V}{\partial T}\right)_P = \frac{1}{kL^3}\left(\frac{\partial\,(kL^3)}{\partial T}\right)_P = \frac{1}{kL^3}3kL^2\left(\frac{\partial L}{\partial T}\right)_P = \frac{3}{L}\left(\frac{\partial L}{\partial T}\right)_P, \tag{3.24}$$

so that

$$\boxed{\alpha_V = 3\alpha_L.} \tag{3.25}$$

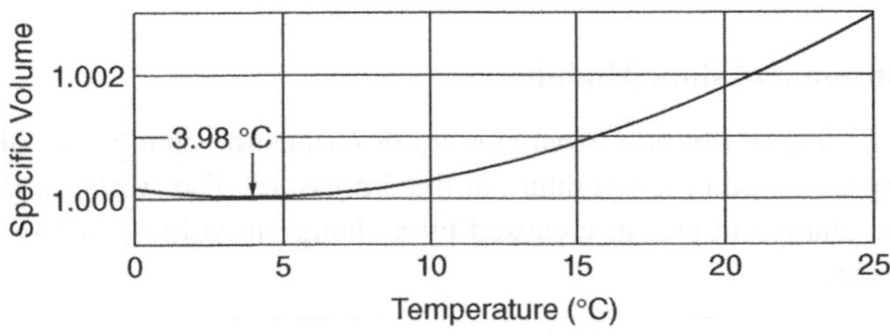

Figure 3.2 *Specific Volume of Water in kilograms per liter*

Simple Solid Model

The parameters a_o, b_o, and c_o in the simple solid model are related to the values of α_L, B_T, and c_V^M in lists of material properties. According to (1.15), the equation of state for a simple solid is $P = a_oT - b_o \ln(V/V_o)$. Differentiating this with respect to V gives us the isothermal bulk modulus,

$$B_T = -V\left(\frac{\partial P}{\partial V}\right)_T = -V\frac{-b_o}{V} = b_o. \tag{3.26}$$

To obtain the volume coefficient of thermal expansion, we rewrite the equation of state as $V = V_o e^{(a_oT-P)/b_o}$ and differentiate with respect to T,

$$\alpha_V = \frac{1}{V}\left(\frac{\partial V}{\partial T}\right)_P = \frac{a_o}{b_o}. \tag{3.27}$$

According to (3.19), the constant-volume specific heat is $c_V^M = c_o$. From these relationships it follows that the model parameters a_o, b_o, and c_o are related to the material properties α_L, B_T, and c_V^M by

$$a_o = \alpha_V B_T, \qquad b_o = B_T, \qquad c_o = c_V^M. \tag{3.28}$$

As a_o, b_o, and c_o are constants, the material properties are independent of temperature and pressure in the simple solid model. Although this is not exactly true, in many applications involving solids and some liquids it is a useful simplifying approximation. As lists of material properties often give only one value of each property, we frequently have no choice but to treat them as constants.

Example 3.3 Specific heat of iron

The simple solid model can be used to estimate the constant-volume specific heat of iron from the constant-pressure specific heat. It follows from (3.28) and the result for $C_P - C_V$ in (3.20) that[1]

$$C_P - C_V = (a_o/b_o)^2 b_o TV = \alpha_V{}^2 B_T TV.$$

Dividing this by the mass M and introducing the density $\rho = M/V$ indicates that

$$c_P^M - c_V^M = \frac{\alpha_V{}^2 B_T T}{\rho}.$$

By using the data in Table 1.2 and assuming that T is room temperature (300 K), we find that

$$c_P^M - c_V^M = \frac{(35.7\times 10^{-6}\ \text{K}^{-1})^2(160\times 10^9\ \text{Pa})(300\ \text{K})}{7880\ \text{kg m}^{-3}}\left(\frac{1\ \text{kcal}}{4184\ \text{J}}\right)$$
$$= 0.0019\ \text{kcal kg}^{-1}\ \text{K}^{-1}.$$

[1] Although derived for the simple solid model, the result $C_P - C_V = \alpha_V{}^2 B_T TV$ is valid for PVT systems of all kinds, as shown in Section 11.3.

Since the value of the constant-pressure specific heat is $c_P^M = 0.106$ kcal kg^{-1} K^{-1}, the constant-volume specific heat of iron is

$$c_V^M = (0.106 - 0.0019)\,\text{kcal kg}^{-1}\,\text{K}^{-1} = 0.104\,\text{kcal kg}^{-1}\,\text{K}^{-1}.$$

The difference $c_P^M - c_V^M$ for iron is less than 2% the value of c_P^M. Such small differences are typical of solids. In contrast, the differences for gases are much larger. As the value of c_V for diatomic gases such as air is $2.5R$ (see (15.35)), the difference $c_P - c_V = R$ is 40% the value of c_V.

3.3 Relationships between Properties

The general relationships between the properties of systems are more important than the results for specific systems. Many of these relationships are obtained by employing the chain rule for differentiating composite functions. Although students regularly use the chain rule when working with explicitly given functions, they sometimes are less familiar with its use with functions expressed in generic form. For example, finding the derivative of $\sqrt{x^2 + y^2}$ with respect to x is straightforward, but finding the derivative of $F(x, g(x, z))$ with respect to x may be a less familiar exercise.

Composite Functions

Functions are often combined to create new functions. For example, the internal energy $U(T) = nu_o + nc_V T$ can be combined with the equation of state $T(P, V) = PV/(nR)$ to create the *composite function* $\overline{U}(P, V) = nu_o + c_V PV/R$, where the bar on $\overline{U}$ emphasizes that $\overline{U}(P, V)$ is a different function of its independent variables than $U(T)$. We use the same letter for $\overline{U}$ and U, as both functions refer to the same physical property. A more general composite function formed by combining $F(t, u, \ldots)$ with $t = t(x, y, \ldots)$, $u = u(x, y, \ldots)$, etc. is

$$\overline{F}(x, y, \ldots) = F(t(x, y, \ldots), u(x, y, \ldots), \ldots). \tag{3.29}$$

Chain Rule

The partial derivative of a composite function is a sum of products. According to the *chain rule*, the partial derivative with respect to x of the above function is

$$\frac{\partial \overline{F}(x, y, \ldots)}{\partial x} = \frac{\partial F(t, u, \ldots)}{\partial t}\frac{\partial t(x, y, \ldots)}{\partial x} + \frac{\partial F(t, u, \ldots)}{\partial u}\frac{\partial u(x, y, \ldots)}{\partial x} + \cdots. \tag{3.30}$$

The following example is included to help illustrate the meaning of the terms in this somewhat abstract formula.

Example 3.4 Chain rule

Find the partial derivative with respect to x of the composite function $\overline{F}(x, y)$ obtained by combining $F(t, u) = \sqrt{u} \sin t$ with the functions $u(x, y) = x^2 + y^2$ and $t(x, y) = xy$.

Solution. The explicit form of the composite function is obtained by substituting $u(x, y)$ and $t(x, y)$ into $F(t, u)$, which gives

$$\overline{F}(x, y) = \sqrt{x^2 + y^2} \sin(xy).$$

By straightforward differentiation, we obtain

$$\frac{\partial \overline{F}(x, y)}{\partial x} = \frac{1}{2} \frac{2x}{\sqrt{x^2 + y^2}} \sin(xy) + \sqrt{x^2 + y^2} \cos(xy)\, y.$$

To show that this result is also implied by the general formula in (3.30), we need the following derivatives:

$$\frac{\partial F(t, u)}{\partial u} = \frac{1}{2} \frac{1}{\sqrt{u}} \sin t, \qquad \frac{\partial F(t, u)}{\partial t} = \sqrt{u} \cos t,$$

$$\frac{\partial u(x, y)}{\partial x} = 2x, \qquad \frac{\partial t(x, y)}{\partial x} = y.$$

Substituting these results into (3.30) gives

$$\frac{\partial \overline{F}(x, y)}{\partial x} = \left(\frac{1}{2} \frac{1}{\sqrt{u}} \sin t \right) (2x) + \left(\sqrt{u} \cos t \right) y.$$

By replacing u and t in this with the explicit expressions for $u(x, y)$ and $t(x, y)$, we find that

$$\frac{\partial \overline{F}(x, y)}{\partial x} = \frac{1}{2} \frac{1}{\sqrt{x^2 + y^2}} \sin(xy)(2x) + \sqrt{x^2 + y^2} \cos(xy)\, y,$$

which is the same as the result obtained by straightforward differentiation.

Different symbols should be used for different mathematical functions, even when they refer to the same physical property. To see why, consider the composite function

$$\overline{F}(x, z) = F(x, y(x, z)), \tag{3.31}$$

which is the special case of (3.29) in which the arguments $(x, y, \cdots)$ are replaced by (x, z) and the function $u(x, \cdots)$ is simply x. According to the chain rule, the partial derivative of $\overline{F}(x, z)$ with respect to x is

$$\frac{\partial \overline{F}(x, z)}{\partial x} = \frac{\partial F(x, y)}{\partial x} + \frac{\partial F(x, y)}{\partial y} \frac{\partial y(x, z)}{\partial x}, \tag{3.32}$$

where y is assigned the value specified by $y(x, z)$. Notice what happens when we fail to identify the variables that are fixed: Instead of (3.32) we would have

$$\frac{\partial F}{\partial x} = \frac{\partial F}{\partial x} + \frac{\partial F}{\partial y} \frac{\partial y}{\partial x}. \tag{3.33}$$

This equation appears to imply that $\dfrac{\partial F}{\partial y} \dfrac{\partial y}{\partial x} = 0$. As a general statement, the implication is *false*!

Composite functions of the type in (3.31) are important in thermodynamics. To make the partial derivatives appear more like they do in thermodynamics, we represent the variables by capital letters and give the derivatives in thermodynamic notation. Then, the composite function is

$$\overline{F}(X, Z) = F(X, Y(X, Z)). \tag{3.34}$$

The partial derivative in (3.32) is

$$\boxed{\left(\frac{\partial F}{\partial X}\right)_Z = \left(\frac{\partial F}{\partial X}\right)_Y + \left(\frac{\partial F}{\partial Y}\right)_X \left(\frac{\partial Y}{\partial X}\right)_Z,} \tag{3.35}$$

and the derivative with respect to Z is

$$\boxed{\left(\frac{\partial F}{\partial Z}\right)_X = \left(\frac{\partial F}{\partial Y}\right)_X \left(\frac{\partial Y}{\partial Z}\right)_X.} \tag{3.36}$$

Notice that a bar is not needed on the functions F in (3.35) or (3.36), because the notation makes it clear that the independent variables of the function F on the left are X and Z, while the independent variables on the function F on the right are X and Y.

The existence of more than two independent variables is indicated in thermodynamic notation by additional subscripts, as in

$$\frac{\partial F(X, Y, Z)}{\partial X} = \left(\frac{\partial F}{\partial X}\right)_{YZ} \quad \text{and} \quad \frac{\partial^2 F(X, Y, Z)}{\partial X \partial Y} = \left(\frac{\partial^2 F}{\partial X \partial Y}\right)_Z. \tag{3.37}$$

An important property of partial derivatives is their independence from the order of differentiation, as indicated by

$$\frac{\partial}{\partial Y}\left(\frac{\partial F(X, Y)}{\partial X}\right) = \frac{\partial}{\partial X}\left(\frac{\partial F(X, Y)}{\partial Y}\right) \tag{3.38}$$

and

$$\boxed{\left(\frac{\partial^2 F}{\partial Y \partial X}\right) = \left(\frac{\partial^2 F}{\partial X \partial Y}\right).} \tag{3.39}$$

Functions $y(x)$ and $x(y)$ are the inverses of each other if

$$y(x(y)) = y. \tag{3.40}$$

For example, the functions $y(x) = 1/(x - 1)$ and $x(y) = 1 + (1/y)$ are inverses,

$$y(x(y)) = \frac{1}{x(y) - 1} = \frac{1}{(1 + 1/y) - 1} = y. \tag{3.41}$$

Functions of two independent variables can be inverses when one of the variables is fixed. The function $y(x, z)$ is the inverse of $x(y, z)$ if

$$y(x(y, z), z) = y. \tag{3.42}$$

Using the chain rule to differentiate both sides of this with respect to y with z fixed gives

$$\frac{\partial y(x,z)}{\partial x}\frac{\partial x(y,z)}{\partial y} = 1 \quad \text{or} \quad \left(\frac{\partial y}{\partial x}\right)_z\left(\frac{\partial x}{\partial y}\right)_z = 1. \tag{3.43}$$

This relationship is easily visualized. Choose a value of z and hold it fixed. Plot $y(x, z)$ as a line on a graph with the y-axis pointing up and the x-axis pointing to the right. Rotate the graph 90° counter-clockwise, so the x-axis points up and the y-axis points to the left. Then, flip the graph left to right, so the y-axis points to the right. The final graph represents $x(y, z)$. The slope $(\partial y/\partial x)_z$ of the line at some point on the original graph is the limiting value of $\Delta y/\Delta x$, while the slope $(\partial x/\partial y)_z$ of the line on the final graph is the limiting value of $\Delta x/\Delta y$. Thus, $(\partial x/\partial y)_z(\partial y/\partial x)_z = 1$. Changing x, y, and z to capitals gives

$$\boxed{\left(\frac{\partial Y}{\partial X}\right)_Z\left(\frac{\partial X}{\partial Y}\right)_Z = 1.} \tag{3.44}$$

Provided $(\partial Y/\partial X)_Z \neq 0$, this implies that

$$\boxed{\left(\frac{\partial X}{\partial Y}\right)_Z = \frac{1}{\left(\frac{\partial Y}{\partial X}\right)_Z}.} \tag{3.45}$$

Compressibility

The *isothermal compressibility* K_T describes the change in volume caused by a change in pressure at fixed temperature,

$$\boxed{K_T = \frac{-1}{V}\left(\frac{\partial V}{\partial P}\right)_T.} \tag{3.46}$$

K_T describes how easy it is to compress a substance. It is an immediate consequence of (3.45), (3.46), and the definition of the bulk modulus B_T in (3.21) that

$$K_T = \frac{1}{B_T} \quad \text{and} \quad K_T B_T = 1. \tag{3.47}$$

Surfaces and Partial Derivatives

A relationship $z = z(x, y)$ can be expressed as $x = x(y, z)$ by solving $z = z(x, y)$ for x (provided it is single valued). For example, the relationship $P = nRT/V$ can be expressed as $T = PV/(nR)$. Both $z(x, y)$ and $x(y, z)$ are represented by the same surface. By combining the function $z(x, y)$ that gives the value of z at some point with the function $x(y, z)$ which gives the value of x at that point, we can form the composite function $x(y, z(x, y))$, which gives the value of x at the point as a function of x and y. Since the value of x is x, it follows that

$$x = x(y, z(x, y)). \tag{3.48}$$

Since $(\partial x/\partial y)_x = 0$ when x is fixed, differentiating both sides of (3.48) with respect to y with x fixed gives

$$0 = \left(\frac{\partial x}{\partial y}\right)_z + \left(\frac{\partial x}{\partial z}\right)_y \left(\frac{\partial z}{\partial y}\right)_x, \tag{3.49}$$

By multiplying this by $(\partial y/\partial x)_z$ and using (3.43), we find that

$$\left(\frac{\partial y}{\partial x}\right)_z \left(\frac{\partial x}{\partial z}\right)_y \left(\frac{\partial z}{\partial y}\right)_x = -1. \tag{3.50}$$

The minus sign may be unexpected. The symbols ∂x, ∂y, and ∂z can sometimes be treated as small displacements that can be cancelled between numerator and denominator. Such cancellations are *not* justified here, because the symbol in the numerator refers to a different displacement than the corresponding symbol in the denominator. The notation specifies that the ∂y in the numerator represents a displacement in the x direction, while the ∂y in the denominator represents a displacement with x fixed.

The existence of a surface is implicit in equation (3.50). Its significance is illustrated by the part of the surface of the unit sphere $x^2 + y^2 + z^2 = 1$ shown in **Figure 3.3**. Three different lines are drawn through point $(x, y, z) = \left(\frac{2}{3}, \frac{2}{3}, \frac{1}{3}\right)$. Line 1 is at the intersection of the sphere with the plane perpendicular to the z-axis at $z = \frac{1}{3}$. Lines 2 and 3 are at the intersections of the sphere with planes perpendicular to the x-axis and the y-axis, respectively. **Figure 3.4** shows lines 1, 2, and 3 when plotted in two dimensions. The slopes of the tangents to the lines in the three graphs are equal to the derivatives $(\partial y/\partial x)_z$, $(\partial z/\partial y)_x$, and $(\partial x/\partial z)_y$ evaluated at $(x, y, z) = \left(\frac{2}{3}, \frac{2}{3}, \frac{1}{3}\right)$. Their values are $(\partial y/\partial x)_z = -1$, $(\partial z/\partial y)_x = -2$, and $(\partial x/\partial z)_y = -1/2$. The product of the derivatives is negative, since all three lines have negative slopes. Specifically, $(-1)(-2)(-1/2)_z = -1$.

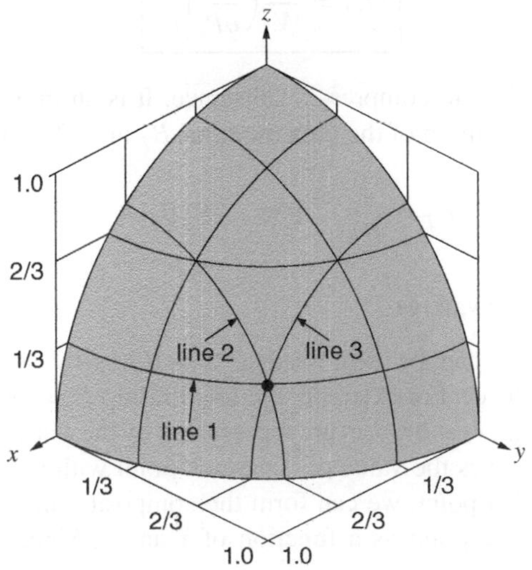

Figure 3.3 *Surface of a Unit Sphere*

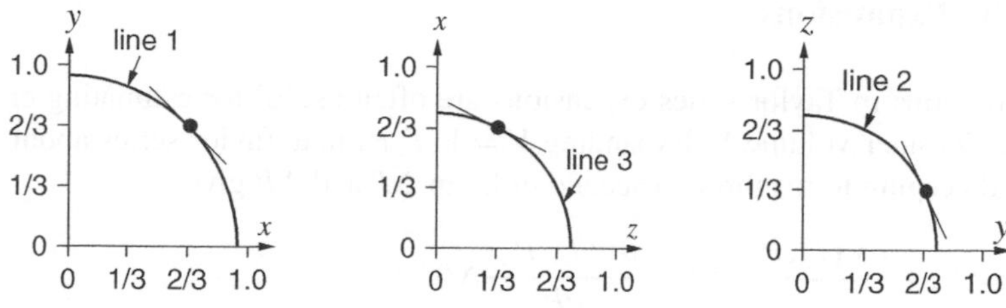

Figure 3.4 *Lines Tangent to a Unit Sphere in Different Directions*

The relationship in (3.50) with *x*, *y*, and *z* changed to capitals is

$$\left(\frac{\partial Y}{\partial X}\right)_Z\left(\frac{\partial X}{\partial Z}\right)_Y\left(\frac{\partial Z}{\partial Y}\right)_X = -1. \tag{3.51}$$

Notice that each of the three variables *X*, *Y*, and *Z* appears once in each partial derivative, either as the function differentiated, the variable that is varied, or the variable that is fixed. Also, each symbol ∂X, ∂Y, and ∂Z occurs once in a numerator and once in a denominator.

The relationship in (3.49) is also useful. By replacing (x, y, z) with (X, Z, Y) and rearranging terms, it becomes

$$\left(\frac{\partial X}{\partial Z}\right)_Y = -\left(\frac{\partial X}{\partial Y}\right)_Z\left(\frac{\partial Y}{\partial Z}\right)_X. \tag{3.52}$$

It is useful to contrast this with the result in (3.36), which states that

$$\left(\frac{\partial F}{\partial Z}\right)_X = \left(\frac{\partial F}{\partial Y}\right)_X\left(\frac{\partial Y}{\partial Z}\right)_X. \tag{3.53}$$

Notice that a different variable is held fixed in each of the partial derivatives in (3.52) but not in (3.53). Also, there are only three variables (*X*, *Y*, and *Z*) in (3.52), while there are four variables (*X*, *Y*, *Z*, and *F*) in (3.53). Finally, there is a minus sign in (3.52) but not in (3.53).

Temperature Derivative of Pressure

The relationships between partial derivatives imply useful relationships between physical properties. By replacing (X, Y, Z) with (P, V, T), the relationship in (3.52) specifies that

$$\left(\frac{\partial P}{\partial V}\right)_T\left(\frac{\partial V}{\partial T}\right)_P = -\left(\frac{\partial P}{\partial T}\right)_V. \tag{3.54}$$

By multiplying and dividing this by *V* and using the definitions of B_T and α_V in (3.21), it follows that

$$\left(\frac{\partial P}{\partial T}\right)_V = -V\left(\frac{\partial P}{\partial V}\right)_T\frac{1}{V}\left(\frac{\partial V}{\partial T}\right)_P = B_T\alpha_V. \tag{3.55}$$

Thus, the change in pressure caused by a change in temperature at fixed volume can be found by multiplying the bulk modulus by the coefficient of thermal expansion.

3.4 Series Expansions

The leading terms in Taylor series expansions are often useful for estimating changes in properties. Consider volume V. Expanding $V = V(T, P)$ in a Taylor series about the state (T_o, P_o) and keeping terms through second order in ΔT and ΔP give

$$V(T,P) = V_o + \frac{\partial V(T_o,P_o)}{\partial T}\Delta T + \frac{\partial V(T_o,P_o)}{\partial P}\Delta P$$
$$+\frac{\partial^2 V(T_o,P_o)}{\partial T^2}\frac{\Delta T^2}{2} + \frac{\partial^2 V(T_o,P_o)}{\partial T\partial P}\Delta T\Delta P + \frac{\partial^2 V(T_o,V_o)}{\partial P^2}\frac{\Delta P^2}{2} + \cdots, \quad (3.56)$$

where $V_o = V(T_o, P_o)$, $\Delta T = T - T_o$, and $\Delta P = P - P_o$. By introducing $\Delta V = V(T, P) - V_o$ and neglecting all but the lowest powers of ΔT and ΔP, we obtain

$$\Delta V \approx \frac{\partial V(T_o,P_o)}{\partial T}\Delta T + \frac{\partial V(T_o,P_o)}{\partial P}\Delta P. \quad (3.57)$$

The terms on the second line of (3.56) can be used to estimate the accuracy of the approximation in (3.57). It follows from the definitions of α_V and K_T that

$$\Delta V \approx V_o\alpha_V^o\Delta T - V_o K_T^o\Delta P, \quad (3.58)$$

where the coefficients α_V^o and K_T^o are the values of α_V and K_T in state (T_o, P_o).

Example 3.5 Change in volume

Estimate the relative change in the volume of a piece of aluminum that is dropped to the bottom of Lake Baikal in Russia. At 1620 m (over a mile) it is the deepest lake in the world. Assume that the temperature of the aluminum decreases by 15 °C as it sinks.

Solution. An expression for estimating the relative change is obtained by dividing both sides of (3.58) by V_o and using $K_T = 1/B_T$,

$$\Delta V/V_o = +\alpha_V^o\Delta T - \frac{\Delta P}{B_T^o}.$$

The increase in pressure caused by an increase Δd in depth is $\Delta P = \rho g\Delta d$, so that

$$\Delta P = (1000\ \text{kg m}^{-3})(9.8\ \text{m s}^{-2})(1620\ \text{m}) = 15.9\times10^6\ \text{Pa}.$$

By using the values of B_T^o and α_V^o in Table 1.2, we find that

$$\frac{-\Delta P}{B_T^o} = -\frac{15.9\times10^6\ \text{Pa}}{75\times10^9\ \text{Pa}} = -0.000212$$

and

$$+\alpha_V^o\Delta T = (69.6\times10^{-6}\ \text{K}^{-1})(-15\ \text{K}) = -0.001044.$$

Thus, the relative change in volume is

$$\frac{\Delta V}{V_o} = -0.001044 - 0.000212 = -0.00126 = -0.13\%.$$

Notice that the change resulting from the 15 °C change in temperature is almost five times greater than the change caused by the increase in pressure.

When expressed in thermodynamic notation, the approximate change in volume in (3.57) becomes

$$\Delta V \approx \left(\frac{\partial V}{\partial T}\right)_P \Delta T + \left(\frac{\partial V}{\partial P}\right)_T \Delta P. \tag{3.59}$$

This is similar in appearance to the following relationship between differentials:

$$dV = \left(\frac{\partial V}{\partial T}\right)_P dT + \left(\frac{\partial V}{\partial P}\right)_T dP. \tag{3.60}$$

Equation (3.59) gives a useful interpretation of (3.60) if the differential on the left is exact, which dV is. However, when the differential on the left is inexact, like the differential for heat $đQ$, the coefficients on the right are in general not partial derivatives. Differentials will not be used until the meanings exact and inexact are reviewed in Chapter 9.

3.5 Summary

It is important to distinguish between relationships that describe processes and relationships between state functions. The first law of thermodynamics asserts for processes which begin and end in equilibrium that

$$Q = \Delta U + W, \tag{3.61}$$

where $\Delta U = U_f - U_0$ is the difference between the initial and final values of the internal energy. The following are some the important state functions of PVT systems:

$$P = P(T, V), \quad U = U(V, T), \quad H = U + PV, \tag{3.62}$$

$$B_T = -V\left(\frac{\partial P}{\partial V}\right)_T = -v\left(\frac{\partial P}{\partial v}\right)_T, \quad K_T = \frac{1}{B_T} = -\frac{1}{V}\left(\frac{\partial V}{\partial P}\right)_T, \tag{3.63}$$

$$\alpha_V = \frac{1}{V}\left(\frac{\partial V}{\partial T}\right)_P = \frac{1}{v}\left(\frac{\partial v}{\partial T}\right)_P, \quad \left(\frac{\partial P}{\partial T}\right)_V = B_T\alpha_V, \tag{3.64}$$

$$C_V = nc_V = M\,c_V^M = \lim \left.\frac{Q}{\Delta T}\right|_{V=const} = \left(\frac{\partial U}{\partial T}\right)_V, \tag{3.65}$$

$$C_P = nc_P = M\,c_P^M = \lim \left.\frac{Q}{\Delta T}\right|_{P=const} = \left(\frac{\partial H}{\partial T}\right)_P. \tag{3.66}$$

For ideal gases we have

$$PV = nRT, \quad U = U(T), \quad c_P = c_V + R. \tag{3.67}$$

The internal energy of an ideal gas with constant specific heats is such that

$$U = n(u_o + c_V T) \quad \text{and} \quad \Delta U = nc_V\Delta T. \tag{3.68}$$

When pressure is constant and uniform throughout a system, the work done W and the heat accepted Q are

$$W = P(V_f\text{-}V_0) \quad \text{and} \quad Q = H_f - H_0 = \Delta H, \tag{3.69}$$

where H is enthalpy. The restriction to uniform pressure arises from the use of $W = \int P(V)dV$ to find the work done.

Problems

Simplify answers so there are no fractions in a numerator or denominator. For example, $\dfrac{a + x/b}{c + y/d}$ should be simplified to $\dfrac{d(ab + x)}{b(cd + y)}$ or $\dfrac{abd + dx}{bcd + by}$.

3.1 One mole of an ideal gas with $c_V = 2.5R$ at 0.0°C is heated at constant pressure. How much heat is required to double its volume?

3.2 Heat is transferred to an ideal gas with $c_V = 2.5R$ in a constant-pressure process. What fraction of the heat introduced is used to increase its internal energy? What fraction is used to do work?

3.3 The purpose of this problem is illustrate the meanings of

$$\overline{F}(x,z) = F(x, y(x,z)), \tag{A}$$

$$\frac{\partial \overline{F}(x,z)}{\partial z} = \frac{\partial F(x,y)}{\partial y}\frac{\partial y(x,z)}{\partial z}, \tag{B}$$

$$\frac{\partial \overline{F}(x,z)}{\partial x} = \frac{\partial F(x,y)}{\partial x} + \frac{\partial F(x,y)}{\partial y}\frac{\partial y(x,z)}{\partial x}. \tag{C}$$

Consider the functions $F(x,y) = x^2 - y^2$ and $y(x,z) = xz^2$ and use conventional notation. (Your answers to parts (b) and (d) should be consistent.)

(a) Find the explicit expression for $\overline{F}(x,z)$ by substituting the above expressions for $F(x,y)$ and $y(x,z)$ into (A) and simplify.

(b) Find $\dfrac{\partial \overline{F}(x,z)}{\partial x}$ and $\dfrac{\partial \overline{F}(x,z)}{\partial z}$ by differentiating the explicit function from part (a).

(c) Find explicit expressions for $\dfrac{\partial F(x,y)}{\partial x}$, $\dfrac{\partial F(x,y)}{\partial y}$, $\dfrac{\partial y(x,z)}{\partial x}$, and $\dfrac{\partial y(x,z)}{\partial z}$.

(d) Use (B) and (C) and the results of part (c) to find $\dfrac{\partial \overline{F}(x,z)}{\partial x}$ and $\dfrac{\partial \overline{F}(x,z)}{\partial z}$.

3.4 **(a)** Find explicit expressions for an ideal gas for the partial derivatives

$$\left(\frac{\partial P}{\partial V}\right)_T, \quad \left(\frac{\partial V}{\partial T}\right)_P, \quad \text{and} \quad \left(\frac{\partial T}{\partial P}\right)_V.$$

(b) Use the results of part (a) to evaluate the product $\left(\dfrac{\partial P}{\partial V}\right)_T\left(\dfrac{\partial V}{\partial T}\right)_P\left(\dfrac{\partial T}{\partial P}\right)_V$.

(c) Express the definitions of $\alpha_V(T,P)$, $K_T(T,P)$, and $B_T(T,V)$ in terms of the indicated independent variables, where each property depends on only two independent variables.

3.5 Consider a gas whose equation of state is

$$P = \frac{nRT}{V} + \frac{n^p b}{V^2} + \frac{n^q cT}{V^3},$$

where b, c, p, and q are intensive constants.

(a) What are the values of p and q? (Use the intensive characteristic of P.)

(b) Find $B_T(T,V)$.

(c) Since expressing the above equation in the form $V = V(T,P)$ is not practical, a explicit expression for $\alpha_V(T,P)$ cannot be obtained from $\alpha_V = V^{-1}(\partial V/\partial T)_P$. Find the expression for α_V in terms of the generic functions $(\partial P/\partial T)_V$ and $(\partial P/\partial V)_T$.

(d) Use the result of part (c) and the above explicit expression for the equation of state to find the *explicit* expression for $\alpha_V(T,V)$. (To be independent of the size, material properties such as α_V should be intensive.)

(e) Express α_V as an explicit function of T and the specific volume $v = V/n$, and show that the resulting expression for $\alpha_V(T, v)$ is intensive.

3.6 The temperature of a piece of lead is increased from 20 to 35 °C while its volume is held constant by increasing the hydrostatic pressure. If the initial pressure is 1.0 atm, what is the final pressure?

4

Processes in Gases

It is now time to do something with the terms and concepts developed in the previous chapters, and applications to gases are a good place to start. After finding the characteristics of several types of equilibrium processes, such as isothermal, adiabatic, and isobaric, we investigate temperature changes in the atmosphere and the energy efficiency of heat engines.

4.1 Ideal Gases

Although idealizations are required for real processes to be treated as equilibrium processes, the idealizations are often quite accurate and allow us to represent the changes that occur by lines in state space. In many applications, real gases are described with useful accuracy as ideal gases with constant specific heats. These are the gases we consider. Their equations of state and internal energies are

$$PV = nRT \quad \text{and} \quad U = n(u_0 + c_V T). \tag{4.1}$$

As indicated in (3.8), u_0 is an integration constant and can be set equal to zero. As the internal energy of an ideal gas depends on temperature only, its change during any process is

$$\Delta U = U_f - U_0 = nc_V(T_f - T_0) = nc_V \Delta T, \tag{4.2}$$

where T_0 and T_f are the initial and final temperatures. Although the constant-volume specific heat c_V appears in this, the result applies to any process that starts and ends in equilibrium. As found in (3.14), the constant-pressure specific heat is

$$c_P = c_V + R, \tag{4.3}$$

which implies that c_P is constant if c_V is constant. A useful quantity is the *ratio of specific heats*

$$\gamma = \frac{c_P}{c_V} = \frac{C_P}{C_V}, \tag{4.4}$$

Thermodynamics and Statistical Mechanics: An Integrated Approach, First Edition.
Robert J Hardy and Christian Binek.

which is constant when the specific heats are constant. The result in (4.3) suggests that specific heats be expressed in terms of the universal gas constant. Useful values for diatomic gases such as air are $c_V = 2.5R$, $c_P = 3.5R$, and $\gamma = 1.4$.

Constant-Pressure Processes

The heat transferred Q and the work done W in a constant-pressure (isobaric) process according to (1.27) and (2.4) are

$$Q = nc_P\Delta T, \quad W = P\Delta V, \quad \Delta U = nc_V\Delta T. \tag{4.5}$$

Constant-Volume Processes

Since no work is done by a PVT system when its volume is constant, it follows from (4.2) and the first law, $Q = \Delta U + W$, that

$$Q = \Delta U = nc_V\Delta T \quad \text{and} \quad W = 0. \tag{4.6}$$

Isothermal Processes

Since the internal energy U of an ideal gas depends on temperature only, U is constant during an isothermal (constant temperature) process, which implies that $\Delta U = 0$. It then follows from the first law that $Q = W$. By using $P = nRT/V$, the work done in a isothermal process becomes (see (2.3))

$$W = \int_{V_0}^{V_f} P(V)dV = nRT\int_{V_0}^{V_f}\frac{dV}{V} = nRT\ln\left(\frac{V_f}{V_0}\right). \tag{4.7}$$

Hence,

$$Q = W = nRT\ln\left(\frac{V_f}{V_0}\right) \quad \text{and} \quad \Delta U = 0. \tag{4.8}$$

Adiabatic Processes

No heat is transferred into or out of a system at any time during an adiabatic process. To impose this restriction, we divide the process into many small sub-processes and require that $Q = 0$ in each sub-process. When the changes in the pressure and volume are small, the work done is

$$W = \int PdV \approx P\Delta V, \tag{4.9}$$

so that the first law $Q = \Delta U + W$ becomes

$$Q = \Delta U + P\Delta V. \tag{4.10}$$

Since $PV = nRT$ and $\Delta U = nc_V\Delta T$, the requirement $Q = 0$ implies that

$$nc_V\Delta T + \left(\frac{nRT}{V}\right)\Delta V = 0. \tag{4.11}$$

When considering specific processes, volume V is a function of a single variable. Hence, $V = V_{ad}(T)$ for an adiabatic process. When ΔV and ΔT are small, we have

$$\Delta V \approx \frac{dV_{ad}(T)}{dT}\Delta T. \tag{4.12}$$

By substituting this into (4.11) and rearranging terms, we obtain

$$\frac{1}{V_{ad}(T)}\frac{dV_{ad}(T)}{dT} = -\frac{c_V}{R}\frac{1}{T}, \tag{4.13}$$

which is exact for limitingly small sub-processes.

The solution to this ordinary differential equation is obtained by finding the indefinite integral of both sides, which gives

$$\ln V_{ad}(T) = -\left(\frac{c_V}{R}\right)\ln T + \text{const.} \tag{4.14}$$

It is convenient to rewrite this by using $\gamma - 1 = R/c_V$, which follows from $\gamma = c_P/c_V$ and $c_P - c_V = R$. Multiplying through by $\gamma - 1$ gives

$$(\gamma - 1)\ln V_{ad}(T) + \ln T = \text{const,} \tag{4.15}$$

where the constant term is different than in (4.14). From this and the properties of logarithms it follows that

$$\boxed{TV^{\gamma-1} = \text{const,}} \tag{4.16}$$

where the constant term is again different. This tells us that the value of $TV^{\gamma-1}$ remains unchanged everywhere along a line that represents an adiabatic equilibrium process in an ideal gas.

To find the relationship between P and V in an adiabatic process, we substitute the equation of state in the form $T = PV/nR$ into (4.16),

$$\left(\frac{PV}{nR}\right)V^{\gamma-1} = PV^{\gamma}\left(\frac{1}{nR}\right) = \text{const.} \tag{4.17}$$

Combining nR with the other constant gives

$$\boxed{PV^{\gamma} = \text{const.}} \tag{4.18}$$

Similarly, substituting $V=nRT/P$ into (4.16) gives

$$T\left(\frac{nRT}{P}\right)^{\gamma-1} = \frac{T^{\gamma}(nR)^{\gamma-1}}{P^{\gamma-1}} = \text{const.} \tag{4.19}$$

Combining constants in this gives

$$\boxed{T^{\gamma}/P^{\gamma-1} = \text{const.}} \tag{4.20}$$

To find the work done in an adiabatic equilibrium process, we use the initial pressure and volume P_0 and V_0 to specify the value of the constant in (4.18), which then becomes

$$PV^{\gamma} = P_0V_0^{\gamma}. \tag{4.21}$$

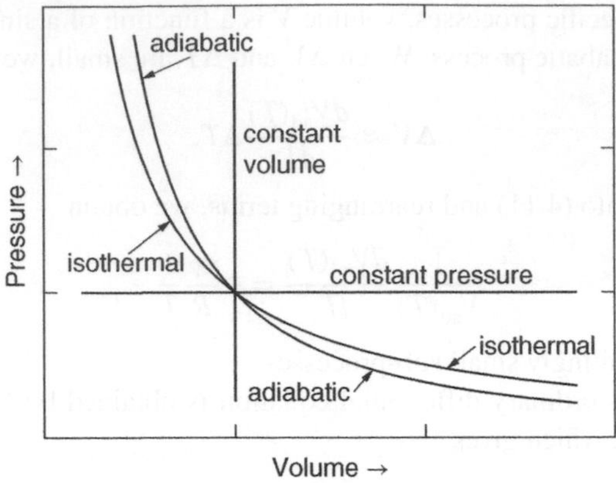

Figure 4.1 Equilibrium Processes in an Ideal Gas

Table 4.1 Equilibrium processes in an ideal gas with constant specific heats

Process	Q	ΔU	W
Constant-volume	$nc_V(T_f - T_0)$	$nc_V(T_f - T_0)$	0
Constant-pressure	$nc_P(T_f - T_0)$	$nc_V(T_f - T_0)$	$P(V_f - V_0)$
Isothermal	$nRT\ln(V_f/V_0)$	0	$nRT\ln(V_f/V_0)$
Adiabatic	0	$nc_V(T_f - T_0)$	$(P_0V_0^\gamma/(1-\gamma))(V_f^{1-\gamma} - V_0^{1-\gamma})$

Hence, $P = P_0V_0^\gamma/V^\gamma$. Substituting this into the integral for work gives

$$W = \int_{V_0}^{V_f} P(V)dV = P_0V_0^\gamma \int_{V_0}^{V_f} \frac{dV}{V^\gamma}, \tag{4.22}$$

so that

$$W = \frac{-P_0V_0^\gamma}{\gamma - 1}\left(\frac{1}{V_f^{\gamma-1}} - \frac{1}{V_0^{\gamma-1}}\right). \tag{4.23}$$

The different types of processes are plotted on a *PV*-diagram in **Figure 4.1**. There are no arrows on the lines as the direction of the processes, which are reversible, has not been specified. Note that the line representing an adiabatic process is steeper than the line representing an isothermal process. The expressions for Q, ΔU, and W for equilibrium processes in an ideal gas with constant specific heats are summarized in Table 4.1.

4.2 Temperature Change with Elevation

The cooler temperatures we experience on trips to the mountains are manifestations of adiabatic processes. To understand why, we visualize the atmosphere as a collection of localized air masses that are moved about by winds and convection currents. As the movements are rapid compared to the time needed for significant heat transfers, the processes in the various air masses are essentially adiabatic. When air rises, it expands and does work on the surrounding air. The energy needed to do the work comes from internal energy, which causes the temperature to decrease. Conversely, when an air mass descends, work is done on it, which causes its internal energy and temperature to increase.

The system used to model the motion of the atmosphere is a localized mass M of air that extends from elevation h to elevation $h + \Delta h$ in a vertical cylinder of cross-sectional area A. Assume that any accelerations of the system are negligible, so that the sum of the vertical components of the forces on the mass is zero. The upwards force exerted on the bottom of the system caused by the pressure of the air below it is $P(h)A$. The downward force on the top of the system is $P(h + \Delta h)A$, and the downward force of gravity (weight) is Mg. Adding these up gives

$$P(h)A - P(h + \Delta h)A - Mg = 0. \tag{4.24}$$

Air can be treated as an ideal gas. As a result, the volume of our system is related to its temperature and pressure by $PV = nRT$, which implies that $1 = (V/n)(P/RT)$. Multiplying this by M and expressing the volume V of the air mass as $A\Delta h$ gives

$$M = A\Delta h\left(\frac{M}{n}\right)\left(\frac{P}{RT}\right). \tag{4.25}$$

Substituting this into (4.24) and rearranging terms gives

$$\frac{P(h + \Delta h) - P(h)}{\Delta h} = -\left(\frac{M}{n}\right)\frac{g}{R}\frac{P(h)}{T(h)}, \tag{4.26}$$

where A has cancelled out and the dependence of P and T on h has been made explicit. Making Δh limitingly small gives

$$\frac{dP(h)}{dh} = -\left(\frac{M}{n}\right)\frac{g}{R}\frac{P(h)}{T(h)}. \tag{4.27}$$

The above result was obtained by treating air as an ideal gas, but no use was made of the adiabatic nature of the motion of the atmosphere. Since changes in the elevation of an air mass can be treated as adiabatic processes, it follows from (4.20) that

$$\frac{T(h)^{\gamma}}{P(h)^{\gamma-1}} = \text{const} \quad \text{or} \quad \gamma \ln T(h) - (\gamma - 1)\ln P(h) = \text{const}. \tag{4.28}$$

Differentiating this gives

$$\frac{\gamma}{T(h)}\frac{dT(h)}{dh} - \frac{\gamma - 1}{P(h)}\frac{dP(h)}{dh} = 0 \quad \text{or} \quad \frac{dP(h)}{dh} = \frac{\gamma}{\gamma - 1}\frac{P(h)}{T(h)}\frac{dT(h)}{dh}. \tag{4.29}$$

By substituting this value for $dP(h)/dh$ into (4.27) and multiplying by $(\gamma - 1)/\gamma$, we find that

$$\frac{dT(h)}{dh} = -\frac{\gamma - 1}{\gamma}\left(\frac{M}{n}\right)\frac{g}{R}. \tag{4.30}$$

M/n is the mass per mole, which equals the molecular weight in grams per mole.

For the atmosphere the parameters in (4.30) are $\gamma = 1.4$, $M/n = 28.8\ \text{g mol}^{-1}$, and $g = 9.8\ \text{m s}^{-2}$. Substituting these into (4.30) gives

$$\frac{dT(h)}{dh} = -\frac{0.4}{1.4}(0.0288\ \text{kg mol}^{-1})\frac{9.8\ \text{m s}^{-2}}{8.314\ \text{J (mol}\cdot\text{K)}^{-1}} = -9.70\times10^{-3}\ \text{K m}^{-1}. \quad (4.31)$$

That is, the temperature drops by 9.7 °C with every kilometer increase in elevation. In terms of feet and Fahrenheit temperature, this corresponds to a drop of 5.3 degrees Fahrenheit for every thousand feet.

4.3 Cyclic Processes

In a *cyclic process* a system periodically returns to its initial state, and each repeat of the process is one cycle. As the initial and final states are identical, each cycle leaves the system's state unchanged. Hence, all state functions have the same value at the end of a cycle that they had at the beginning. In particular, $U_f = U_0$, so that

$$\Delta U = U_f - U_0 = 0. \quad (4.32)$$

It then follows from the first law that

$$Q = W. \quad (4.33)$$

Thus, in a cyclic process, the heat Q equals the work W. It is not required that the system be in equilibrium throughout the process, but the existence of the state function U requires that it be in equilibrium at the point where the cycle begins and ends.

A cyclic process that is also an equilibrium process is represented by a closed contour (line) in the system's state space. In *PVT* systems, the work done in one cycle is equal to the area enclosed on a *PV*-diagram, as illustrated in **Figure 4.2**. The arrows indicate the direction of the process, and W is positive when the contour encloses the area in a clockwise direction, as in the figure. The integral $\int PdV$ that gives the work done during the part of

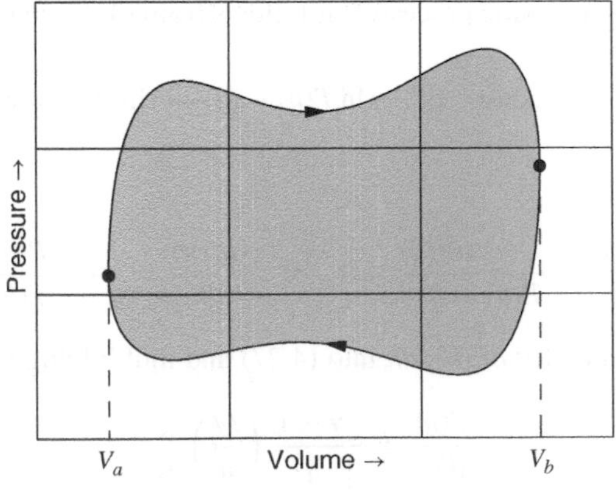

Figure 4.2 *PV-Diagram of a Cyclic Equilibrium Process*

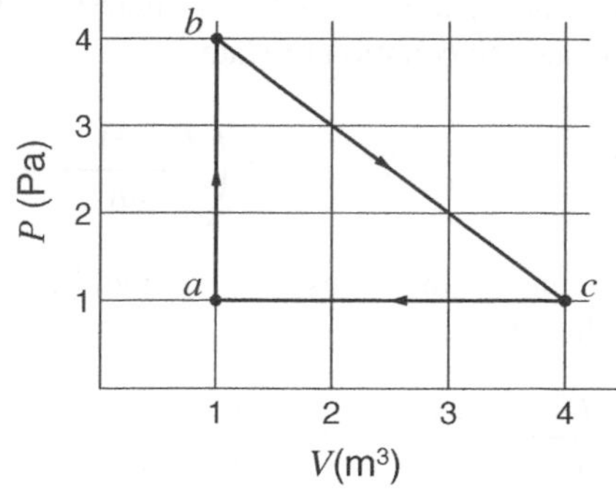

Figure 4.3 *PV-Diagram of a Cyclic Process*

the process from state a to state b is equal to the area under the upper part of the contour, *i.e.*, the area between the upper line, the horizontal axis, and the vertical lines between the axis and points a to b. The integral for the part of the process from state b back to state a subtracts the area under the lower part of the contour, so that the total work done is equal to the shaded area. The work done during one cycle of processes represented by counter-clockwise contours is the negative of the area enclosed.

Example 4.1 A cyclic process

Consider the process in an ideal gas diagrammed in **Figure 4.3**. Assume that $c_V = 2.5R$ and verify the information given in the table.

Process	W	ΔU	$Q = \Delta U + W$
$a \to b$	(1) 0.0	(5) 7.5 J	(9) 7.5 J
$b \to c$	(2) 7.5 J	(6) 0.0	(10) 7.5 J
$c \to a$	(3) −3.0 J	(7) −7.5 J	(11) −10.5 J
$a \to b \to c \to a$	(4) 4.5 J	(8) 0	(12) 4.5 J

From the graph we see that $P_a = 1.0\,\text{Pa}$ and $V_a = 1.0\,\text{m}^3$. The product P_aV_a is the area of each square on the graph, which is $1.0\,\text{Pa}\,\text{m}^3 = 1.0\,\text{Nm} = 1.0\,\text{J}$. As W is plus or minus the area under the line that represents the process, the work done can be found from the figure, which indicates that the works done in the first three items in the table are respectively 0.0, 7.5 J, and −3.0 J. The work done in one cycle (item 4) is the sum of the works done in the first three items, which equals 4.5 J. The change in the internal energy in process $a \to b$ (item 5) is found by using the equation of state $PV = nRT$ and the values $P_b = 4P_a$, $V_b = V_a$, and $c_V = 2.5R$,

$$\Delta U_{a\to b} = U_b - U_a = nc_V(T_b - T_a) = n(2.5R)\left(\frac{(4P_a)\,V_a}{nR} - \frac{P_aV_a}{nR}\right) = 7.5P_aV_a = 7.5\,\text{J}.$$

The change in the internal energy in process $b \to c$ (item 6) is found by using the temperatures in states b and c and the equation of state,

$$\Delta U_{b\to c} = U_c - U_b = nc_V(T_c - T_b) = nc_V\left(\frac{P_a(4V_a)}{nR} - \frac{(4P_a)V_a}{nR}\right) = 0.$$

The change in the internal energy in process $c \to a$ (item 7) is found by using $P_c = P_a$ and $V_c = 4V_a$, which implies that

$$\frac{nRT_c}{V_c} = \frac{nRT_a}{V_a} \quad \text{and} \quad \frac{T_c}{T_a} = \frac{V_c}{V_a} = 4.$$

It then follows that

$$\begin{aligned}\Delta U_{c\to a} &= U_a - U_c = nc_V(T_a - T_c) = n(2.5R)(T_a - 4T_a)\\ &= 2.5nRT_a(-3) = -7.5P_aV_a = -7.5\,\text{J}.\end{aligned}$$

As internal energy U is a state function, it is unchanged during a complete cycle (item 8), so that $\Delta U = 0$. After the columns for W and ΔU are completed, the first law $Q = \Delta U + W$ can be used to find the heats transferred in the final column (items 9, 10, 11, & 12). Hence, their values are 7.5, 7.5, −10.5, and 4.5 J.

We can check our calculations by checking that the value of Q in a cycle (item 12) equals the sum of its values in the sub-processes (items 9,10, & 11), which it does. The sum of the changes ΔU in the sub-process (items 5, 6, & 7) must equal the change for the cycle (item 8), which it does. Also, as $Q = W$ in a cyclic process, the value of W for each cycle (item 4) must equal the value of Q (item 12), which it does.

4.4 Heat Engines

In everyday speech, the word engine refers to a practical device, such as an automobile engine. The term heat engine is used more abstractly in thermodynamics. *A heat engine is a cyclic process in a system or device that absorbs heat $|Q_{\text{in}}|$, rejects heat $|Q_{\text{out}}|$, and does a positive amount of work W.* Although abstractly defined, heat engines are useful for understanding the performance of real devices. Because of the absolute value signs, $|Q_{\text{in}}|$ and $|Q_{\text{out}}|$ are positive. As the sign convention for heat associates a plus sign with $|Q_{\text{in}}|$ and a minus sign with $|Q_{\text{out}}|$, the total heat transferred to the system in one cycle is

$$Q = |Q_{\text{in}}| - |Q_{\text{out}}| = W > 0, \tag{4.34}$$

where the second equal sign follows from the characteristic $Q = W$ of a cyclic process.

To help understand the concept of a heat engine, consider a fossil fueled or nuclear fueled electric power plant. A block diagram of a fossil fueled plant is shown in **Figure 4.4**. In a nuclear plant, the heat transferred to the steam engine comes from a nuclear reaction. The heat engine is the part of the power plant (the steam engine) that accepts heat from the burning fuel or nuclear reaction, rejects heat to some body of water or the atmosphere, and does useful work (via the rotating shaft) by powering a generator that produces electrical

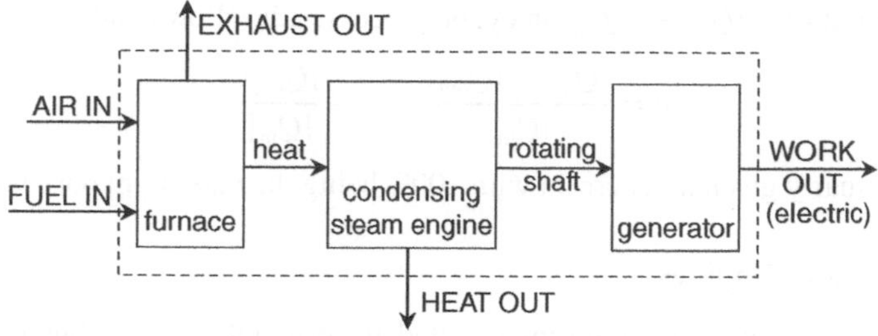

Figure 4.4 *Electric Power Plant*

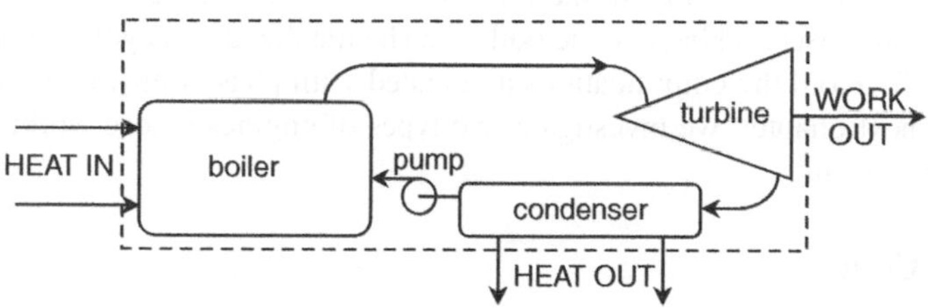

Figure 4.5 *Condensing Steam Engine*

energy.[1] The heat engine is different from the engine of everyday speech, which refers to the entire apparatus for converting the chemical energy in the fuel into useful work.

A simplified diagram of the steam engine in a fossil fueled plant is given in **Figure 4.5**. The dashed lines in the figure represent a *control surface*, which is an abstract surface that separates the system being analyzed from its surroundings. Everything inside the control surface is the heat engine. Heat $|Q_{in}|$ is transferred to the water in the boiler from the burning of fuel in a fire box, which is external to the system. This heat is used to change liquid water into steam, which then passes to the turbine, where it expands and pushes on moving turbine blades. The rotating shaft of the turbine supplies the work W that turns a generator, which is external to the system. In the condenser, the expanded steam from the turbine transfers heat $|Q_{out}|$ to a cooling medium. As heat is transferred from the steam, it changes back to liquid water and is returned to the boiler by the pump.

Energy Efficiency

An efficiency is the ratio of something desired divided by whatever is required to obtain the desired result. As the purpose of a heat engine is to obtain useful work from the energy transferred to it as heat, the *energy efficiency* of a heat engine is

$$e = \frac{\text{work out}}{\text{heat in}} = \frac{W}{|Q_{in}|}. \tag{4.35}$$

[1] The steam engines formerly used on railroads are *not* good examples of heat engines, as they did not operate on closed cycles. Instead of transferring heat to and from a working substance that remained within the device, they are open systems which take in water from their tenders and reject steam from their stacks along with the combustion products from the burning fuel.

Since $W = Q$ and $Q = |Q_{in}| - |Q_{out}|$ in cyclic processes, it follows that

$$e = \frac{|Q_{in}| - |Q_{out}|}{|Q_{in}|} = 1 - \frac{|Q_{out}|}{|Q_{in}|}. \tag{4.36}$$

Efficiency is usually given as a percent with 100% being the maximum possible efficiency.

Cyclic Processes as Engines

Although practical engines are quite complicated, the principles underlying their operation can be understood by analyzing the changes that occur to the device's working substance as it circulates through the engine. In a steam engine, the working substance is water in its liquid and vapor phases. The transformations of the water as it passes from the boiler to the turbine to the condenser and back to the boiler can be idealized as a cyclic equilibrium process in water. To avoid the complications associated with phase transitions, which are the subject of the next chapter, we investigate two types of engines whose working substance can be idealized as air.

Gas Turbine Cycle

Gas turbines are internal combustion engines used in propeller-driven aircraft, in ships, and to generate electrical power. As diagrammed in **Figure 4.6**, a gas turbine takes in air from the atmosphere, increases the pressure of the air in a compressor, and adds fuel to it. The fuel-air mixture is then burned, and the hot gases from the burner pass through two turbines. The first turbine supplies the work required to drive the compressor, and the second does useful work. In a turbo-prop aircraft, this work is used to rotate the propeller. A jet engine is also a gas turbine but the useful work is then used to increase the kinetic energy of the air, so that the exhaust gases have greater momentum than the intake gases.

Unlike a condensing steam engine, which is a closed system, a gas turbine exchanges matter with its surroundings and thus is an open system. However, they can be idealized as closed systems by replacing the burner with a heat exchanger and returning the gas from the turbines to the compressor after cooling it in a second heat exchanger, as diagrammed in **Figure 4.7**.

A *PV*-diagram for the *gas turbine cycle* for an ideal gas with $\gamma = 1.4$ is shown in **Figure 4.8**. This cycle is also known as a *Brayton cycle* and a *Joule cycle*. It combines two adiabatic processes and two constant pressure processes. To understand the relationship between the real engine and the diagrammed cyclic process, consider a small quantity of air as it passes through the real device. The adiabatic process $a \to b$ in the figure

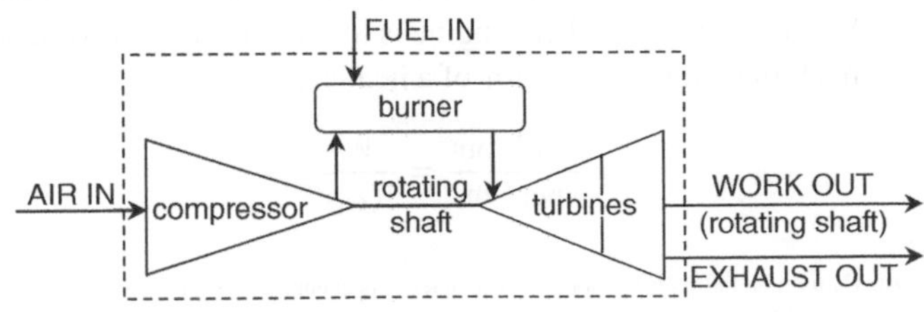

Figure 4.6 *Fuel Burning Gas Turbine*

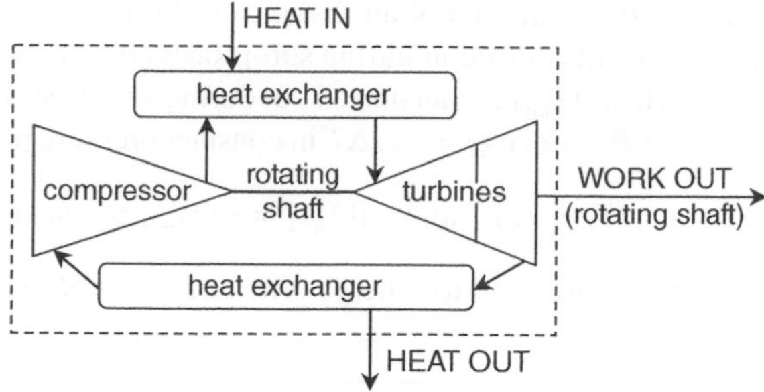

Figure 4.7 *Gas Turbine with Heat Exchangers*

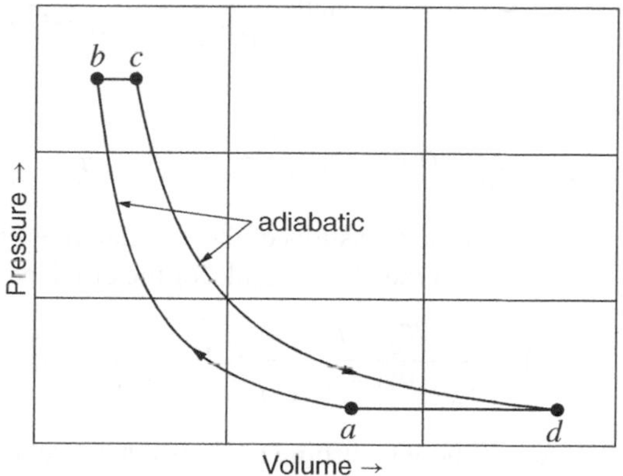

Figure 4.8 *PV-Diagram of the Gas Turbine Cycle*

represents the compression of the air in the compressor, the constant pressure process $b \to c$ represents the heating of the air by the burning of the fuel, and the adiabatic process $c \to d$ represents the expansion of the gas in the turbines. The gas enters the compressor as air, which is a mixture of about 80% nitrogen and 20% oxygen, while the exhaust gases are a mixture of nitrogen, oxygen, carbon dioxide, water vapor, and other combustion products. The idealized cycle ignores the small changes to the number of moles and the specific heat that results from burning the fuel. It replaces the real process in the burner by one that transfers heat to the air without changing its composition. In the real engine, the exhaust gases leave at atmospheric pressure, and the intake air enters at atmospheric pressure. We can imagine that the exhaust gases, idealized as hot air, are cooled at atmospheric pressure and returned to the compressor. This imagined cooling is represented by the constant pressure process $d \to a$.

Example 4.2 Gas turbine efficiency

Find the energy efficiency of the gas turbine cycle and express the answer as a function of the ratio of the pressures of the constant pressure processes.

Solution. The system is a small amount of air that passes through the cycle diagrammed in **Figure 4.8**. Heat $|Q_{in}|$ is transferred in during sub-process $b \to c$, which is at the high pressure $P_h = P_b = P_c$. Heat $|Q_{out}|$ is transferred out during sub-process $d \to a$, which is at low pressure $P_l = P_d = P_a$. Since $Q = nc_p\Delta T$ in constant pressure processes, we have

$$|Q_{in}| = Q_{b\to c} = nc_P(T_c - T_b) \quad \text{and} \quad |Q_{out}| = -Q_{d\to a} = -nc_P(T_a - T_d).$$

Substituting these into the expression for energy efficiency in (4.36) gives

$$e = 1 - \frac{|Q_{out}|}{|Q_{in}|} = 1 - \frac{nc_P(T_d - T_a)}{nc_P(T_c - T_b)} = 1 - \frac{T_d - T_a}{T_c - T_b}.$$

As $(T/P^{(\gamma-1)/\gamma}) = \text{const}$ for the two adiabatic processes (see (4.20)), we have

$$\frac{T_a}{P_l^{(\gamma-1)/\gamma}} = \frac{T_b}{P_h^{(\gamma-1)/\gamma}} \quad \text{and} \quad \frac{T_c}{P_h^{(\gamma-1)/\gamma}} = \frac{T_d}{P_l^{(\gamma-1)/\gamma}}.$$

Rearranging these gives

$$T_b = r_P^{(\gamma-1)/\gamma} T_a, \quad \text{and} \quad T_c = r_P^{(\gamma-1)/\gamma} T_d,$$

where $r_P = P_h/P_l$ is the ratio of the pressures of the constant pressure processes. Substituting the values of T_b and T_c into the above result for the efficiency gives

$$e = 1 - \frac{T_d - T_a}{r_P^{(\gamma-1)/\gamma}(T_d - T_a)} = 1 - \frac{1}{r_P^{(\gamma-1)/\gamma}}. \tag{4.37}$$

Thus, the energy efficiency of the gas turbine cycle is determined by the *pressure ratio* r_P. Higher pressure ratios give greater energy efficiency.

Otto Cycle

The operation of an automobile engine is idealized by the cyclic processes called the Otto cycle, named for Nikolaus Otto, the inventor of the four-stroke gasoline engine. A *PV*-diagram of the cycle for a gas with $\gamma = 1.4$ is shown in **Figure 4.9**. Two adiabatic processes are combined with two constant volume processes. To understand the relationship of the Otto cycle to a gasoline engine, consider the gas in one of the engine's cylinders at the beginning of the compression stroke. This gas is our system. The cycle starts in state a, at which point the gas is a mixture of air and vaporized fuel. The adiabatic process $a \to b$ in the diagram represents the compression of the gas as the piston moves up during the compression stroke and reduces the volume of the cylinder from a maximum value of $V_{max} = V_a = V_d$ to a minimum value of $V_{min} = V_b = V_c$. A spark plug then ignites the fuel-air mixture, which rapidly burns and causes a rapid increase in temperature and pressure. The burning is sufficiently rapid that it can be idealized by the constant volume process $b \to c$. The expansion of the system (the burnt gases) from volume V_{min} to volume V_{max} as the piston moves down in the power stroke is represented by adiabatic process $c \to d$. The cylinder's exhaust valve opens at the end of the power stroke and the burnt gases are forced out as the piston moves up during the exhaust stroke. The exhaust valve closes and the intake valve opens when the volume is minimized. A fresh mixture of fuel and air is then

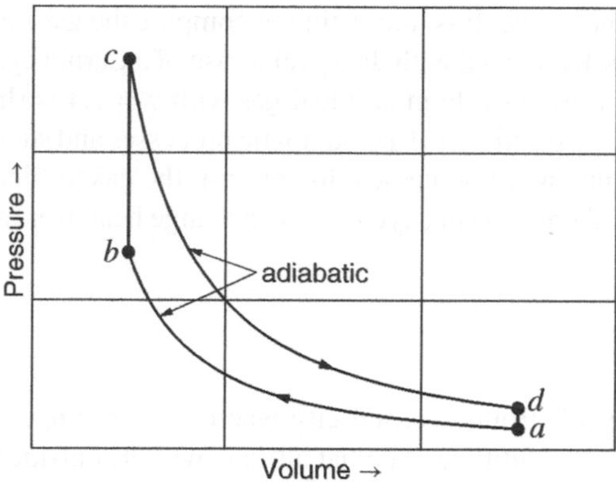

Figure 4.9 *PV-Diagram of the Otto Cycle*

drawn in as the piston moves down during the intake stroke. The purpose of the exhaust and intake strokes is to replace the burnt gases at volume V_{max} with a fresh fuel-air mixture of the same volume. The replacement is represented by the constant volume process $d \rightarrow a$. The cylinder is now filled with a fresh fuel-air mixture and the cycle is repeated.

It is left as a problem to show that the energy efficiency of an Otto cycle in an ideal gas with constant c_V is

$$e = 1 - (1/r_V)^{\gamma-1}, \tag{4.38}$$

where $r_V = V_{max}/V_{min}$, which is the *compression ratio* of the engine. It follows from this that the energy efficiency of a gasoline engine can be increased by increasing its compression ratio until factors not considered here prevent it from being further increased.

Heat engines both accept and reject heat and they reject it at a lower temperature than they accept it. As will be shown in Chapter 7, when there is a range of temperatures at which heat can be exchanged, the most efficient engine accepts heat at the highest available temperature and rejects heat at the lowest temperature. This ideal type of engine operates

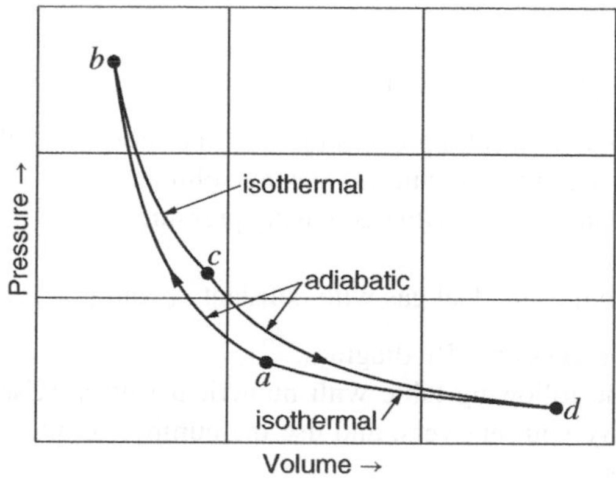

Figure 4.10 *Carnot Cycle in an Ideal Gas*

in what is called a Carnot cycle. It is interesting to compare the gas turbine and Otto cycles diagrammed in Figures 4.7 and 4.8 with the special case of a Carnot cycle that uses the same working substance. A Carnot cycle in an ideal gas with $\gamma = 1.4$ is shown in **Figure 4.10**, where sub-processes $a \to b$ and $c \to d$ are adiabatic processes and sub-processes $b \to c$ and $d \to a$ are constant temperature processes. In contrast, the gas turbine and Otto cycles are idealizations of practical engineering cycles that exchange heat over ranges of temperature.

Problems

4.1 An airplane takes off from a coastal city where the air temperature is 20 °C (68 °F) and climbs to cruising altitude. At what altitude will the outside temperature be 0 °C? (Assume that there is no temperature inversion like those which occur in the Los Angeles area.)

4.2 Consider a cyclic equilibrium process made up of following sub processes:
$a \to b$ is isothermal with increasing volume;
$b \to c$ is at constant volume with decreasing pressure;
$c \to a$ is adiabatic.

(a) Sketch the process on a *PV* diagram.

(b) Complete the following table with numerical values. (Use a separate piece of paper.) Justify your answers, and use the numbers in the table to identify your justifications.

sub process	Q	ΔU	W
$a \to b$	3600 J	(3)	(7)
$b \to c$	(1)	(4)	(8)
$c \to a$	0	(5)	−1200 J
Complete cycle	(2)	(6)	(9)

(c) Find the energy efficiency of the process.

4.3 Consider a cyclic equilibrium process made of the following sub-processes:
$a \to b$ is at constant pressure with increasing volume.
$b \to c$ is at constant volume with decreasing pressure.
$c \to a$ is adiabatic.
Assume the system is an ideal gas with constant c_V and $\gamma = 1.5$.

(a) Sketch the process on a *PV* diagram.

(b) Complete the following table with numerical values. (Use a separate piece of paper). Justify your answers, and use the numbers in the table to identify your justifications.

State	P	V	T
a	20×10^5 Pa	$0.3\,\text{m}^3$	400 K
b	(1)	$0.6 \times 0.6\text{m}^3$	(4)
c	(2)	(3)	(5)

4.4 A cyclic equilibrium process in n moles of an ideal gas with $c_V = 2.5R$ is formed of three sub-processes:
$a \to b$ is a constant pressure doubling of the volume;
$b \to c$ is at constant volume with decreasing pressure;
$c \to a$ is adiabatic.
Assume that $V_b = V_c = 2V_a$.

(a) Sketch the process on a PV diagram.
(b) Find the heat absorbed in sub-process $a \to b$ in terms of T_a, n, and R.
(c) Find the heat rejected in sub-process $b \to c$ in terms of T_a, n, and R.
(d) Find the energy efficiency of the cycle. Give a numerical answer.

4.5 Find the final temperature when air is compressed adiabatically from an initial temperature of 293 K (20 °C):

(a) When the compression ratio V_f/V_0 is $1/10$ (as in gasoline engines).
(b) When the compression ratio V_f/V_0 is $1/20$ (as in diesel engines).

4.6 **(a)** Show that the energy efficiency of the air-standard Otto cycle is $e = 1 - (1/r_V)^{\gamma-1}$, where $r_V = V_{max}/V_{min}$.
(b) Find numerical values of e for $\gamma = 1.4$ when $r_V = 10$ and when $r_V = 20$.

5

Phase Transitions

The freezing of water and the melting of ice are familiar phase transitions, but the existence of several high pressure phases of ice is less well known. A phase is a homogenous part of a system that is separated by a boundary from other phases, and different types of systems have different phases. For example, *PVT* systems have solid, liquid, and gaseous phases, while magnetic systems have paramagnetic and ferromagnetic phases (and other magnetically ordered phases). The phase diagrams for *PVT* and magnetic systems help illustrate the nature of phase transitions, and the van der Waals model gives useful insights into the liquid-vapor transition.

5.1 Solids, Liquids, and Gases

As mentioned earlier, fluids flow to conform to the shape of their container, and both liquids and gases are fluids. A gas can expand to fill a container of arbitrary size, while a liquid can expand only to a certain amount before some of it vaporizes and the system becomes a two-phase mixture of liquid and gas. The discussion in this chapter is limited to single-component systems, which are systems that contain molecules of a single type.[1]

Phase Diagrams

The plot of pressure versus temperature in **Figure 5.1** is a *phase diagram* of a system like water. Points on the curves represent states in which two phases can coexist in equilibrium. Points not on a curve represent states with a single phase present. At temperatures below the critical point, a gas is called a *vapor*. The liquid and vapor phases are in equilibrium along the *vaporization curve*, the solid and liquid phases are in equilibrium along the *fusion curve*, and the solid and vapor (gas) phases are in equilibrium along the *sublimation curve*.

When two phases are present in thermal equilibrium, the temperature is uniquely determined by the pressure, and vice versa. The Celsius temperature scale is based on this fact.

[1] Multi-component systems are discussed in Chapter 13.

Thermodynamics and Statistical Mechanics: An Integrated Approach, First Edition.
Robert J Hardy and Christian Binek.

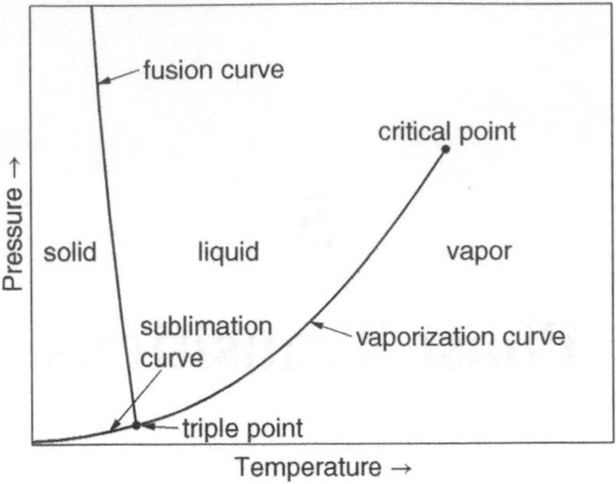

Figure 5.1 *Phase Diagram of a PVT System*

It assigns 0 °C to the temperature at which ice and water are in equilibrium at one standard atmosphere (101.325 kPa) and assigns 100 °C to the temperature at which steam (water vapor) and liquid water are in equilibrium at one standard atmosphere. These two temperatures are called the *ice point* and the *steam point*.

Triple Point

The point on the phase diagram where the three two-phase curves meet is a *triple point*. It represents the state in which the three phases can coexist in equilibrium. A triple point has a unique temperature and unique pressure. The Kelvin temperature scale is defined by the temperature of the triple point of water.[2] It is the scale with the value of zero at absolute zero and has the same size temperature increment as the Celsius scale. Since the Celsius temperature of absolute zero is −273.15 °C and the triple point of water is 0.01 °C, the Kelvin scale is the absolute temperature scale that assigns temperature 273.16 K to the triple point of water. Some important properties of water are given in Table 5.1. Notice that the pressure at the triple point is much less than one standard atmosphere.

Many substances have more than one solid phase, with the different phases having different macroscopic properties and different crystal structures. Some of the high pressure phases of water are shown in **Figure 5.2**. The vaporization curve is not shown because of

Table 5.1 *Properties of water*

Triple point: $P_{tr} = 612\,\text{Pa}$, $T_{tr} = 273.16\,\text{K}$ (exactly)
Critical point: $P_c = 22.12\,\text{MPa}$, $T_c = 647.4\,\text{K}$, $v_c = 3.17\times10^{-3}\,\text{m}^3\text{kg}^{-1}$
Steam point: $v_{steam} = 1.673\,\text{m}^3\text{kg}^{-1}$, $v_{liq} = 1.044\times10^{-3}\,\text{m}^3\text{kg}^{-1}$
Ice point: $v_{ice} = 1.091\times10^{-3}\,\text{m}^3\,\text{kg}^{-1}$, $v_{liq} = 1.000\times10^{-3}\,\text{m}^3\,\text{kg}^{-1}$
Latent heat of fusion at 1 atm: $L_f = 80\,\text{kcal}\,\text{kg}^{-1}$
Latent heat of vaporization at 1 atm: $L_v = 539\,\text{kcal}\,\text{kg}^{-1}$

[2] Temperature scales are discussed in more detail in Section 7.5 and Chapter 8.

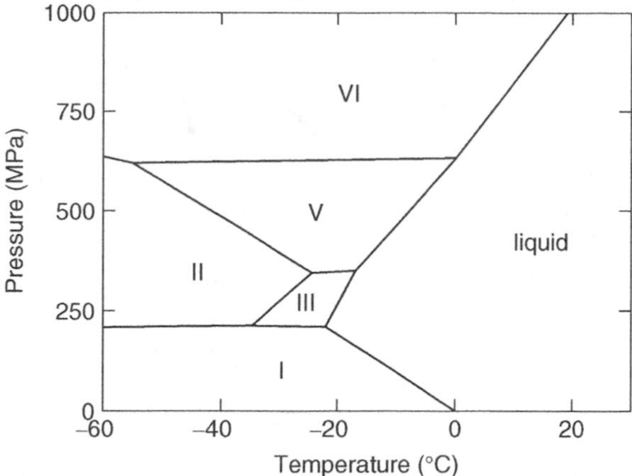

Figure 5.2 *High Pressure Phases of Water*

the high pressures being plotted. The different solid phases, or ices, are identified by Roman numerals. The only ice that exists at atmospheric pressure is ice-I, which is the only ice that is less dense than water. Notice the large number of triple points, *i.e.*, points with a unique temperature and pressure at which three phases can be present in equilibrium. For example, the temperature and pressure at triple point for ice-I, ice-II, and ice-III are $T = -34.7\ °\text{C}$ and $P = 212.9\,\text{MPa}$.

Critical Point

The point on the phase diagram where the vaporization curve ends is the *critical point*. To understand its significance, consider an equilibrium process in which the point representing a system moves along the vaporization curve. Below the critical point, the liquid and gas (vapor) are separated by a surface with the gas above and the liquid below. Beyond the critical point the surface disappears, and the qualitative distinction between liquid and gas ceases to exist. This implies it is possible to change a liquid to a gas in a process in which only one phase is present at all times. Such a process is represented by the line $a \to d \to c$ in **Figure 5.3**. The system is a liquid in state a and a gas in state c. By passing above the critical point the system remains in a single phase as it goes from c through d to c. This is in contrast with what happens in the constant pressure process represented by the line $a \to b \to c$. The system is in a single phase (liquid) as the temperature increases from point a to point b. When it reaches point b, the system has two phases present as it changes from all liquid to all gas with no change in temperature. As the temperature increases beyond point b to point c, the system is again in a single phase.

Notice that the slope of the fusion curve is negative in Figure 5.1 but is positive in Figure 5.3. A positive slope is characteristic of a system whose volume decreases upon freezing. Most substances are of this type. A fusion curve with a negative slope is characteristic of substances, such as water, that expand upon freezing.[3]

[3] The relationship between the slopes of two-phase curves and the associated change in volume is described by Clapeyron's equation, which is discussed in Section 11.4.

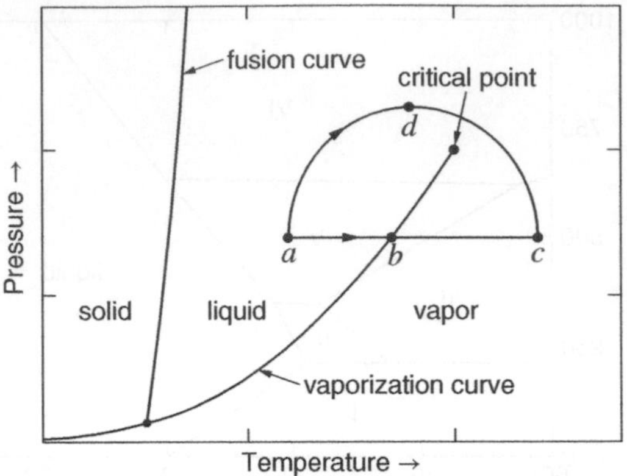

Figure 5.3 *Two Processes that Transform a Liquid into a Vapor*

Equation of State Surface

An *equation of state surface* for a substance whose volume decreases upon freezing is shown in **Figure 5.4**. A few isotherms are included to help us visualize it as three dimensional. Notice that the critical point remains a point when viewed in three dimensions, that the triple point becomes a line, and that the two-phase curves become regions on the equation of state surface. The projection of the surface onto the *PV* plane yields the *PV*-diagram in **Figure 5.5**. Each point on the *PV*-diagram and on the equation of state surface represents a different state of the system. In contrast, a point on one of the two-phase curves in a phase diagram, such as Figure 5.3, represents many states. The process represented by the isothermal line $b1 \rightarrow b2$ in Figure 5.5 shows the sequence of states that a system passes through in the transition represented by the single point b in Figure 5.3. As heat is transferred to the system, it passes through the states between $b1$ and $b2$, and its volume increases as the amount of gas increases relative to the amount of liquid. Along the line $b1 \rightarrow b2$ the

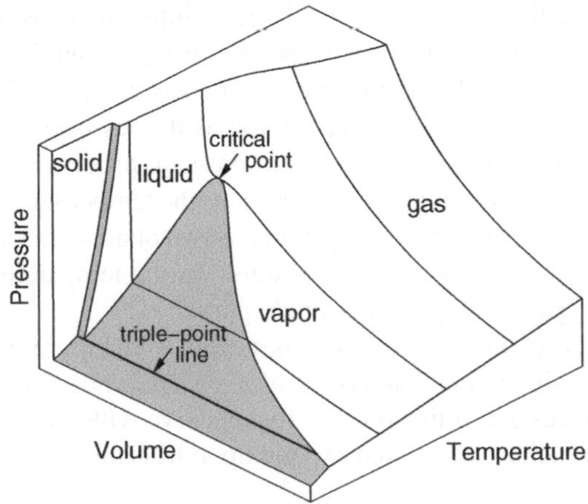

Figure 5.4 *Equation of State of a PVT System*

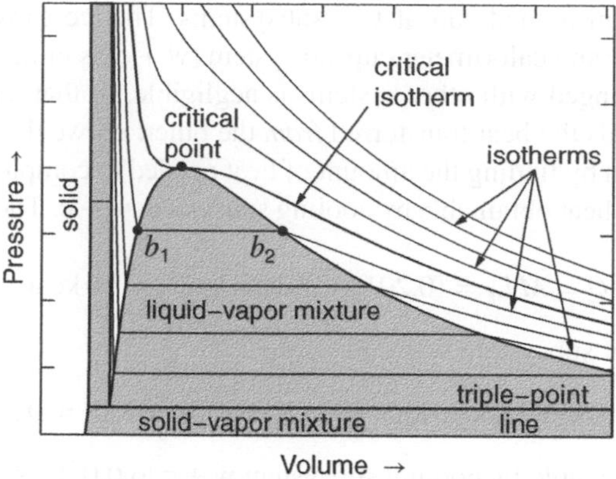

Figure 5.5 *Isotherms on a PV-Diagram*

specific volume of the liquid is the same as in state $b1$ and the specific volume of the gas is the same as in state $b2$. Notice that the pressure decreases with increasing volume much more rapidly along an isotherm in the liquid phase than it does along the same isotherm in the gas phase. This difference is reflected in the value of the isothermal bulk modulus $B_T = -V(\partial P/\partial V)_T$, which is much greater for the liquid than for the gas.

5.2 Latent Heats

The amount of heat absorbed or released when a substance changes phase at constant temperature, called a *latent heat*, depends on the size of the system and the phases involved. For example, heat is required to transform a system from a liquid at state $b1$ in Figure 5.5 to a vapor (gas) at state $b2$. The heat Q required to transform the system from one phase to another without changing its temperature or pressure is

$$Q = ML, \tag{5.1}$$

where L is a *specific latent heat* and $M$is the mass of the system. Specific latent heats are defined so their values are positive. The *specific latent heat of fusion* L_f is the heat per unit mass absorbed when a system changes from a solid to a liquid (melting), which is also the heat per unit mass given off when the liquid freezes. The *specific latent heat of vaporization* L_v is the heat per unit mass absorbed when a liquid vaporizes to become a gas, which is the heat per unit mass given off when the gas condenses. The latent heats L_f and L_v for water are given in Table 5.1. The *specific latent heat of sublimation* L_s is the heat absorbed when a solid sublimes, *i.e.*, when it changes directly to a gas without passing through a liquid phase.

Example 5.1 Calorimetry with a phase change

A 0.20 kg piece of ice is placed in a calorimeter cup containing 0.50 kg of water at 27 °C, and the combined system is allowed to come to thermal equilibrium. Find the final state of the system. The heat capacity of the calorimeter cup is 0.030 kcal K^{-1}.

Solution. The system is made up of two subsystems. The ice subsystem is initially at 0.0 °C and the water and calorimeter cup subsystem (w + c) is initially at 27 °C. Presumably, the heat exchanged with other systems is negligible, so that the heat transferred *to* one subsystem equals the heat transferred *from* the other. As we do not know how much ice remains, we start by finding the amount of heat needed to completely melt the ice and compare it with the heat obtainable by cooling subsystem w + c. The heat needed to melt all of the ice is

$$Q = ML_f = (0.20\,\text{kg})(80\,\text{kcal}\,\text{kg}^{-1}) = 16\,\text{kcal}.$$

The heat capacity of subsystem w + c is

$$C_P^{\text{w+c}} = (0.50\,\text{kg})(1.0\,\text{kcal}\,\text{kg}^{-1}\text{K}^{-1}) + 0.030\,\text{kcal}\,\text{K}^{-1} = 0.53\,\text{kcal}\,\text{K}^{-1}.$$

Hence the heat obtainable by cooling subsystem w + c to 0.0 °C is

$$Q = C_P^{\text{w+c}}(T_f - T_0) = (0.53\text{kcal}\,\text{K}^{-1})(0.0\,^\circ\text{C} - 27\,^\circ\text{C}) = -14.3\,\text{kcal}.$$

As the 14.3 kcal obtainable from subsystem w + c is less than the 16 kcal needed to melt the ice, some ice will remain. Presumably, the pressure of the system is close enough to 1.0 atm that the ice and water in the final state are at 0.0 °C. Since final state of the system is not uniquely determined by its temperature and pressure, we need to determine how the system is distributed between the two phases. To do this, we set the heat transferred *to* the ice subsystem equal to the heat transferred *from* subsystem w + c. As found above, this is 14.3 kcal. Hence, $M_{\text{ice}}L_f = 14.3\,\text{kcal}$, so that the amount of ice that has melted is $M_{\text{ice}} = 14.3\,\text{kcal}/(80\,\text{kcal}\,\text{kg}^{-1}) = 0.179\,\text{kg}$. As $0.20\,\text{kg} - 0.179\,\text{kg} = 0.021\,\text{kg}$ and $1.1\,\text{kg} + 0.179\,\text{kg} = 0.679\,\text{kg}$, the final state of the system has 0.021 kg of ice, 0.679 kg of liquid water, and the calorimeter cup in equilibrium at 0.0 °C.

Enthalpy of Transformation

Latent heats are defined for processes at constant temperature and pressure. We found in Chapter 3 that the heat transferred to a system at constant pressure is equal to the change in its enthalpy (see (3.11)). Thus, the heat $Q_{A\to B}$ needed to change a system *from* phase A *to* phase B is

$$Q_{A\to B} = \Delta H = H_B(T,P) - H_A(T,P), \tag{5.2}$$

where H_A and H_B are the enthalpies of two phases. It then follows from (5.1) that

$$L_{A\to B} = h_B(T,P) - h_A(T,P), \tag{5.3}$$

where $h = H/M$ is the specific enthalpy. That is, *the specific latent heat for a phase transition equals the difference between the specific enthalpies of the two phases.*

The above result can be used to relate the latent heat of sublimation to the latent heats of fusion and vaporization. Consider three points near the triple point in Figure 5.1 or 5.3 with point (T_f, P_f) on the fusion curve, point (T_v, P_v) on the vaporization curve, and point (T_s, P_s) on the sublimation curve. Then,

$$L_f = L_{\text{sol}\to\text{liq}} = h_{\text{liq}}(T_f, P_f) - h_{\text{sol}}(T_f, P_f), \tag{5.4}$$

$$L_v = L_{\text{liq}\to\text{vap}} = h_{\text{vap}}(T_v, P_v) - h_{\text{liq}}(T_v, P_v), \tag{5.5}$$

and

$$-L_s = L_{\text{vap}\to\text{sol}} = h_{\text{sol}}(T_s, P_s) - h_{\text{vap}}(T_s, P_s), \tag{5.6}$$

where sol, liq, and vap respectively designate the solid, liquid, and gas phases. There is a minus sign in the last equation, as L_s is the heat per unit mass needed for a sol → vap transition, not a vap → sol transition. Move the three points along the two-phase curves toward the triple point, so that points (T_f, P_f), (T_v, P_v), and (T_s, P_s) converge at the triple point (T_{tr}, P_{tr}). When the three points converge, the sum of the right hand sides of (5.4) through (5.6) add up to zero. The sum of the left hand sides must also be zero, so that $L_f + L_v - L_s = 0$. Thus, at the triple point we have

$$L_s = L_f + L_v. \tag{5.7}$$

For water $L_f = 79\,\text{kcal}\,\text{kg}^{-1}$ and $L_v = 596\,\text{kcal}\,\text{kg}^{-1}$ at the solid-liquid-gas triple point, so that $L_s = 675\,\text{kcal}\,\text{kg}^{-1}$. Notice that these values for L_f and L_v differ from those in Table 5.1. This is because the values of the specific latent heats vary as one moves along the associated two-phase curves.

5.3 Van der Waals Model

A relatively simple representation of the properties of fluids is given by the model based on the *van der Waals equation of state* developed by Johannes van der Waals. It states that

$$\left(P + \frac{a}{v^2}\right)(v - b) = RT, \tag{5.8}$$

or equivalently

$$P = \frac{RT}{v - b} - \frac{a}{v^2}, \tag{5.9}$$

where $v = V/n$ is specific volume. The parameters a and b are adjusted to fit the properties of specific substances. The equation of state tends to the ideal gas equation $Pv = RT$ at low densities, which are characterized by $|b| << v$ and $|a| << Pv^2$. The model nicely illustrates the nature of liquid-vapor phase transition.

Several of the isotherms of the van der Waals equation of state are shown in **Figure 5.6**, where the *critical isotherm* is the isotherm through the *critical point*. The straight constant pressure lines below the critical isotherm represent mixtures of liquid and vapor. (Where to draw them is discussed in Chapter 13.) The curved isotherms between where they touch the straight liquid-vapor isotherms represent states with a single phase present. These single phase states are not true equilibrium states because the system is more stable as a two-phase mixture.

Each of the curved isotherms below the critical point has a maximum and a minimum point. As the temperature increases, these points approach each other and become a single point at the critical isotherm. This single point is a point of inflection, which is a point

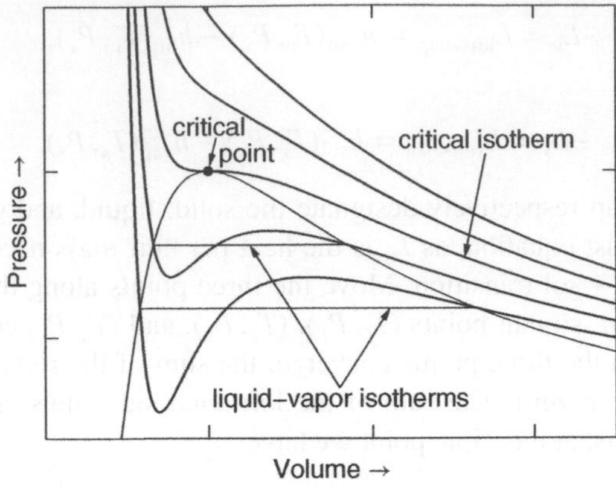

Figure 5.6 *Isotherms of a van der Waals Fluid*

where both the first and second derivatives vanish. Hence, the temperature T_c and specific volume v_c at the critical point are determined by

$$\frac{\partial P(T_c, v_c)}{\partial v} = 0 \quad \text{and} \quad \frac{\partial^2 P(T_c, v_c)}{\partial v^2} = 0. \tag{5.10}$$

Evaluating these equations for the van der Waals model gives

$$-\frac{RT_c}{(v_c - b)^2} + \frac{2a}{v_c^3} = 0 \quad \text{and} \quad \frac{2RT_c}{(v_c - b)^3} - \frac{6a}{v_c^4} = 0, \tag{5.11}$$

which imply that

$$v_c = 3b \quad \text{and} \quad T_c = \frac{8a}{27bR}. \tag{5.12}$$

The critical pressure P_c, obtained by substituting (5.12) into (5.9), is

$$P_c = \frac{RT}{(v_c - b)} - \frac{a}{v_c^2} = \frac{a}{27b^2}. \tag{5.13}$$

To choose parameters to fit the properties of specific substances, we use (5.12) and (5.13) to express a and b as functions of P_c and T_c,

$$a = \frac{27(RT_c)^2}{64P_c} \quad \text{and} \quad b = \frac{RT_c}{8P_c}. \tag{5.14}$$

Comparison with Experiment

To facilitate the comparison with experiment, we introduce the dimensionless *compressibility factor* $z = Pv/RT$, whose value for ideal gases is one ($z = 1$). At the critical point its value is

$$z_c = \frac{P_c v_c}{RT_c}. \tag{5.15}$$

Table 5.2 Critical properties and van der Waals parameters

	P_c (MPa)	T_c (K)	v_c (L mol^{-1})	z_c	a (J L mol^{-2})	b (L mol^{-1})
He	0.229	5.30	0.0577	0.300	0.0036	0.0241
Ne	2.62	44.5	0.0416	0.295	0.0220	0.0176
Ar	4.86	151	0.0752	0.291	0.137	0.0323
H_2	1.30	34.2	0.0650	0.297	0.0263	0.0274
N_2	3.50	126	0.0900	0.300	0.132	0.0375
O_2	5.04	154	0.0744	0.292	0.138	0.0318
CO_2	7.50	304	0.0956	0.283	0.360	0.0422

1 L = 1 liter = 10^{-3} m^3

As z_c is determined by three experimental parameters (P_c, T_c, v_c) but only two of them (P_c and T_c) are used to parameterize the model, comparing experimental values of z_c with the model value is a simple way of testing the model. The value of z_c obtained from the model values of P_c, T_c, and v_c is

$$z_c = \frac{P_c v_c}{RT_c} = \left(\frac{a}{27b^2}\right)(3b)\frac{1}{R}\left(\frac{27Rb}{8a}\right) = \frac{3}{8} = 0.375. \tag{5.16}$$

The van der Waals model predicts that the compressibility factor at the critical point is 0.375 for all fluids.

The experimental values for P_c, T_c, and v_c given in Table 5.2 were used to calculate the experimental compressible factors z_c and the model parameters a and b. Notice that the values of z_c are closer to each other than to the model value of 0.375. As the experimental compressibility factors are about 80% of the model value, we conclude that the van der Waals model is only moderately accurate.

Reduced Variables

The van der Waals equation of state can be simplified by giving the values of P, T, and v relative to their critical values. To do this, use (5.12) and (5.13) to express a and b as

$$a = 3P_c v_c^2 \quad \text{and} \quad b = \frac{1}{3}v_c. \tag{5.17}$$

From these and (5.16) it follows that

$$R = \frac{8}{3}\frac{P_c v_c}{T_c}. \tag{5.18}$$

Substituting these values for a, b, and R into the van der Waals equation of state in (5.9) and dividing by P_c gives

$$\frac{P}{P_c} = \frac{8v_c}{T_c}\frac{T}{3v - v_c} - \frac{3v_c^2}{v^2}. \tag{5.19}$$

The values of P, T, and v relative to their critical values, called *reduced variables*, are

$$\widetilde{P} = \frac{P}{P_c}, \quad \widetilde{T} = \frac{T}{T_c}, \quad \widetilde{v} = \frac{v}{v_c}. \tag{5.20}$$

In terms of reduced variables, the van der Waals equation of state is

$$\boxed{\widetilde{P} = \frac{8\widetilde{T}}{3\widetilde{v} - 1} - \frac{3}{\widetilde{v}^2}.} \tag{5.21}$$

This equation is universal in the sense that it does not depend on the parameters a and b, which characterize specific fluids. This suggests that all fluids behave similarly when P, T, and v are scaled relative to their critical values. The values of z_c in Table 5.2 also suggest this conclusion. Although the suggestion is not strictly accurate, it is an excellent approximation for many fluids.[4]

5.4 Classification of Phase Transitions

So far, the solid, liquid, and gaseous phases of *PVT* systems have been used to describe the nature of phase transitions. Other materials have different types of phases. To mention a few, magnetic materials have paramagnetic and ferromagnetic phases. Some dielectric materials possess a ferroelectric phase. Alloys, which are mixtures of different metals, have phases whose stability depends on the relative concentrations of the ingredient. Some metals have a superconducting phase at low temperatures, and liquid helium has a superfluid phase.

Classification

Phase transitions are classified as first order or second order. At a *first order phase transition*, the specific heat and other properties change discontinuously, and a latent heat is associated with the transition. The freezing and melting of water and its vaporization and condensation at temperatures below the critical point are familiar first order transitions. *Second order phase transitions* occurs at critical points. The properties of systems change continuously at a second order transition and there is no latent heat of transformation. They are also called *continuous phase transitions*. Systems behave in unusual ways at a critical point and their detailed description involves mathematical singularities.

Magnetic Systems

The paramagnetic-ferromagnetic transition in magnetic materials is an important example of a second order transition. In analogy to *PVT* systems, we can think of magnetic materials as $\mathcal{HMT}$ systems, where $\mathcal{H}$ is the magnetic field and $\mathcal{M}$ is the magnetization. **Figure 5.7** shows the typical dependence of $\mathcal{M}$ on temperature T and field $\mathcal{H}$ in systems with a single

[4] The similarity of the behavior of different fluids is the subject of the *principle of corresponding states*, which is outside the scope of this book.

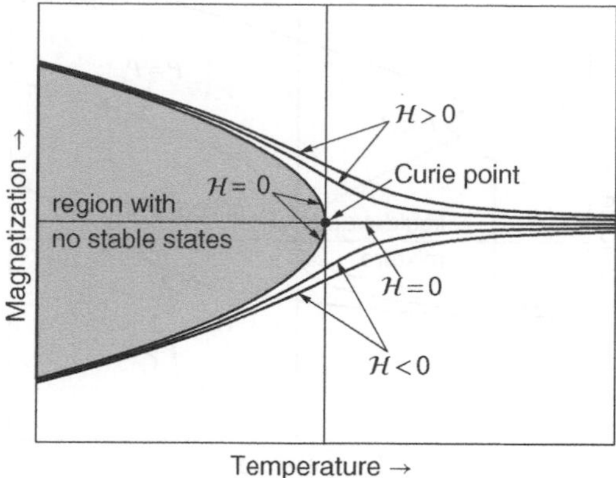

Figure 5.7 *Magnetization of a Ferromagnetic System*

magnetic domain.[5] The magnetization has two possible orientations ("up" and "down") determined by the shape of the sample and its microscopic structure. The horizontal line across the center of the figure represents states of zero magnetization ($\mathcal{M} = 0$), points above it represent the "up" orientation ($\mathcal{M} > 0$), and points below it represent the "down" orientation ($\mathcal{M} < 0$). The *critical point* of the system is called the *Curie point*.

At temperatures above the Curie point, the system is in a paramagnetic phase, and the magnetization changes continuously from positive to negative as the field is changed from positive to negative. In weak fields, the magnetization is proportional to the magnetizing field, and there is no phase transition.

At temperatures below the Curie point, the edges of the shaded region in the figure represent states with zero $\mathcal{H}$-field but non-zero magnetization. As the field is changed from small positive values to small negative values, the magnetization $\mathcal{M}$ changes discontinuously from "up" to "down". The two orientations are the two ferromagnetic phases of the system, and the phase transition is first order.

To see what is special about the critical point, consider a process in which the magnetic field is zero and the temperature is lowered from above the Curie point. The point that represents the state of the system starts at the right and moves along the zero magnetization line until it arrives at the Curie point, where the system changes from its paramagnetic phase to one of the two ferromagnetic phases. Whether the magnetization enters the "up" or "down" phase is determined by the small random fluctuations present in real systems. The system is intrinsically unstable, and the transition is second order. The slope of the zero-field magnetization curve diverges as the Curie point is approached from the ferromagnetic phases.

The behaviors of $\mathcal{HMT}$ and *PVT* systems near their critical point are both similar and different. **Figure 5.8** shows three isobars (constant pressure lines) of a *PVT* system, which can be compared with the constant field lines in the previous figure. Both $\mathcal{HMT}$ and *PVT* systems possess two phases below the critical temperature and one phase above it.

[5] Obtaining a single magnetic domain in most materials requires a small sample. For iron the sample is required to have dimensions of the order of ten nanometers.

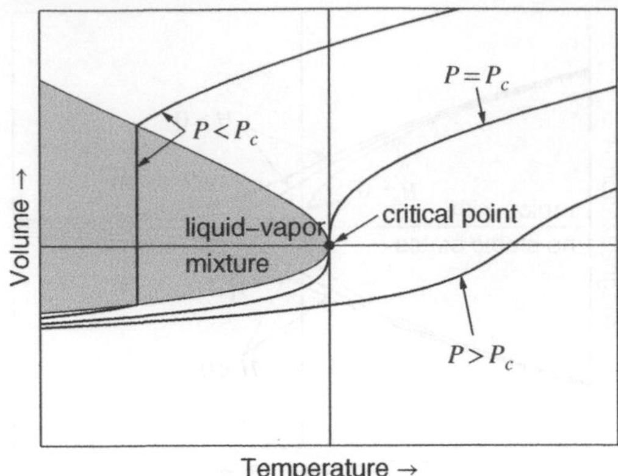

Figure 5.8 *Isobars of a PVT System Near the Critical Point*

Unlike the magnetic system, *PVT* systems have states in the shaded region of the figure. In the shaded region, the line labeled $P < P_c$ represents states in which both phases are present, while $\mathcal{HMT}$ systems have no stable states in the shaded region in Figure 5.7. The slope of the constant pressure line labeled $P = P_c$ diverges at the critical point, just as the slope of the zero-field line diverges at the Curie point. The instability at the critical point in *PVT* systems is manifested by the phenomenon of *critical opalescence* in which a normally transparent fluid becomes translucent.

Problems

5.1 A 200 g piece of ice at −10 °C is placed in 500 g of water at 50 °C. What is the final temperature? Assume the heat transferred to the container is negligible.

5.2 Solid carbon dioxide is known as *dry ice*. The solid and gaseous phases are in equilibrium at −78.5 °C and 1.0 atm, and the latent heat of sublimation is 137 kcal kg^{-1}. How much dry ice at −78.5 °C must be changed to a gas at 0 °C to freeze 1 kg of water at 0 °C? The specific heat of CO_2 gas is 0.205 kcal kg^{-1} K^{-1}.

5.3 A 1000 W heater is used to heat 1.0 kg of ice initially at −20 °C until it is transformed into steam at 120 °C and 1.0 atm. How many minutes does it take:

(a) To raise the temperature of the ice to 0 °C?
(b) To melt the ice?
(c) To heat the water to 100 °C?
(d) To boil the water?
(e) To heat the steam to 120 °C?

The constant-pressure specific of the steam is 0.452 kcal kg^{-1} K^{-1}. Other data are available in Tables 1.2 and 5.1.

5.4 Consider one mole of gas at a temperature and volume of 295.0 K and 24.50 L and test the accuracy of the ideal gas equation of state:

(a) Use the ideal gas equation of state to find the pressure accurate to four significant figures.

(b) Use the van der Waals equation of state to find the pressure accurate to four significant figures when the gas is N_2, O_2, and CO_2.

6

Reversible and Irreversible Processes

As many of the techniques for analyzing equilibrium processes are not applicable to nonequilibrium processes, it is important to distinguish between them. Reversible and irreversible processes are associated with equilibrium and nonequilibrium behavior respectively. The ability to identify a process as reversible is aided by having good examples of behavior that is *not* reversible. To this end, the behavior of some simple electric circuits consisting of resistors, capacitors, and batteries are discussed. We also describe an elementary irreversible phenomenon, the decay of a localized hot spot.

6.1 Idealization and Reversibility

Idealization

Some idealization is usually required before real processes can be treated as reversible. Idealizations are used in all fields of physics. For example, textbook descriptions of projectile motion assume that the gravitational force is independent of height, even though we know that the pull of gravity decreases as the square of the distance from the center of the earth. The situations described by idealizations do not exactly represent what is occurring, but represent situations that can be approached arbitrarily closely. Idealizations are not falsehoods, which are imaginary things that are impossible. For example, neglecting the effects of air resistance on projectile motion is an idealization that can be approached by performing experiments in an evacuated chamber with lower and lower pressures. In contrast, a massive object moving at twice the speed of light is a falsehood. Not only does it not occur (except in science fiction), but the situation cannot even be approached. Idealizations are approximations. When air resistance is neglected, the forces on the projectile have been approximated. Idealizations and approximations are made because they greatly simplify the analysis and because they prevent insignificant details from obscuring the essential characteristics of the phenomena involved.

Thermodynamics and Statistical Mechanics: An Integrated Approach, First Edition.
Robert J Hardy and Christian Binek.

Reversibility

The idealization of a reversible process is extremely useful. In the strictest sense, almost all real processes are irreversible. Experience tells us that macroscopic motion ultimately stops unless it is sustained by some external means. A swinging pendulum comes to rest. A spinning top slows down and tumbles. Macroscopic motion is almost always accompanied by irreversible phenomena that tend to cause the motion to cease. Nevertheless, there are many situations in which irreversible effects are sufficiently small that they can be neglected, and the processes involved can be treated as reversible.

The concept of reversibility is illustrated by imagining that we have made a video recording of some process or physical phenomenon and then run the video backwards. The phenomenon is reversible if you *cannot* tell the direction in which the video is running without knowing how it was made. For example, if you threw a ball from right to left and recorded its motion without showing either the thrower or the receiver, you would be unable to distinguish the reversed video from a similar video in which a ball had been thrown from left to right. The motion of the ball is reversible. In contrast, the behavior of a cup of coffee dropped on the floor is irreversible. The video run backwards would show the broken parts of the cup and the wet mess becoming an intact cup filled with coffee. As such things never happen, you would have no difficulty telling whether the video was running forwards or backwards.

A process can be idealized as reversible if it can be made to evolve in the reversed direction by a small change in the constraints on the system. The effect of a small change in constraints is illustrated by the behavior of a mixture of ice and water that is maintained at a constant pressure of one atmosphere. In equilibrium, the temperature of the mixture would be exactly 0.0 °C. If we place the system in contact with a heat reservoir with a temperature the slightest bit less than 0.0 °C, heat will be transferred *from* the system *to* the reservoir and some of the water will freeze. If we place the system in contact with a reservoir with a temperature the slightest bit greater than 0.0 °C, the direction of the heat transfer will be reversed and some of the ice will melt. If ΔU, Q, and W are the change in internal energy, heat transferred, and work done in a reversible process, their values in the reversed process are $-\Delta U$, $-Q$, and $-W$.

6.2 Nonequilibrium Processes and Irreversibility

Equilibrium processes and reversible processes have different defining characteristics. In an equilibrium process, the system changes in response to changes in the constraints on it and stops changing when the constraints stop changing, so that it is effectively in equilibrium at all times. In a reversible process, a system can be made to evolve in the reversed direction by an infinitesimal change in the constraints on it. In practice, equilibrium processes are reversible and nonequilibrium processes are irreversible.

Any process in which the system is *not* in equilibrium at all times is a nonequilibrium process. As systems in equilibrium do not change spontaneously, any system that continues to evolve after the constraints on it stop changing is experiencing a nonequilibrium process. The processes involved in the calorimetry experiments in Examples 1. 2 and 5.1 and in the dissipation of mechanical energy by friction in Example 2.2 are examples of nonequilibrium processes.

In the calorimetry experiments, an object at one temperature is placed in a calorimeter cup containing water at another temperature, and the combined system is then allowed to come to thermal equilibrium. The constraint was the initial separation (adiabatic isolation) of the object from the water and calorimeter cup. As the combined system continued to change after the constraint stopped changing, the system was not in equilibrium at all times. The process was a nonequilibrium process. It was also irreversible. The reversed video of the experiments would show the equilibrated calorimeter spontaneously evolving into a situation in which the object has a different temperature than the water. That never happens.

Free Expansions

In the free expansion experiment discussed in Section 2.4, a gas that was initially confined to one container was allowed to expand into a second, initially evacuated container. A free expansion is a nonequilibrium process. The gas, which is the system, is in equilibrium at the start and finish of the process, but not at times in between. Opening the valve between the two containers changes the constraints on the system. After the valve is opened the gas flows without external assistance. A free expansion is also an irreversible process. A reversed video would show gas that is initially in two containers suddenly flowing into one of them and leaving the other one empty. That never happens. Furthermore, as the pressure and local temperature have different values at different places from the moment the valve is opened until the gas has come to mechanical and thermal equilibrium, the process *cannot* be represented by a line on a PV-diagram or other representation of the state space of the gas.

The first law $Q = \Delta U + W$ is a general principle that applies to both equilibrium and nonequilibrium processes. In reversible processes the work done by a PVT system is $W = \int PdV$, but this is generally *not* true for irreversible processes. The formula $W = \int PdV$ is only valid when the volume change is caused by the forces exerted by pressure P. Even though a large change in volume occurs in the free expansion experiment discussed in Section 2.4, we concluded that no work was done because the volume change was not caused by the gas exerting a force on part of the surroundings that moved. Nevertheless, we could still conclude from $Q = \Delta U + W$ and the results $W = 0$ and $Q = 0$ that $\Delta U = 0$. Free expansion experiments also illustrate the value of global reasoning. By taking a global point of view we did not need to know how the pressure of the gas was changing in time and space as it flowed into the evacuated container.

Throttling Processes

Another example of an irreversible process is a throttling process in which gas is forced through a constriction with no transfer of heat. The porous plug experiment diagrammed in **Figure 6.1** is an example of such a process. Gas is continuously introduced at the left end of a long tube, passes through a porous plug, and exits at a lower pressure at the right. The porous plug is insulated to prevent any heat transfer, and the velocity of the flow, which is constant, is sufficiently low that the associated kinetic energy is negligible. Consider the process during which the sample of gas between x_1 and x_3 at initial time t_0 moves so that it is located between x_2 to x_4 at the final time t_f. The pressure of the gas to the left of the plug is P_a, and V_a is the volume between x_1 and x_2. The pressure on the right side of the plug is P_b, and V_b is the volume between x_3 and x_4.

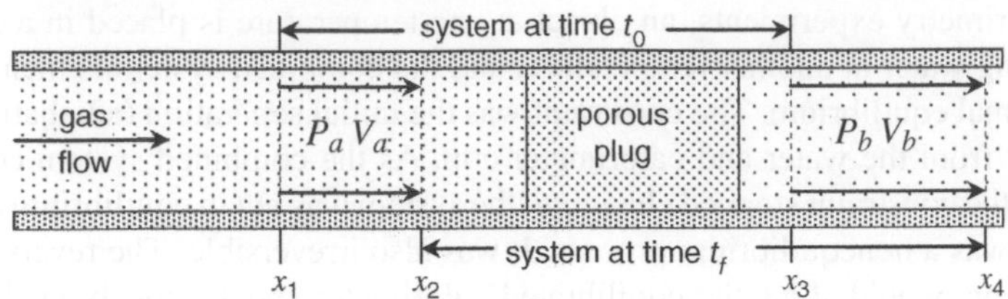

Figure 6.1 *Porous Plug Experiment*

During the process, the gas at the left does work $P_a V_a$ on the system, while the system does work $P_b V_b$ on the gas at the right. Thus, the total work done *by* the system is $W = -P_a V_a + P_b V_b$. As there is no heat transfer, Q is zero. As the properties of the system between x_2 and x_3 are unchanged by the process, the change in the system's internal energy is $\Delta U = U_b - U_a$, where U_a is the internal energy of the part of the system that was initially to the left of x_2 but is not there at the final time, and U_b is the internal energy of the part of the system to the right of x_3 that was not there initially. Applying the first law $Q = \Delta U + W$ to the process tells us that

$$0 = (U_b - U_a) + (-P_a V_a + P_b V_b) \tag{6.1}$$

or

$$U_a + P_a V_a = U_b + P_b V_b. \tag{6.2}$$

As enthalpy H is defined as $H = U + PV$, the above result states that

$$H_a = H_b. \tag{6.3}$$

Thus, a throttling process is a constant enthalpy process.

The change in temperature that results from a throttling process is called the *Joule–Thompson effect*. It is used commercially for cooling gases. At room temperature, all gases except hydrogen and helium can be cooled by a throttling process. If the change in pressure is not too great, a good estimate for the change in temperature is

$$\Delta T = \left(\frac{\partial T}{\partial P}\right)_H \Delta P. \tag{6.4}$$

The derivative $\mu_{JT} = (\partial T/\partial P)_H$ is the *Joule–Thompson coefficient*. As the coefficient μ_{JT} is zero for ideal gases, the effect depends on the amount by which the properties of real gases deviate from those of ideal gases. To see that $(\partial T/\partial P)_H$ is zero for an ideal gas, we need an expression for the enthalpy $H = U + PV$. Since $PV = nRT$ and $U = U(T)$, its enthalpy is

$$H = U(T) + nRT, \tag{6.5}$$

which is independent of pressure. Hence, the derivative $\partial T/\partial P$ is zero.

Since no heat is transferred to or from the system, the increase in volume in the throttling experiment is an adiabatic expansion, as it is in a free expansion experiment. In both experiments the expansion is an *irreversible* adiabatic process. This contrasts with the *reversible* adiabatic expansions discussed in Chapter 4, in which the temperature decreases as the volume increases.

6.3 Electrical Systems

Interesting examples of reversible and irreversible behavior are found in simple electrical circuits. A convenient way to think about electrical systems is diagrammed in **Figure 6.2**. The system being analyzed is the contents of the black box. The battery, switch, and wires in the surroundings are introduced to facilitate the calculation of the work W. The purpose of the heat reservoir is to maintain the system at a constant temperature, which facilitates the calculation of the heat Q.

The *electric potential difference* (or voltage) V_{el} between two points is the work per unit charge required to move electric charge from one point to the other. The battery in the figure supplies the energy that moves charge through the system. The work done against electric forces by the battery to move charge q through the system from terminal 1 to terminal 2 is qV_{el}. As this energy is transferred *to* the black box, the *electrical work* done *by* the system is

$$W = -qV_{el}. \tag{6.6}$$

(Subscript el distinguishes voltage V_{el} from volume V.) V_{el} is positive when the potential (voltage) at terminal 1 is greater than at terminal 2, *i.e.*, when $V_{el,1} > V_{el,2}$.

The *electric current* I (in amperes) is the charge per unit time that flows through any cross section of the wire. When the current is constant, the current is

$$I = q/\Delta t, \tag{6.7}$$

where q is the amount of charge that passes through the cross section in time interval Δt. When the current flows in the direction indicated in the figure, the electrical work done *by* the system is

$$W = -V_{el}I\Delta t. \tag{6.8}$$

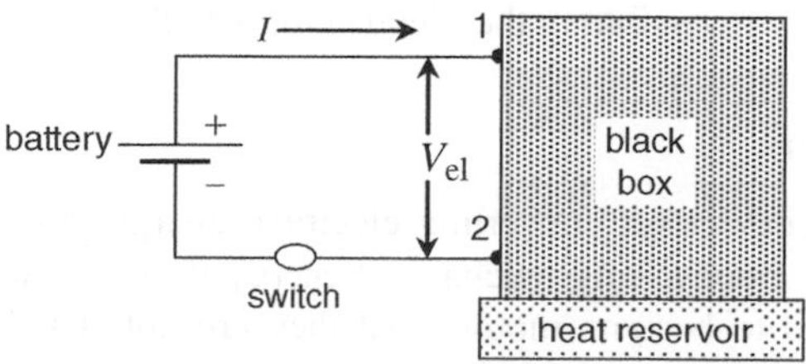

Figure 6.2 *An Electrical System in a Black Box*

If the current varies with time, the work done is

$$W = -\int_{t_0}^{t_f} V_{el}(t)I(t)\,dt, \tag{6.9}$$

where $I(t) = dq(t)/dt$.

Resistors

When the system inside the black box is a network of resistors, the process is *irreversible*. The potential difference and current are related by Ohm's law $V_{el} = IR$, where R is the effective resistance between the terminals. The electrical work done *by* the resistors is

$$W = -I^2 R \Delta t. \tag{6.10}$$

Resistors are used as heaters in both laboratory and everyday applications. For the contents of the black box to remain unchanged, heat must be continuously transferred to the heat reservoir.

Capacitors

When the system inside the black box is a capacitor, the process is *reversible*, provided the resistance of the capacitors is negligible. The internal energy of a capacitor is $U = \frac{1}{2}C{V_{el}}^2 = q^2/2C$, where C is the capacitance and q is the magnitude of the charge on each capacitor plate. When an initially uncharged capacitor is charged, the electrical work done *on* the capacitor is $q^2/2C$. When a charged capacitor is completely discharged, the electrical work done *by* the capacitor is $W = q^2/2C$. No heat is transferred during the charging or discharging of a capacitor with negligible resistance, so that $Q = 0$.

Batteries

When the system inside the black box is a battery, the process can be reversible or irreversible, depending on the battery. An ideal rechargeable battery is *reversible*, but many small batteries (AA, AAA, etc.) are *irreversible*. When current flows into the battery at the terminal with the higher voltage, a rechargeable battery converts the incoming electrical energy to chemical energy. In the same situation, a nonrechargeable battery converts the incoming energy into thermal energy and causes the battery's temperature to increase. When current flows out of the terminal with the higher voltage, a battery transform chemical energy into electrical energy. As all batteries have some internal resistance, rechargeable batteries are only strictly reversible in the idealization which neglects that resistance.

Motors and Generators

An electric motor is a device for converting electrical energy into mechanical energy. A generator is a device for converting mechanical energy into electrical energy. Some generators can be reversed to become motors, but they are not strictly reversible, as some dissipation of energy is associated with their operation whether operating as generators or as motors.

The following example discusses the heat dissipated by a generator and emphasizes the meanings of the symbols Q, ΔU, and W. The example also illustrates the technique of breaking a problem into smaller ones, analyzing the smaller problems, and then combining the intermediate results to obtain the desired information.

Example 6.1 Generator

A water-proof generator is placed in an ice-water bath, as diagrammed in **Figure 6.3**. The apparatus is initially in equilibrium, after which the generator delivers 250 W of electric power to an external load for 1.0 h. After the generator stops and the apparatus has returned to equilibrium, it is found that 0.25 kg of ice has melted. Find the work done by the rotating shaft supplying the mechanical energy to the generator.

Solution. We start by dividing the apparatus into three subsystems: subsystem ice is the initially frozen water, subsystem liq is the initially liquid water, and subsystem gen is the generator. As no information is given to the contrary, we assume that no heat is exchanged between the combination of subsystems and their surroundings. Presumably, the pressure of the atmosphere is constant, so that we can assume that the temperature is the same at the start and finish of the process. First, consider subsystem ice. As 0.25 kg of the ice has melted, the heat transferred *to* the subsystem is

$$Q_{\text{ice}} = (0.25\,\text{kg})(80\ \text{kcal kg}^{-1})(4184\,\text{J}\ \text{kcal}^{-1}) = 83.7\ \text{kJ},$$

where $80\ \text{kcal kg}^{-1}$ is the latent heat of fusion of water. Next, consider subsystem liq. As its temperature and pressure are unchanged, its state is unchanged. This implies that its internal energy and volume are unchanged. Hence, $\Delta U_{\text{liq}} = 0$ and $W_{\text{liq}} = 0$. Since the heat transferred *to* the subsystem is $Q_{\text{liq}} = \Delta U_{\text{liq}} + W_{\text{liq}}$, it follows that $Q_{\text{liq}} = 0$. Subsystem liq exchanges heat with the other two subsystems, so that $Q_{\text{liq}} = -Q_{\text{gen}} - Q_{\text{ice}}$, where Q_{gen} is the heat transferred *to* the generator. It then follows from $Q_{\text{liq}} = 0$ and the value of Q_{ice} that

$$Q_{\text{gen}} = -Q_{\text{ice}} = -83.7\,\text{kJ}.$$

Finally, consider the generator and analyze the effect of the process on it. After the generator has cooled to the temperature of the ice-water mixture, its state is the same as it was initially. As its state is unchanged, its internal energy is unchanged, so that $\Delta U_{\text{gen}} = 0$. The first law then tells us that $Q_{\text{gen}} = W_{\text{gen}}$. There are two contributions to W_{gen}: (i) mechanical work is done *on* the generator by the rotating shaft and (ii) electrical work is done *by* the generator through the electrical energy delivered through the wires. The

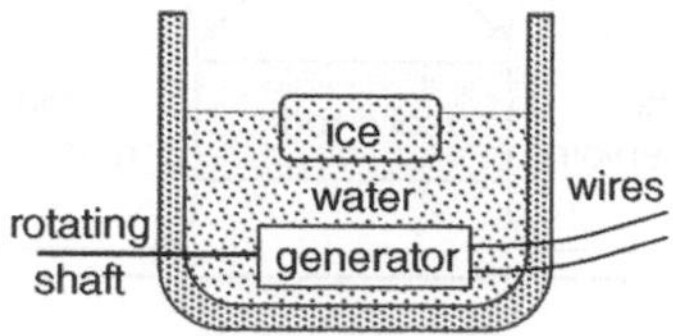

Figure 6.3 An Electrical Generator Immersed in Water

total work done by the generator is

$$W_{gen} = -|W_{shaft}| + |W_{electric}|,$$

where the absolute value signs are used to emphasize the signs of the contributions. The electrical work done by the generator is

$$|W_{electric}| = (250\ \text{W})(1.0\,\text{h}) = 900\,\text{kJ}.$$

It then follows from $Q_{gen} = W_{gen}$ and the value of Q_{gen} that $W_{gen} = -83.7\,\text{kJ}$. Combining these results indicates that the work done by the rotating shaft is

$$|W_{shaft}| = 900\,\text{kJ} - (-83.7\,\text{kJ}) = 984\,\text{kJ}.$$

The values of Q, ΔU, and W for the generator and subsystem *liq* were found in the example, but only Q was found for subsystem *ice*. ΔU_{ice} and W_{ice} were not needed. Notice that W_{ice} is not zero. As the specific volume of frozen water decreases as it melts, a small amount of work is done *by* the atmosphere, so that the work done *by* the subsystem is negative. It then follows from $Q_{ice} = \Delta U_{ice} + W_{ice}$ that ΔU_{ice} is slightly larger than Q_{ice}.

6.4 Heat Conduction

So far, our principal tool for analyzing nonequilibrium processes has been the first law, which is a statement about the overall effect. To go beyond such global considerations, we look at *transport phenomena*, which are phenomena that involve the transfer of energy, mass, and momentum and include heat conduction, diffusion, and viscous flow. The parameters used to determine the temporal and spatial evolution of these phenomena – called *transport coefficients* – are *thermal conductivity*, *diffusivity*, and *viscosity*. Before deriving the diffusion equation that describes the evolution of the local temperature in solids, we consider a steady state example of heat conduction.

To have a definite situation to visualize, consider the apparatus diagrammed in **Figure 6.4**. A solid rod of uniform cross section is maintained in a steady state by transferring heat Q *to* the rod in time Δt from the hot reservoir and transferring heat *from* the rod at the same rate to the cold reservoir. The rate of heat transfer is $Q/\Delta t$, which is proportional to the

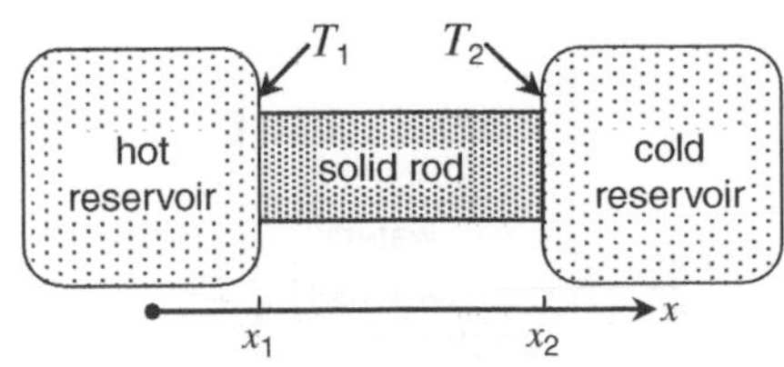

Figure 6.4 *Heat Conduction Experiment*

Table 6.1 Thermal conductivities

	Aluminum	Copper	Iron	Lead	Glass
K (W m^{-1} K^{-1})	237	401	80.2	36.3	1.1

cross-sectional area A and the temperature difference $T_2 - T_1$ and inversely proportional to $x_2 - x_1$,

$$\frac{Q}{\Delta t} = -KA\frac{T_2 - T_1}{x_2 - x_1}. \tag{6.11}$$

The proportionality constant K is the *thermal conductivity* of the material. Some typical values of the thermal conductivity are given in Table 6.1. Although the rod is in a steady state, it is not in thermal equilibrium because heat must be continuously added and removed to maintain the temperature gradient. If the rod were suddenly isolated, it would spontaneously evolve until it comes to equilibrium with a uniform temperature.

Example 6.2 Heat Conduction

Compare the rate of heat transfer through a sheet of aluminum (Al) with the rate of transfer through a sheet of glass of same thickness and shape.

Solution. A reasonable way to make such a comparison is to find the ratio of the rates of heat transfer $Q/\Delta t$ for the two materials. As the sheets are of the same thickness and shape, the values of $x_2 - x_1$ and A are the same. For the comparison to be meaningful, the values of $T_2 - T_1$ also need to be the same. It then follows from (6.11) that the ratio of $Q/\Delta t$ for the two materials is equal to the ratio of their thermal conductivities,

$$\frac{(Q/\Delta t)_{\rm Al}}{(Q/\Delta t)_{\rm glass}} = \frac{K_{\rm Al}}{K_{\rm glass}} = \frac{237\,{\rm W\ m^{-1}\ K^{-1}}}{1.1\,{\rm W\ m^{-1}\ K^{-1}}} = 215.$$

This result tells us, for example, that water being heated in a pot placed on a hot plate will come to a boil faster in an aluminum pot than in a glass container of the same thickness and shape.

Time and Space Dependence

Equation (6.11) is a special case of a more general relationship, called *Fourier's law*, which for one-dimensional spatial variations is

$$\boxed{S(x,t) = -K\frac{\partial T(x,t)}{\partial x}.} \tag{6.12}$$

Fourier's law is an empirical relationship which asserts that the heat flux is proportional to the temperature gradient. The heat flux $S(x,t)$ is the rate at which heat is transferred through a surface per unit area. The heat flux in the situation diagramed in **Figure 6.4** is $S = Q/(A\,\Delta t)$ and is the same everywhere along the rod. Heat is transferred from the left to the right when the temperature is higher at the left.

The equation for describing the temporal and spatial evolution of the local temperature is obtained by combining Fourier's law with the continuity equation for energy conservation. Continuity equations are associated with conserved quantities that are distributed in space. For one-dimensional variations the *continuity equation* for energy conservation is

$$\frac{\partial u(x,t)}{\partial t} = -\frac{\partial S(x,t)}{\partial x}, \tag{6.13}$$

where $u(x,t)$ is internal energy per unit volume. $S(x,t)$ is the *heat flux* and $u(x,t)$ is the energy density. The significance of (6.13) can be seen by multiplying it by the cross-sectional area of the rod and integrating over an interval of length Δx. This gives

$$A\int_x^{x+\Delta x} \frac{\partial u(x,t)}{\partial t} dx = -A\int_x^{x+\Delta x} \frac{\partial S(x,t)}{\partial x} dx = -A[S(x+\Delta x,t) - S(x,t)]. \tag{6.14}$$

As the integral is over x, the partial derivative with respect to t can be taken outside the integral, so that the first integral in (6.14) becomes $\partial U/\partial t$, where $U = A\int u(x,t)dx$ is the internal energy of the material in the interval between x and $x+\Delta x$. Substituting this into (6.14) gives

$$\frac{\partial U}{\partial t} = AS(x,t) - AS(x+\Delta x,t). \tag{6.15}$$

This result states that the rate of change of the internal energy is equal to the rate at which heat is transferred in at the left edge of the interval minus the rate at which heat is transferred out at its right edge.

Before the continuity equation can be used to determine the dependence of the local temperature T on x and t, we need to relate $\partial u(x,t)/\partial t$ to $\partial T(x,t)/\partial t$. We limit our considerations to situations in which the properties of the rod vary only slightly from uniform background values, so that the internal energy U in (6.15) becomes

$$U = U_0 + (\partial U/\partial T)_V(T - T_0) = U_0 + M c_V^M (T - T_0), \tag{6.16}$$

where U_0 and T_0 are the background values of U and T. The mass of the material in the interval between x and $x+\Delta x$ is M and its specific heat is c_V^M. The energy per unit volume is

$$u = U/V = (U_0/V) + \rho c_V^M (T - T_0), \tag{6.17}$$

where $V = A\,\Delta x$ is the volume of the interval and $\rho = M/V$ is its mass density. When Δx is made limitingly small, the time derivative of the energy density becomes

$$\frac{\partial u(x,t)}{\partial t} = \rho c_V^M \frac{\partial T(x,t)}{\partial t}. \tag{6.18}$$

It follow from Fourier's law that

$$\frac{\partial S(x,t)}{\partial x} = -K\frac{\partial^2 T(x,t)}{\partial^2 x}. \tag{6.19}$$

By substituting (6.18) and (6.19) into the continuity equation in (6.13), we obtain

$$\boxed{\frac{\partial T(x,t)}{\partial t} = D\frac{\partial^2 T(x,t)}{\partial x^2},} \tag{6.20}$$

where $D = K/(\rho c_V^M)$. D is called the *thermal diffusivity*. Equation (6.20) is the one-dimensional form of the *diffusion equation* for local temperature. It is a linear partial differential equation whose solution $T(x, t)$ is uniquely determined by the initial distribution and the boundary conditions. The diffusion equation describes the spontaneous evolution of nonuniform temperature distributions.

Decay of a Localized Hot Spot

The evolution of a localized hot spot is an interesting example of nonequilibrium behavior. We investigate the evolution of a temperature distribution with the general shape of a hot spot created by the rapid heating of a small region of a long thin rod. The disturbance is assumed to be localized at $x = 0$ in a rod oriented parallel to the x-axis, and the rod is sufficiently long that its ends (boundaries) are effectively at $x = \pm\infty$. The solution to the diffusion equation is

$$T(x, t) = T_0 + \frac{A}{\sqrt{t + t_0}} \exp\left(\frac{-x^2}{4D\left(t + t_0\right)}\right), \tag{6.21}$$

where T_0 is the uniform background temperature and D is the thermal diffusivity. The constants A and t_0 characterize the shape of the initial temperature distribution, *i.e.*, the distribution at $t = 0$. The spatial dependence of $T(x, t)$ is given by the Gaussian function e^{-x^2/w^2}, where $w = \sqrt{4D(t + t_0)}$. Since $e^{-x^2/w^2} = 1$ at $x = 0$ and drops to $e^{-1} = 0.368$ at $x = +w$ and $x = -w$, a reasonable parameter for characterizing the width of the region with an elevated temperature is $2w = 4\sqrt{D(t + t_0)}$.

Figure 6.5 shows the evolution of the temperature disturbance $T(x, t) - T_0$. The solution describes a hot spot with an initial width of $4\sqrt{Dt_0}$ and a maximum temperature increase of $A/\sqrt{t_0}$. As the amount of energy in the hot spot is finite, the disturbance tends to zero as the boundaries are approached. As a Gaussian goes to zero at $x = \pm\infty$, the trial solution has the desired behavior. To verify that the solution satisfies the diffusion equation, we need to find the first time derivative and second spatial derivative of $T(x, t)$. They are

$$\frac{\partial T(x, t)}{\partial t} = \left(\frac{-1}{2\left(t + t_0\right)} + \frac{x^2}{4D_T(t + t_0)^2}\right) \frac{A}{\sqrt{t + t_0}} \exp\left(\frac{-x^2}{4D\left(t + t_0\right)}\right) \tag{6.22}$$

and

$$\frac{\partial^2 T(x, t)}{\partial x^2} = \left(\frac{-1}{2D\left(t + t_0\right)} + \frac{x^2}{4D^2(t + t_0)^2}\right) \frac{A}{\sqrt{t + t_0}} \exp\left(\frac{-x^2}{4D\left(t + t_0\right)}\right). \tag{6.23}$$

It is easily seen that the first time derivative equals D times the second spatial derivative, as is required by the diffusion equation.

The figure shows how the initial peak in the temperature drops and broadens. At long times ($t >> t_0$) the width of the peak increases as the square root of time. Although real hot spots are not likely to be exactly Gaussian in shape, their time evolution will be similar to that in the figure. The tendency of local temperature to spontaneously decay to a uniform value is characteristic of processes described by the diffusion equation. Also, the evolution of the hot spot is obviously irreversible. The reverse of the process diagrammed in the

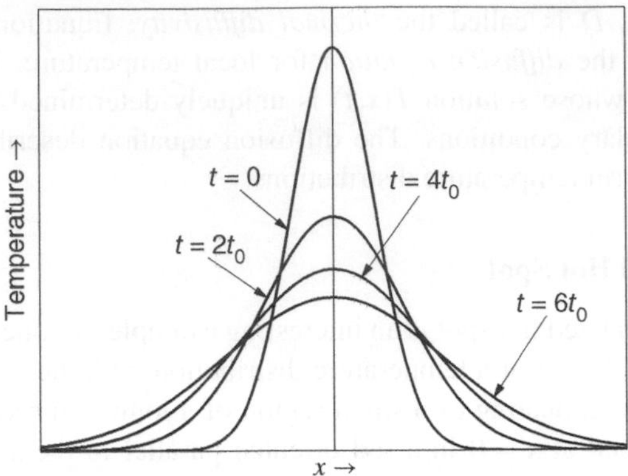

Figure 6.5 *Evolution of a Localized Hot Spot*

figure would have a rod at uniform temperature T_0 spontaneously proceeding through the sequence of temperature distributions shown in the figure in reversed order, *i.e.*, from the broad distribution at $t = 6t_0$ to the peaked distribution for $t = 0$.

The primary purpose for introducing Fourier's law, the continuity equation, and the diffusion equations was to present a simple example of nonequilibrium behavior. Only their one-dimensional forms have been given. Their three dimensional forms are

$$\boldsymbol{S}(\boldsymbol{r}, t) = -K\nabla T(\boldsymbol{r}, t), \tag{6.24}$$

$$\frac{\partial u(\boldsymbol{r}, t)}{\partial t} = -\nabla \cdot \boldsymbol{S}(\boldsymbol{r}, t), \tag{6.25}$$

$$\frac{\partial T(\boldsymbol{r}, t)}{\partial t} = D\nabla^2 T(\boldsymbol{r}, t). \tag{6.26}$$

Problems

6.1 The internal energy function for a van der Waals gas with constant c_V is

$$U(T, v) = n(c_V T - a/v),$$

where v is the specific volume. Find the change in temperature $T_f - T_0$ that occurs in a free expansion when the volume changes from V_0 to $2V_0$. Assume that the initial specific volume is $20\,\mathrm{L\,mol^{-1}}$. Use $c_V = 2.5R$ and the data in Table 5.2 to find the temperature change for nitrogen gas (N_2).

6.2 Containers 1 and 2 have rigid insulating walls and are connected by a valve that is initially closed. Both containers are filled with the same gas, which is an ideal gas with constant c_V. Initially:

$$n_1 = 0.10 \text{ moles}, \quad n_2 = 0.20 \text{ moles}, \quad T_1 = 250\,\mathrm{K}, \quad T_2 = 350\,\mathrm{K}.$$

The valve is opened and the gases in the containers are allowed to mix and come to equilibrium. Consider the system to be all of the gas in both containers.

(a) What are the values of Q and W?
(b) What is its final temperature?

6.3 A bottled beverage at 0 °C has ice in it. You want to melt the ice by placing it in water at 20 °C. Estimate how much ice will be melted in one minute. Assume that the glass walls of the bottle are 2.5 mm thick and that 300 cm^2 of the bottle are in contact with the water.

6.4 A small water proof electric motor is being tested. The motor is in a container with some ice and water, which are initially in equilibrium. (The situation is similar to Figure 6.3 except that the rotating shaft, the generator, and wires are replaced respectively by wires, a motor, and a shaft.)
60 watts of power are delivered to the motor for 40 s while the motor turns a shaft which causes a 25 kg mass to be raised vertically by 8.0 m. The contents of the container then come to equilibrium.
The motor is the system in parts (a), (b), and (c).

(a) What is the value of W for the process?
(b) What is the value of ΔU for the process?
(c) What is the value of Q for the process?
(d) How many grams of ice were melted during the process?

6.5 The walls of a vacuum chamber are punctured so that air ($\gamma = 1.4$) can flow in. What is the final temperature of the air in the chamber? Assume the temperature of the atmosphere is 300 K? (This is not a free expansion experiment. The system to analyze is the air that occupies the chamber in the final state.)

Part II
Foundations of Thermodynamics

The science of thermodynamics is developed in Part II as a self-contained theory of macroscopic thermal phenomena based on the first and second laws of thermodynamics. The first law, introduced in Chapter 2, is a statement of the principle of energy conservation. The second law is a carefully phrased assertion of the observation that most natural phenomena are irreversible. We can transform any amount of mechanical or electrical energy into heat by using frictional forces or electrical resistance, but these processes cannot be reversed. If we could reverse them, we could power our vehicles and generate electricity by extracting energy from our environment in the form of heat. Because of the restriction imposed by the second law, society's energy needs cannot be satisfied in this way. In addition to indicating how to use energy efficiently, thermodynamics establishes exact general relationships, which are valuable in the analysis of the properties of materials. One of the strengths of thermodynamics is that these relationships are independent of any assumptions about the microscopic structure of matter.

Well-formulated theories associate observable properties with mathematical concepts, which can then be used to deduce results that may not otherwise be apparent. The existence of entropy and of thermodynamic potentials are such results. The tendency of entropy to increase whenever irreversible behavior has occurred and our ability to summarize a system's equilibrium properties in a single potential are powerful tools for the analysis of thermodynamic behavior.

Thermodynamics and Statistical Mechanics: An Integrated Approach, First Edition.
Robert J Hardy and Christian Binek.

7

Second Law of Thermodynamics

After reviewing the fundamental concepts introduced in Part I, the statements of the second law of Clausius and Kelvin are introduced. The logical development of thermodynamics then proceeds with the introduction of Carnot cycles and the definition of absolute temperature scales. Since they are the type of cyclic process with the highest energy efficiency, Carnot cycles have practical implications for the efficiencies of engines, air conditioners, and heat pumps.

7.1 Energy, Heat, and Reversibility

When viewed microscopically, a system's constituent particles have kinetic energy associated with their motion and potential energy associated with their interactions. Thus, energy at the atomic level is a function of the momentum and position variables of every one of a system's constituent particles. In contrast, the internal energy of a system is a function of the small number of thermodynamics coordinates (two for PVT systems) needed to specify its equilibrium state.

First Law of Thermodynamics

The first law is a statement of the principle of energy conservation that equates the change in the internal energy state function to the energy transferred to or from a system as heat or work. As stated in Chapter 2, the law asserts *that all systems have an internal energy state function U and that energy can be transferred between systems either as heat Q or as work W.* The first law is summarized by the equation

$$\Delta U = Q - W. \tag{7.1}$$

By convention, Q is positive when energy is transferred *to* the system and is negative when energy is transferred *from* it. W is positive when work is done *by* the system and is negative

Thermodynamics and Statistical Mechanics: An Integrated Approach, First Edition.
Robert J Hardy and Christian Binek.

when work is done *on* it. Systems that cannot exchange energy with other systems in the form of heat are *adiabatically isolated* systems, and systems without adiabatically isolated parts are *thermally interacting* systems.

Temperature

Temperature is an intensive state function that describes the tendency of a system (or subsystem) *to exchange heat with other systems*. When brought into thermal contact, systems at different temperatures spontaneously exchange heat, while systems at the same temperature are in thermal equilibrium with each other. Although it is conceivable that the exchange of heat could be determined by more than one property, it is fundamental that a single property – temperature – is sufficient. The temperatures associated with the interfaces where heat is exchanged are local temperatures. When the system involved is in thermal equilibrium, these local temperatures are the same as the system's equilibrium temperature.

In everyday life, the terms hotter and colder imply higher and lower temperatures. Since the basis for the commonly used temperature scales has not been established, we need to define hotter and colder without reference to temperature. When two systems exchange heat, the system whose internal energy is decreasing is the *hotter* system and the system whose energy is increasing is the *colder* system.

Heat Reservoirs

One goal of thermodynamics is to assign values to temperature in as general a way as possible. To do this we need to be able to assert that a system has a definite tendency to exchange heat without referring to the readings on a specific thermometer. Heat reservoirs are the idealizations used to do this. *A heat reservoir is a system with a fixed tendency to accept and reject heat*.

An important example of a heat reservoir is the *triple point cell*. It is a closed vessel that contains ice, liquid water, and water vapor in thermal equilibrium at the triple point of water (described Section 5.1). When heat is transferred *to* the cell, its temperature and pressure remain unchanged as some ice melts and/or some liquid vaporizes. When heat is transferred *from* the cell, the phase changes are reversed. Another practical heat reservoir consists of the liquid and vapor phases of a substance in equilibrium at constant pressure. Its temperature corresponds to a point on the vaporization curve shown in Figure 5.1. When heat is transferred to or from the system, some liquid changes to vapor or vice versa. Reservoirs at different temperatures can be obtained by varying the pressure.

Reversible Processes

The direction in which things evolve is a fundamental concern of the second law of thermodynamics. *A reversible process is one whose direction in time can be reversed.* In practice, reversible processes are equilibrium processes. In an equilibrium process, the system is in thermal equilibrium at all times. Although reversible and equilibrium processes have different defining characteristics, processes with the characteristics of one type have the characteristics of the other. Because of the fundamental importance of reversibility in the following discussion, such processes will be referred to as reversible processes and will be represented by lines in state space.

A process is considered to be reversible if it can be made to evolve in the reversed direction by an infinitesimal change in the constraints on the system. If ΔU, Q, and W are the change in internal energy, the heat transferred, and the work done in a reversible process, their values in the reversed process are $-\Delta U$, $-Q$, and $-W$.

7.2 Cyclic Processes

Thought experiments are useful aids for investigating and interpreting physical theories. The experiments can in principle be performed, but it is usually not practical to do so, since they tend to be highly idealized. Their purpose is to emphasize basic concepts. Cyclic processes are frequently used in the thought experiments of thermal physics.

A cyclic process periodically returns the system to the same condition that it had earlier, and each repeat is a cycle. If the system is in equilibrium at the start and finish of a cycle, its thermodynamic state is unchanged. If its state is unchanged, its state functions are unchanged. As a result, $\Delta P = 0$, $\Delta V = 0$, $\Delta T = 0$, and $\Delta U = 0$. It follows from $\Delta U = 0$ and the first law that

$$Q = W. \tag{7.2}$$

Engines and refrigerators are examples of systems (also called devices) that operate cyclically.

Heat Engines

A cyclic device that exchanges heat with its surroundings and does a positive amount of work is a heat engine. It can be a complicated device, such as a real condensing steam engine, a specialized laboratory instrument, or a conceptual system used in a thought experiment. As discussed in Chapter 4, some engines can be represented by closed contours in the state space of their working substance.

It is convenient to introduce a special type of engine, a 2-*temperature engine*, that accepts heat from its surroundings at one temperature and rejects heat at one other temperature. This is different from the usual situation for practical engines, which tend to accept and reject heat over a range of temperatures. A real engine can be modified to a 2-temperature device by supplying heat from a reservoir at the hottest temperature utilized by the engine and rejecting heat to a reservoir at the coldest temperature. Suitable heat conducting material can be used to transfer heat between the engine and the reservoirs in such a way that the temperatures at which heat is exchanged by the engine are unchanged. The modified device formed by combining the real engine with the heat conducting material is a 2-temperature engine.

Refrigerators

A refrigerator is a cyclic device that accepts heat at a colder temperature and rejects heat at a hotter temperature. Mechanical or electrical energy is needed to operate a refrigerator, *i.e.*, work is done *on* it. It can be a complicated device, like the mechanism in a household refrigerator, a laboratory instrument, or a conceptual device. For example, the mechanism in a household refrigerator transfers heat from inside, where it is colder, to the room, where it is hotter, and requires electrical energy to operate. If the temperature of the contents of a refrigerator and the temperature of the air in the room are approximately constant, they can be idealized as heat reservoirs.

A *2-temperature refrigerator* accepts heat from a single colder reservoir, rejects heat to a single hotter reservoir, and has work done on it. Practical refrigerators are usually not 2-temperature devices, but like practical engines they can be combined with suitable heat conducting material to become 2-temperature refrigerators.

Diagrams

It is helpful to represent heat engines and refrigerators by diagrams that emphasize their purpose. **Figure 7.1** shows diagrams of a general heat engine, a 2-temperature engine, and a 2-temperature refrigerator. As the devices are cyclic, the energy transferred in during a cycle is equal to the energy transferred out. This characteristic is suggested by the pipes in the diagrams. The amount of water that goes into a collection of pipes equals the amount of water that comes out. Water is neither created nor destroyed in plumbing. It is conserved. In engines and refrigerators, it is energy that is conserved. The horizontal lines in the diagrams of the 2-temperature devices represent heat reservoirs. By convention, the hotter reservoir is represented by the higher line. Subscripts h and c indicate hotter and colder. The symbol Θ is used for temperature. The symbol T is reserved for absolute temperatures, which are not defined until Section 7.7.

The directions of the energy transfers are indicated by arrows. Notice that the arrows for a refrigerator are in the opposite directions of those for an engine. The amounts of energy transferred are indicated by $|Q_{in}|$, $|Q_{out}|$, and $|W|$, where absolute value signs are placed on the symbols so the diagrams can be interpreted without reference to the sign conventions for heat and work. The symbol Q is used to represent the total amount of heat transferred *to* a device in one cycle, so that $Q = |Q_{in}| - |Q_{out}|$.

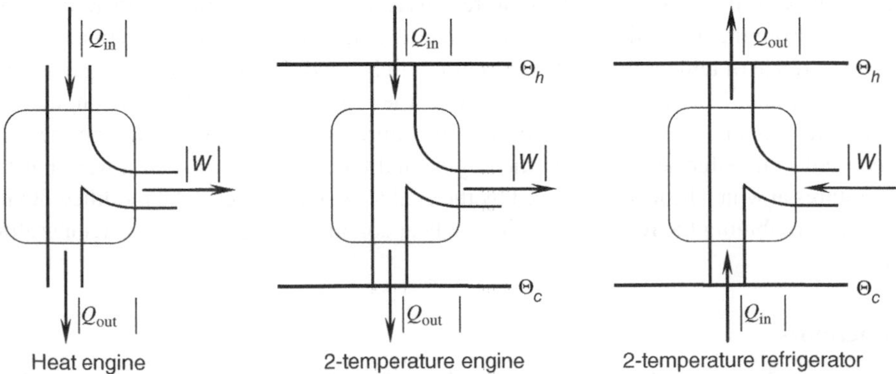

Figure 7.1 Cyclic Devices

7.3 Second Law of Thermodynamics

Although the irreversible aspects of reversible processes can be neglected in many applications, experience tells us that natural phenomena are ultimately irreversible. Before this observation can be used to make physically significant predictions, it needs to be expressed more precisely. This is done by the second law of thermodynamics as formulated by Rudolf Clausius and William Thomson, who is also known as Lord Kelvin.

Clausius Statement of the Second Law
There is no process whose only effect is to accept heat from a colder reservoir and transfer it to a hotter one.

The word "only" is crucial to the Clausius statement. Refrigerators accept heat from a colder reservoir and transfer heat to a hotter reservoir, but that is not all they do, as electrical or mechanical energy is required to operate them. When objects at different temperatures are placed in thermal contact, energy is spontaneously transferred *from* the hotter object *to* the colder object as heat. The Clausius statement tells us that, no matter how clever we are or how complicated a device we create, it is impossible to create a device whose *only* effect is to transfer energy *to* the hotter object *from* the colder object.

Kelvin Statement of the Second Law
There is no process whose only effect is to accept heat from a single heat reservoir and transform it entirely into work.

The word "only" is also crucial to the Kelvin statement. Heat engines accept heat from a reservoir and transform it into work, but that is not all they do. They also transfer heat to a reservoir at a lower temperature. The opposite of accepting energy as heat and converting it to work is accepting energy as work and converting it to heat. There is no restriction on the latter process, which is what friction and electrical resistance do. Friction converts energy into internal energy by exerting a force through a distance, *i.e.*, by doing work. The internal energy is often transferred elsewhere as heat. Similarly, a resistor changes electrical energy into internal energy, which can then be transferred to other objects as heat.

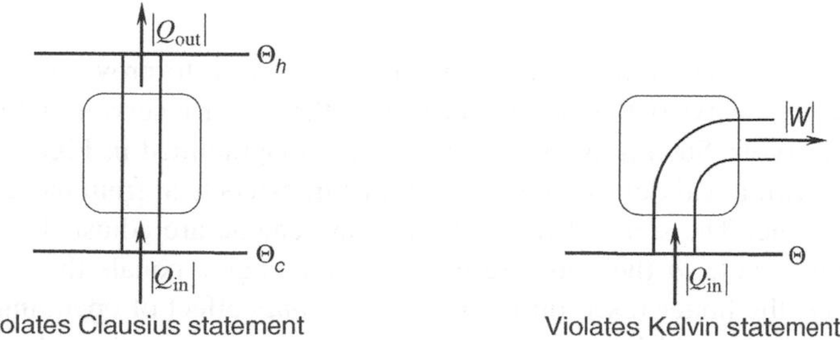

Figure 7.2 *Devices that Violate the Second Law*

Hypothetical devices that violate the second law of thermodynamics are diagrammed in **Figure 7.2**. The only effect of operating the device on the left is to transfer heat from the colder reservoir to the hotter one, which violates the Clausius statement of the second law. The only effect of operating the device on the right is to accept heat from one reservoir and do work, which violates the Kelvin statement.

For designers of economical machines, the ideal heat engine would be one that changes 100% of the heat available into mechanical or electrical energy. The Kelvin statement of the second law asserts that it is impossible to construct such an engine. The ideal household refrigerator would be one that transfers heat from inside the box to the warmer air in the room without using any electrical of mechanical energy. The Clausius statement asserts that it is impossible to construct such a refrigerator. The Kelvin statement is often the more useful because it refers to a single heat reservoir instead of two. There is no need to consider one statement more fundamental than the other, since we can prove the two statements are equivalent.

Proofs

A useful procedure for obtaining physically significant results is the *method of indirect proof*. A statement is made about the physical world. It is then shown that, if the negation of the statement were true, one of our assumptions about the physical world must be false. Or, if our assumptions are valid, the contradiction must be false and the original statement is true. We assume that the first and second laws of thermodynamics are valid. If a hypothetical cyclic process can be combined with a real process to make a device that is functionally equivalent to one of the impossible devices diagrammed in Figure 7.2, the hypothetical process represents an impossibility.

The goal of thermodynamic theory is to describe as wide a range of thermal phenomena as possible. For the theory to be complete, it must include facts that we accept without question because of their familiarity. Such facts should either be postulates of the theory or be obtainable from them. The most important postulates of thermodynamics are the first and second laws. The diagram of a 2-temperature engine in Figure 7.1 assumes without proof that an engine accepts heat *from* the hotter reservoir and rejects heat *to* the colder reservoir. The following example obtains this fact as a consequence of the second law.

Example 7.1 Hotter and colder

Prove that 2-temperature engines accept heat from the hotter reservoir and reject heat to the colder reservoir.

Solution. The contradiction of the statement that we seek to prove is: At least one 2-temperature engine exists that accepts heat from the colder reservoir and rejects heat to the hotter reservoir. Such a hypothetical engine is diagrammed in **Figure 7.3**, where it is combined with a real device (a metal rod) that transfers heat from the hotter reservoir to the colder one. The sizes of the metal rod and engine are adjusted so that the heat transferred per cycle *to* the hotter reservoir by the engine equals the heat transferred per cycle *from* the hotter reservoir by the rod. The *only* effect of operating the resulting composite device would be to accept heat from one reservoir (the colder one) and do work, so that the composite device is functionally equivalent to the second device in

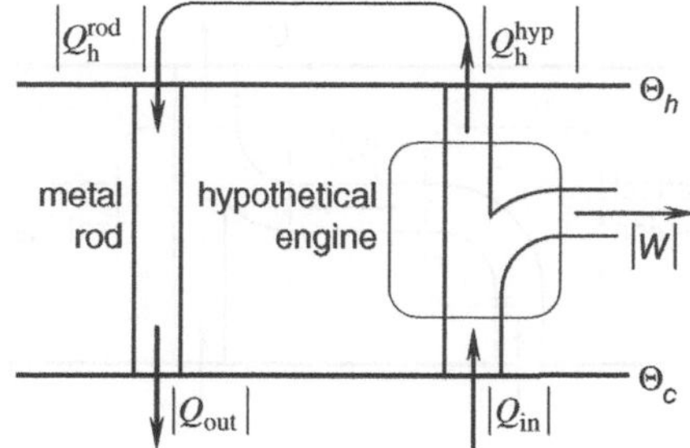

Figure 7.3 *Composite Device that Violates the Second Law*

Figure 7.2, which violates the Kelvin statement of the second law. Since we have assumed that the Kelvin statement is valid, the statement at the beginning of the paragraph must be false. Thus, 2-temperature engines accept heat *from* the hotter reservoir and reject heat *to* the colder reservoir.

As engines accept heat from the hotter reservoir and reject it to the colder reservoir, we have $|Q_{in}| = |Q_h|$ and $|Q_{out}| = |Q_c|$,where subscripts h and c indicate hotter and colder. It then follows from the first law and $\Delta U = 0$ that

$$|W| = |Q_{in}| - |Q_{out}| = |Q_h| - |Q_c|. \tag{7.3}$$

Because of the absolute value signs, this equation is also valid for refrigerators.

When proving statements by rejecting contradicting possibilities, it is essential that all possibilities be considered. In the above example, an apparent possibility is an engine that transfers heat between two reservoirs at the *same* temperature. However, in the conceptual arguments being made, a heat reservoir is anything and everything that accepts and rejects heat at one temperature, so that having two reservoirs at the same temperature is an excluded possibility.

Equivalence of the Clausius and Kelvin Statements

The Clausius and Kelvin statements are not both needed as postulates, since either one can be deduced from the other. To show this, we first prove that, *if the Kelvin statement of the second law is false, then the Clausius statement is false*. If the Kelvin statement is false, a device must exist that is functionally equivalent to the second device in Figure 7.2. The problem is to show that, if such a device is possible, then a device like the one on the left of Figure 7.2 is also possible. We do this by creating the composite device diagrammed in **Figure 7.4**, which is equivalent to combines the hypothetical device in Figure 7.2 which violates the Kelvin statement with a real refrigerator. Their sizes are adjusted so that the work needed to operate the refrigerator exactly equals the work done by the hypothetical device. The *only* effect of operating the resulting composite device is to accept heat from the colder reservoir and transfer it to the hotter reservoir, which makes it functionally equivalent to the device in Figure 7.2, which violates the Clausius statement. Hence, *if the Kelvin statement is false, then the Clausius statement is false.*

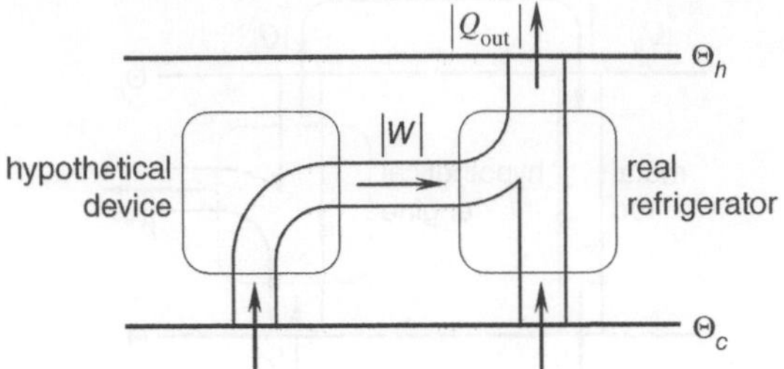

Figure 7.4 Composite Device that Violates the Second Law

It is left as a problem to prove that, *if the Clausius statement of the second law is false, then the Kelvin statement is false*. After proving this, we still have two possibilities: Specifically, both statements could be true or both could be false. Logic alone is not capable of selecting between these two possibilities. If it were, the statements of the second law would not be significant statements about the real world. The Clausius and Kelvin statements are justified by observations of the real world.

7.4 Carnot Cycles

A *Carnot engine* named after Sadi Carnot is an idealized system that is cyclic and reversible and interacts with two heat reservoirs. Although the term "engine" usually refers to the complete device for transforming heat into work, our focus here will be on the working substances on which the devices operate. For example, gas is the working substance in a gas turbine, and water in its liquid and vapor phases is the working substance in a steam engine.[1] A Carnot cycle is made up two isothermal and two adiabatic processes:

1. Heat is transferred *to* the system (working substance) from the hotter reservoir at temperature Θ_h in an isothermal process.
2. The system is cooled to the temperature Θ_c of the colder reservoir in an adiabatic process.
3. Heat is transferred to the colder reservoir *from* the system in an isothermal process.
4. The system is returned to its initial state at temperature Θ_h in an adiabatic process.

Carnot engines are important because they are the most efficient type of 2-temperature engine and because their efficiency is independent of the properties of the working substance. When a Carnot engine is operated in reverse, it becomes a Carnot refrigerator. Carnot cycles can be either engines ($W > 0$) or refrigerators ($W < 0$). Their essential characteristics are summarized as follows:

> ***Carnot Cycle***
> *A Carnot cycle is a cyclic process that accepts heat at one temperature, rejects heat at another temperature, and is reversible.*

[1] Heat engines are discussed in more detail in Section 4.3.

The word "reversible" is the crucial feature that distinguishes a Carnot cycle from other 2-temperature cycles. The cycle involves reversible heat transfers, which is an idealization that is approached by having very small temperature gradients where heat is exchanged. Since reducing temperature gradients reduces the rate of heat transfer, devices whose operation approximates a Carnot cycle tend to operate very slowly and thus are not useful for everyday applications. Nevertheless, their existence has practical implications because of the limits they establish on the energy efficiencies. As defined in (4.35), *energy efficiency* is

$$e = \frac{|W|}{|Q_{\text{in}}|}, \tag{7.4}$$

where $|Q_{\text{in}}|$ is the heat transferred to the device per cycle and $|W|$ is the work done. As $|Q_{\text{in}}| = |Q_h|$ and $|W| = |Q_h| - |Q_c|$, it follows that

$$e = \frac{|Q_h| - |Q_c|}{|Q_h|} = 1 - \frac{|Q_c|}{|Q_h|}, \tag{7.5}$$

where $|Q_h|$ and $|Q_c|$ are the heats exchanged at the hot and cold reservoirs.

Carnot's Theorem

No engine operating between two heat reservoirs is more efficient than a Carnot engine.

The proof of the theorem proceeds by considering the composite device diagrammed in **Figure 7.5**, where X represents a general 2-temperature engine that may or may not be reversible. The complete device combines engine X operating normally and a Carnot engine operating in reverse, which makes it a Carnot refrigerator. The sizes of the devices are adjusted so the work produced by X equals the work needed to operate the refrigerator.

There are two possibilities: either $e^X \le e^C$ or $e^X > e^C$. We now show that, if $e^X > e^C$ were true, the composite device would violate the Clausius statement of the second law. If the inequality $e^X > e^C$ were true, it would follow from the definition of energy efficiency that

$$\frac{|W|}{|Q_h^X|} > \frac{|W|}{|Q_h^C|}, \tag{7.6}$$

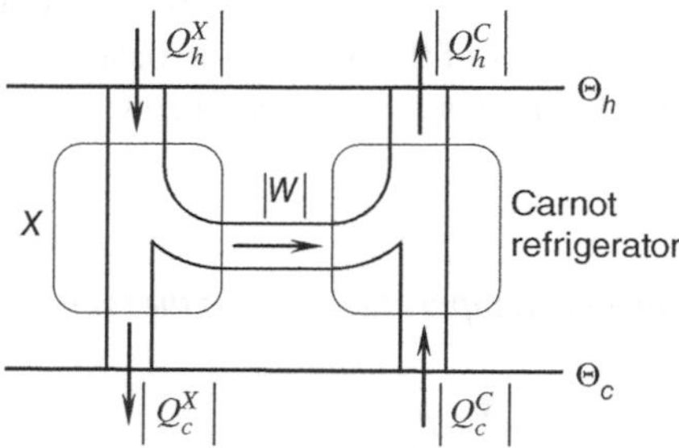

Figure 7.5 Composite Device Made from an Engine and a Carnot Refrigerator

which implies that

$$|Q_h^C| > |Q_h^X| \quad \text{and} \quad |Q_h^C| - |Q_h^X| > 0. \tag{7.7}$$

As can be seen from the figure, the composite device rejects heat $|Q_h^C| - |Q_h^X|$ to the hotter reservoir. This fact combined with (7.7) implies that heat is transferred *to* the hotter reservoir. Since the heat would have to come from the colder reservoir, the only effect of operating the composite device would be to transfer heat from the colder reservoir to the hotter one. Since that would violate the Clausius statement of the second law, the possibility $e^X > e^C$ is rejected. The remaining possibility is

$$e^X \leq e^C, \tag{7.8}$$

which effectively summarizes Carnot theorem.

> ***Corollary to Carnot's Theorem***
> *All Carnot engines operating between the same two heat reservoirs have the same efficiency.*

The proof is easy. Consider two Carnot engines C and $\overline{C}$. Depending on which Carnot engine is considered to be engine X, applying Carnot's theorem implies that $e^{\overline{C}} \leq e^C$ and that $e^C \leq e^{\overline{C}}$. From these it follows that $e^C = e^{\overline{C}}$, and the corollary is proved.

7.5 Absolute Temperature

The defining characteristic of heat reservoirs is their fixed tendency to exchange heat. Since all Carnot engines operating between the same two reservoirs have the same energy efficiency e^C, we can use their efficiency to assign temperatures in a way that does not depend on the properties of any system or type of system. It follows from the definition of energy efficiency that

$$e^C = \frac{|W|}{|Q_h^C|} = \frac{|Q_h^C| - |Q_c^C|}{|Q_h^C|} = 1 - \frac{|Q_c^C|}{|Q_h^C|}, \tag{7.9}$$

where $|Q_h^C|$ and $|Q_c^C|$ are the heats exchanged at the hotter and colder reservoirs by a Carnot engine. As Carnot engines are reversible and e^C is the same for all Carnot engines, the ratio $|Q_c^C/Q_h^C|$ is the same for all Carnot devices (engines or refrigerators) operating between the same two reservoirs. The uniqueness of this ratio is the basis for the definition of *absolute temperatures*, also known as *thermodynamic temperatures*.

Definition

A temperature scale is an absolute temperature scale if the temperatures T_1 and T_2 assigned to two reservoirs are such that

$$\boxed{\frac{T_2}{T_1} = \frac{|Q_2^C|}{|Q_1^C|},} \tag{7.10}$$

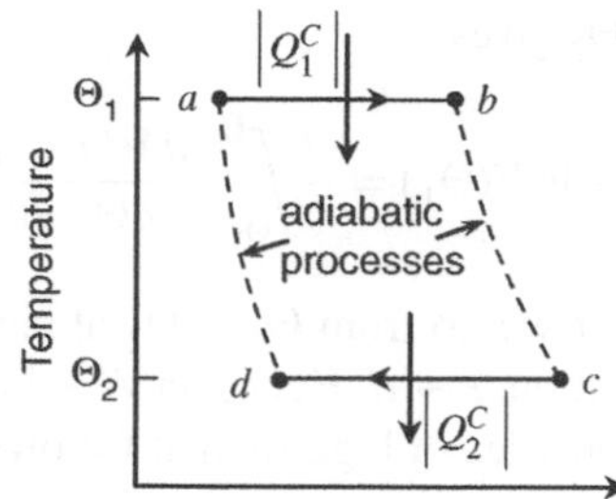

Figure 7.6 Generalized Diagram of a Carnot Cycle

where $|Q_1^C|$ *and* $|Q_2^C|$ *are the heats exchanged by a Carnot cycle operating between the reservoirs.*

By convention, the symbol T is reserved for temperatures measured on an absolute scale. For the above definition to be useful, we need practical methods for assigning temperatures consistent with (7.10). To show how this can be done without actually constructing a Carnot device, we consider *PVT* systems whose internal energies depend on temperature only. Although this is a characteristic of ideal gases, we cannot assume that $PV = nRT$ since we have not yet established how absolute temperatures T are measured. Instead, we assume that the product of pressure times volume is a function of temperature, but make no assumption about the temperature scale used to specify the temperatures designated by Θ. As a result, the internal energy and equation of state of the system are

$$U = A(\Theta) \quad \text{and} \quad PV = B(\Theta), \tag{7.11}$$

where $A(\Theta)$ and $B(\Theta)$ are unspecified functions of Θ. A Carnot engine based on the properties of this system is diagrammed in **Figure 7.6**, where Θ_1 and Θ_2 are the temperatures of the reservoirs and the dashed curves represent adiabatic processes.

As internal energy U depends on temperature only, its change ΔU vanishes during the isothermal processes $a \to b$ and $c \to d$, so that the first law specifies that $Q = W$, where $W = \int PdV$. It then follows from $PV = B(\Theta)$ that the heat transferred at temperature Θ_1 is

$$Q_1^C = \int_{V_a}^{V_b} \frac{B(\Theta_1)}{V} dV = B(\Theta_1)(\ln V_b - \ln V_a) = B(\Theta_1) \ln \frac{V_b}{V_a}. \tag{7.12}$$

Similarly, the heat transferred in process $c \to d$ is $Q_2^C = B(\Theta_2)\ln(V_d/V_c)$. Since $Q_2^C > 0$ and $Q_2^C < 0$, the ratio of the heats exchanged is

$$\frac{|Q_2^C|}{|Q_1^C|} = \frac{B(\Theta_2)\ln(V_c/V_d)}{B(\Theta_1)\ln(V_b/V_a)}. \tag{7.13}$$

We next consider the adiabatic processes $b \to c$ and $d \to a$. Since no heat is transferred during adiabatic processes, the first law specifies that $\Delta U + W = 0$. For small changes $\Delta\Theta$ and ΔV this becomes $(dU/d\Theta)\Delta\Theta + P\Delta V = 0$ and $\Delta V/\Delta\Theta$ becomes $dV/d\Theta$. Then, by using $dU/d\Theta = dA/d\Theta$ and $P = B/V$, which follow from (7.11), we find show that

$$\frac{1}{V(\Theta)} \frac{dV(\Theta)}{d\Theta} = -\frac{dA(\Theta)}{d\Theta} \frac{1}{B(\Theta)}. \tag{7.14}$$

Integrating over Θ from Θ_1 to Θ_2 gives

$$\ln V(\Theta_2) - \ln V(\Theta_1) = -\int_{\Theta_1}^{\Theta_2} \frac{dA(\Theta)}{d\Theta}\frac{1}{B(\Theta)}d\Theta, \tag{7.15}$$

which is independent of whether we go from Θ_1 to Θ_2 along the line $b \to c$ or along the line between a and d in the direction $a \to d$. Hence, $\ln V_c - \ln V_b = \ln V_d - \ln V_a$. By rearranging terms and using the properties of logarithm, it follows that

$$\ln V_c - \ln V_d = \ln V_b - \ln V_a \quad \text{and} \quad \ln\left(\frac{V_c}{V_d}\right) = \ln\left(\frac{V_b}{V_a}\right). \tag{7.16}$$

Combining this with (7.13) implies

$$\frac{|Q_2^C|}{|Q_1^C|} = \frac{B(\Theta_2)}{B(\Theta_1)}. \tag{7.17}$$

Comparing this with (7.10) indicates that $B(\Theta)$ is proportional to absolute temperature T. By using nR as the proportionality constant, the comparison indicates that $B(\Theta) = nRT$, where R is an unspecified constant, where the number of moles n is included so that $B(\Theta)$ is extensive.

Combining $B(\Theta) = nRT$ with the equation of state $PV = B(\Theta)$ yields the familiar ideal gas equation $PV = nRT$, which can be rearranged as $T = PV/nR$. This gives us a method for measuring absolute temperatures: We would use a thermometer that utilizes the properties of a gas which closely approximates an ideal gas. The temperature T of the thermometer would then be found by measuring the pressure P and volume V of n moles of a gas and equating T to the value of PV/nR. As discussed in the next chapter, absolute temperatures measured in this way are called *ideal gas temperatures*.

As the ratio of the heats exchanged is the same for all Carnot devices, absolute temperatures can be measured by any device that assigns temperatures to heat reservoirs so the ratio of temperatures equals the ratio of heats. Since the absolute values of Q_2^C and Q_1^C are used in (7.10), the ratio of temperatures is always greater than zero, which implies that absolute temperatures are either all positive or all negative. The positive option is used. Since Celsius and Fahrenheit temperatures can be negative, they are clearly not absolute temperatures. The smallest absolute temperature is $T = 0$, which is the zero point on all absolute temperature scales. The special significance of this lowest (coldest) temperature – called *absolute zero* – is discussed in Sections 8.2 and 10.6.

Kelvin Temperatures

Since the ratio T_2/T_1 is unchanged when both T_2 and T_1 are multiplied by the same factor, absolute temperatures are only defined up to a multiplicative constant. A unique scale can be obtained by specifying the temperature of a single heat reservoir. The Kelvin temperature scale is the absolute temperature scale obtained by specifying the temperature of the triple point of water and requiring that the temperature difference between the ice point and the steam point is the same as on the Celsius scale. Since the temperature of absolute zero on the Celsius scale is −273.15 °C and the temperature of the triple point of water is 0.01 °C, the temperature at the triple point is 273.16 K. *The Kelvin temperature scale is the absolute scale that assigns* 273.16 K *to the temperature of the triple point of water.*

Another absolute scale, used by engineers, is the *Rankine temperature scale*. Just as the Kelvin scale has the same temperature intervals as the Celsius scale, the Rankine scale has the same intervals as the Fahrenheit scale. As the temperature difference between the ice point and the steam point is 100° on the Celsius scale and 180° on the Fahrenheit scale $(212-32=180)$, the values of temperature differences on the Fahrenheit scale are larger than on the Celsius scale by a factor of 1.8 (180/100=9/5). Hence, temperatures on Rankine and Kelvin scales are related by $T^R = 1.8\,T^K$.

7.6 Applications

Carnot engines and refrigerators give us useful insights into the operation of practical devices. The key to this is the definition of absolute temperatures in terms of the heats exchanged at two heat reservoirs.

Refrigerators and Heat Pumps

A parameter is needed for characterizing the performance of refrigerators. Since we desire to remove heat $|Q_c|$ *from* the cold interior and want to minimize the energy $|W|$ needed to operate the refrigerator, the optimal refrigerator is one that maximizes the ratio $|Q_c|/|W|$. This ratio is usually greater than one. If we called it efficiency, we would have efficiencies greater than 100%, which is contrary to convention. Instead, it is called a *coefficient of performance*. The appropriate coefficient for devices whose purpose is to transfer heat $|Q_c|$ *from* a colder place is

$$\mathrm{COP}_c = \frac{|Q_c|}{|W|}. \tag{7.18}$$

As the purpose of air conditioners is to transfer heat *from* inside a building when it is cooler than outside, COP_c is the appropriate coefficient of performance for air conditioners.

The *heat pumps* used for heating buildings are technically refrigerators but are called heat pumps because their purpose is to deliver heat to a hotter place. The appropriate coefficient of performance is

$$\mathrm{COP}_h = \frac{|Q_h|}{|W|}, \tag{7.19}$$

where $|Q_h|$ is the heat transferred *to* the building.

Refrigerators, air conditioners, and heat pumps are diagrammed in **Figure 7.7**, where the lines with arrows are simplified representations of the pipes used in the earlier diagrams. The best possible coefficient of performance is achieved by a Carnot refrigerator. (Justifying this assertion is left as a problem.) By using $|W| = |Q_h| - |Q_c|$, the best coefficient of performance for refrigerators and air conditioners becomes

$$\mathrm{COP}_c^C = \frac{|Q_c^C|}{|W^C|} = \frac{|Q_c^C|}{|Q_h^C| - |Q_c^C|} = \frac{1}{|Q_h^C|/|Q_c^C| - 1}. \tag{7.20}$$

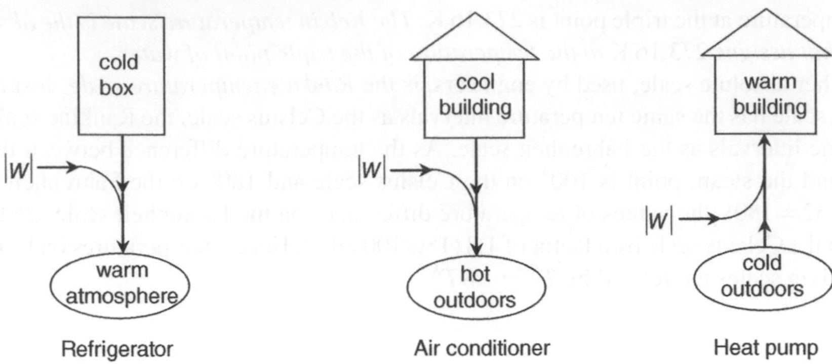

Figure 7.7 *Practical Devices*

Using the definition of absolute temperature in (7.10) to express this in terms of temperatures gives

$$\mathrm{COP}_c^C = \frac{1}{T_h/T_c - 1} = \frac{T_c}{T_h - T_c}. \tag{7.21}$$

Similarly, the best coefficient of performance for heat pumps is

$$\mathrm{COP}_h^C = \frac{|Q_h^C|}{|W^C|} = \frac{|Q_h^C|}{|Q_h^C| - |Q_c^C|} = \frac{1}{1 - |Q_c^C|/|Q_h^C|}, \tag{7.22}$$

so that

$$\mathrm{COP}_h^C = \frac{1}{1 - T_c/T_h} = \frac{T_h}{T_h - T_c}. \tag{7.23}$$

Both coefficients indicate that the smaller the temperature difference $T_h - T_c$ between the hotter and colder locations, the greater the coefficient of performance.

Example 7.2 Heat pumps

Find the ratio of the energy needed to heat a building with electric resistance heaters to the energy needed to heat it with a Carnot heat pump. Assume that the temperature is 20 °C inside and 0 °C outside.

Solution. Let $|Q_{20}|$ represent the heat required to maintain the building at 20 °C. An electric heater changes electrical energy $|W^{\mathrm{el}}|$ directly into heat, so that

$$|W^{\mathrm{el}}| = |Q_{20}|.$$

By using this, the coefficient of performance of the heat pump becomes

$$COP_h = \frac{|Q_{20}|}{|W^{h\text{-}p}|} = \frac{|W^{\mathrm{el}}|}{|W^{h\text{-}p}|},$$

where $|W^{h\text{-}p}|$ is the energy input to the heat pump. As the coefficient of performance is $\mathrm{COP}_h = T_h/(T_h - T_c)$, the ratio of the energy needed to heat the building with resistance heaters to the energy needed with a Carnot heat pump is

$$\frac{|W^{el}|}{|W^{h\text{-}p}|} = \frac{273 + 20}{(273 + 20) - 273} = \frac{293}{20} = 14.7.$$

The example tells us that resistance heaters require 14.7 times more energy than the best possible heat pump when it is 0 °C outside. The performance of commercially available heat pumps is typically much less than that of ideal Carnot devices. Unfortunately, the coefficients of performance of commercially available devices are not expressed as dimensionless numbers but are commonly expressed as ratios of two numbers with different units of energy.

Engines

The purpose of heat engines is to accept heat and do work. Designers of heat engines would like to transform as much as possible of the available heat into useful work and thereby minimize the heat rejected. Carnot's theorem states that the efficiency of an engine cannot exceed the efficiency of a Carnot engine ($e^X \leq e^C$). It follows from the expression for e^C in (7.9) and the definition of absolute temperature in (7.10) that the efficiency of a Carnot engine is

$$e^C = 1 - \frac{|Q_c^C|}{|Q_h^C|} = 1 - \frac{T_c}{T_h}. \tag{7.24}$$

By combining this with the expression for energy efficiency $e = 1 - |Q_c|/|Q_h|$ and the relationship $e^X \leq e^C$ in (7.8), we obtain

$$1 - \frac{|Q_c|}{|Q_h|} \leq 1 - \frac{T_c}{T_h}, \tag{7.25}$$

which implies that

$$|Q_c| \geq \frac{T_c}{T_h}|Q_h|. \tag{7.26}$$

This inequality imposes a lower bound to the amount of heat $|Q_c|$ that must be rejected at the cold reservoir and indicates that we can reduce the amount of heat rejected either by increasing the temperature T_h of the hot reservoir or by decreasing the temperature T_c of the cold reservoir. This fundamental restriction on our ability to transform heat into work is a consequence of the second law and supplements the restriction (energy conservation) imposed by the first law.

Example 7.3 Gas turbine cycle

Find the energy efficiency of the gas turbine cycle and compare it with the energy efficiency of a Carnot cycle operating between the same maximum and minimum temperatures. Assume that an ideal gas (air) enters the compressor at temperature $T_a = 300\,\text{K}$, that the pressure ratio is $r_P = P_b/P_a = 10$, and that the change in temperature in the burner is $T_c - T_b = 500\,\text{K}$.

Solution. Gas turbines are described in Section 4.3, and a diagram of the idealized cycle is shown in **Figure 7.8**. The energy efficiency e^{gt} of a gas turbine was expressed as a

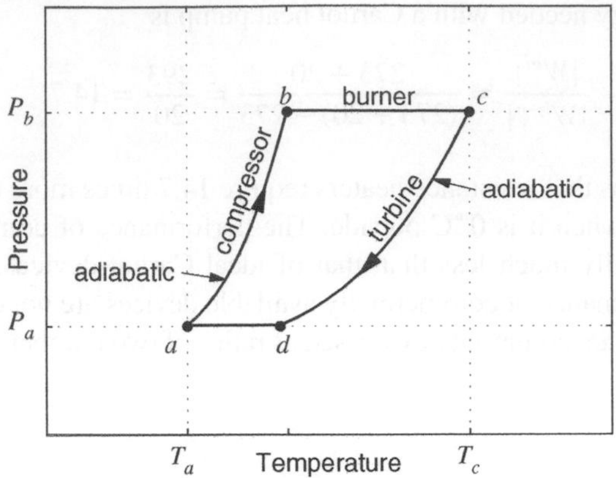

Figure 7.8 Gas Turbine Cycle

function of the pressure ratio $r_P = P_b/P_a$ in Example 4.2. Substituting $r_P = 10$ into the result in (4.37) gives

$$e^{\text{gt}} = 1 - \frac{1}{r_P^{(\gamma-1)/\gamma}} = 1 - \left(\frac{1}{10}\right)^{0.4/1.4} = 48\%. \tag{7.27}$$

The temperature of the gas increases from T_a to T_b in process $a \to b$. As given in (4.20), the relationship between temperature and pressure in an adiabatic process is $T^\gamma/P^{\gamma-1} =$ const. Applying this to process $a \to b$ gives

$$\frac{T_b}{P_b^{(\gamma-1)/\gamma}} = \frac{T_a}{P_a^{(\gamma-1)/\gamma}}.$$

By rearranging terms and using $\gamma = 1.4$ and $P_b/P_a = 10$, the ratio of temperatures becomes

$$\frac{T_b}{T_a} = \left(\frac{P_b}{P_a}\right)^{(\gamma-1)/\gamma} = r_P^{(\gamma-1)/\gamma} = 10^{(0.4/1.4)} = 1.931. \tag{7.28}$$

The maximum temperature in the cycle, which is T_c, occurs where the gas exits the burner. By using $T_c - T_b = 500\,\text{K}$ and $T_a = 300\,\text{K}$, we find that

$$T_c = T_c - T_b + \frac{T_b}{T_a}T_a = 500\,\text{K} + (1.931)(300\,\text{K}) = 1080\,\text{K}.$$

The minimum temperature, which is T_a, occurs where gas enters the compressor. By using (7.24), the efficiency of a Carnot engine operating between these maximum and minimum temperatures is

$$e^C = 1 - \frac{T_a}{T_c} = 1 - \frac{300}{1080} = 72\%.$$

This Carnot efficiency is significantly greater than the 48% efficiency of the idealized gas turbine cycle.

It is interesting to see the effect of decreasing the temperature change in the burner on the efficiency of a gas turbine. To do this, we express both the Carnot and gas turbine efficiencies in terms of T_a and T_b and the temperature change in the burner $(T_c - T_b)$. The Carnot efficiency is

$$e^C = 1 - \frac{T_a}{T_c} = 1 - \frac{T_a}{T_b + (T_c - T_b)}. \quad (7.29)$$

The gas turbine efficiency is found by combining $e^{gt} = 1 - 1/r_P^{(\gamma-1)/\gamma}$ with the equality $r_P{}^{(\gamma-1)/\gamma} = T_b/T_a$ from (7.28). The result is

$$e^{gt} = 1 - \frac{T_a}{T_b}. \quad (7.30)$$

Comparing these expressions indicates that the gas turbine efficiency approaches the optimal Carnot efficiency as the temperature change in the burner deceases, a decrease that can be achieved by decreasing the amount of heat transferred in the burner. However, reducing the amount of heat transferred also reduces the amount of useful work produced, which means reducing power output of the device. As a result, the goal of achieving greater efficiency is in conflict with the goal of obtaining more power.

Problems

7.1 What temperature drop in the water in the oceans would be needed to supply the energy used in the USA in one year if we could violate the Kelvin statement of the second law and obtain all of that energy from the oceans as heat without any other effect? Use US engineering units throughout. Do not change to metric units. In engineering practice $Q = wc_P^w \Delta T$, where w is weight and c_P^w has units of Btu lb^{-1} °F^{-1}. Data: 1 mile = 5280 ft; volume of the oceans, 3.29 × 10^8 mile^3; density of sea water, 64 lb ft^{-3}; specific heat of sea water, c_P^w = 1.0 Btu lb^{-1} °F^{-1}; the U.S. used approximately 9 × 10^{16} Btu of energy annually from 2004 to 2010.

7.2 Prove that, if the Clausius statement of the second law is false, then the Kelvin statement is false. Use diagrams of the type introduced in the chapter.

7.3 Find the maximum possible coefficient of performance of a household refrigerator which keeps its contents at 3 °C and operates in a room at 22 °C.

7.4 **(a)** Prove that no 2-temperature refrigerator operating between reservoirs at hotter and colder temperatures T_h and T_c has a greater coefficient of performance than a Carnot refrigerator. Use diagrams of the type introduced in the chapter.
(b) Find the minimum amount of electrical energy needed to create 0.50 kg of ice from water at 0 °C in a household freezer that rejects heat to a room at 20 °C.

7.5 Show that a heat pump with a coefficient of performance COP$_h$ greater than a heat pump that operates on a Carnot cycle would violate Clausius' statement of the second law.

7.6 Consider the possibility of a "thermal battery," *i.e.*, a device for storing mechanical or electrical energy. Use two large containers as heat reservoirs in which one container contains ice and water in equilibrium while the other contains steam and water in equilibrium, and both are held at a pressure of 1 atm. Energy is stored by using a Carnot cycle to transfer heat from the colder reservoir to the hotter reservoir. How much ice is created and how much water is vaporized when 1.0 kilowatt hour of energy is stored?

7.7 Show that the efficiency of the Otto cycle (described in Section 4.4) approaches the efficiency e^C of a Carnot cycle when the energy released by burning the fuel-air mixture is decreased. Assume the compression ratio r_V remains unchanged.

8

Temperature Scales and Absolute Zero

We are familiar with temperatures through their use in everyday life (weather, cooking, etc.) but may not be familiar with the problems associated with measuring it. To appreciate these problems, we need to consider how temperatures are measured by thermometers. In general terms, a thermometer is a device (system) with an easily measured property that is determined by the tendency with which it exchanges heat with other systems. The property measured in laboratory thermometers is often a voltage or resistance. The property measured in the familiar liquid-in-glass thermometers is the length of a column of mercury or other liquid. The device consists of liquid in a glass bulb connected to a long narrow tube along which the liquid expands or contracts as its temperature changes. Marks are made on the tube at 100 equally spaced intervals between the locations of the end of the column of liquid when in equilibrium at the ice point (0 °C) and when in equilibrium at the steam point (100 °C). The marks are numbered from 0 to 100 and the temperature is indicated by the number adjacent to the end of the column. However, as different types of liquid and glass expand differently, the temperatures measured depend on the materials used. Consequently, thermometers made with different types of liquid and glass effectively define different temperature scales. The differences in the measured temperatures are typically not significant in everyday situations but can be significant in scientific work.

8.1 Temperature Scales

To facilitate the discussion of temperature scales, different temperature scales will be indicated by superscripts, and the temperatures of different systems (heat reservoirs) will be indicated by subscripts. For example, the temperature of a system of ice and water in equilibrium at one standard atmosphere (the ice point) is $\Theta^C_{\text{ice·pt}} = 0$ on the Celsius scale and

Thermodynamics and Statistical Mechanics: An Integrated Approach, First Edition.
Robert J Hardy and Christian Binek.

$\Theta^F_{\text{ice·pt}} = 32$ on the Fahrenheit scale. Following convention, the symbol T is reserved for temperatures measured on an absolute scale.

Gas Thermometers

In the previous chapter, we found that temperatures measured with a thermometer based on the properties of an ideal gas are absolute temperatures. We now investigate how to correct for the differences between the properties of real and ideal gases. As suggested by the ideal gas equation of state, $PV = nRT$, the temperature Θ^{gas} measured by a gas thermometer is proportional to PV/n, where P is pressure, V is volume, and n is the number of moles. In a constant volume thermometer, the property measured is pressure. The thermometer (a container of gas) is calibrated by bringing it into thermal equilibrium with a selected reference reservoir and measuring the pressure, which we denote by P_{tr}. (The subscript tr is used because the triple point of water is commonly used as the reference reservoir.) The temperature of system X is measured by bringing the thermometer into thermal equilibrium with it and measuring its pressure P_X. Then, the *gas temperature* of the system is

$$\Theta^{\text{gas}}_X = \frac{\Theta_{\text{tr}}}{P_{\text{tr}}} P_X. \tag{8.1}$$

We can set Θ_{tr} equal to 273.16 in anticipation of the Kelvin temperature scale.

Gas thermometers that use different gases yield slightly different temperatures. This can be seen in **Figure 8.1**, which shows the gas temperatures $\Theta^{\text{gas}}_{\text{st·pt}}$ at the steam point of water as determined when oxygen, hydrogen, and helium gases are used. The data points (squares) are for different densities $\rho = n/V$ and correspond to different values of P_{tr}.

The straight lines in the figure represent (idealized) best fits to the data. The range of values for $\Theta^{\text{gas}}_{\text{st·pt}}$ is small, and the difference between the values for different gases decreases as the density decreases. The convergence of lines indicates that values independent of the gas used can be obtained by extrapolating the data to zero density. Since the behavior

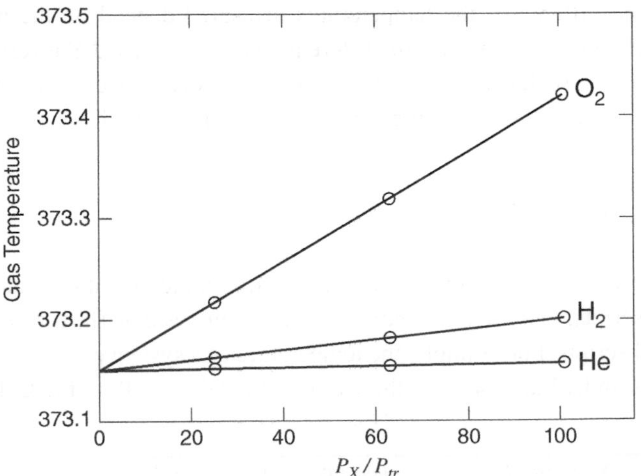

Figure 8.1 *Gas Temperatures at the Steam Point*

of real gases approaches that of ideal gases at low densities, the temperatures obtained by the extrapolation are *ideal gas temperatures* Θ^{ig}. The procedure for measuring Θ^{ig} is summarized by

$$\Theta^{ig} = \Theta_{tr} \lim_{\rho \to 0} \left[\frac{P}{P_{tr}} \right]. \tag{8.2}$$

As ideal gas temperatures are absolute temperatures, they can be used to calibrate thermometers that measure temperatures on the Kelvin scale, or other absolute temperature scales. Different absolute temperature scales differ by a multiplicative constant. The Celsius temperature scale, which is not an absolute temperature scale, differs from the Kelvin scale by an additive constant ($\Theta^C = T^K + 273.15$).

International Temperature Scale

The temperatures obtained with gas thermometers are only one way to obtain *thermodynamic temperatures*, which is the name of scales with the properties of an absolute scale. When high accuracy is required, the *International Temperature Scale* is used. This scale is an equipment calibration standard established by international agreement and consists of a set of fixed points and methods for interpolation between them. Practical thermometers are calibrated by comparison with these fixed points. One of the fixed points is the triple point of water whose isotopic composition is the same as that of ocean water. The reference to ocean water is needed because water is composed of molecules of different isotopes of hydrogen and oxygen, and variations in the relative amounts of the different isotopes cause small variations in high precision measurements of the triple point temperature.

The International Temperature Scale places the zero of the Celsius scale 0.01° below the triple point, so that 0 °C is the same as 273.15 K. Since the freezing and boiling of water are physical phenomena, their temperatures must be determined experimentally. The results for the freezing and boiling temperatures of water at one standard atmosphere of pressure are 0.000089 and 99.9839 °C, respectively – results that are very close to but not exactly equal to 0 and 100 °C.

8.2 Uniform Scales and Absolute Zero

Are absolute temperature scales uniform? That is, can the same significance be attributed to the temperature intervals at all temperatures? The key to the question is the meaning of "same significance." Consider how we might check that different length intervals on an ordinary ruler are the same. We could place a piece of paper next to the ruler, draw lines on the paper next to adjacent marks on the ruler, such as the marks at 1 and 2 cm, and verify that the lines on the paper can be lined up with adjacent marks everywhere on the ruler. The procedure for checking temperature intervals is less straightforward.

Thought Experiment

The significance of the intervals on an absolute temperature scale is illustrated by the thought experiment diagrammed in **Figure 8.2**. The horizontal lines in the figure represent

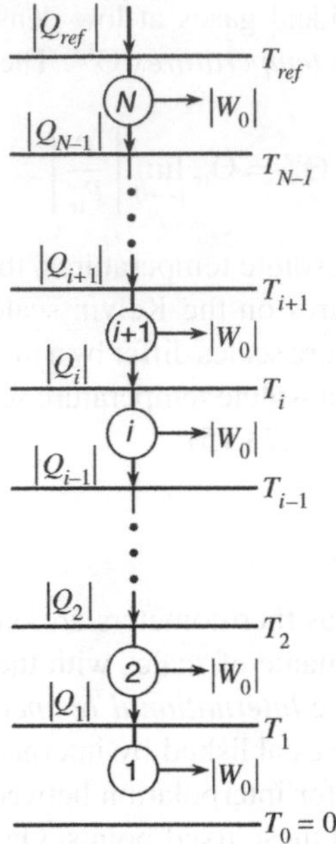

Figure 8.2 *Diagram of the Thought Experiment*

heat reservoirs and the circles represent Carnot engines operating between adjacent reservoirs. The experiment proceeds by selecting $N + 1$ heat reservoirs as follows:

1. Select a reference heat reservoir and assign a specific temperature T_{ref} to it.
2. Select N other heat reservoirs colder than the reference reservoir and label them from $i = 0$ to $i = N - 1$ with smaller values of i indicating colder reservoirs.
3. Starting with the reference reservoir, operate Carnot engines between adjacent reservoirs, and adjust them so the heat rejected by engine i equals the heat accepted by engine $i - 1$.
4. Choose the reservoirs for $i < N$ so that the work done by each engine is

$$W_0 = \frac{1}{N}|Q_{\text{ref}}|, \tag{8.3}$$

where Q_{ref} is the energy accepted as heat from the reference reservoir.

Let T_i represent the absolute temperature of reservoir i. It then follows from the definition of absolute temperature in (7.10) that

$$\frac{T_i}{T_{i+1}} = \frac{|Q_i^C|}{|Q_{i+1}^C|}. \tag{8.4}$$

By expressing the ratio T_i/T_{ref} as a product of ratios, we find that

$$\frac{T_i}{T_{\text{ref}}} = \frac{T_i}{T_{i+1}}\frac{T_{i+1}}{T_{i+2}}\cdots\frac{T_{N-2}}{T_{N-1}}\frac{T_{N-1}}{T_{\text{ref}}} = \frac{|Q_i^C|}{|Q_{i+1}|}\frac{|Q_{i+1}^C|}{|Q_{i+2}|}\cdots\frac{|Q_{N-2}^C|}{|Q_{N-1}^C|}\frac{|Q_{N-1}^C|}{|Q_{\text{ref}}|} = \frac{|Q_i^C|}{|Q_{\text{ref}}|}. \tag{8.5}$$

This indicates that the collection of engines between the reference reservoir and reservoir i is effectively a single Carnot engine, which allows us to express the temperature difference between reservoirs i and $i + 1$ as

$$T_{i+1} - T_i = T_{\text{ref}} \left(\frac{T_{i+1}}{T_{\text{ref}}} - \frac{T_i}{T_{\text{ref}}} \right) = T_{\text{ref}} \left(\frac{|Q_{i+1}^C|}{|Q_{\text{ref}}|} - \frac{|Q_i^C|}{|Q_{\text{ref}}|} \right) = \frac{T_{\text{ref}}}{|Q_{\text{ref}}|} (|Q_{i+1}^C| - |Q_i^C|). \quad (8.6)$$

Conservation of energy requires that $|Q_{i+1}^C| - |Q_i^C| = W_o$. Since $W_o = (1/N)|Q_{\text{ref}}|$, it follows that

$$T_{i+1} - T_i = \frac{T_{\text{ref}}}{|Q_{\text{ref}}|} W_0 = \frac{1}{N} T_{\text{ref}}. \quad (8.7)$$

Thus, the temperature difference $T_{i+1} - T_i$ is the same at all temperatures.

The thought experiment indicates that equal absolute temperature intervals have the same significance in the sense that Carnot engines operating between reservoirs with the same temperature difference produce the same amount of work. For example, a collection of heat reservoirs with uniform temperature differences on the Kelvin scale could be constructed by choosing the reference reservoir to be water at its triple point, assigning T_{ref} the value of 273.16, and setting $N = 27316$. The temperature difference between adjacent reservoirs would be 0.01 K, and a unit temperature interval on the Kelvin scale would be the temperature difference between reservoirs i and $i + 100$.

Absolute Zero

As the ratio of two absolute temperatures T_1/T_2 equals the ratio $|Q_1^C|/|Q_2^C|$ of the heats exchanged by a Carnot cycle, the ratio T_1/T_2 is always positive. Hence, as mentioned earlier, absolute temperatures are either all positive or all negative. It would be consistent to make them all negative, but the positive choice is conventionally made. If there are no negative temperatures, the lowest absolute temperature is zero.[1]

The thought experiment starts by accepting energy Q_{ref} from a reference reservoir and letting each engine transfer energy Q_{ref}/N to the surroundings. As a result, there is no energy remaining after N engines. Specifically, there is no energy remaining for transfer to the heat reservoir at temperature T_0. Then, as indicated by (8.4), the temperature of the final reservoir is $T_0 = 0$. This lower temperature is called *absolute zero*. Notice that engine 1 in the figure would transform heat entirely into work, which would violate the Kelvin statement of the second law, which indicates that absolute zero only exists in a limiting sense.

Absolute zero is the lowest (coldest) absolute temperature to which energy can be transferred as heat. Its existence is a consequence of the first law through energy conservation and the second law through the definition of absolute temperature scales. The significance of absolute zero as the lowest possible energy is consistent with microscopic theory, which identifies absolute zero with the ground state energy of a system, *i.e.*, with the quantum mechanical state with the lowest energy.

---

[1] One may have heard of negative temperatures, which are discussed in Appendix 21.A. Negative temperatures are *not* absolute temperatures. A system with a negative temperature is hotter than all system with positive temperatures.

8.3 Other Temperature Scales

Just as the numerical value of a distance in kilometers differs by a multiplicative constant from its value in miles, different absolute temperature scales differ by a multiplicative constant. The Celsius and Fahrenheit scales, which are not absolute scales, differ from each other by both an additive constant and a multiplicative constant. However, these are not the only ways in which temperature scales can differ.

In everyday life, the concepts of hotter and colder are closely associated with higher and lower temperatures, where higher and lower mean numerically greater than and less than, respectively. This association is certainly valid with the Kelvin, Celsius, and Fahrenheit temperatures. Nevertheless, it is not required by the defining characteristics of temperature, which require (i) that systems in equilibrium with each other have the same temperature and (ii) that systems with different initial temperatures spontaneously exchange heat when brought into thermal contact (see Section 1.4).

To see the kind of possibility that is allowed by the defining characteristics of temperature but not by the second law, consider heat reservoirs A, B, and C at different temperatures Θ_A, Θ_B, and Θ_C. To simplify the discussion, use arrows ($\rightarrow$) as an abbreviation for "is hotter than" and consider the possibility that $A \rightarrow B$, $B \rightarrow C$, and $C \rightarrow A$. If this hypothetical possibility occurred in nature, we could *not* consistently associate hotter and colder with higher and lower temperatures. Nevertheless, the possibility is consistent with the characteristics of temperature stated above. The defining characteristics associate the inequality $\Theta_A \neq \Theta_B$ with systems that would exchange heat, but $\Theta_A \neq \Theta_B$ does not distinguish between $\Theta_A < \Theta_B$ and $\Theta_A > \Theta_B$. Specifically, if hotter systems are associated with higher temperatures, the hypothetical possibility would imply that $\Theta_A > \Theta_B$, $\Theta_B > \Theta_C$, and $\Theta_C > \Theta_A$, and these three inequalities are inconsistent with the properties of real numbers!

As the hypothetical possibility does not occur in nature, its non-occurrence should be derivable from the postulates of thermodynamics. If C is hotter than A, *i.e.*, if $C \rightarrow A$, we could transfer heat Q_0 from C to A by bringing them into thermal contact. Then, if $A \rightarrow B$, we could transfer heat Q_0 from A to B, which would leave A in its original state, so that the only effect of the process would be to transfer heat from C to B. If we also had $B \rightarrow C$, which implies that C is colder than B, we would violate Clausius' statement of the second law, which asserts that there is no process whose only effect is to accept heat from a colder reservoir and transfer it to a hotter one. Hence, the second law rules out the possibility that $A \rightarrow B$, $B \rightarrow C$, and $C \rightarrow A$.

The association of hotter and colder (or colder and hotter) with higher and lower temperatures is valid. The following theorem tells us that different temperature scales are related by monotonic functions. A function $f(x)$ is monotonically increasing if $a > b$ implies $f(a) > f(b)$ and is monotonically decreasing if $a > b$ implies $f(a) < f(b)$. For differentiable functions, the derivative of a monotonically increasing function is everywhere positive, $df(x)/dx > 0$, while the derivative of a monotonically decreasing function is everywhere negative, $df(x)/dx < 0$. Lines that represent the two types of monotonic function are shown in **Figure 8.3** along with a line representing a function that is not monotonic. Monotonic functions have inverses, so that if $f(x)$ is monotonic and $y = f(x)$, then a function $g(y)$ exists such that $x = g(y)$. Nonmonotonic functions do not have inverses, as the relationship between x and y is not one-to-one. For example, the nonmonotonic function in the figure has two values of x for each value of y.

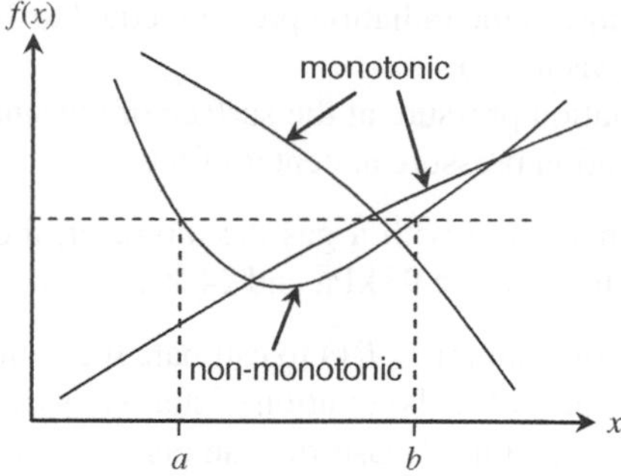

Figure 8.3 *Types of Functions*

> ***Theorem***
> *If Θ^A is a valid temperature scale and $f(x)$ is a monotonic function, then $\Theta^B = f(\Theta^A)$ is also a valid temperature scale.*

The theorem follows from the fact that, when $f(x)$ is monotonic, the ordering of temperatures on the Θ^A scale is either preserved or completely reversed on the Θ^B scale. Monotonically increasing functions preserve the ordering, while monotonically decreasing functions reverse it. In both cases, a single value of Θ^B corresponds to each value of Θ^A, and vice versa. Because of this, systems with the same temperature on one scale have the same temperature on the other. In contrast, when f(x) is *not* monotonic, the equality of temperatures on the Θ^B scale does *not* necessarily imply equality on the Θ^A scale. The nonmonotonic function in the figure illustrates this because $f(x)$ has the same value at two values of x.

A temperature scale in which the ordering from hotter to colder corresponds to the ordering from smaller to larger temperatures is used extensively in statistical mechanics. The temperatures on this scale – called *inverse temperatures* – are related to absolute temperatures T by

$$\beta = \frac{1}{k_B T}, \tag{8.8}$$

where k_B is Boltzmann's constant. Since the function $f(x) = 1/(kx)$ is monotonically decreasing, the β-scale is a valid temperature scale and associates hotter systems with lower temperatures.

Problems

8.1 The pressure exerted by electromagnetic radiation in thermal equilibrium (called black body radiation) is

$$P = \frac{1}{3}K_u T^4,$$

where $K_u = 7.56 \times 10^{-16}\,\mathrm{J\,K^{-4}\,m^{-3}}$.

(a) At what temperature is the radiation pressure equal to 10^{-8} Pa. (In the laboratory 10^{-8} Pa is a high vacuum.)
(b) Estimate the radiation pressure at the surface of the sun. (See Table 1.1.)
(c) Estimate the radiation pressure at center of the sun.

8.2 While measuring temperature with a gas thermometer, the pressure P and specific volume $v = V/n$ are found to be 98 kPa and 24×10^{-3} m^3 mole^{-1}.

(a) Use the equation of state $Pv = R\Theta$ to estimate the value of Θ.
(b) Assume that the van der Waals equation of state $(P + a/v^2)(v - b) = RT$ gives the correct Kelvin temperature T. Use the van der Waals equation of state to derive an expression that relates the difference $T - \Theta$ to P and v.
(c) Find the difference $T - \Theta$ for the values of P and v given above when the gas in the thermometer is helium. (See Table 5.2 for values for a and b.)

8.3 Define a temperature scale Θ^{ln} by the equation $\Theta^{\text{ln}} = 321 \times \ln(T^K/273)$ where T^K is the Kelvin temperature.

(a) Verify that the temperature Θ^{ln} agrees within three significant figures with the Celsius scale at the ice point and the steam point.
(b) Find the temperatures Θ^{ln} of the three highest and three lowest temperatures given in Table 1.1.
(c) What temperature Θ^{ln} corresponds to absolute zero?

9

State Space and Differentials

Much of the power of thermodynamics is obtained through the analysis of equilibrium processes, *i.e.*, processes that can be described by lines in state space, and differentials are a useful tool for describing them. A differential represents a small amount of something, but not just any small amount. The differentials of thermodynamics are the kind of infinitesimals that are summed in line integrals. As the student may not be familiar with some of their important characteristics, we review their properties and emphasize the similarity of the mathematics employed to that used in mechanics, electricity, and magnetism. We begin by outlining the general properties of multi-dimensional spaces.

9.1 Spaces

Let D represent the number of dimensions of the space being investigated. The spaces most familiar to us are the two dimensional surfaces we write on ($D = 2$) and the three dimensional space we live in ($D = 3$). The theory of relativity combines time with the three spatial dimensions to create four dimensional space-time ($D = 4$). These are examples of metric spaces, *i.e.*, spaces for which a concept of length exists. Although the spaces formed by the thermodynamic coordinates of systems are not metric spaces, our familiarity with metric spaces contributes to our understanding of thermodynamic state spaces. The idea of representing states by points in state space and equilibrium processes by lines, also called curves and contours, is applicable to all systems. The state spaces of *PVT* systems are two dimensional ($D = 2$).

The concepts needed for representing equilibrium processes – point, line, scalar field, vector field, and line integral – are familiar from other branches of physics. We review their definitions for spaces with an arbitrary number of dimensions and indicate both detailed and abbreviated notations. The use of bold face type indicates that single symbols like $\boldsymbol{r}$ and $\boldsymbol{F}$ represent quantities with D components.

Thermodynamics and Statistical Mechanics: An Integrated Approach, First Edition.
Robert J Hardy and Christian Binek.

Point

A *point* in a D-dimensional space is specified by giving the values of its D *coordinates*: $(x_1, x_2, \ldots, x_D)$ or $\boldsymbol{r}$.

Line

The parametric representation of a *line* is given by D functions $\mathrm{x}_i(t)$ of a single *parameter* t and a range of values for the parameter:

$$[x_1 = \mathrm{x}_1(t), x_2 = \mathrm{x}_2(t), \ldots, x_D = \mathrm{x}_D(t), \ \ t_0 \leq t \leq t_f] \quad \text{or} \quad [\boldsymbol{r} = \boldsymbol{r}(t), \ \ t_0 \leq t \leq t_f]. \tag{9.1}$$

The Cartesian coordinates of a point in ordinary 3-space are commonly written as (x, y, z) or (x_1, x_2, x_3). For *PVT* systems the coordinates may be (P, V), (P, T), (T, V), or some other set of two state variables. The trajectory of a particle in real space is an example of a line.

Scalar Field

A *scalar field* is specified by a single function of the D coordinates:

$$\Phi(x_1, x_2, \cdots, x_D) \quad \text{or} \quad \Phi(\boldsymbol{r}). \tag{9.2}$$

Vector Field

A *vector field* is specified by its D *components*, which are functions of the coordinates:

$$[F_1(x_1, \ldots, x_D), \ \ F_2(x_1, \ldots, x_D), \ \ldots, F_D(x_1, \ldots, x_D)] \quad \text{or} \quad F(\boldsymbol{r}). \tag{9.3}$$

Scalar fields assign a value to every point in space and are sometimes referred to as *scalars, functions*, or *potentials*. The electrostatic potential is one example. The gravitational potential $-GM/r$ is another. The state functions $P(T, V)$ and $U(T, V)$ are examples of scalar fields from thermodynamics. The electric and magnetic fields $\mathcal{E}(\boldsymbol{r})$ and $\mathcal{B}(\boldsymbol{r})$ are examples of vector fields from electricity and magnetism. Vector fields are implicit in the differentials used in thermodynamics.

Line Integral

A *line integral* is a one-dimensional definite integral formed from a vector field and a line:

$$\int_{t_0}^{t_f} \left[\sum_{i=1}^{D} F_i \left(x_1(t), \ldots, x_D(t) \right) \frac{dx_i(t)}{dt} \right] dt \quad \text{or} \quad \int_L \boldsymbol{F}(\boldsymbol{r}) \cdot d\boldsymbol{r}. \tag{9.4}$$

Other definitions of a line integral are effectively equivalent to this. The abbreviated notation for line integrals makes use of *dot product* notation:

$$\boldsymbol{A} \cdot \boldsymbol{B} = \sum_{i=1}^{D} A_i B_i. \tag{9.5}$$

A line integral is a special case of an integral of a function of one variable,

$$\int_a^b f(t)dt = \lim_{N\to\infty} \sum_{n=1}^{N} f(t_n)\,\Delta t_n. \tag{9.6}$$

The resulting integral is a single number (not a function), and the variable t is called a *dummy variable*. The thing that is special about a line integral is the integrand, which is the quantity inside square brackets in (9.4). To have a line integral, we need both a vector field and a line. In the special case of a line parallel to one of the coordinate axes, the integral becomes an ordinary one dimensional integral. As an illustration, consider a line parallel to the x_1-axis, which implies for $i \neq 1$ that $x_i(t)$ is a constant so that its derivative is $dx_i/dt = 0$. Then, by changing the integration variable from t to x_1, the integral in (9.4) becomes

$$\int_{x_0}^{x_f} F_1(x_1(t), \dots, x_D)\frac{dx_1}{dt}dt = \int_{x_0}^{x_f} F_1(x_1, \dots, x_D)dx_1. \tag{9.7}$$

Line integrals appear in electricity and magnetism in Faraday's law for the electric field $\mathcal{E}$ and Ampere's law for the magnetic field $\mathcal{B}$, which are

$$\oint \mathcal{E}(\boldsymbol{r}) \cdot d\boldsymbol{r} = -\left(\frac{d\Phi_{\mathcal{B}}}{dt}\right) \tag{9.8}$$

and

$$\oint \mathcal{B}(\boldsymbol{r}) \cdot d\boldsymbol{r} = \mu_0 I_{\text{tot}}, \tag{9.9}$$

where the circles on the integral signs indicate integrations are around closed contours. $\Phi_{\mathcal{B}}$ is the magnetic flux through the contour in Faraday's law, and I_{tot} is the total current flowing through the contour in Ampere's law. As discussed in Section 4.2, closed contours are used in thermodynamics to represent cyclic equilibrium processes.

Intuitive reasoning based on our understanding of ordinary 3-space can often be used to evaluate simple line integrals. However, it may be helpful to evaluate a simple line integral by referring to the underlying concepts. Consider the application of Faraday's law to the fields inside a solenoid. **Figure 9.1** shows an end view of a solenoid. The circle represents the wires wrapped around the solenoid, and the current in the wires flows in the clockwise direction. The resulting magnetic field $\mathcal{B}$ is directed into the page and is shown in the

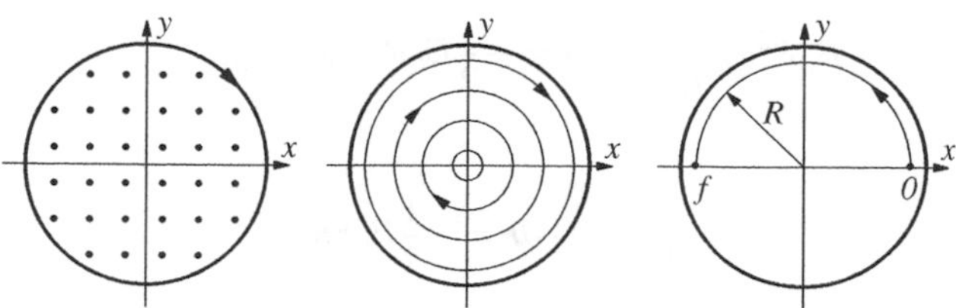

Figure 9.1 *End View of a Solenoid: Magnetic Field $\mathcal{B}$, Electric Field $\mathcal{E}$, and Integration Contour*

first diagram. The magnitude of the B-field is assumed to be increasing at a constant rate $\dot{B}_0$, so that the resulting electric field $\mathcal{E}$ is like that shown in the second diagram. The Cartesian components of this electric field are

$$\mathcal{E}_x(x,y,z) = \frac{1}{2}\dot{B}_0 y, \quad \mathcal{E}_y(x,y,z) = -\frac{1}{2}\dot{B}_0 x, \quad \mathcal{E}_z(x,y,z) = 0. \tag{9.10}$$

Example 9.1 A line integral

Evaluate the line integral $\int_L \mathcal{E}(\boldsymbol{r}) \cdot d\boldsymbol{r}$ for the solenoidal electric field in (9.10) and the line shown in **Figure 9.1(c)**.

Solution. The contour is a semicircle of radius R traversed in the counter-clockwise direction and is given by

$$\left[x = R\cos\omega_0 t, y = R\sin\omega_0 t, z = 0, \ \ 0 \le t \le \pi/\omega_0\right], \tag{9.11}$$

where ω_0 is a constant, $t_0 = 0$, and $t_f = \pi/\omega_0$. The line integral is evaluated by using definition (9.4) and standard mathematical relationships:

$$\int_L \mathcal{E}(\boldsymbol{r}) \cdot d\boldsymbol{r} = \int_{t_0}^{t_f} \mathcal{E}(\boldsymbol{r}(t)) \cdot \frac{d\boldsymbol{r}(t)}{dt}\,dt = \int_{t_0}^{t_f} \left[\mathcal{E}_x(x(t), y(t), z(t))\frac{dx(t)}{dt}\right.$$
$$\left. + \mathcal{E}_y(x(t), y(t), z(t))\frac{dy(t)}{dt} + \mathcal{E}_z(x(t), y(t), z(t))\frac{dz(t)}{dt}\right] dt.$$

Introducing the components $\mathcal{E}_x$, $\mathcal{E}_y$, and $\mathcal{E}_z$ from (9.10) gives

$$\int_L \mathcal{E}(r) \cdot dr = \frac{1}{2}\dot{B}_0 \int_{t_0}^{t_f} \left[y(t)\left(\frac{dx(t)}{dt}\right) - x(t)\left(\frac{dy(t)}{dt}\right) + 0\right] dt.$$

By using the above description of the line in (9.11), it follows that

$$\int_L \mathcal{E}(\boldsymbol{r}) \cdot d\boldsymbol{r} = \frac{1}{2}\dot{B}_0 \int_0^{\pi/\omega_0} [(R\sin\omega_0 t)(-R\omega_0\sin\omega_0 t) - (R\cos\omega_0 t)(R\omega_0\cos\omega_0 t)]\,dt$$
$$= -\frac{1}{2}\dot{B}_0 R^2\omega_0 \int_0^{\pi/\omega_0} [\sin^2\omega_0 t + \cos^2\omega_0 t]\,dt$$
$$= -\frac{1}{2}\dot{B}_0 R^2\omega_0 \int_0^{\pi/\omega_0} dt = -\frac{1}{2}\dot{B}_0 R^2\omega_0(\pi/\omega_0).$$

Thus,

$$\int_L \mathcal{E}(\boldsymbol{r}) \cdot d\boldsymbol{r} = -\frac{1}{2}\dot{B}_0 \pi R^2.$$

If the integral had been taken around a full circle of radius R (instead of a semicircle), the parametric equations for the line would be similar except that t_f would be $2\pi/\omega_0$ (instead of

π/ω_0). The value of the integral would then be multiplied by two. As the resulting contour is closed, a circle is placed on the integral sign, so that

$$\oint \mathcal{E}(\boldsymbol{r}) \cdot d\boldsymbol{r} = -\dot{B}_0 \pi R^2. \tag{9.12}$$

This result is consistent with Faraday's law, as $B\pi R^2$ is the magnitude of the magnetic flux through the closed contour.

It follows from definition (9.4) and the rules for changing integration variables that *the values of line integrals are independent of the parameterization of the line.* As an example of this, notice that the line sketched in the third diagram of Figure 9.1 is described both by (9.11) and by

$$[x = R - v_o t, y = \sqrt{R^2 - (R - v_o t)^2}, z = 0, \quad t_0 \le t \le t_f], \tag{9.13}$$

where v_0 is a constant, $t_0 = 0$, and $t_f = 2R/v_0$. Both parameterizations of the line yield the same value for the line integral.

When time is used to parameterize a line, the fact that line integrals are independent of parameterization has important implications in thermodynamics: *The work done W and the heat transferred Q during an equilibrium process are independent of the rate at which the process is carried out.* This independence from rates is very different from mechanics, where velocities and accelerations are of central importance. Although line integrals are independent of the parameterization of the line, they do depend on the direction in which the line is traversed. Reversing the direction of the line multiplies the integral by -1. This dependence is reflected in the multiplication of W and Q by -1 when a process is reversed.

9.2 Differentials

Infinitesimals, which are quantities that can be made arbitrarily small, are commonly used in relationships which become exact when a limit is taken, as in the definitions of derivatives and integrals. The differentials used in thermodynamics are special types of infinitesimals. Three types are used: coordinate differentials, parameter differentials, and differential forms. Although they are mathematically different, the three types have similar physical interpretations and will simply be called differentials in later chapters.

Coordinate and Parameter Differentials

A *coordinate differential* dx_i represents a small change in the ith coordinate. In a D dimensional space, there are D coordinate differentials:

$$(dx_1, dx_2, \dots, dx_D) \quad \text{or} \quad d\boldsymbol{r}. \tag{9.14}$$

A set of coordinate differentials can be associated with a particular point on a line by

$$dx_i = \frac{dx_i(t)}{dt} dt \;\; (i = 1, 2, \dots, D) \quad \text{or} \quad d\boldsymbol{r} = \frac{d\boldsymbol{r}(t)}{dt} dt, \tag{9.15}$$

where $x_1(t), \dots, x_D(t)$ are the parametric functions that describe the line. The *parameter differential dt* represents a small non-zero change in parameter t.

Differential Forms

A *differential form* $đA$ is a sum of products, sometimes called a dot product, formed with a vector field and a set of coordinate differentials:

$$đA = \sum_{i=1}^{D} F_i(x_1, \ldots, x_D)dx_i \quad \text{or} \quad đA = F(\mathbf{r}) \cdot d\mathbf{r}. \tag{9.16}$$

The bar on $đ$ identifies $đA$ as a differential form. In thermodynamics, it is natural to refer to the vector component F_i that multiplies coordinate differential dx_i as the coefficient of dx_i.

Differential forms and line integrals are closely related. To see this, consider a line L described by the parametric functions $\mathbf{r}(t) = (x_1(t), \cdots, x_D(t))$, and divide the interval $[t_0, t_f]$ into N sub-intervals by values $t_1, t_2, \ldots, t_{N-1}$. Introduce N differential forms $đA_n$, use (9.15) to associate coordinate differentials with these points, and set the parameter differentials dt equal to $\Delta t_n = t_n - t_{n-1}$. In successively less abbreviated notations, the sum of the N differential forms becomes

$$\begin{aligned}\sum_{n=1}^{N} đA_n &= \sum_{n=1}^{N}\left[\sum_{i=1}^{D} F_i\left(x_1, \ldots, x_D\right) dx_i\right] \\ &= \sum_{n=1}^{N}\left[\sum_{i=1}^{D} F_i\left(x_1\left(t_n\right), \ldots, x_D(t_n)\right)\frac{dx_i(t_n)}{dt}\right]\Delta t_n.\end{aligned} \tag{9.17}$$

From this, the definition of a definite integral in (9.6), and the definition of a line integral in (9.4), it follows that

$$\lim_{N\to\infty}\sum_{n=1}^{N} đA_n = \int_{t_0}^{t_f}\left[\sum_{i=1}^{D} F_i\left(x_1(t), \ldots, x_D(t)\right)\frac{dx_i(t)}{dt}\right]dt = \int_L F(\mathbf{r}) \cdot d\mathbf{r}. \tag{9.18}$$

Thus, just as $f(t)\Delta t$ is the quantity summed in a definite integral, *a differential form is the quantity summed in a line integral*. The above result suggests the following concise notation for line integrals:

$$\int_L \mathbf{F}(\mathbf{r}) \cdot d\mathbf{r} = \int_L đA. \tag{9.19}$$

Equations Involving Differentials

The same set of coordinate differentials $(dx_1, dx_2, \cdots, dx_D)$ is used throughout any equation that relates differentials, and equations involving differentials are valid independent of the coordinate differentials used. This allows us to treat the coordinate differentials as a collection of arbitrary numbers. The underlying meaning of an equation involving differentials is found by expressing the symbols $đA$, $đB$, etc. in terms of the vector fields and coordinate differentials implicit in the abbreviated notation. As an illustration, consider the equation

$$đA = \Phi(\mathbf{r})\,đB. \tag{9.20}$$

Let $\boldsymbol{F}(\boldsymbol{r})$ and $\boldsymbol{G}(\boldsymbol{r})$ be the vector fields implicit in $đA$ and $đB$. When $\boldsymbol{F}(\boldsymbol{r})$ and $\boldsymbol{G}(\boldsymbol{r})$ are expressed in terms of their components, equation (9.20) becomes

$$\sum_{i=1}^{D} F_i(x_1, \dots, x_D)\,dx_i = \Phi(x_1, \dots) \sum_{i=1}^{D} G_i(x_1, \dots, x_D)\,dx_i. \tag{9.21}$$

Since this is true for all sets of coordinate differentials, it must be true for the set

$$(dx_1, dx_2, \dots, dx_D) = (0, \dots, 0, dx_j, 0, \dots, 0). \tag{9.22}$$

Substituting (9.22) into (9.21) and omitting the terms equal to zero gives

$$F_j(x_1, \dots, x_D) = \Phi(x_1, \dots, x_D) G_j(x_1, \dots, x_D). \tag{9.23}$$

By successively setting j equal to $(1, 2, \dots, D)$, we find that the relationship $đA = \Phi(\boldsymbol{r})\,đB$ between differentials implies an analogous relationship between vector fields,

$$\boldsymbol{F}(\boldsymbol{r}) = \Phi(\boldsymbol{r})\boldsymbol{G}(\boldsymbol{r}). \tag{9.24}$$

When the dependence on $\boldsymbol{r}$ is left implicit, as is often done, equations (9.20) and (9.24) become

$$đA = \Phi\,đB \quad \text{and} \quad \boldsymbol{F} = \Phi\boldsymbol{G}. \tag{9.25}$$

By setting $\Phi = 1$, it follows that $đA = đB$ implies that $\boldsymbol{F} = \boldsymbol{G}$, and vice versa.

Because of the association of differentials with vector fields, thc types of relationships that are valid for vector fields are also valid for differentials. Specifically, new differentials can be obtained by multiplying by a function (as shown above) or by adding or subtracting differentials. Specifically, if the vector fields associated with $đA$, $đB$, and $đC$ are $\boldsymbol{U}$, $\boldsymbol{V}$, and $\boldsymbol{W}$, respectively, then $đC = đA + đB$ implies that $\boldsymbol{W} = \boldsymbol{U} + \boldsymbol{V}$, and vice versa.

As the differential $đA$, $đB$, etc. are associated with vector fields, it is reasonable to ask why differentials are used in thermodynamics instead of vector fields. It is because of the abstract nature of thermodynamic state space. In ordinary 3-space, the three components of a vector field can be interpreted as an entity with magnitude and direction, but analogous concepts in do not exist for state space.

9.3 Exact Versus Inexact Differentials

The general expression for a differential form in a space with coordinates (x, y, z) is

$$đA = F_x(x, y, z)\,dx + F_y(x, y, z)\,dy + F_z(x, y, z)\,dz. \tag{9.26}$$

The most familiar differential form for many students is the *differential of a function*. For example, the differential of $\Phi(x, y, z)$ is

$$d\Phi = \left(\frac{\partial\Phi}{\partial x}\right)_{yz} dx + \left(\frac{\partial\Phi}{\partial y}\right)_{zx} dy + \left(\frac{\partial\Phi}{\partial z}\right)_{xy} dz, \tag{9.27}$$

where $d\Phi$ is an estimate for the change in $\Phi(x, y, z)$ that results from small changes dx, dy, and dz in the independent variables. The bar on $đ$ can be omitted on the differential of a function. In general, there is no function $\Phi(x, y, z)$ whose partial derivatives equal a given set of functions F_x, F_y, and F_z. When a function Φ does exist, the functions F_x, F_y, and F_z are given by

$$F_x(x, y, z) = \left(\frac{\partial \Phi}{\partial x}\right)_{yz}, \quad F_y(x, y, z) = \left(\frac{\partial \Phi}{\partial y}\right)_{zx}, \quad F_z(x, y, z) = \left(\frac{\partial \Phi}{\partial z}\right)_{zx}. \tag{9.28}$$

A useful property of partial derivatives is their independence from the order of differentiation,

$$\left(\frac{\partial^2 \Phi}{\partial x \partial y}\right)_z = \left(\frac{\partial^2 \Phi}{\partial y \partial x}\right)_z, \quad \left(\frac{\partial^2 \Phi}{\partial y \partial z}\right)_x = \left(\frac{\partial^2 \Phi}{\partial z \partial y}\right)_x, \quad \left(\frac{\partial^2 \Phi}{\partial z \partial x}\right)_y = \left(\frac{\partial^2 \Phi}{\partial x \partial z}\right)_y. \tag{9.29}$$

Combining these equalities with (9.28) implies that

$$\left(\frac{\partial F_x}{\partial y}\right)_z = \left(\frac{\partial F_y}{\partial x}\right)_z, \quad \left(\frac{\partial F_y}{\partial z}\right)_x = \left(\frac{\partial F_z}{\partial y}\right)_x, \quad \left(\frac{\partial F_z}{\partial x}\right)_y = \left(\frac{\partial F_x}{\partial z}\right)_y. \tag{9.30}$$

These conditions – called *exactness conditions* – must be satisfied for the differential $đA$ to be the differential of a function. A differential form $đA$ that satisfies the exactness conditions is an *exact differential*, sometimes called a *perfect differential*. A differential in ordinary 3-space is exact if the curl of the associated vector field vanishes. The Cartesian components of the curl of $\boldsymbol{F} = (F_x, F_y, F_z)$ are

$$(\nabla \times \boldsymbol{F})_x = \left(\frac{\partial F_z}{\partial y} - \frac{\partial F_y}{\partial z}\right)_x, \quad (\nabla \times \boldsymbol{F})_y = \left(\frac{\partial F_x}{\partial z} - \frac{\partial F_z}{\partial x}\right)_y,$$

$$(\nabla \times \boldsymbol{F})_z = \left(\frac{\partial F_y}{\partial x} - \frac{\partial F_x}{\partial y}\right)_z. \tag{9.31}$$

When the exactness conditions are satisfied, the x, y, and z components are zero, so that $\nabla \times \boldsymbol{F} = 0$.

Exact Differential

A differential $đA = \boldsymbol{F} \cdot d\boldsymbol{r}$ is *exact* if for all i and j it is true that

$$\frac{\partial F_i(x_1, \ldots, x_D)}{\partial x_j} = \frac{\partial F_j(x_1, \ldots, x_D)}{\partial x_i}. \tag{9.32}$$

This is the general form of the *exactness conditions*. The conditions for $i \neq j$ are referred to as the requirement that *cross derivatives* are equal. A differential form that does not satisfy the exactness conditions is an *inexact differential*.

Differential of a Function

When generalized to spaces of D dimensions, the *differential of a function* is a differential whose vector field has components that are the partial derivatives of some scalar field.

Specifically, the differential of the function $\Phi(x_1, \cdots, x_D)$ is

$$d\Phi = \sum_{i=1}^{D} \frac{\partial \Phi(x_1, \ldots, x_D)}{\partial x_i} dx_i \quad \text{or} \quad d\Phi = \nabla\Phi \cdot d\boldsymbol{r}, \tag{9.33}$$

where the abbreviated notation $\nabla\Phi$ is suggested by the gradient notation used in vector analysis. By using (9.33) the differential of a product of two functions is

$$d(FG) = F\,dG + G\,dF. \tag{9.34}$$

As partial derivatives are independent of the order of differentiation, it follows from (9.32) that in general the *differential of a function is an exact differential.*

The bar on $đ$ can be omitted when the symbols $đA$, $đB$, etc. represent exact differentials or differentials of functions. As most differentials are not exact, the bar should be kept until we are sure its omission is justified. The special properties of exact differentials are summarized in the following theorem, which is especially important in thermodynamics. It tells us that an exact differential is the differential of a state function, a property that is used in the next chapter to demonstrate the existence of entropy.

Exact Differential Theorem

If any of the following four statements is true, the other three are also true:

1. *$đA$ is the differential of a function.*
2. *$đA$ is exact.*
3. *$\oint đA = 0$ for all closed contours.*
4. *$\int_L đA$ is independent of the line L between points $\boldsymbol{r}_0$ and $\boldsymbol{r}_f$.*

The theorem is proved by showing that statements 2 through 4 follow from the preceding statement and that statement 1 follows from statement 4, so that if any one of them is true, the other three are also true. As indicated earlier, it follows from the independence of partial derivatives on the order of differentiation that $đA$ is exact (statement 2) if it is the differential of a function. That statement 3 is true when $đA$ is exact follows from Stokes' theorem, which is used in electromagnetic theory to obtain Maxwell's equation for $\nabla \times \mathcal{E}$ from Faraday's law. Stokes' theorem states that[1]

$$\oint \boldsymbol{F}(\boldsymbol{r}) \cdot d\boldsymbol{r} = \int_S d\boldsymbol{S} \cdot \nabla \times \boldsymbol{F}(\boldsymbol{r}), \tag{9.35}$$

i.e., the line integral of a vector field $\boldsymbol{F}$ around a closed contour equals the surface integral of the curl of the field over a surface bounded by the contour. As indicated in (9.31), it follows from the exactness of $đA = \boldsymbol{F} \cdot d\boldsymbol{r}$ that $\nabla \times \boldsymbol{F} = 0$. It then follows from Stokes' theorem that $\oint đA = \oint \boldsymbol{F} \cdot d\boldsymbol{r} = 0$. Since no restriction was placed on the closed contour, the result is true for all closed contours.

[1] The theorem as given is for three-dimensional spaces. Results for two-dimensions are obtained by assuming that F_x and F_y are independent of z and that $F_z = 0$.

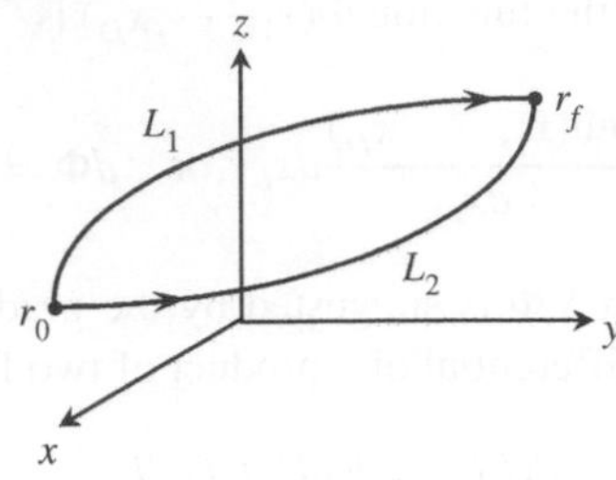

Figure 9.2 *Two Lines Between the Same End Points*

To show that $\oint đA = 0$ implies that the line integral of $đA$ along a line between two points is independent of the line connecting them, we consider two lines L_1 and L_2 between points $\boldsymbol{r}_0$ and $\boldsymbol{r}_f$, as diagrammed in **Figure 9.2**. A closed contour is obtained by going from $\boldsymbol{r}_0$ to $\boldsymbol{r}_f$ along line L_1 and returning to $\boldsymbol{r}_0$ along line $\overline{L}_2$, where $\overline{L}_2$ represents L_2 when traversed in the reversed direction. Since traversing a line in the reversed direction multiplies a line integral by -1, it follows that

$$\oint đA = \int_{L_1} đA + \int_{\overline{L}_2} đA = \int_{L_1} đA - \int_{L_2} đA. \tag{9.36}$$

From this and $\oint đA = 0$, we obtain

$$\int_{L_1} đA = \int_{L_2} đA. \tag{9.37}$$

As the only restriction imposed on L_1 and L_2 is that they connect $\boldsymbol{r}_0$ to $\boldsymbol{r}_f$, the line integral $\int_L đA$ is independent of the line between the two points.

To show that $đA$ is the differential of a function (statement 1) when $\int_L đA$ is independent of the line between $\boldsymbol{r}_0$ and $\boldsymbol{r}_f$, we define a function $\Phi(\boldsymbol{r})$ and then show that its differential of $\Phi(\boldsymbol{r})$ is $đA = \boldsymbol{F} \cdot d\boldsymbol{r}$. The function $\Phi(\boldsymbol{r})$ is

$$\Phi(\boldsymbol{r}) = A_0 + \int_L đA = A_0 + \int_L \boldsymbol{F}(\boldsymbol{r}) \cdot d\boldsymbol{r}, \tag{9.38}$$

where A_0 is an arbitrary constant and L is a line between points $\boldsymbol{r}_0$ and $\boldsymbol{r}$. No subscript is placed on the final point $\boldsymbol{r}$, which can be any point in the space. To illustrate the ideas involved, consider a three-dimensional space with $\boldsymbol{r}_0 = (x_0, y_0, z_0)$, $\boldsymbol{r} = (x, y, z)$, and $\boldsymbol{F} = (F_x, F_y, F_z)$. Assume that $\int_L đA$ is independent of the line L between points $\boldsymbol{r}_0$ and $\boldsymbol{r}$ and consider lines made up of segments which are parallel to the coordinate axes. First, consider a line that goes from point (x_0, y_0, z_0) to point (x, y_0, z_0), then to point (x, y, z_0), and finally to point (x, y, z), so that

$$\Phi(x, y, z) = A_0 + \int_{x_0}^{x} F_x(x, y_0, z_0)dx + \int_{y_0}^{y} F_x(x, y, z_0)dy + \int_{z_0}^{z} F_x(x, y, z)dz. \tag{9.39}$$

As the only dependence on z is in the upper limit of the last integral, it follows that

$$\left(\frac{\partial \Phi}{\partial z}\right)_{xy} = F_z(x, y, z). \tag{9.40}$$

By using similar reasoning, we find that

$$\left(\frac{\partial \Phi}{\partial x}\right)_{yz} = F_x(x, y, z) \quad \text{and} \quad \left(\frac{\partial \Phi}{\partial y}\right)_{zx} = F_y(x, y, z). \tag{9.41}$$

It follows from (9.40) and (9.41) that the differential $đA = F_x dx + F_y dy + F_z dz$ is

$$đA = \left(\frac{\partial \Phi}{\partial x}\right)_{yz} dx + \left(\frac{\partial \Phi}{\partial y}\right)_{zx} dy + \left(\frac{\partial \Phi}{\partial z}\right)_{xy} dz. \tag{9.42}$$

Thus, $đA = F_x dx + F_y dy + F_z dz$ is the differential of a function. Furthermore, the function $\Phi(\boldsymbol{r})$ defined in (9.38) is a function whose differential is $đA$. Because of the arbitrariness of $\boldsymbol{r}_0$ and $\boldsymbol{A}_0$, the function is not unique.

9.4 Integrating Differentials

The exact differential theorem tells us that every exact differential is the differential of some function. Finding that function is called integrating the differential. In this section we show how indefinite integrals can be used to do that. Calculus tells us that the one dimensional indefinite integral of the derivative of $f(x)$ is

$$\int \frac{df(x)}{dx} dx = f(x) + C, \tag{9.43}$$

where C is an arbitrary constant.

Consider an exact differential in a space with coordinates (x, y),

$$dA = F_x(x, y)\, dx + F_y(x, y)\, dy. \tag{9.44}$$

Since dA is exact, it is the differential of a function $A(x, y)$ and the components of the vector field are the partial derivatives of $A(x, y)$,

$$F_y(x, y) = \frac{\partial A(x, y)}{\partial x} \quad \text{and} \quad F_y(x, y) = \frac{\partial A(x, y)}{\partial y}. \tag{9.45}$$

To find $A(x, y)$, we hold y constant and integrate $F_x(x, y)$ over x,

$$\int F_x(x, y)\, dx = \int \frac{\partial A(x, y)}{\partial x} dx = A(x, y) + [-g(y) - h], \tag{9.46}$$

where the term in square brackets corresponds to the constant C in (9.43). As the integrand $F_x(x, y)$ depends on y, the constant is a function of y. For convenience, we have written it as $-g(y) + h$, where h is a constant independent of both x and y. Solving (9.46) for $A(x, y)$ gives

$$A(x, y) = \int F_x(x, y)\,dx + g(y) + h. \tag{9.47}$$

Similar reasoning with the roles of x and y exchanged gives

$$A(x, y) = \int F_y(x, y)\,dy + f(x) + k. \tag{9.48}$$

It should be emphasized that the integral of the exact differential dA in (9.44) is *not* in general equal to the following sum of one-dimensional indefinite integrals:

$$\int F_x(x, y)\,dx + \int F_y(x, y)\,dy. \tag{9.49}$$

This is easily seen by considering the function

$$B(x, y) = xy + \text{const}, \tag{9.50}$$

whose differential is $dB = y\,dx + x\,dy$. Substituting $F_x(x, y) = y$ and $F_y(x, y) = x$ into (9.49) and performing the indicated integrations gives

$$f(x, y) = \int y\,dx + \int x\,dy = 2xy + \text{const}. \tag{9.51}$$

Because of the factor of 2 in this, $f(x, y)$ is definitely *not* the function $B(x, y)$ in (9.50). This is because differential forms are the quantities summed in line integrals, not ordinary indefinite integrals.

Example 9.2 Integrating a differential

Find a function whose differential is

$$đA = (ax + cxy^2)\,dx + (by + cx^2y)\,dy, \tag{9.52}$$

where a, b, and c are constants.

Solution. Five steps should be followed when integrating a differential:

Step (1). Verify that the differential is exact. For this example, checking the equality of cross-derivatives gives

$$\frac{\partial(ax + cxy^2)}{\partial y} \stackrel{?}{=} \frac{\partial(by + cx^2y)}{\partial x} \quad \text{or} \quad 2cxy \stackrel{?}{=} 2cxy. \tag{9.53}$$

As both equalities are valid, differential $đA$ is indeed exact.

Step (2). Evaluate the indefinite integral of one component of the vector field and use (9.47) or (9.48) to find $A(x, y)$. For example, using (9.47) gives

$$A(x, y) = \int (ax + cxy^2)\,dx + g(y) + h = \frac{1}{2}ax^2 + \frac{1}{2}cx^2y^2 + g(y) + h. \tag{9.54}$$

Step (3). Find $A(x, y)$ by evaluating the indefinite integral of the other component. Using (9.48) gives

$$A(x, y) = \int (by + cx^2y)dy + f(x) + k = \frac{1}{2}by^2 + \frac{1}{2}cx^2y^2 + f(x) + k. \tag{9.55}$$

Step (4). Use the results of steps (2) and (3) to determine the unknown function $g(y)$ or $f(x)$. For example, the result of step (3) contains $f(x)$, which depends on x but not on y. The only such term in the result of step (2) is $\frac{1}{2}ax^2$. Hence, $f(x) = \frac{1}{2}ax^2$. Substituting this into (9.55) gives

$$A(x, y) = \frac{1}{2}ax^2 + \frac{1}{2}by^2 + \frac{1}{2}cx^2y^2 + k. \tag{9.56}$$

Step (5). Check that the result obtained is correct, which is done by verifying that the partial derivatives of the function obtained are the vector components of the given differential. Differentiating (9.56) gives

$$\frac{\partial A(x, y)}{\partial x} = \frac{\partial\left(\frac{1}{2}ax^2 + \frac{1}{2}by^2 + \frac{1}{2}cx^2y^2 - k\right)}{\partial x} = ay + cxy^2. \tag{9.57}$$

and

$$\frac{\partial A(x, y)}{\partial y} = \frac{\partial\left(\frac{1}{2}ax^2 + \frac{1}{2}by^2 + \frac{1}{2}cx^2y^2 - k\right)}{\partial y} = by + cx^2y. \tag{9.58}$$

These are indeed the x and y components of the differential $đA$ in (9.52). Hence, $đA$ is the differential of the function $A(x, y)$.

The above procedure can be extended to spaces with more than two dimensions, but it becomes more involved. For example, if $D = 3$, verifying the exactness of the differential requires checking three cross derivatives, instead of one, and the arbitrary constants become functions of two variables, not one.

9.5 Differentials in Thermodynamics

Exact Differentials

The differentials of the state functions of thermodynamics are exact. For example, the differential of the state function for volume is

$$dV = \left(\frac{\partial V}{\partial T}\right)_P dT + \left(\frac{\partial V}{\partial P}\right)_T dP. \tag{9.59}$$

As defined in (3.21) and (3.46), the partial derivatives of volume are related to the coefficient of thermal expansion α_V and the compressibility K_T by

$$\alpha_V = \frac{1}{V}\left(\frac{\partial V}{\partial T}\right)_P \quad \text{and} \quad K_T = \frac{-1}{V}\left(\frac{\partial V}{\partial P}\right)_T. \tag{9.60}$$

Hence,

$$dV = V\alpha_V dT - V K_T dP. \tag{9.61}$$

This equation is similar to the expression for the change in volume ΔV in (3.58), which was obtained by truncating a series expansion of $V(T, P)$. The difference is that (3.58) is an approximate relationship between ΔV, ΔT, and ΔP, while the relationship between dV, dT, and dP in (9.61) is exact. The similarity of the expressions for dV and ΔV indicates that exact differentials can be interpreted as representing small amounts of something.

Inexact Differentials

The lines that represent equilibrium processes can be divided into short segments which represent sub-processes. The heat transferred and the work done during these sub-processes are given by inexact differentials $đQ$ and $đW$. The values of Q, W, and ΔU for the complete process are the sums of the values of $đQ$, $đW$, and dU for the sub-processes. As the sub-processes are made limitingly small, the sums become line integrals:

$$Q = \int_L đQ, \quad W = \int_L đW, \quad \Delta U = \int_L dU. \tag{9.62}$$

Combining (9.62) with the first law $Q = \Delta U + W$ gives

$$\int_L đQ = \int_L dU + \int_L đW. \tag{9.63}$$

For this to be true for all lines L, *i.e.*, for all equilibrium processes, the differentials must satisfy the equation

$$\boxed{đQ = dU + đW,} \tag{9.64}$$

which summarizes the first law in investigations of equilibrium processes.

PVT Systems

We start by using P and V as the thermodynamic coordinates of a *PVT* system. The differential for work is then related to the coordinate differentials by

$$đW = F_P(P, V)dP + F_V(P, V)dV, \tag{9.65}$$

and we need to find the functional forms of the vector components $F_P(P, V)$ and $F_V(P, V)$. As work is a mechanical concept, we look to mechanics for the answer. The work done by a *PVT* system is (see (2.3))

$$W = \int_{V_0}^{V_f} P(V)dV, \tag{9.66}$$

which is a one dimensional definite integral. To express it as a line integral, we introduce the line

$$[P = P(t), V = V(t), \quad t_0 < t \leq t_f], \tag{9.67}$$

where $P(t)$ and $V(t)$ describe the time dependence of the pressure and volume during some process. The function $P(V)$ in (9.66) is related to the function $P(t)$ in (9.67) by $P(V(t)) = P(t)$. By using $V(t)$ to change the integration variable in (9.66) from V to t, we obtain

$$W = \int_{t_0}^{t_f} P(V(t))\frac{dV(t)}{dt}\,dt = \int_{t_0}^{t_f} P(t)\frac{dV(t)}{dt}\,dt, \tag{9.68}$$

where the limits in (9.66) are related to those in (9.68) by $V_0 = V(t_0)$ and $V_f = V(t_f)$. To give this integral the form of a line integral, we introduce a term that vanishes and rewrite (9.68) as

$$W = \int_{t_0}^{t_f} \left[0\frac{d0\,(t)}{dt} + P(t)\frac{dV(t)}{dt}\right] dt. \tag{9.69}$$

This has the form of the line integral

$$W = \int_L [0dP + PdV] = \int_L đW. \tag{9.70}$$

Thus, $đW = 0dP + PdV$, so that the components of the vector field in the differential for work are

$$F_P(P, V) = 0 \quad \text{and} \quad F_V(P, V) = P. \tag{9.71}$$

When the vanishing pressure component is omitted, the differential for work in a *PVT* system becomes

$$\boxed{đW = PdV.} \tag{9.72}$$

Example 9.3 Differential for work

Is the differential for work in a *PVT* system exact or inexact?

Solution. As $D = 2$, the only cross derivative to check is

$$\left(\frac{\partial F_P}{\partial V}\right)_P \stackrel{?}{=} \left(\frac{\partial F_V}{\partial P}\right)_V.$$

Substituting the vector components in (9.71) into this gives

$$\left(\frac{\partial 0}{\partial V}\right)_P \stackrel{?}{=} \left(\frac{\partial P}{\partial P}\right)_V \quad \text{or} \quad 0 \stackrel{?}{=} 1.$$

This is false, hence $đW$ is *inexact*.

Since internal energy U is a state function, its differential is

$$dU = \left(\frac{\partial U}{\partial P}\right)_V dP + \left(\frac{\partial U}{\partial V}\right)_P dV. \tag{9.73}$$

The differential for heat is obtained from the first law expressed as $đQ = dU + đW$. Substituting (9.72) and (9.73) into this gives

$$đQ = \left(\frac{\partial U}{\partial P}\right)_V dP + \left[\left(\frac{\partial U}{\partial V}\right)_P + P\right] dV. \tag{9.74}$$

When the coordinates of state space are transformed, the vector fields and coordinate differentials are also transformed. The notation for differentials makes this easy to do. Simply treat the original coordinate differentials as differentials of the functions that relate the old coordinates to the new coordinates.

Transformation to Coordinates (*T*, *V*)

When the coordinates are (T, V), the differential of the internal energy is

$$dU = \left(\frac{\partial U}{\partial T}\right)_V dT + \left(\frac{\partial U}{\partial V}\right)_T dV. \tag{9.75}$$

As the pressure component of the differential $đW = PdV$ is zero when the coordinates are (P, V) and as dV is one of the coordinate differentials when the coordinates are (T, V), the differential for work is unaffected by the transformation from (P, V) to (T, V). Substituting dU and $đW$ into $đQ = dU + đW$ gives

$$đQ = dU + PdV = \left(\frac{\partial U}{\partial T}\right)_V dT + \left[\left(\frac{\partial U}{\partial V}\right)_T + P\right] dV. \tag{9.76}$$

This can be simplified by using the constant-volume heat capacity $C_V = (\partial U/\partial T)_V$ from (3.5). Thus,

$$\boxed{đQ = C_V(T, V)dT + \left[\left(\frac{\partial U}{\partial V}\right)_T + P(T, V)\right] dV.} \tag{9.77}$$

Transformation to Coordinates (*T*, *P*)

When the coordinates are changed from (P, V) to (T, P), the coordinate V is given in terms of the new coordinates by the equation of state $V = V(T, P)$. Its differential is

$$dV = \left(\frac{\partial V}{\partial T}\right)_P dT + \left(\frac{\partial V}{\partial P}\right)_T dP. \tag{9.78}$$

Since P is a coordinate both before and after the transformation, dP remains a coordinate differential. Substituting dV into the differential for work in (9.72) gives

$$đW = PdV = P\left(\frac{\partial V}{\partial T}\right)_P dT + P\left(\frac{\partial V}{\partial P}\right)_T dP. \tag{9.79}$$

The differential of U when the coordinates are T and P is

$$dU = \left(\frac{\partial U}{\partial T}\right)_P dT + \left(\frac{\partial U}{\partial P}\right)_T dP. \tag{9.80}$$

It now follows from $đQ = dU + đW$ that

$$đQ = \left[\left(\frac{\partial U}{\partial T}\right)_P + P\left(\frac{\partial V}{\partial T}\right)_P\right] dT + \left[\left(\frac{\partial U}{\partial P}\right)_T + P\left(\frac{\partial V}{\partial P}\right)_T\right] dP. \tag{9.81}$$

This can be simplified by using $C_P = (\partial H/\partial T)_P$ for the relationship between the constant-pressure heat capacity and the enthalpy H from 3.13. Since $H = U + PV$, the constant-pressure heat capacity is

$$C_P(T,P) = \left(\frac{\partial H}{\partial T}\right)_P = \left[\left(\frac{\partial U}{\partial T}\right)_P + P\left(\frac{\partial V}{\partial T}\right)_P\right]. \tag{9.82}$$

Using this in the above expression for $đQ$ gives

$$\boxed{đQ = C_P(T,P)\,dT + \left[\left(\frac{\partial U}{\partial P}\right)_T + P\left(\frac{\partial V}{\partial P}\right)_T\right] dP.} \tag{9.83}$$

In summary, the general expression for the differential for work in *PVT* systems is $đW = PdV$. When the coordinates are (P, V), the differential dV is a coordinate differential and P is a coordinate. When the coordinates are (T, V), the differential dV is a coordinate differential and P is given by a function $P(T, V)$. When the coordinates are (T, P), the differential dV is the differential of a function and P is a coordinate.

Quotients of Differentials

A differential form can be divided by a coordinate differential. To see the justification for this, consider a line parallel to the x-axis and the differential

$$đA = F_x dx + F_y dy, \tag{9.84}$$

where F_x and F_y are functions of x and y. A line parallel to the x-axis can be obtained by using the x coordinate to parameterize the line and holding the y-coordinate constant $(y(x) = y_0)$. The line is described by

$$[x = x, y = y_0, \quad x_0 \leq x \leq x_f]. \tag{9.85}$$

The coordinate differentials associated with this line are

$$(dx, dy) = \left(\frac{dx\,(x)}{dx}dx, \frac{dy(x)}{dx}dx\right) = (dx, 0), \tag{9.86}$$

where the symbol dx appears both as a coordinate differential and a parameter differential. Since $dy = 0$, substituting these coordinate differentials into (9.84) gives

$$đA = F_x dx. \tag{9.87}$$

As dx is a parameter differential, it is non-zero and can be used as a divisor. Dividing both sides by dx gives

$$\left(\frac{đA}{dx}\right)_y = F_x, \tag{9.88}$$

where the subscript on the parentheses indicates that y is constant along the line. Similarly, by considering a line parallel to the y-axis, we obtain

$$\left(\frac{đA}{dy}\right)_x = F_y. \tag{9.89}$$

Substituting these results into (9.84) gives

$$đA = \left(\frac{đA}{dx}\right)_y dx + \left(\frac{đA}{dy}\right)_x dy. \tag{9.90}$$

From the above, we see that the quotient $(đA/dx)_y$ is an alternate notation for the y-component of a vector field. Notice that both the symbols $đ$ and d appear in (9.90) but the partial symbol ∂ does not. It is essential not to confuse the quotient $(đA/dx)_y$ with the partial derivative $(\partial A/\partial x)_y$. In the special case when $đA$ is exact, a state function $A(x, y)$ exists, and the quotients $(đA/dx)_y$ and $(đA/dy)_x$ can be (and should be) replaced by the partial derivatives $(\partial A/\partial x)_y$ and $(\partial A/\partial y)_x$.

Heat Capacities

The quotient notation for vector components leads to alternate expressions for the heat capacities. It follows from (9.77) and (9.83) that the constant-pressure and constant-volume heat capacities can be expressed as

$$C_V(T, V) = \left(\frac{đQ}{dT}\right)_V \quad \text{and} \quad C_P(T, P) = \left(\frac{đQ}{dT}\right)_P. \tag{9.91}$$

These are suggestive of the empirical definitions of C_P and C_V in Chapter 1 (see (1.25)),

$$C_V = \lim \frac{Q}{\Delta T}\bigg|_{V=\text{const}} \quad \text{and} \quad C_P = \lim \frac{Q}{\Delta T}\bigg|_{P=\text{const}}. \tag{9.92}$$

These empirical definitions are more general than (9.91), because the symbol Q describes the heat transferred in both equilibrium and non-equilibrium processes, while the symbol $đQ$ in (9.91) describes the heat transferred in an equilibrium process.

9.6 Discussion and Summary

Differentials in Electricity

Our understanding of electricity and magnetism and ordinary 3-space contribute to our understanding of differentials. The infinitesimal amount of work done against electric forces as charge q is displaced by $d\boldsymbol{r}$ is given by the differential for work $đW = -q\mathcal{E}(\boldsymbol{r}) \cdot d\boldsymbol{r}$, where $\mathcal{E}(\boldsymbol{r})$ is the electric field. The differential for the work done per unit charge is

$$đV_{\text{el}} = -\mathcal{E}(\boldsymbol{r}) \cdot d\boldsymbol{r}. \tag{9.93}$$

With this differential, *Faraday's law* in (9.8) becomes

$$\oint đV_{\text{el}} = \frac{d\Phi_B}{dt}. \tag{9.94}$$

That is, the total work done per unit of charge as a charge is moved around a closed contour is equal to the rate of change of the magnetic flux Φ_B through the contour.

Example 9.4 Solenoidal electric field

Is the differential for the work done per unit charge exact or inexact when $\mathcal{E}$ is the electric field inside a solenoid?

Solution. The components of the electric field inside a solenoid for a B-field increasing at a constant rate $\dot{B}_0$ are given in (9.10). For this field the differential $đV_{el}$ is

$$đV_{el} = -\mathcal{E} \cdot d\boldsymbol{r} = -\mathcal{E}_x dx - \mathcal{E}_y dy - \mathcal{E}_z dz = -\frac{1}{2}\dot{B}_0 y\,dx + \frac{1}{2}\dot{B}_0 x\,dy + 0\,dz.$$

The exactness conditions in (9.30) require that

$$\left(\frac{\partial \mathcal{E}_x}{\partial y}\right)_z \stackrel{?}{=} \left(\frac{\partial \mathcal{E}_y}{\partial x}\right)_z, \quad \left(\frac{\partial \mathcal{E}_y}{\partial z}\right)_x \stackrel{?}{=} \left(\frac{\partial \mathcal{E}_z}{\partial y}\right)_x, \quad \left(\frac{\partial \mathcal{E}_z}{\partial x}\right)_y \stackrel{?}{=} \left(\frac{\partial \mathcal{E}_x}{\partial z}\right)_y,$$

where the question marks indicate our uncertainly about the validity of the equal signs. By using the explicit expressions for $\mathcal{E}_x$, $\mathcal{E}_y$, and $\mathcal{E}_z$, the conditions become

$$\frac{\partial}{\partial y}\left(\frac{1}{2}\dot{B}_0 y\right) \stackrel{?}{=} \frac{\partial}{\partial x}\left(-\frac{1}{2}\dot{B}_0 x\right), \quad \frac{\partial}{\partial z}\left(-\frac{1}{2}\dot{B}_0 x\right) \stackrel{?}{=} \frac{\partial}{\partial y}0, \quad \frac{\partial}{\partial x}0 \stackrel{?}{=} \frac{\partial}{\partial z}\left(\frac{1}{2}\dot{B}_0 y\right),$$

which simplify to

$$\frac{1}{2}\dot{B}_0 \stackrel{?}{=} -\frac{1}{2}\dot{B}_0, \quad 0 \stackrel{?}{=} 0, \quad 0 \stackrel{?}{=} 0.$$

Since one of these is *not* true (except in the trivial case when $\dot{B}_0 = 0$), the differential $đV_{el}$ is *not* exact.

In electrostatics time derivatives are neglected so that Faraday's law simplifies to $\oint đV_{el} = 0$. From this and the exact differential theorem it follows that $đV_{el}$ is *exact* and that it is the differential of a function, which is the electrostatic potential $V_{el}(\boldsymbol{r})$. The electric field is obtained by differentiation

$$\mathcal{E}(\boldsymbol{r}) = -\nabla V_{el}(\boldsymbol{r}). \tag{9.95}$$

Since $d\Phi_B/dt$ is in general *not* zero, $đW = -q\mathcal{E}(\boldsymbol{r}) \cdot d\boldsymbol{r}$ is in general *not* exact, as we found in the above example, *i.e.*, the electric field is in general *not* derivable from a scalar potential. The electrostatic case and the general case are summarized in Table 9.1. In electrostatics $đV_{el}$ is exact, like dU, dT, etc. In the general case $đV_{el}$ is not exact, like the differentials $đQ$ and $đW$ of thermodynamics.

Table 9.1 *Differentials in electricity*

Electrostatics	General case
$đV_{el}$ is exact	$đV_{el}$ is *inexact*
$\oint đV_{el} = -\oint \mathcal{E}(\boldsymbol{r}) \cdot d\boldsymbol{r} = 0$	$\oint đV_{el} = -\oint \mathcal{E}(\boldsymbol{r}) \cdot d\boldsymbol{r} = (d\Phi_B/dt) \neq 0$
There exists a potential $V_{el}(\boldsymbol{r})$ such that $\mathcal{E}(\boldsymbol{r}) = -\nabla V_{el}(\boldsymbol{r})$.	$\mathcal{E}(\boldsymbol{r})$ is *not* derivable from a scalar potential
$\nabla \times \mathcal{E}(\boldsymbol{r})$ is zero everywhere	$\nabla \times \mathcal{E}(\boldsymbol{r})$ is *not* zero everywhere

Summary

The word differential refers collectively to differential forms, coordinate differentials, and parameter differentials. Physically, a differential is interpreted as a small amount of something. Their mathematical properties are as follows:

- A differential form $đA$ (or dA if exact) is an abbreviated notation for a product of vector components times coordinate differentials. (In thermodynamics vector components are called coefficients of coordinate differentials.)
- Differential forms are the quantities summed in line integrals.
- Differential forms can be added or subtracted.
- A differential form can be multiplied by a function.
- A differential form that is exact can be changed to a coordinate differential by a coordinate transformation and vice versa.
- An exact differential is the differential of a state function.
- A coordinate differential becomes a parameter differential when the associated coordinate parameterizes a line.
- A bar is required on the symbol $đ$ for differential forms that are *not* exact, but is omitted on differentials that are known to be exact.

The distinction between barred and unbarred differentials is more important in thermodynamics than the distinction between differential forms, coordinate differentials, and parameter differentials. The three types are simply called differentials in the following chapters. The exact (unbarred) differentials dP, dV, dT, and dU represent infinitesimal changes in the state function P, V, T, and U. In contrast, the inexact (barred) differentials $đQ$ and $đW$ represent infinitesimal transfers of energy that occur during equilibrium (reversible) processes, *i.e.*, processes that are representable by lines in state space.

Problems

9.1 The magnetic field inside a cylindrical conductor with a uniform current density J in the $+z$ direction is

$$B_x = -(\mu_0 J/2)y, \quad B_y = +(\mu_0 J/2)x, \quad B_z = 0.$$

Evaluate the line integral $\int_L \mathbf{B} \cdot d\mathbf{r}$ for the following lines by integrating with respect to parameters t, s, and u.

(a) $[x = R\cos t,\ y = R\sin t,\ z = z_0.\ 0 \leq t \leq \pi]$. Sketch the line and evaluate.

(b) $[x = R\cos s^3,\ y = R\sin s^3,\ z = z_0.\ 0 \leq s \leq \pi^{1/3}]$. Sketch the line and evaluate. (The purpose of part (b) is to show by example that the value of a line integral is independent of the parameterization of the line.)

(c) $[x = R - u,\ y = 0,\ z = z_0.\ 0 \leq u \leq 2R]$. Sketch the line and evaluate. (The purpose of part (c) is to show that in general the values of line integrals between the same end points are different for different lines.)

9.2 Determine which of the following differentials are exact and which are inexact. Integrate the exact differentials to find the associated scalar potential. Assume the spaces are two dimensional.

(a)

$$đA = (ax + cxy)\,dx + (by + cxy)\,dy,$$

where a, b, and c are constants.

(b) The differential for the electric potential outside a line of charge is

$$đV = \frac{-Kx}{x^2 + y^2}dx + \frac{-Ky}{x^2 + y^2}dy,$$

where $K = \lambda/2\pi\epsilon_0$ and λ is charge per unit length.

(c) A differential for a gas is

$$đF = (2aT + bV^3)dT + (3bTV^2)dV,$$

where a and b are constants.

9.3 Consider the following differential in a space with coordinates x and y:

$$đA = Kx^p y^q dx + Lx^m y^n dy.$$

The constants p, q, K, L, m, and n can be chosen arbitrarily if $đA$ is inexact, but not if $đA$ is exact. Find the values of K, p, and q when the values of L, m, and n are specified and we require that $đA$ be exact. (The answer is equations of the form $K = \cdots$, $p = \cdots$, and $q = \cdots$.)

9.4 Use the expression for an ideal gas to find the line integral $W = \int_L đW$ for the isothermal process

$$[P = P_0 + (P_f - P_0)t,\ T = T_0,\ 0 \le t \le 1],$$

where P_f, P_0, and T_0 are constants. Check that your answer agrees with Table 4.1.

9.5 Consider a *PVT* system for which

$$PV = n(RT + na/V) \ \text{and}\ U = n(cT + bT^2) + n^2a/V,$$

where a, b, and c are constants.

(a) Find $đQ$ for the system with T and V as coordinates. That is, complete the equation

$$đQ = (nc + \ldots)dT + (\ldots)dV.$$

(b) Show that $đQ$ is not exact.

9.6 Assume that the differential of $U(T, V)$ is

$$dU = n\left(c - \frac{nb}{V}\right)dT + \frac{n^2bT}{V^2}dV,$$

where b and c are constants and n is the number of moles.

(a) Verify that dU is exact.

(b) Find $U(T, V)$ and verify that the differential of the function obtained is the original differential.

(c) Find the specific heat $c_V(T, V)$.

9.7 **(a)** Express the differential of the pressure $P(T, V)$ in terms of B_T and α_V. (The definitions of B_T and α_V are given in Chapter 3.)

(b) Assume that $B_T = b_o$ and $\alpha_V = a_o/b_o$, where a_o and b_o are constants. Find $P(T, V)$ by integrating dP. Show that $P = P(T, V)$ is the equation of state for a simple solid in (1.15).

10
Entropy

The Kelvin statement of the second law, Carnot cycles, absolute temperatures, and the properties of differentials are essential for establishing the existence of the entropy state function. Although the differential for heat $đQ$ is not exact, the differential $đQ/T$ is exact, which implies that it is the differential of a function – the state function we call entropy. An important characteristic of entropy is its tendency to increase. That increase indicates that irreversible behavior has occurred, which in turn indicates that an opportunity to do useful work has been lost. The concept of entropy leads to a reformulation of the second law and is essential to the third law of thermodynamics and the restrictions it imposes on the behavior of systems near absolute zero.

10.1 Definition of Entropy

To obtain the concept of entropy, we need to demonstrate that the inexact differential $đQ$ becomes exact when divided by absolute temperature. Before doing that, we point out that the definition of absolute temperature T given in (7.10) can be expressed as

$$\frac{|Q_1^C|}{T_1} = \frac{|Q_2^C|}{T_2}, \tag{10.1}$$

where Q_1^C and Q_2^C are the heats exchanged by Carnot cycles operating between heat reservoirs at temperatures T_1 and T_2. As both Q^C and $đQ$ describe reversible heat transfers, equation (10.1) anticipates the definition of entropy which utilizes the exactness of $đQ/T$. To demonstrate that $đQ/T$ is exact, we show that its line integral around closed contours in state space is zero. We begin by showing how reversible cyclic processes can be interpreted as collections of Carnot cycles.

The closed contour C in the PV-diagram in **Figure 10.1** represents a reversible cyclic process in an ideal gas. The diagram is divided into narrow strips by adiabatic curves, which are connected by short isothermal curves. The result is a collection of small clockwise

Thermodynamics and Statistical Mechanics: An Integrated Approach, First Edition.
Robert J Hardy and Christian Binek.

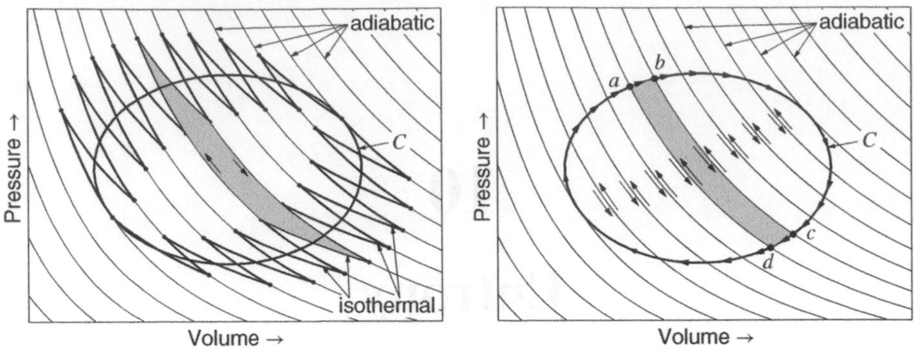

Figure 10.1 *Cyclic Process C as a Sum of Carnot Cycles*

Carnot cycles with the adiabatic curves shared by adjacent cycles.[1] In the second diagram in Figure 10.1 the small cycles are slightly modified so that adjacent adiabatic curves are connected by short segments of contour *C*, which becomes our integration contour. This allows the contour to be interpreted as a sum of the integrals around many small cycles, so that

$$\oint_C \frac{đQ}{T} = \sum_j \oint_j \frac{đQ}{T}, \tag{10.2}$$

where the different cycles are identified by index *j*. The integrals along the adiabatic curves do not contribute to the sum since $đQ = 0$ along them. (Even if the integrals were not zero, they would be cancelled because the integrals are in the opposite direction in the adjacent cycles.) The integral around the contour for each cycle is the sum of integrals along four different lines, two of which vanish. For example, the line integral around the shaded cycle in the figure is

$$\oint_j \frac{đQ}{T} = \int_{a\to b} \frac{đQ}{T} + 0 + \int_{c\to d} \frac{đQ}{T} + 0, \tag{10.3}$$

where the integrals along lines $b \to c$ and $d \to a$ are zero because $đQ$ is zero. The integral along lines $a \to b$ and $c \to d$ are integrals of $đQ$ divided by a temperature that varies slightly along the line. These variations become less and less as the strips are made narrower, so that the temperature at which heat is exchanged along the two short line segments is essentially constant. This implies that

$$\oint_j \frac{đQ}{T} \approx \frac{|Q_{a\to b}|}{T_a} - \frac{|Q_{c\to d}|}{T_c}, \tag{10.4}$$

where the minus sign accounts for the fact that $Q_{c\to d}$ is negative. As the temperatures of the heat transfers are effectively constant, each small cycle becomes a Carnot cycle as the strips are made narrower. It follows from (10.1) that the right hand side of (10.4) equals

[1] When contours with complex shapes are considered, some strips may be crossed in more than two places, in which case more than one Carnot cycle will be associated with the strip.

zero.[2] As the only characteristic of contour C that we have used is that it is closed, we conclude that the line integral of $đQ/T$ around any closed contours in state space is zero,

$$\oint \frac{đQ}{T} = 0. \tag{10.5}$$

This is the crucial result needed for the definition of entropy.

Entropy

According to the exact differential theorem, if the integral of a differential around all closed contours is zero, the differential is exact and is the differential of a function (see Section 9.3). Thus, the result in (10.5) implies that $đQ/T$ is exact and is the differential of a function. We represent this differential by

$$\boxed{dS = \frac{đQ}{T}.} \tag{10.6}$$

The associated function S is called *entropy*. Another consequence of the exact differential theorem is that the change in entropy in processes that takes a system from state 0 to state f is

$$\boxed{\Delta S = S_f - S_0 = \int_{L_{0\to f}} \frac{đQ}{T},} \tag{10.7}$$

where $L_{0\to f}$ is any line that represents a process between those two states. This tells us that the change in entropy is independent of the reversible process used to go from state 0 to state f. We obtain specific values for the entropy by letting 0 represent some fixed reference state o and assigning a value S_o to that state. Then, the entropy of any state f is

$$\boxed{S_f = S_o + \int_{L_{o\to f}} \frac{đQ}{T}.} \tag{10.8}$$

Since the value of S_o is arbitrary, the entropy state function is only determined up to an additive constant. As indicated in Section 10.6, the third law indicates that we can set $S_o = 0$ at absolute zero.

When an inexact differential is made exact by multiplying by a function, the function is called an *integrating factor*. Integrating factors do not exist for most inexact differentials. As a result, the fact that the differential for heat has an integrating factor is physically significant: *The differential $đQ$ that describes the transfer of heat in reversible processes has an integrating factor, and the integrating factor is* $1/T$.

[2] The integrals around the first and last small cycles in the figure also tend to zero. These cycles consist of two lines: One line is an adiabatic curve, while the other is a segment of contour C which becomes both shorter and more like an adiabatic curve as the strips become narrower.

Since the differential for heat is $đQ = dU + PdV$, the differential of entropy is

$$dS = \frac{1}{T}dU + \frac{P}{T}dV. \tag{10.9}$$

As dS is exact, this implies that

$$\frac{1}{T} = \left(\frac{\partial S}{\partial U}\right)_V \text{ and } \frac{P}{T} = \left(\frac{\partial S}{\partial V}\right)_U. \tag{10.10}$$

It then follows from $(\partial S/\partial U)_V = 1/(\partial U/\partial S)_V$ that a system's absolute temperature can be expressed as

$$T = \left(\frac{\partial U}{\partial S}\right)_V. \tag{10.11}$$

10.2 Clausius' Theorem

Although only reversible processes are involved in its definition, the usefulness of entropy for predicting the evolution of systems arises from its behavior in irreversible processes. The crucial result needed for demonstrating the behavior of entropy in all kinds of situations is contained in Clausius' theorem. The theorem states that: *The heats Q_i exchanged by a system at absolute temperatures T_i in a cyclic process of N steps are such that*

$$\boxed{\sum_{i=1}^{N} \frac{Q_i}{T_i} \leq 0.} \tag{10.12}$$

By convention $Q_i > 0$ when heat is transferred to the system in step i, and $Q_i < 0$ when heat is transferred from the system.

The cyclic process in the theorem can be either reversible or irreversible, and it can take place in a complex system or in a relatively simple one like a *PVT* system. The condensing steam engine diagrammed in **Figure 10.2** is an example of a complex system. In an electrical power plant the heat from a fire or nuclear reaction is transferred to the working fluid (water and steam) at a range of temperatures, and waste heat is transferred to a cooling medium at a colder temperature. The device, which is our system, is everything inside the dashed line.

The proof of Clausius' theorem applies the Kelvin statement of the second law to a thought experiment of a very general type. In this experiment heat is exchanged between a cyclic device D and a single heat reservoir and all of the heat transfers are done by Carnot cycles. This is illustrated in **Figure 10.3**, where device D can be something simple like a *PVT* system or something complex like a condensing steam engine. The work done in one cycle *by* the device is W^D, and the absolute temperature of the heat reservoir is T^R. Each cycle consists of N steps, where T_i is the temperature at which heat is transferred to or from the device in step i. The Carnot cycle labeled C_i exchanges heat Q_i^R with reservoir R, exchanges heat Q_i with the device, and does work W_i on the surroundings. There are no

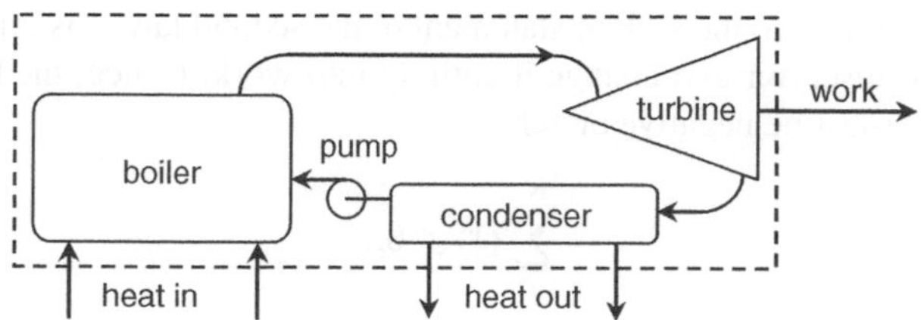

Figure 10.2 Condensing Steam Engine

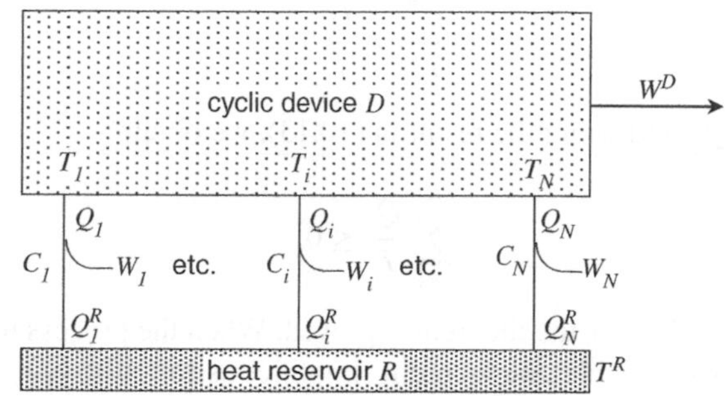

Figure 10.3 A Cyclic Device D Interacting with a Single Heat Reservoir

arrows on the lines representing the Carnot cycles, as they can be engines or refrigerators, depending on the directions of the heat transfers.

Since energy can be transferred either to or from the device, the following sign conventions are important:

$W^D > 0$. Work is done *by* device D.
$W^D < 0$. Work is done *on* device D.
$W_i > 0$. Work is done *by* Carnot cycle C_i.
$W_i < 0$. Work is done *on* Carnot cycle C_i.
$Q_i^R > 0$. Heat is transferred *from* reservoir R.
$Q_i^R < 0$. Heat is transferred *to* reservoir R.
$Q_i > 0$. Heat is transferred *to* device D.
$Q_i < 0$. Heat is transferred *from* device D.

There are four possibilities for each Carnot cycle:

1. $T_i < T^R$ and $Q_i > 0$. C_i is a Carnot engine and $W_i > 0$.
2. $T_i < T^R$ and $Q_i < 0$. C_i is a Carnot refrigerator and $W_i < 0$.
3. $T_i > T^R$ and $Q_i > 0$. C_i is a Carnot refrigerator and $W_i < 0$.
4. $T_i > T^R$ and $Q_i < 0$. C_i is a Carnot engine and $W_i > 0$.

For all four possibilities, the conservation of energy implies that $W_i = Q_i^R - Q_i$.

The only effect of running the cyclic device D in conjunction with the collection of Carnot cycles is to transfer heat $\sum_i Q_i^R$ *from* a single reservoir and do work $W^D + \sum_i W_i$ *on* the surroundings. According to the Kelvin statement of the second law, it is impossible to take heat from a single reservoir and change it entirely into work. Hence, the heat transferred from the reservoir must be negative or zero,

$$\sum_{i=1}^{N} Q_i^R \leq 0. \tag{10.13}$$

Since Q_i and Q_i^R refer to heats exchanged by Carnot cycles, it follows from (10.1) and the above sign conventions for Q_i and Q_i^R that

$$\frac{Q_i^R}{T^R} = \frac{Q_i}{T_i}. \tag{10.14}$$

By solving this for Q_i^R and using the result in (10.13), we obtain

$$\sum_{i=1}^{N} \frac{Q_i}{T_i} \leq 0. \tag{10.15}$$

As this is the same as (10.12), the theorem is proved. When the process to which Clausius' theorem is applied is reversible, the theorem simplifies as follows:

Corollary to Clausius' Theorem

The heats Q_i and temperatures T_i in reversible cyclic processes are such that

$$\boxed{\sum_{i=1}^{N} \frac{Q_i}{T_i} = 0.} \tag{10.16}$$

Since the corollary assumes that the system (device) is reversible, Clausius' theorem can be applied to the reversed system, which implies that

$$\sum_{i=1}^{N} \frac{Q_i^{\text{reversed}}}{T_i} \leq 0. \tag{10.17}$$

When a process is reversed, the heats Q_i^R and Q_i and the works W_i are multiplied by -1, while the temperatures are unchanged. Since $Q_i^{\text{reversed}} = -Q_i^{\text{forward}}$, it follows from (10.17) that

$$-\sum_{i=1}^{N} \frac{Q_i^{\text{forward}}}{T_i} \leq 0. \tag{10.18}$$

Applying Clausius' theorem to the system operating in the forward direction gives

$$\sum_{i=1}^{N} \frac{Q_i^{\text{forward}}}{T_i} \leq 0. \tag{10.19}$$

Since the equal sign is the only possibility consistent with both (10.18) and (10.19), the corollary is proved.

$đQ/T$ is Exact

The corollary to Clausius' theorem can be used to prove that $đQ/T$ is exact. We proceed by investigating reversible cyclic processes in systems with a single temperature when in equilibrium, and represent the process by a closed contour L_{cycle} in the system's state space. The process is divided into N steps, which divide the contour into segments L_i. Let Q_i represent the heat transferred in step i, and let T_i represent the average temperature during that step. As the process is reversible, Q_i equals the line integrals of $đQ$. Then, by using small steps, the sum of the integrals of $đQ/T$ can be approximated as

$$\sum_{i=1}^{N} \int_{L_i} \frac{đQ}{T} \approx \sum_{i=1}^{N} \frac{1}{T_i} \int_{L_i} đQ = \sum_{i=1}^{N} \frac{Q_i}{T_i}, \tag{10.20}$$

which becomes exact as the steps are made limitingly small, *i.e.*, as $N \to \infty$. Taking the limit gives

$$\lim_{N\to\infty} \sum_{i=1}^{N} \frac{Q_i}{T_i} = \oint \frac{đQ}{T}. \tag{10.21}$$

It follows from this and the corollary to Clausius' theorem that

$$\oint \frac{đQ}{T} = 0. \tag{10.22}$$

This is the same as the conclusion obtained in (10.5) by using different reasoning. It follows from (10.22) and the exact differential theorem that $đQ/T$ is exact and is the differential of a function. The function is the *entropy state function* and its differential is $dS = đQ/T$.

10.3 Entropy Principle

We want to show that the entropy of adiabatically isolated systems can only increase or remain the same. To do this, we need to determine how much the entropy has changed, which is done by integrating the differential dS along a line in state space. As real processes are often irreversible, they are not representable by lines in state space. As a result, determining the entropy change during many processes requires that we find a reversible process that produces the same change. To satisfy this requirement, we investigate two part processes like that diagrammed in **Figure 10.4**, where a real process takes the system from state a to state b is followed by a reversible process that returns it to state a. As the complete process is cyclic, Clausius' theorem is applicable.

It is convenient to replace the single sum over i in the summary of the Clausius' theorem in (10.12) by separate sums for the real and the reversible parts of the process, so that the theorem becomes

$$\sum_i \frac{Q_i}{T_i} = \sum_j \frac{Q_j^{\text{real}}}{T_j} + \sum_k \frac{Q_k^{\text{rev}}}{T_k} \le 0, \tag{10.23}$$

where Q_j^{real} is the heat exchanged with the surroundings at temperatures T_j in step j of the real process, and Q_k^{rev} is the heat exchanged at temperatures T_k in step k of the reversible process.

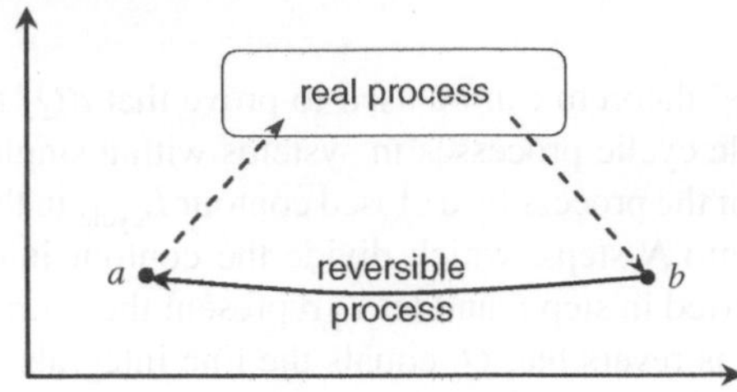

Figure 10.4 *Two-Part Cyclic Process*

We first consider systems with no adiabatically isolated parts and represent the reversible process by line $L_{b\to a}$. The process from b to a is divided into K steps, so that $L_{b\to a}$ is made up of K line segments L_k. The heat transferred to the system during step k is $Q_k = \int_{L_k} đQ$, and the temperature change becomes less and less as the step is made smaller. By considering an infinite number of infinitesimal steps, the sum over k in (10.23) becomes

$$\sum_k \frac{Q_k^{\text{rev}}}{T_k} = \lim_{K\to\infty} \sum_{k=1}^{K} \frac{1}{T_k} \int_{L_k} đQ = \int_{L_{b\to a}} \frac{đQ}{T} = S_a - S_b, \tag{10.24}$$

where (10.8) has been used. We now restrict the real processes considered to those in which no heat is exchanged between the system and its surroundings, *i.e.*, we only consider real processes in which the system is adiabatically isolated. This implies that $Q_j = 0$ in all steps of the real process. Hence,

$$\sum_j \frac{Q_j^{\text{real}}}{T_j} = 0. \tag{10.25}$$

Combining this with (10.23) and (10.24) gives

$$\sum_i \frac{Q_i}{T_i} = S_a - S_b \leq 0. \tag{10.26}$$

Since states a and b are the initial and final states of the real process, which are often represented by 0 and f, the result in (10.26) implies that

$$S_0 \leq S_f, \tag{10.27}$$

where S_0 and S_f are the initial and final entropies of the real process.

Since the entropy change during an adiabatic process from state a to state b is such that $S_a \leq S_b$, reversing a real reversible process changes the relationship to $S_b \leq S_a$. Taken together, the two relationships imply that $S_b = S_a$. Hence: *The entropy of a system is unchanged during reversible adiabatic processes.*

Composite Systems

The above result was obtained for systems with a single equilibrium temperature. To avoid this restriction, we treat systems with adiabatically isolated parts as collections of subsystems without isolated parts. We then restrict the reversible process that returns the system from state b back to state a to processes in which the heat transferred to or from the subsystems is exchanged with the surroundings of the combined system with no heat exchanges between subsystems. This allows us to separately apply the arguments used to derive (10.24) to each subsystem. By using index n to label the subsystems, applying (10.24), and adding up the results, it follows that

$$\sum_k \frac{Q_k^{\text{rev}}}{T_k} = \sum_n \sum_k \frac{Q_{n,k}}{T_{n,k}} = \sum_n (S_{n,a} - S_{n,b}), \tag{10.28}$$

where $S_{n,a}$ and $S_{n,b}$ are the entropies of subsystem n in states a and b. By using this result to represent the reversible part of the cycle in the summary of Clausius' theorem in (10.23), we find that

$$\sum_j \frac{Q_j^{\text{real}}}{T_j} + \sum_n (S_{n,a} - S_{n,b}) \le 0, \tag{10.29}$$

where the heats are exchanged with the surroundings of the combined system. We now restrict the real processes considered to those in which the combined system is adiabatically isolated. It then follows that

$$\sum_n S_{n,a} \le \sum_n S_{n,b}. \tag{10.30}$$

Although the composite system is adiabatically isolated during the real process, there is no restriction on the exchange of heat between the subsystems. Also, there is no restriction on the exchange of energy in the form of work.

The sums in (10.30) indicate that entropy is additive, which is the characteristic of an extensive property (see Section 1.5). Since $đQ$ is extensive (proportional to subsystem size) and T is intensive (independent of size), it follows that $đQ/T$ is extensive. The entropies of the individual subsystems are also extensive, and entropy of the combined (total) systems is

$$S^{\text{tot}} = S_1 + S_2 + \cdots = \sum_n S_n. \tag{10.31}$$

By combining this with the previous result and representing the initial and final states by 0 and f, instead of a and b, we find for adiabatically isolated systems that

$$\boxed{S_0^{\text{tot}} \le S_f^{\text{tot}}.} \tag{10.32}$$

The following principle is obtained by combining this result with the definition of the entropy state function:

Entropy Principle

1. *All thermally interacting systems in thermal equilibrium possess an entropy state function S whose differential is $dS = đQ/T$.*
2. *The entropy of an adiabatically isolated system never decreases.*

Entropy will Increase If It Can

An important aspect of the entropy principle is the idea that entropy will increase if it can. Specifically, if the total entropy of an adiabatically isolated system can increase, it will spontaneously evolve in the direction that causes its entropy to increase.[3] Since there are states available with greater entropy when the entropy is not maximized, we conclude that: *When in equilibrium, the entropy of an adiabatically isolated system has the maximum value consistent with the constraints on it.*

The entropy principle is useful for predicting how systems evolve. Since entropy is an equilibrium state function, the principle is limited to processes that start and end in equilibrium. However, no such limitation applies while the processes are evolving. Although the time evolution of systems is ultimately determined by the behavior of their microscopic constituents, the entropy principle establishes a theory of the evolution of thermodynamic systems that is entirely independent of assumptions about their atomic-level structure.

The Second Law

Our derivation of the entropy principle makes critical use of the second law of thermodynamics as given in Chapter 7. An alternate expression of the second law can be obtained by modifying entropy the principle. The modified principle replaces the first part by the assertion that entropy increases as internal energy increases when no work is done. Symbolically, the assertion specifies that $dS > 0$ if $dU > 0$ provided $đW = 0$. This is true because $đQ$ becomes dU when $đW = 0$, so that $dS = đQ/T$ becomes $dS = dU/T$ which implies that S increases as U increases. That the following statement is an independent expression of the second law is shown in Appendix 10.A:

> ***Entropy statement of the second law***
> 1. *All thermally interacting systems in thermal equilibrium possess an entropy state function that is extensive and increases as internal energy increases when no work is done.*
> 2. *The entropy of an adiabatically isolated system never decreases.*

10.4 Entropy and Irreversibility

Much of the significance of energy comes from its conservation. Entropy is different. Instead of being conserved, it increases as time evolves. Central to its significance are the concepts of reversibility and its absence – irreversibility. When using the entropy principle, it is essential to remember that it refers to the total entropy of an adiabatically isolated system. The entropy of a system can decrease if it is not isolated, provided the decrease is

[3] This concept is implicit in Rudolph Clausius' summary of the first and second laws: "*The energy of the world is constant.*" and "*The entropy of the world tends towards a maximum.*" ("Die Energie der Welt ist konstant." and "Die Entropie der Welt strebt einem Maximum zu.") Clausius, R. (1865) Annalen der Physik, **125** (7), 353–400, ser. 2.

compensated by an increase in the entropy of the other systems involved. For example, the entropy of a hot cup of coffee placed on a table ready to drink decreases as it cools, but the decrease is accompanied by an increase in the entropy of the air in the room. When all of the entropy changes are included, the total entropy has increased.

To prevent the concept of entropy from becoming too abstract, we will find the entropy changes in a few easily visualized situations. When the entropy state function is known, we can find its change by calculating the difference between its initial and final values. When the state function is not known, the change ΔS is found by integrating $đQ/T$ along a line in state space that takes the system between the same initial and final states. As lines in state space represent reversible processes while real processes are frequently irreversible, the reversible process employed often involves a hypothetical situation that has little similarity to the real process. Some idealization of the processes and the systems involved may be necessary to obtain a system that is adiabatically isolated.

Experiments are often performed at constant pressure, which is accomplished by leaving the system open to the atmosphere. In this case, the only significant variable in systems with two thermodynamic coordinates is temperature. It is convenient to have an expression for the change in entropy in these situations. As dP is zero when pressure is constant, the differential for heat becomes $đQ = C_P dT$, so that the line integral in (10.7) becomes an ordinary one-dimensional integral. When the heat capacity C_P is constant, the change in entropy between temperatures T_0 and T_f during a reversible constant-pressure process is

$$\Delta S = \int_L \frac{đQ}{T} = \int_{T_0}^{T_f} \frac{C_P dT}{T} = C_P \int_{T_0}^{T_f} \frac{dT}{T} = C_P \ln \frac{T_f}{T_0}. \tag{10.33}$$

When the heat is transferred at a fixed temperature, as it is for heat reservoirs, the temperature can be taken outside the integration. The resulting change is

$$\Delta S_R = \frac{Q_R}{T_R}, \tag{10.34}$$

where Q_R is the heat transferred *to* a reservoir at temperature T_R.

Example 10.1 Cooling a hot utensil

Find the change in entropy that results when a 2.0 kg cooking utensil made of iron at 80 °C is placed in 5.0 kg (5 L) of water at 15 °C. Assume that any exchange of heat with other objects is negligible.

Solution. Our system is a combination of two subsystems, the iron utensil and the water. After being placed in thermal contact, they come to equilibrium at the same temperature Θ_f, which must be found before the change in entropy can be calculated. It follows from the simplifying assumption that

(heat *to* 5.0 kg of water) = (heat *from* 2.0 kg of iron).

By using $Q = M c_P^M \Delta T$, where C_P is $M c_P^M$ and M is mass, we obtain

$$(5.0\,\text{kg})c_P^{\text{w}}(\Theta_f - 15\,^\circ\text{C}) + (1.0\,\text{kg})c_P^{\text{Fe}}(\Theta_f - 80\,^\circ\text{C}) = 0.$$

Thus,

$$\Theta_f = \frac{c_P^{\text{w}}(5.0\,\text{kg})(15\,^\circ\text{C}) + c_P^{\text{Fe}}(2.0\,\text{kg})(80\,^\circ\text{C})}{c_P^{\text{w}}(5.0\,\text{kg}) + c_P^{\text{Fe}}(2.0\,\text{kg})}.$$

Using the values for the specific heats from Table 1.2 gives

$$\Theta_f = \frac{1.00(5.0)(15\,^\circ\text{C}) + 0.106(2.0)(80\,^\circ\text{C})}{1.00(5.0) + 0.106(2.0)} = 17.64\,^\circ\text{C}.$$

As the cooling of the utensil is irreversible, we need a reversible process that achieves the same amount of cooling, otherwise we have no line in state space along which to integrate $đQ/T$. Consider the following two-part process during which the subsystems are isolated with well-defined equilibrium properties:

1. Reversibly heat the 5.0 kg of water from 10 °C to 17.64 °C.
2. Reversibly cool the 2.0 kg iron from 80 °C to 17.64 °C.

The reversible heating or cooling of an object is an idealization of the process of placing the object in thermal contact with a sequence of heat reservoirs with essentially the same temperature. The total change in entropy is found by adding the entropy changes of the two subsystems. Presumably, the experiment is performed at constant atmospheric pressure, so that (10.33) can be used. The entropy change of the iron is

$$\begin{aligned}\Delta S^{\text{Fe}} &= M^{\text{Fe}} c_P^{\text{Cu}} \ln \frac{T_f}{T_0^{\text{Fe}}} = 2.0\,\text{kg}(0.106\,\text{kcal}\,\text{kg}^{-1}\text{K}^{-1}) \ln \frac{273.15 + 17.64}{273.15 + 80} \\ &= -\,0.04119\,\text{kcal}\,\text{K}^{-1}.\end{aligned}$$

The entropy change of the water is

$$\begin{aligned}\Delta S^{\text{w}} &= M^{\text{w}} c_P^{\text{w}} \ln \frac{T_f}{T_0^{\text{w}}} = 5.0\,\text{kg}(1.00\,\text{kcal}\,\text{kg}^{-1}\,\text{K}^{-1}) \ln \frac{273.15 + 17.64}{273.15 + 15} \\ &= +\,0.04567\,\text{kcal}\,\text{K}^{-1}.\end{aligned}$$

The total change in entropy is

$$\Delta S^{\text{tot}} = \Delta S^{\text{Fe}} + \Delta S^{\text{w}} = -41.19\,\text{cal}\,\text{K}^{-1} + 45.67\,\text{cal}\,\text{K}^{-1} = 4.48\,\text{cal}\,\text{K}^{-1}.$$

The increase in the entropy of the water is greater than the decrease in iron. This is because the temperature in the denominator of $đQ/T$ in (10.33) is smaller for the water than for the iron.

As no heat was exchanged with other objects, the increase in entropy in the example is consistent with the second law. The entropy of one subsystem (iron) decreased, while the entropy of the other (water) increased. Notice that the entropy change of each subsystem is significantly greater than the total entropy change. This is typical. Since subtracting two numbers of approximately the same magnitude causes a loss of accuracy, it is often necessary to keep more that the usual number of significant figures in the intermediate steps of entropy calculations.

Example 10.2 Dry friction

In Example 2.3 an iron block was moved back and forth against a stationary block. Because of the frictional forces acting between the blocks, 32 kJ of work was done by an external force and the temperature of the blocks increased by 36 K. Estimate the change in entropy that occurred. (The combined mass of the blocks is 2.0 kg.)

Solution. Assume that no heat is transferred between the environment and our system (the blocks), so that the process is adiabatic. It is reasonable to neglect the small amount of $P\Delta V$ work done on the atmosphere by the expansion of the iron. Since the external force moving the block back and forth does work *on* the system, the work done *by* the system is -32 kJ. The same change in the system's temperature can also be achieved by heating the blocks reversibly by bringing them into contact with a sequence of heat reservoirs with temperatures infinitesimally greater than that of the blocks. As the heat capacity C_P of iron is essentially constant over the range of temperatures involved, we can use (10.33) to estimate the change in entropy. By using the value of c_P^M for iron in Table 1.2 and assuming the blocks were initially at room temperature (300 K), the change in entropy becomes

$$\Delta S = M c_P^M \ln \frac{T_f}{T_0} = (2\,\text{kg})(444\,\text{J}\,\text{kg}^{-1}\,\text{K}^{-1}) \ln \frac{300+36}{300} = 101\,\text{J}\,\text{K}^{-1}.$$

Since the process in the example is adiabatic, the positive value of ΔS is consistent with the second law. The dissipation of mechanical energy by friction is the quintessential example of an irreversible process. The process takes place every time you apply the brakes in your car. Another example is the dissipation of electrical energy in a resistor. Instead of rubbing the iron blocks against each other, we could have allowed an electric current to flow through the blocks so they acted like a resistor. By dissipating 320 W of electric power for a period of 100 s, we could transfer the same amount of energy to the system electrically that was transferred mechanically in the example, and the same increase in entropy would result.

Irreversibility

A collection of objects that only exchange heat with each other can be treated as adiabatically isolated. When its entropy remains unchanged in a process, the process is reversible. When the process is irreversible, the increase in the system's entropy can be used as a quantitative indicator of the irreversibility that has occurred.

10.5 Useful Energy

An increase in entropy indicates a loss of useful energy. Since energy is conserved, *i.e.*, cannot be created or destroyed, one might ask why people are concerned about energy shortages. Of course, we know that the concern is not about energy in general but about energy that is available for performing useful tasks. For example, on cold days there is a large amount of energy in the environment outside, but it is not available for warming the air indoors until we have a heat pump and some energy with which to operate it. (Heat pumps were discussed in Section 7.6.) Heat pumps are usually powered by electrical energy, which in thermodynamics is energy transferred as work. Since the second law restricts our ability to convert heat into work but sets no limit on the conversion of work into heat, energy that can be transferred as work is more useful than energy that is transferred as heat. The design of devices that use a minimum amount of energy or fuel to perform useful tasks, such as heating a building or powering an automobile, are practical challenges, challenges that are met by avoiding irreversible behavior, *i.e.*, by minimizing the creation of entropy.

When designing devices (engines) whose purpose is to deliver work, it is usually desirable to increase the amount of work done, which increases the value of W. Since the work done *by* devices that accept energy while performing a useful task is negative ($W < 0$), decreasing the energy needed to operate the device will bring the value of W closer to zero. Thus, in both situations it is desirable to increase the value of W. In this section, we show that the energy available for doing useful work is increased by reducing irreversible behavior. Specifically, we use the properties of entropy to demonstrate that: *The work delivered in any process that takes a system between specific initial and final states is maximized when the process employed is reversible. Furthermore, all reversible processes between two specific states deliver the same amount of work.*

To verify the above statement, we idealize the interactions between systems and their environment as shown in **Figure 10.5**. Process X in the figure represents a general process in which the system's change can be either reversible or irreversible, while its change in the other process is reversible. Both processes take the system from the same initial state 0 to the same final state f. The interactions of the system with its environment in process X are represented by reversible exchanges of heat at temperature T_i with heat reservoirs R_i and by energy exchanges in the form of work with mechanical device M. The same reservoirs R_i are included in process *rev* and the same amounts of heat are exchanged by them. However, all heat exchanges in the reversible process (designated *rev*) are made through Carnot cycles, which operate between a single auxiliary reservoir at temperature T_A and the system and each reservoir R_i. The collection of system and reservoirs in each diagram is adiabatically isolated.

We now compare the entropy changes in the two processes. Since the states 0 and f are the same in both processes, the entropy change ΔS_{sys} of the system is the same. By letting

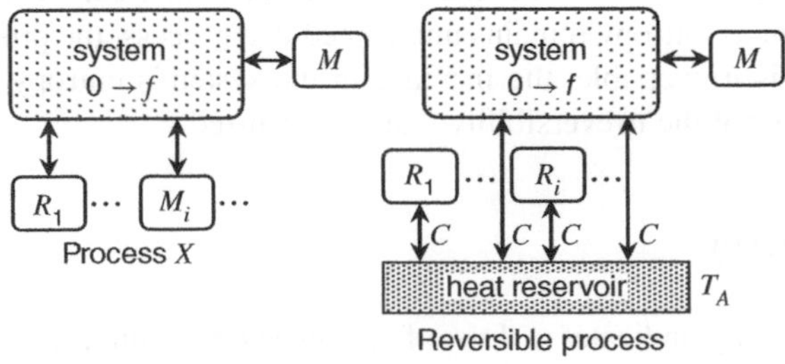

Figure 10.5 Processes that Take a System from State 0 to State f

Q_i^X represent the heat transferred *to* reservoir R_i *from* the system in general process X, the entropy change of the reservoir given in (10.34) is Q_i^X/T_i. Since the entropy never decreases in an adiabatically isolated system, the total entropy change in process X is

$$\Delta S_{\text{sys}} + \sum_i \frac{Q_i^X}{T_i} \geq 0. \tag{10.35}$$

Since the same reservoirs R_i and the same amounts of heat are exchanged in both processes, the entropy change of reservoir R_i in process *rev* is also Q_i^X/T_i. The entropy change in the auxiliary reservoir is Q_A/T_A. As the system's change is reversible and the Carnot cycles are reversible, the entropy remains unchanged in process *rev*,

$$\Delta S_{\text{sys}} + \sum_i \frac{Q_i^X}{T_i} + \frac{Q_A}{T_A} = 0. \tag{10.36}$$

Combining this with (10.35) implies that $-(Q_A/T_A) \geq 0$. The heat $Q_{0\to f}^{\text{rev}}$ transferred to the system and to the other reservoirs is the negative of the heat Q_A transferred to the auxiliary reservoir, so that

$$Q_{0\to f}^{\text{rev}} + \sum_i Q_i^X = -Q_A. \tag{10.37}$$

Since $T_A > 0$, it follows from $-(Q_A/T_A) \geq 0$ that

$$Q_{0\to f}^{\text{rev}} + \sum_i Q_i^X \geq 0. \tag{10.38}$$

The heat transferred to the system in process X is $Q_{0\to f}^X = -\sum_i Q_i^X$. Combining this with the previous result implies that

$$Q_{0\to f}^{\text{rev}} - Q_{0\to f}^X \geq 0. \tag{10.39}$$

As process X can be reversible, this result can be used to compare two reversible processes: *rev*-1 and *rev*-2. By identifying them successively as process X, we obtain $Q_{0\to f}^{\text{rev-2}} \geq Q_{0\to f}^{\text{rev-1}}$ and $Q_{0\to f}^{rev\text{-}1} \geq Q_{0\to f}^{rev\text{-}2}$ which together imply that $Q_{0\to f}^{\text{rev-1}} - Q_{0\to f}^{\text{rev-2}} = 0$. Thus, the equal sign in (10.39) applies when process X is reversible.

When process X is irreversible, the heat transferred to the system in the reversible process is greater than in the irreversible process ($Q_{0\to f}^{\text{rev}} > Q_{0\to f}^X$). Because a smaller fraction of the energy needed to change the system's internal energy comes from mechanical device M. By using the first law $Q = \Delta U + W$, it follows that

$$Q_{0\to f}^{\text{rev}} - Q_{0\to f}^X = (\Delta U_{\text{sys}} + W_{0\to f}^{\text{rev}}) - (\Delta U_{\text{sys}} + W_{0\to f}^X) = W_{0\to f}^{\text{rev}} - W_{0\to f}^X. \tag{10.40}$$

By combining (10.40) with (10 39), we find that

$$\boxed{W_{0\to f}^{\text{rev}} \geq W_{0\to f}^X,} \tag{10.41}$$

where $W_{0\to f}^{\text{rev}}$ and $W_{0\to f}^X$ are works done in process *rev* and process X. Since the equal sign applies when process X is reversible, we conclude that all reversible processes deliver the same amount of work.

The above result tells us that the work done by a system as it goes from state 0 to state f is greatest when the process involved is reversible. As almost all activities involving transfers of energy change the state of some system, the result indicates that the amount of energy available for performing useful tasks is maximized by minimizing the creation of entropy, which is accomplished by avoiding irreversible behavior.

Example 10.3 Work from cooling water

Find the maximum amount of work that can be obtained from one liter of water at 100 °C by cooling it to the temperature of a room at 20 °C.

Solution. We begin by finding the maximum work obtainable from any object at temperature T_0 which interacts with a heat reservoir at temperature T_R. For simplicity, assume that its heat capacity C_0 and its volume V are constant. To find the change in its internal energy as the object cools from temperature T_0 to temperature T_R, we bring it into thermal contact with the heat reservoir and allow heat to be transferred to the reservoir irreversibly. Since no work is done when V is constant, the change in internal energy is

$$\Delta U = U_f - U_0 = C_0(T_R - T_0).$$

To obtain the maximum amount of work from the object, we connect it to a device (engine) that accepts heat from it and rejects heat to the reservoir. We adiabatically isolate the combination of object and reservoir and transfer the heat reversibly. The work done by the device (whose state remains unchanged) equals the energy removed from the object minus the heat Q_R transferred to the reservoir,

$$W = (U_0 - U_f) - Q_R = C_0(T_0 - T_R) - Q_R.$$

Since the maximum amount of energy available for doing work is obtained by minimizing the creation of entropy, we need to know the entropy changes involved. The entropy change of the reservoir is $\Delta S_R = Q_R/T_R$. The entropy change of the object is

$$\Delta S_{\text{obj}} = \int_{T_0}^{T_R} \frac{đQ}{T} = C_o \int_{T_0}^{T_R} \frac{dT}{T} = C_o(\ln T_R - \ln T_0) = C_o \ln \frac{T_R}{T_0}.$$

As entropy is unchanged in reversible adiabatic processes, it follows that $\Delta S_{\text{obj}} + \Delta S_R = 0$. By using this and the above results, it follows that

$$Q_R = T_R \Delta S_R = -T_R \Delta S_{\text{obj}} = -C_o T_R \ln \frac{T_R}{T_0}.$$

By combining this with the above the expression for W, the maximum amount of work obtainable is[4]

$$W_{\text{max}} = C_o \left[(T_0 - T_R) + T_R \ln \frac{T_R}{T_0} \right].$$

Since the mass M of one liter of water is 1.0 kg, its heat capacity is $C_0 = Mc_P^M = 4.184\,\text{kJ}\,\text{K}^{-1}$. (Presumably the water is cooled at atmospheric pressure with negligible

[4] It is left as a problem to show that the same result can be obtained by using Carnot engines to transfer heat from the system to the reservoir.

change in volume.) By converting temperatures (100 and 20 °C) to $T_0 = 373\,\mathrm{K}$ and $T_R = 293\,\mathrm{K}$, the maximum amount of work obtainable from one liter of water by cooling it by 80 °C is

$$W_{\max} = (4.184\,\mathrm{kJ\,K^{-1}})\left[(373\,\mathrm{K} - 293\,\mathrm{K}) + 293\,\mathrm{K}\ln\frac{293}{373}\right] = 39\,\mathrm{kJ}.$$

To put this result into perspective, we mention that 39 kJ is sufficient energy to operate a 100 W electrical appliance for 390 s (6.5 min). This implies that by allowing one liter of boiling water cool to room temperature, we have caused 39 kJ of energy that was available for doing useful work to become unavailable. In other words, 39 kJ of energy has been wasted. For comparison, this is less than the energy needed to heat the water from 20 to 100 °C, which is $(4.184\,\mathrm{kJ\,K^{-1}})(80\,\mathrm{K}) = 335\,\mathrm{kJ}$.

10.6 The Third Law

Because of the T in the denominator of $đQ/T$, we should expect entropy to behave in a special way as T approaches absolute zero. That behavior is the subject of the third law. The principle underlying the third law was proposed in 1905 by Walther Nernst, and is sometimes called *Nernst's heat theorem*. It is elegantly expressed in the *Planck statement* of the third law:

> ***Third Law of Thermodynamics***
> *The entropy of a system is zero at absolute zero.*

The essential idea of the third law is that a system's entropy becomes independent of its thermodynamic coordinates other than temperature as absolute zero is approached. The definition of entropy in (10.8) contains an arbitrary reference state o and an arbitrary constant S_o. The third law tells us that we can use absolute zero as the reference state and set S_o equal to zero.

To find the implications of the third law, we consider a system with two thermodynamic coordinates. Let temperature T be one of them and for generality represent the other by X. Entropy is then given by a function $S(T, X)$. For PVT systems, we could let X be V or P. Adiabatic demagnetization experiments have been especially important in the study of low temperature phenomena, in which case we would let X represent the strength of the magnetic field $\mathcal{H}$.

Consider a reversible process that takes the system from state b at temperature T_b to state 0 at absolute zero while X is held constant. The differential for heat at constant X is $đQ = C_X dT$, where $C_X = (đQ/dT)_X$ is the heat capacity. When the line considered represents a constant X process in TX-space, the line integral simplifies to an ordinary one dimensional integral, so that the entropy change in (10.7) becomes

$$\Delta S = S(0, X) - S(T_b, X) = \int_{L_{b\to 0}} \frac{đQ}{T} = \int_{T_b}^{0} \frac{C_X}{T} dT = -\int_0^{T_b} \frac{C_X}{T} dT. \tag{10.42}$$

The third law tells us that we can set $S(0,X) = 0$, so that

$$S(T_b, X) = \int_0^{T_b} \frac{C_X}{T} dT. \tag{10.43}$$

The convergence of this integral requires that: *Heat capacities tend to zero as absolute zero is approached.* When a system has more than two thermodynamic coordinates, X can be interpreted as an abbreviation for all of the coordinates except T.

Another consequence of the third law is the concept of *absolute entropy*. The arbitrariness in the definition of entropy represented by the additive constant S_o in (10.8) is removed by using absolute zero as the reference state and setting $S_o = 0$. The resulting definition of the entropy in any other state a is

$$\boxed{S_b = \int_{L_{o \to a}} \frac{đQ}{T}.} \tag{10.44}$$

10.7 Unattainability of Absolute Zero

The unattainability statement is another formulation of the third law: *It is impossible to reduce the temperature of a system to absolute zero in any finite number of operations.*

The relationship of the two statements of the third law can be seen by imaging how the temperature of a paramagnetic system might be lowered. To illustrate the procedure, we represent the dependence of its entropy on T and $\mathcal{H}$ near absolute zero by

$$S(T, \mathcal{H}) = g(\mathcal{H}) + f(\mathcal{H}) T^{\gamma}, \tag{10.45}$$

where the functions $g(\mathcal{H})$and $f(\mathcal{H})$ are independent of temperature and γ is a constant. To prevent the entropy from diverging at absolute zero, we assume that $\gamma > 0$. The constant-field heat capacity $C_{\mathcal{H}}(T)$ is

$$C_{\mathcal{H}}(T) = T\left(\frac{\partial S}{\partial T}\right)_{\mathcal{H}} = f(\mathcal{H}) \gamma T^{\gamma}. \tag{10.46}$$

To make $C_{\mathcal{H}}(T)$ positive, we assume that $f(\mathcal{H}) > 0$. Since the Planck statement of the third law specifies that the entropy $S(T, \mathcal{H})$ becomes independent of $\mathcal{H}$ as T approaches zero, the Planck statement implies that $g(\mathcal{H})$ vanishes, so that $S(T, \mathcal{H}) = f(\mathcal{H}) T^{\gamma}$.

The two diagrams in **Figure 10.6** illustrate the implications of the unattainability statement of the third law. The solid lines represent $S(T, \mathcal{H})$ at two values of the magnetic field $\mathcal{H}$ when $\gamma = 1$. The locations of the constant field lines for $\mathcal{H}$=0 and $\mathcal{H}$>0 indicate that the entropy decreases as the magnetic field increases, which is characteristic of paramagnetic systems. The dotted lines represent reversible processes. The vertical lines (constant T lines) represent isothermal processes. Since $đQ = TdS$ implies that $đQ = 0$ when $dS = 0$, the horizontal lines (constant S lines) represent adiabatic processes.

In the diagram at the left the term $g(\mathcal{H})$ depends on the strength of the magnetic field. If this were the case, absolute zero could be reached in two operations. We could divide a

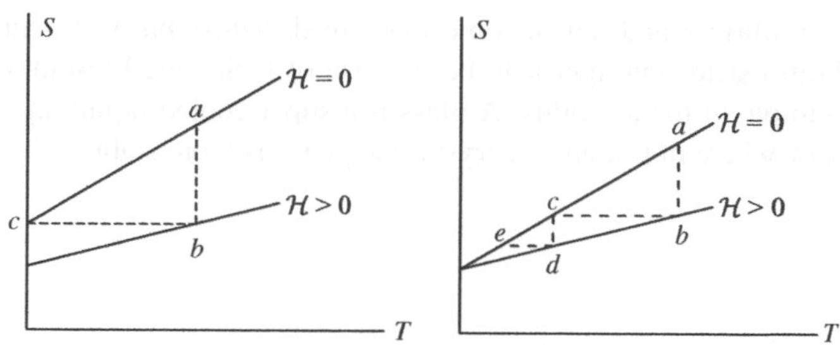

Figure 10.6 *Entropy Near Absolute Zero*

sample of paramagnetic material into two parts, a small part and a much larger part. The small part is the system being cooled, while the larger part becomes a heat reservoir. Both parts are initially at temperature T_a and field $\mathcal{H} = 0$. In the first operation, the field on the small system is slowly increased while the field on the reservoir is maintained at zero. This causes the system to evolve isothermally to state b. In the next operation, the system is isolated from the reservoir and the field is slowly brought back to zero. This causes the system to evolve adiabatically to state c. As a result, we could place state c at absolute zero by appropriately choosing the strength of the non-zero field. Since this would violate the unattainability statement of the the law, the assumption that the temperature independent term in (10.45) depends on the magnetic field is invalid.

In the diagram at the right the term $g(\mathcal{H})$ is assumed to be constant, so that $g(\mathcal{H})$ can be replaced by a constant g_o. Equation (10.45) then becomes $S(T,\mathcal{H}) = g_o + f(\mathcal{H})T^\gamma$, which causes the constant field lines to converge at $T = 0$. As a result, the temperature changes in adiabatic processes ($b \to c$, $d \to e$, etc.) become less and less as the starting temperatures (T_b, T_d, etc.) become smaller. Specifically, the entropy change between a state with $\mathcal{H}>0$ and a state with $\mathcal{H}=0$ is

$$\Delta S = f(0)T_f{}^\gamma - f(\mathcal{H})T_0{}^\gamma, \tag{10.47}$$

and $\Delta S = 0$ because the process is adiabatic and reversible. Solving for the final temperature T_f gives

$$T_f = T_0\left[\frac{f(\mathcal{H})}{f(0)}\right]^{1/\gamma}. \tag{10.48}$$

Since the entropy decreases with increasing magnetic field *i.e.*, $f(\mathcal{H}) < f(0)$, the final temperature T_f is less than the initial temperature T_0, but it is *not* zero. Consequently, we cannot reach $T = 0$ in a finite number of operations (processes). Hence, the unattainability statement is obeyed when the term $g(\mathcal{H})$ in (10.45) is a constant. By setting the constant equal to zero, the entropy becomes $S(T,\mathcal{H}) = f(\mathcal{H})T^\gamma$, which is the same as the result implied by the Planck statement.

Like all scientific laws, the third law summarizes the properties of the real world and is subject to experimental verification. It is generally accepted to be valid. Nevertheless, because processes at very low temperatures are typically very sluggish, experiments to verify the third law are not simple. When applying the law, it is essential that the system be in thermal equilibrium and not in some long lived *metastable state*. Such states are not true equilibrium states but appear to be because of their long life-times. Glasses are an important example of metastability. A glass is a super cooled liquid, *i.e.*, a liquid at a temperature below where in principle a crystalline phase is more stable.

Problems

10.1 Use the entropy principle to derive the Clausius statement of the second law.

10.2 Find the entropy change that occurs when 2.0 kg of water at 90 °C is mixed with 3.0 kg of water at 20 °C.

10.3 A 0.50 kg piece of iron at 100 °C is dropped into a bucket containing a mixture of ice and water. Some ice remains after the iron, ice, and water come to equilibrium. Find the total change in entropy.

10.4 It is possible to obtain super-cooled water, which is water below its normal freezing temperature by carefully cooling very pure water. Super-cooled water is in a meta-stable state which without warning can spontaneously change to an ice-water mixture in a true equilibrium state. The super-cooling process is reversible, but the spontaneous change is not. What is the change in entropy when 1.0 g of water at −8.0 °C spontaneously changes to an ice-water mixture?

10.5 Find the work obtainable by using a sequence of Carnot engines to cool an object initially at temperature T_0 to the temperature T_R of a heat reservoir. Assume the volume and heat capacity of the object are constant. (Compare with Example 10.3.)

10.6 The oceans contain a surface layer of water that is warmer than the water beneath it. It has been proposed that useful work (electrical energy) can be obtained by cooling the surface layer to the temperature of the water deeper down. Assume that the temperature of the surface layer is $T_0 = 285$ K (12 °C) and the temperature of colder water deeper down (the reservoir) is $T_R = 275$ K (2 °C). Estimate the thickness of the surface layer that must be cooled to supply the energy used (useful work) in the United States in one year. Use Example 10.3 or the result of the previous problem. It is interesting to compare with the answer to Problem 7.1. Data: oceans cover 71% of the surface of earth; radius of earth, 6370 km; density of sea water, 1025 kg m^{-3}; specific heat of sea water, $c_P^M = 3990\,\mathrm{J\,kg^{-1}\,K^{-1}}$. The USA uses approximately 9×10^{16} Btu annually.

Appendix 10.A

Entropy Statement of the Second Law

The entropy statement of the second law was obtained in Section 10.3 from the Kelvin statement by following the logic of the Clausius-Kelvin formulation of thermodynamics. An alternate formulation uses the entropy statement as the fundamental expression of the physical principle expressed by the law.[5] As given in Section 10.3, the entropy statement of the second law asserts that:

1. *All thermally interacting systems in thermal equilibrium possess an entropy state function that is extensive and increases as internal energy increases when no work is done.*
2. *The entropy of an adiabatically isolated system never decreases.*

The purpose of this appendix is to demonstrate the physical equivalence of the two formulations of thermodynamics by showing that the entropy statement of the second law is independent of the earlier Kelvin statement (or Clausius statement). We do this by deducing the Kelvin statement from the entropy statement. To illustrate the ideas involved, we consider *PVT* systems and begin with internal energy U and volume V as thermodynamic coordinates, so that the system's entropy is given by a function $S(U, V)$. As no work is done when volume is constant, the assertion that entropy increases as internal energy increases when no work is done implies that $(\partial S/\partial U)_V > 0$ and allows us to re-express the relationship $S = S(U, V)$ as $U = U(S, V)$. We begin by deriving the relationships $T = (\partial U/\partial S)_V$ and $dS = đQ/T$ from the entropy statement without using any results obtained from the earlier statements of the law.

Temperature

As the entropy statement of the law does not presume the existence of a state function with the characteristics of absolute temperature, we need to verify that such a function exists. As summarized in Section 7.1, temperature is an intensive state function with the characteristics (i) that systems in thermal contact have the same temperature when in equilibrium with each other, and (ii) that systems with different initial temperatures exchange heat when brought into thermal contact. We start by observing that $(\partial U/\partial S)_V$ is intensive: Both U and S are extensive, and dividing one extensive variable by another yields an intensive variable.

To find the significance of $(\partial U/\partial S)_V$ for systems in thermal contact, we consider two systems identified by subscripts 1 and 2. So that changes in their internal energies can be associated with heat transfers, we consider processes in which no work is done and assume that no heat is exchanged with other systems. These restrictions imply that the combination of systems is adiabatically isolated and that the volumes V_1 and V_2 are constant. As no work is done, the total internal energy $U_1 + U_2$ is constant, which implies that $\Delta U_1 = -\Delta U_2$. The initial entropy of the combination is

$$S_0^{\text{tot}} = S_1(U_{1,0}, V_{1,0}) + S_2(U_{2,0}, V_{2,0}), \tag{10.A.1}$$

where $U_{1,0}$, $V_{1,0}$, $U_{2,0}$, and $V_{2,0}$ represent U_1, V_1, U_2, and V_2 in the initial state. The entropies of the other states in the processes considered are

$$S_f^{\text{tot}} = S_1(U_{1,0} + \Delta U_1, V_{1,0}) + S_2(U_{2,0} - \Delta U_1, V_{2,0}). \tag{10.A.2}$$

[5] The alternate formulation was developed by Lazlo Tisza, *Generalized Thermodynamics*, (M.I.T. Press, Cambridge, 1966) and is the basis of the text by Callen H.B. (1985) *Thermodynamics and an Introduction to Thermostatistics*, 2nd edn, John Wiley & Sons, Inc., New York.

We now assume that the individual systems are initially adiabatically isolated and see what happens when they are brought into thermal contact. Consider the possibility that a small amount of energy ΔU_1 is transferred. Then, by expanding the change in entropy in powers of ΔU_1 and neglecting higher order terms, we obtain

$$S_f^{\text{tot}} = S_0^{\text{tot}} + \left[\left(\frac{\partial S_1}{\partial U_1}\right)_{V_1} - \left(\frac{\partial S_2}{\partial U_2}\right)_{V_2}\right]_0 \Delta U_1. \tag{10.A.3}$$

where the derivatives are evaluated in the initial state.

As mentioned in Chapter 10, it is implicit in the assertion that "the entropy of an adiabatically isolated system never decreases" that entropy will increase if it can. This implies that a spontaneous exchange of heat will occur whenever it causes the entropy to increase. As spontaneous changes do not occur when systems are in equilibrium, the entropy in the initial state must be at a maximum when the systems are initially in equilibrium with each other, which implies that entropy S_f^{tot} cannot be greater than S_0^{tot}, which implies that

$$\left(\frac{\partial S_1}{\partial U_1}\right)_{V_1} - \left(\frac{\partial S_2}{\partial U_2}\right)_{V_2} = 0. \tag{10.A.4}$$

By using the mathematical relationship $(\partial X/\partial Y)_Z = 1/(\partial Y/\partial X)_Z$, we find that

$$\left(\frac{\partial U_1}{\partial S_1}\right)_{V_1} = \left(\frac{\partial U_2}{\partial S_2}\right)_{V_2}. \tag{10.A.5}$$

Thus, two systems in thermal contact have the same value of $(\partial U/\partial S)_V$ when in equilibrium with each other.

The direction of the heat transfer when the systems are not in equilibrium is found by observing that positive and negative values of ΔU_1 indicate heat transfers *to* system 1 and system 2, respectively. It follows from (10.A.3) that the entropy will increase ($S_f^{\text{tot}} > S_0^{\text{tot}}$) when both ΔU_1 and the quantity in brackets are positive. As entropy will increase if it can, it follows that heat will be transferred to system 1 when $(\partial S_1/\partial U_1)_{V_1} > (\partial S_2/\partial U_2)_{V_2}$, which can be re-expressed as

$$\left(\frac{\partial U_1}{\partial S_1}\right)_{V_1} < \left(\frac{\partial U_2}{\partial S_2}\right)_{V_2}. \tag{10.A.6}$$

Thus, heat is transferred to the system with smaller values of $(\partial U/\partial S)_V$. Similarly, the entropy will increase when both ΔU_1 and the term in brackets are negative, which leads to the inequality

$$\left(\frac{\partial U_1}{\partial S_1}\right)_{V_1} > \left(\frac{\partial U_2}{\partial S_2}\right)_{V_2}. \tag{10.A.7}$$

As $\Delta U_1 < 0$ indicates a heat transfer *to* system 2, we again find that heat is transferred to the system with smaller value of $(\partial U/\partial S)_V$. Since non-zero values of ΔU_1 lead to an increase in the entropy when the initial values of $(\partial U/\partial S)_V$ are not the same, we obtain the second characteristic of temperature, *i.e.*, that systems with different initial values of $(\partial U/\partial S)_V$ exchange heat when brought into thermal contact. Furthermore, the heat transfers are from the system with the larger value of $(\partial U/\partial S)_V$ to the system with the smaller value.

Also, $(\partial U/\partial S)_V$ is positive, which follows from $(\partial S/\partial U)_V > 0$. Before we can show that $(\partial U/\partial S)_V$ is an absolute temperature, an expression for the differential for heat is needed.

Differential for Heat

To find the differential $đQ$, we consider processes represented by lines in state space, which are reversible and have characteristics described by differentials. We consider processes in adiabatically isolated systems and use S and V as thermodynamic coordinates. As the direction of travel along a line in state space can be reversed, the entropy is constant in adiabatic processes, which implies that $dS = 0$; otherwise the entropy would decrease in one of the directions. The differential of energy is

$$dU = \left(\frac{\partial U}{\partial S}\right)_V dS + \left(\frac{\partial U}{\partial V}\right)_S dV. \tag{10.A.8}$$

When $dS = 0$, this becomes $dU = (\partial U/\partial V)_S dV$. As the differential for heat vanishes for adiabatic processes, the relationship $đQ = dU + PdV$ indicates that

$$dU = -PdV. \tag{10.A.9}$$

When considering all kinds of reversible processes, the first law specifies that

$$đQ = dU + PdV = \left(\frac{\partial U}{\partial S}\right)_V dS + \left(\frac{\partial U}{\partial V}\right)_S dV + PdV. \tag{10.A.10}$$

Since the final two terms cancel because of (10.A.9), the general expression for the differential for heat is

$$đQ = (\partial U/\partial S)_V dS. \tag{10.A.11}$$

Absolute Temperature

We now show that $(\partial U/\partial S)_V$ is an absolute temperature. Absolute temperatures were defined in Section 7.5 by reference to Carnot cycles. (The definition of absolute temperatures does not depend on Carnot's theorem, which used the Clausius statement of the second law.) As diagrammed in **Figure 10.A.1**, a Carnot cycle is a reversible cyclic process made up of two isothermal processes and two adiabatic processes. Heat $|Q_1^C|$ is

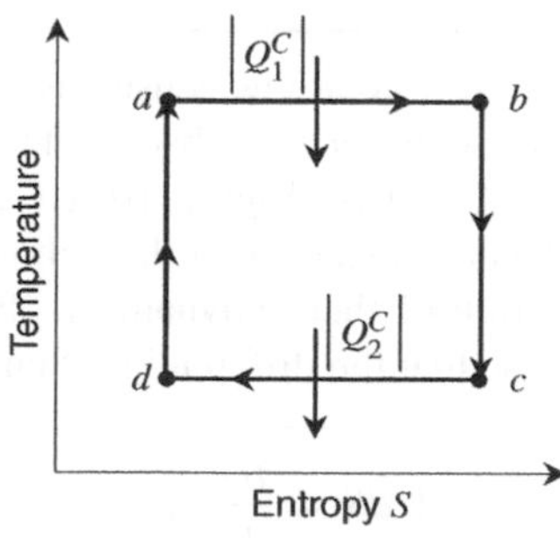

Figure 10.A.1 Temperature-Entropy Diagram of a Carnot Cycle

transferred *to* the Carnot cycle (engine) *from* heat reservoir 1 at temperature $(\partial U/\partial S)_1$ in processes $a \to b$, and heat $|Q_2^C|$ is transferred *from* the engine *to* heat reservoir 2 at temperature $(\partial U/\partial S)_2$ in processes $c \to d$. As entropy is constant in reversible adiabatic processes, the constant entropy processes $b \to c$ and $d \to a$ are adiabatic processes.

Absolute temperature scales are defined by the requirement that

$$\frac{T_1}{T_2} = \frac{|Q_1^C|}{|Q_2^C|}, \tag{10.A.12}$$

where $|Q_1^C|$ and $|Q_2^C|$ are the heats exchanged by a Carnot cycle operating between two heat reservoirs. To show that $(\partial U/\partial S)_1$ and $(\partial U/\partial S)_2$ are absolute temperatures, we must show that their ratio equals the ratio of the heats exchanged. Then, following the convention that only absolute temperatures are represented by T, the derivatives $(\partial U/\partial S)_1$ and $(\partial U/\partial S)_2$ can be represented by T_1 and T_2.

Since $đQ = (\partial U/\partial S)_V dS$ and $(\partial U/\partial S)_1$ is constant, the heat transferred during process $a \to b$ is

$$Q_1^C = \int_{a\to b} đQ = (\partial U/\partial S)_1 \int_{a\to b} dS = (\partial U/\partial S)_1 (S_b - S_a). \tag{10.A.13}$$

Similarly, the heat transferred in process $c \to d$ is $Q_2^C = (\partial U/\partial S)_2 (S_d - S_c)$. Since $S_b > S_a$ and $S_c > S_d$, it follows that Q_1^C is positive and Q_2^C is negative. When absolute values are taken, we find that

$$\frac{|Q_1^C|}{|Q_2^C|} = \frac{(\partial U/\partial S)_1 (S_b - S_a)}{(\partial U/\partial S)_2 (S_c - S_d)}. \tag{10.A.14}$$

Because processes $b \to c$ and $d \to a$ are the constant entropy (adiabatic) processes, it follows that $S_b = S_c$ and $S_d = S_a$, so that $S_b - S_a = S_c - S_d$. Using this in (10.A.14) gives

$$\frac{|Q_1^C|}{|Q_2^C|} = \frac{(\partial U/\partial S)_1}{(\partial U/\partial S)_2}. \tag{10.A.15}$$

Since the above result is valid for any two heat reservoirs, we conclude that *the temperature* $(\partial U/\partial S)_V$ *is an absolute temperature*. Thus,

$$\boxed{T = \left(\frac{\partial U}{\partial S}\right)_V.} \tag{10.A.16}$$

When the entropy statement of the second law is used as the fundamental expression of the law, this equation becomes the definition of absolute temperature T. In this approach to thermodynamics, the analysis based on Figure 10.A.1 is the verification that using (10.A.16) as the definition of T is consistent with the definition of absolute temperature in the Clausius–Kelvin formulation of thermodynamics. By substituting (10.A.16) into (10.A.11), we obtain $đQ = TdS$, so that the differential of entropy is

$$\boxed{dS = \frac{đQ}{T}.} \tag{10.A.17}$$

Kelvin Statement

The entropy statement of the second law was found in Section 10.3, where its derivation made use of the second law of thermodynamics as expressed in the Kelvin statement. If the Kelvin statement can be derived as a consequence of the entropy statement (which it can), any results that follow from the Kelvin statement can also be obtained from the entropy statement. The two statements then become alternate expressions of the physical content of the second law.

To derive the Kelvin statement from the entropy statement, we analyze a hypothetical process that violates the Kelvin statement by using the result $dS = đQ/T$. Specifically, we consider a process whose only effect is to accept heat *from* a single reservoir and transform it entirely into work. As the temperatures of heat reservoirs are constant, the entropy change of the reservoir is

$$\Delta S_{\text{res}} = \int dS = \int \left(\frac{đQ}{T}\right) = \left(\frac{1}{T}\right) \int đQ = \frac{Q}{T}. \tag{10.A.18}$$

Since a transfer of heat *from* the reservoir is indicated by a negative value of Q, the entropy change of the reservoir is negative ($\Delta S_{res} < 0$). Since the only effect of the process is to transform heat into work, the device that transforms the heat into work is left unchanged. Since the entropy of the reservoir will have decreased, the entropy of an adiabatically isolated system of reservoir and device will have decreased. Such a decrease violates the entropy statement of the second law: Hence, any process whose only effect is to accept heat from a single reservoir and transform it entirely into work is impossible, which is precisely what the Kelvin statement asserts.

11

Consequences of Existence of Entropy

The existence of entropy was demonstrated in the previous chapter. We now ask how a system's entropy can be determined from measurable properties. In PVT systems the measurable properties are the heat capacity and equation of state. An important consequence of the existence of the entropy state function is the exact general relationships that it establishes between the properties of systems. We also find that the uniformity of temperature and pressure in systems in equilibrium can be understood as a consequence of the tendency of entropy to increase.

11.1 Differentials of Entropy and Energy

The differentials of entropy and internal energy establish a relationship between the internal energy and pressure that leads to a useful relationship between a system's entropy and its heat capacity and the equation of state. When T and V are used as the thermodynamic coordinates of a PVT system, the differential of internal energy is

$$dU = \left(\frac{\partial U}{\partial T}\right)_V dT + \left(\frac{\partial U}{\partial V}\right)_T dV. \tag{11.1}$$

Combining this with $đQ = dU + PdV$ implies that

$$đQ = \left(\frac{\partial U}{\partial T}\right)_V dT + \left[\left(\frac{\partial U}{\partial V}\right)_T + P\right] dV. \tag{11.2}$$

Since $dS = đQ/T$, the differential of entropy is

$$dS = \frac{đQ}{T} = \frac{1}{T}\left(\frac{\partial U}{\partial T}\right)_V dT + \frac{1}{T}\left[\left(\frac{\partial U}{\partial V}\right)_T + P\right] dV. \tag{11.3}$$

Thermodynamics and Statistical Mechanics: An Integrated Approach, First Edition.
Robert J Hardy and Christian Binek.

Since dS is an exact differential, it follows from the equality of cross derivatives (see (9.32)) that

$$\frac{\partial}{\partial V}\left\{\frac{1}{T}\left(\frac{\partial U}{\partial T}\right)_V\right\} = \frac{\partial}{\partial T}\left\{\frac{1}{T}\left[\left(\frac{\partial U}{\partial V}\right)_T + P\right]\right\}. \tag{11.4}$$

Carrying out the indicated differentiations gives

$$\frac{1}{T}\frac{\partial^2 U}{\partial V \partial T} = \frac{-1}{T^2}\left[\left(\frac{\partial U}{\partial V}\right)_T + P\right] + \frac{1}{T}\left[\frac{\partial^2 U}{\partial T \partial V} + \left(\frac{\partial P}{\partial T}\right)_V\right]. \tag{11.5}$$

As partial derivatives are independent of the order of differentiation, the terms containing second derivatives cancel. By rearranging terms and multiplying through by T^2, we find that

$$\left(\frac{\partial U}{\partial V}\right)_T = T\left(\frac{\partial P}{\partial T}\right)_V - P. \tag{11.6}$$

By using $C_V = (\partial U/\partial T)_V$ and the above result, the differential of internal energy becomes

$$dU = C_V dT + \left[T\left(\frac{\partial P}{\partial T}\right) - P\right] dV. \tag{11.7}$$

Combining this with $đQ = dU + PdV$ and $dS = đQ/T$ gives

$$dS = \frac{1}{T}\left\{C_V dT + \left[T\left(\frac{\partial P}{\partial T}\right)_V - P\right] dV\right\} + \frac{PdV}{T}, \tag{11.8}$$

which simplifies to

$$dS = \frac{C_V}{T} dT + \left(\frac{\partial P}{\partial T}\right)_V dV. \tag{11.9}$$

Equations (11.7) and (11.9) can be used to determine the internal energy and entropy from the heat capacity and equation of state. This is significant, because the experimental determination of $P(T, V)$ and $C_V(T, V)$ is relatively straightforward.

Since dS is the differential of a function, we have

$$dS = \left(\frac{\partial S}{\partial T}\right)_V dT + \left(\frac{\partial S}{\partial V}\right)_T dV. \tag{11.10}$$

From this and (11.9) it follows that

$$\left(\frac{\partial S}{\partial V}\right)_T = \left(\frac{\partial P}{\partial T}\right)_V \quad \text{and} \quad \left(\frac{\partial S}{\partial T}\right)_V = \frac{C_V}{T}. \tag{11.11}$$

Thus,

$$C_V = T\left(\frac{\partial S}{\partial T}\right)_V. \tag{11.12}$$

It is left as a problem to show that C_P equals $T(\partial S/\partial T)_P$. The definitions of the constant-volume and constant-pressure heat capacities and the formulas for them are summarized in Table 11.1.

Table 11.1 Heat capacities for PVT systems

Constant-volume	$C_V = \lim_{Q\to 0} \frac{Q}{\Delta T}\Big\vert_{V=\text{const}} = \left(\frac{\partial U}{\partial T}\right)_V = \left(\frac{đQ}{dT}\right)_V = T\left(\frac{\partial S}{\partial T}\right)_V$
Constant-pressure	$C_P = \lim_{Q\to 0} \frac{Q}{\Delta T}\Big\vert_{P=\text{const}} = \left(\frac{\partial H}{\partial T}\right)_P = \left(\frac{đQ}{dT}\right)_P = T\left(\frac{\partial S}{\partial T}\right)_P$

Adiabatic Bulk Modulus

The differential for heat $đQ$ describes the transfer of heat in reversible processes. When $đQ$ is zero, $dS = đQ/T$ is also zero, which implies that the entropy is constant. Since no heat is exchanged when $đQ$ is zero, it follows that *reversible adiabatic processes are isentropic processes, i.e.*, constant entropy processes. When matter is compressed adiabatically, the change in pressure caused by a change in volume is given by the *adiabatic bulk modulus*

$$B_S = -V\left(\frac{\partial P}{\partial V}\right)_S, \tag{11.13}$$

where the entropy is fixed. Similarly, the change in volume caused by a change in pressure is given by the *adiabatic compressibility*,

$$K_S = -\frac{1}{V}\left(\frac{\partial V}{\partial P}\right)_S = \frac{1}{B_S}. \tag{11.14}$$

A common example of an isentropic process is the rapid compression or expansion of a gas, liquid, or solid. When the change in volume is sufficiently rapid, there is insufficient time for a significant exchange of heat to occur.

11.2 Ideal Gases

Internal Energy

An ideal gas was defined in Chapter 1 as a *PVT* system for which

$$PV = nRT \quad \text{and} \quad U = U(T). \tag{11.15}$$

It was stated without proof that the second of these relationships is a consequence of the first, a statement that is easily proved now that we have a relationship between U and P: It follows from $P = nRT/V$ that

$$(\partial P/\partial T)_V = nR/V. \tag{11.16}$$

Substituting this into (11.6) gives

$$\left(\frac{\partial U}{\partial V}\right)_T = T\left(\frac{\partial P}{\partial T}\right)_V - P = T\frac{nR}{V} - \frac{nRT}{V} = 0. \tag{11.17}$$

The vanishing of the derivative $(\partial U/\partial V)_T$ implies that U is independent of V. Hence, the internal energy depends on T only, which is summarized by writing $U = U(T)$.

When U is independent of volume, the heat capacity $C_V = (\partial U/\partial T)_V$ becomes

$$C_V(T) = \frac{dU(T)}{dT}. \tag{11.18}$$

It follows from this that the internal energy function can be obtained from $C_V(T)$ by ordinary one-dimensional integration. In many situations the dependence of C_V on T is sufficiently weak that it can be treated as a constant.

Entropy

Explicit expressions for the entropy of an ideal gas with constant C_V are easily found by integrating the differential dS using the procedures outlined in Section 9.4. By introducing the specific heat $c_V = C_V/n$ and using $(\partial P/\partial T)_V = nR/V$, the expression for dS in (11.9) becomes

$$dS = \frac{nc_V}{T}\,dT + \frac{nR}{V}\,dV. \tag{11.19}$$

Since c_V is constant, checking the equality of cross derivatives gives

$$\left(\frac{\partial\left(nc_V/T\right)}{\partial V}\right)_T \stackrel{?}{=} \left(\frac{\partial\left(nR/V\right)}{\partial T}\right)_V, \tag{11.20}$$

which reduces to $0 = 0$. Hence, the differential in (11.19) is indeed exact. The indefinite integral of the coefficient of dT in (11.19) is

$$S(T, V) = \int \frac{nc_V}{T}\,dT = nc_V \ln T + g(V) + h. \tag{11.21}$$

The indefinite integral of the coefficient of dV is

$$S(T, V) = \int \frac{nR}{V}\,dV = nR \ln V + f(T) + k. \tag{11.22}$$

Comparing these expressions indicates that $g(V) = nR \ln V$ and $f(T) = C_V \ln T$. Thus, the entropy of an ideal gas with constant specific heat is

$$\boxed{S(T, V) = nc_V \ln T + nR \ln V + k.} \tag{11.23}$$

To check this result, we find its derivatives with respect to T and V,

$$\left(\frac{\partial S}{\partial T}\right)_V = \frac{nc_V}{T} \quad \text{and} \quad \left(\frac{\partial S}{\partial V}\right)_T = \frac{nR}{V}, \tag{11.24}$$

which are the coefficients of dT and dV in (11.19), as they should be.

Now that we have an entropy state function, we can find the change in entropy by calculating the difference between its initial and final values. This is illustrated by the free expansion of the gas. As described in Section 2.4, a free expansion experiment is performed by opening a valve which allows the gas in one container to expand into a second container that was initially evacuated. No heat is exchanged in the process and no work done, so that there is no change in the internal energy. By using (11.23), the entropy change in an expansion from volume V_0 to volume V_f is

$$\Delta S = S(T_f, V_f) - S(T_0, V_0)$$
$$= [nc_V \ln T_f + nR \ln V_f + k] - [nc_V \ln T_0 + nR \ln V_0 + k]. \qquad (11.25)$$

Since the internal energy of an ideal gas is a function of temperature only, the final temperature is the same as the initial temperature ($T_f = T_0$), so that

$$\Delta S = nR \ln(V_f/V_0). \qquad (11.26)$$

Since the volume has increased ($V_f > V_0$), the entropy has increased, which tells us that the process is irreversible.

The expression for $S(T, V)$ in (11.23) has two shortcomings: (i) An expression for entropy should be conspicuously extensive. (ii) The arguments of functions other than powers (such as $T^m V^n$) should be dimensionless and intensive. These shortcomings are overcome by the choice of the integration constant k. We select a reference state o and specify the temperature T_o, volume V_o, and entropy per mole s_o in that state. Since $s_o = S(T_o, V_o)/n$, it follows that $ns_o = nc_V \ln T_o + nR \ln V_o + k$, so that

$$k = -nc_V \ln T_o - nR \ln V_o + ns_o. \qquad (11.27)$$

Substituting this into (11.23), and rearranging terms gives

$$\boxed{S(T, V) = nc_V \ln \frac{T}{T_o} + nR \ln \frac{V}{V_o} + ns_o.} \qquad (11.28)$$

The entropy of an ideal gas as a function of T and P is obtained by using the equation of state $V = nRT/P$ to replace the V in (11.28). After taking the logarithm of nRT/P, we obtain

$$S(T, P) = nc_V \ln(T/T_o) + nR(\ln nR + \ln T - \ln P - \ln V_o) + ns_o. \qquad (11.29)$$

The two terms containing $\ln T$ can be combined by using the relationship $c_P = c_V + R$ from (3.14). By introducing the pressure $P_o = nRT_o/V_o$ in the state o, we obtain

$$\boxed{S(T, P) = nc_P \ln \frac{T}{T_o} - nR \ln \frac{P}{P_o} + ns_o.} \qquad (11.30)$$

Entropy as a function of P and V is obtained by writing the equation of state as $T = PV/nR$ and using it to replace the T in (11.28), which gives

$$S(P, V) = nc_V(\ln P + \ln V - \ln nR - \ln T_o) + nR \ln\left(\frac{V}{V_o}\right) + ns_o. \qquad (11.31)$$

By using $P_o = nRT_o/V_o$ and $c_P = c_V + R$, it follows that

$$\boxed{S(P, V) = nc_V \ln \frac{P}{P_o} + nc_P \ln \frac{V}{V_o} + ns_o.} \qquad (11.32)$$

Notice that the above expressions for $S(T, V)$ and $S(T, P)$ are not consistent with the third law, because they do not yield values independent of V and P as T goes to zero. Instead, they diverge at $T = 0$. This inconsistency is the result of extending our model beyond its range of validity. The above expressions were derived for systems with constant C_V, which is inconsistent with the requirement that heat capacities become zero at absolute zero.

Example 11.1 B_S for an ideal gas

Find the adiabatic bulk modulus of an ideal gas with constant c_V and compare it with the isothermal bulk modulus B_T.

Solution. By using the relationships between partial derivatives, the expression for the adiabatic bulk modulus in (11.13) becomes

$$B_S = -V\left(\frac{\partial P}{\partial V}\right)_S = V\left(\frac{\partial P}{\partial S}\right)_V\left(\frac{\partial S}{\partial V}\right)_P = V\frac{(\partial S/\partial V)_P}{(\partial S/\partial P)_V}.$$

It now follows from the expression for $S(P, V)$ in (11.32) that

$$B_S = V\frac{(nc_P/V)}{(nc_V/P)} = \frac{c_P}{c_V}P.$$

By using the ratio of specific heats $\gamma = c_P/c_V$, the adiabatic bulk modulus becomes

$$B_S = \gamma P. \tag{11.33}$$

The isothermal bulk modulus is

$$B_T = -V\left(\frac{\partial P}{\partial V}\right)_T = -V\left(\frac{-nRT}{V^2}\right) = P.$$

Since $\gamma > 1$, the adiabatic bulk modulus is greater than the isothermal bulk modulus.

Both bulk moduli describe a change in pressure caused by a change in volume. Since the transfer of heat needed to keep the temperature constant takes time, the isothermal bulk modulus is appropriate for processes that are relatively slow. The adiabatic bulk modulus is used for processes that take place sufficiently rapidly that any heat transfer is negligible. Since sound waves involve rapid changes in pressure, the speed of sound is determined by the adiabatic bulk modulus. In gases, the speed of sound is $v_{\text{snd}} = \sqrt{B_S/\rho}$, where ρ is the mass density of the gas.

11.3 Relationships Between C_V, C_P, B_T, B_S, and α_V

Five of the measurable properties of *PVT* systems are summarized in Table 11.2. These properties are not independent. The values of one heat capacity and one bulk modulus can be found from the values of the other heat capacity, the other bulk modulus, and the coefficient of thermal expansion. This has practical implications. For solids and liquids it is usually easier to measure heat capacities at constant pressure than at constant volume. Also, speed of sound measurements are often used to find the bulk modulus, which yields

Table 11.2 Properties of PVT systems

Coefficient of thermal expansion	Isothermal bulk modulus	Adiabatic bulk modulus	Constant-volume heat capacity	Constant-pressure heat capacity
$\alpha_V = \frac{1}{V}\left(\frac{\partial V}{\partial T}\right)_P$	$B_T = -V\left(\frac{\partial P}{\partial V}\right)_T$	$B_S = -V\left(\frac{\partial P}{\partial V}\right)_S$	$C_V = T\left(\frac{\partial S}{\partial T}\right)_V$	$C_P = T\left(\frac{\partial S}{\partial T}\right)_P$

adiabatic values. As an illustration, imagine that we have values for α_V, B_T, and C_V from a statistical mechanical calculation and want to compare them with experimental values, when the available values are for B_S and C_P. In this case we could use the formulas derived below to convert the data.

Two relationships between these five material properties are obtained by differentiating the composite function formed from the entropy $S(T, V)$ and the equation of state $V = V(T, P)$,

$$\overline{S}(T, P) = S(T, V(T, P)). \tag{11.34}$$

Using the chain rule to find the derivative with respect to temperature gives

$$\left(\frac{\partial S}{\partial T}\right)_P = \left(\frac{\partial S}{\partial T}\right)_V + \left(\frac{\partial S}{\partial V}\right)_T\left(\frac{\partial V}{\partial T}\right)_P. \tag{11.35}$$

Multiplying through by T and using the expressions for C_P and C_V in Table 11.2 gives

$$C_P = C_V + T\left(\frac{\partial P}{\partial T}\right)_V\left(\frac{\partial V}{\partial T}\right)_P. \tag{11.36}$$

Re-expressing $(\partial P/\partial T)_V$ as $-(\partial P/\partial V)_T(\partial V/\partial T)_P$ gives

$$C_P = C_V + T\left[-\left(\frac{\partial P}{\partial V}\right)_T\left(\frac{\partial V}{\partial T}\right)_P\right]\left(\frac{\partial V}{\partial T}\right)_P. \tag{11.37}$$

It now follows from the expressions for α_V and B_T in Table 11.2 that

$$\boxed{C_P = C_V + TVB_T{\alpha_V}^2.} \tag{11.38}$$

Differentiating (11.34) with respect to pressure gives

$$\left(\frac{\partial S}{\partial P}\right)_T = \left(\frac{\partial S}{\partial V}\right)_T\left(\frac{\partial V}{\partial P}\right)_T. \tag{11.39}$$

By using the relationships between partial derivatives, this becomes

$$\left[-\left(\frac{\partial S}{\partial T}\right)_P\left(\frac{\partial T}{\partial P}\right)_S\right] = \left[-\left(\frac{\partial S}{\partial T}\right)_V\left(\frac{\partial T}{\partial V}\right)_S\right]\left(\frac{\partial V}{\partial P}\right)_T. \tag{11.40}$$

Re-expressing $(\partial T/\partial P)_S$ as $1/(\partial T/\partial P)_S$ and rearranging terms gives

$$\frac{(\partial S/\partial T)_P}{(\partial S/\partial T)_V} = \left(\frac{\partial P}{\partial T}\right)_S\left(\frac{\partial T}{\partial V}\right)_S\left(\frac{\partial V}{\partial P}\right)_T. \tag{11.41}$$

Combining the first two derivatives on the right side and inverting $(\partial V/\partial P)_T$ gives

$$\frac{(\partial S/\partial T)_P}{(\partial S/\partial T)_V} = \left(\frac{\partial P}{\partial V}\right)_S \left(\frac{\partial V}{\partial P}\right)_T = \frac{(\partial P/\partial V)_S}{(\partial V/\partial P)_T}. \tag{11.42}$$

From this and the expressions for C_V, C_P, B_T, and B_S, it follows that

$$\boxed{\frac{C_P}{C_V} = \frac{B_S}{B_T}.} \tag{11.43}$$

Equations (11.38) and (11.43) can be combined to show that

$$\boxed{B_S = B_T + \frac{TVB_T{}^2\alpha_V{}^2}{C_V}.} \tag{11.44}$$

11.4 Clapeyron's Equation

Another consequence of the existence of entropy is our ability to predict the slopes of two-phase curves on phase diagrams. The phase diagram (*PT*-diagram) for a substance that expands upon freezing is shown in **Figure 11.1**, which is the same as Figure 5.1. When two phases coexist is equilibrium, there is one value of P for each value of T, which implies that a function $P_{2ph}(T)$ exists, so that each two-phase curve is described by an equation like

$$P = P_{2\text{ph}}(T). \tag{11.45}$$

We proceed by using the relationship between internal energy and pressure in (11.6). Substituting (11.45) into that relationship for a system with two phases present gives

$$\left(\frac{\partial U}{\partial V}\right)_T = T\frac{dP_{2\text{ph}}(T)}{dT} - P_{2\text{ph}}(T). \tag{11.46}$$

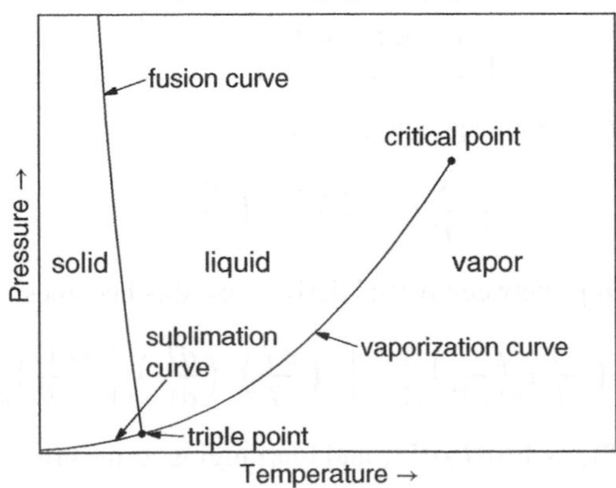

Figure 11.1 *Phase Diagram of a PVT System*

The total mass M of the system is

$$M = M_1 + M_2, \tag{11.47}$$

where the phases are identified by subscripts 1 and 2. The volume V and internal energy U are

$$V = M_1 \mathrm{v}_1(T) + M_2 \mathrm{v}_2(T) \tag{11.48}$$

and

$$U = M_1 u_1(T) + M_2 u_2(T), \tag{11.49}$$

where $\mathrm{v}_1(T)$ and $\mathrm{v}_2(T)$ are the volumes per unit mass of the phases and $u_1(T)$ and $u_2(T)$ are the internal energies per unit mass. The total mass M is fixed, so that $M_1 = M - M_2$. Hence,

$$V = (M - M_2)\mathrm{v}_1(T) + M_2 \mathrm{v}_2(T) = V(T, M_2) \tag{11.50}$$

and

$$U = (M - M_2)u_1(T) + M_2 u_2(T) = U(T, M_2), \tag{11.51}$$

where the independent variables are T and M_2. When T is held constant, the derivative of U with respect to V can be re-expressed as

$$\left(\frac{\partial U}{\partial V}\right)_T = \left(\frac{\partial U}{\partial M_2}\right)_T \left(\frac{\partial M_2}{\partial V}\right)_T = \frac{(\partial U/\partial M_2)_T}{(\partial V/\partial M_2)_T}. \tag{11.52}$$

By using (11.50) and (11.51) to evaluate the derivatives with respect to M_2, we find that

$$\left(\frac{\partial U}{\partial V}\right)_T = \frac{u_2(T) - u_1(T)}{\mathrm{v}_2(T) - \mathrm{v}_1(T)}. \tag{11.53}$$

Combining this and (11.46) gives

$$\frac{u_2(T) - u_1(T)}{\mathrm{v}_2(T) - \mathrm{v}_1(T)} = T\frac{dP_{2\text{ph}}(T)}{dT} - P_{2\text{ph}}(T), \tag{11.54}$$

which can be rearranged as

$$\frac{dP_{2\text{ph}}}{dT} = \frac{1}{T}\frac{(u_2 + P_{2\text{ph}}\mathrm{v}_2) - (u_1 + P_{2\text{ph}}\mathrm{v}_1)}{\mathrm{v}_2 - \mathrm{v}_1}. \tag{11.55}$$

It is left implicit that v_1, v_2, u_1, u_2, and $P_{2\text{ph}}$ are functions of T.

The specific enthalpy h of a phase is

$$h = \frac{H}{M} = \frac{(U + PV)}{M} = u + P\mathrm{v}. \tag{11.56}$$

Since the pressure of both phases is $P_{2\text{ph}}$, it follows that

$$(u_2 + P_{2\text{ph}}\mathrm{v}_2) - (u_1 + P_{2\text{ph}}\mathrm{v}_1) = h_2 - h_1 = L_{1\to 2}, \tag{11.57}$$

where $L_{1\to2}$ is the latent heat of the phase transition. As discussed in Section 5.2, $L_{1\to2}$ equals the difference between the specific enthalpies of the two phases. Substituting (11.57) into (11.55) gives

$$\boxed{\frac{dP_{2\text{ph}}}{dT} = \frac{1}{T}\frac{L_{1\to2}}{v_2 - v_1}.} \tag{11.58}$$

This is *Clapeyron's equation* for the slope of a two-phase coexistence curve on a phase diagram.

When subscripts 1 and 2 are the liquid and vapor phases, $L_{1\to2}$ is the latent heat of vaporization ($L_{1\to2} > 0$). Since the specific volume of the vapor is greater than that of the liquid ($v_2 > v_1$), Clapeyron's equation indicates that the slope of the vaporization curve is positive, as it is in Figure 11.1. When phases 1 and 2 are the solid and liquid phases, $L_{1\to2}$ is the latent heat of fusion, and the change in specific volume $v_2 - v_1$ becomes $v_{\text{liq}} - v_{\text{sol}}$, which can be positive or negative. Since ordinary ice (Ice I) has a higher specific volume (lower density) than liquid water, $v_{\text{liq}} - v_{\text{sol}}$ is negative, so that Clapeyron's equation indicates that the slope of fusion curve is negative, as it is in Figure 11.1. Water is unusual in that most substances have a lower specific volume (higher density) in the solid phase, so that $v_{\text{liq}} - v_{\text{sol}}$ is positive, which indicates that the slope of the fusion curve is positive. As Ice-I is the only ice that is less dense than the water, it is the only ice whose fusion curve has a negative slope, as is shown in Figure 5.2.

11.5 Maximum Entropy, Equilibrium, and Stability

Maximum Entropy

The second law establishes a maximum principle which asserts that an adiabatically isolated system in equilibrium has the maximum entropy consistent with the constraints on it. To be able to use the principle, some variability must remain in a system after the constraints are imposed. The principle is applied by expressing entropy as a function of *free* and *fixed* variables. The fixed variables are often controlled experimentally, while the free variables are determined by the maximum principle.

Equilibrium Conditions

The equilibrium and stability conditions for *PVT* systems illustrate how the maximum principle can be used to obtain useful results. To apply the maximum principle, we investigate processes in an adiabatically isolated system of fixed size and shape. Adiabatic isolation implies that no heat is transferred ($Q = 0$). Since the system's size is fixed, its volume is fixed ($\Delta V = 0$). When there is no change in volume, no work is done ($W = 0$). The first law then specifies that the internal energy does not change ($\Delta U = Q - W = 0$).

When U and V are thermodynamic coordinates, the entropy is given by a function $S(U, V, \cdots)$. When U and V are the only independent variables, the assertion that entropy is maximized has no predictive value. To obtain a function with free variables, we

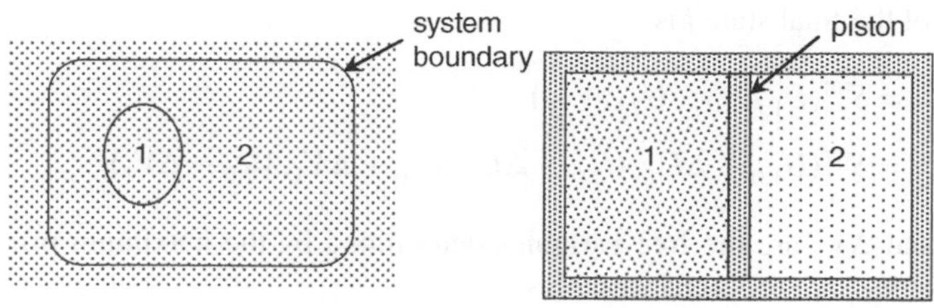

Figure 11.2 *Two-Part Systems*

divide the system into subsystems, as illustrated in **Figure 11.2**. In the first diagram the adiabatically isolated system is part of a larger piece of matter. Subsystem 1 is a region of arbitrary size within it and subsystem 2 is the remainder of the isolated system. Subsystem 1 can change size and exchange heat with subsystem 2 but can not exchange matter with it. In the second diagram the adiabatically isolated system is a fluid inside a perfectly insulating container with rigid walls. The fluid is divided into subsystems by a piston that is free to move and conduct heat. The rigid walls and free piston illustrate the type of process being considered, but are not essential. What is essential is that the internal energy and volume of the combined system are fixed, which can occur even when not externally enforced.

We can obtain conclusions of general validity by investigating what happens to the entropy as a consequence of imagined (but impossible) changes. So that equilibrium state functions can be used, we restrict ourselves to processes in which the subsystems are individually in equilibrium at the start and finish of the process and investigate what happens if they are not in equilibrium with each other.

Since entropy is extensive, the total entropy S^{tot} of the combined system is the sum of the entropies of the subsystems,

$$S^{\text{tot}}(U_1, V_1, U_2, V_2) = S_1(U_1, V_1) + S_2(U_2, V_2), \tag{11.59}$$

where U_1, U_2, V_1, and V_2 are the thermodynamics coordinates of the combined system. The values of the coordinates in initial state 0 are $(U_{1,0}, V_{1,0}, U_{2,0}, V_{2,0})$, and the entropy in state 0 is

$$S_0^{\text{tot}} = S^{\text{tot}}(U_{1,0}, V_{1,0}, U_{2,0}, V_{2,0}). \tag{11.60}$$

The internal energy and volume of the total system are sums of subsystem values,

$$U^{\text{tot}} = U_1 + U_2 \quad \text{and} \quad V^{\text{tot}} = V_1 + V_2. \tag{11.61}$$

We are interested in processes in which U^{tot} and V^{tot} are constant, so that the changes to U and V in one subsystem are the negative of the changes in the other. Hence,

$$\Delta U_1 = -\Delta U_2 \quad \text{and} \quad \Delta V_1 = -\Delta V_2, \tag{11.62}$$

where $\Delta U_1 = U_{1,f} - U_{1,0}$ and $\Delta U_2 = U_{2,f} - U_{2,0}$ and similarly for the volume changes. The hypothetical processes being considered are characterized by ΔU_1 and ΔV_1.

The entropy of the final state f is

$$\begin{aligned} S_f^{\text{tot}} &= S^{\text{tot}}(U_{1,f}, V_{1,f}, U_{2,f}, V_{2,f}) \\ &= S^{\text{tot}}(U_{1,0} + \Delta U_1, U_{2,0} - \Delta U_1, V_{1,0} + \Delta V_1, V_{2,0} - \Delta V_1), \end{aligned} \tag{11.63}$$

where ΔU_1 and ΔV_1 are the free variables determined by the maximum principle. $U_{1,0}$, $U_{2,0}$, $V_{1,0}$, and $V_{2,0}$ are the fixed variables.

To find what changes are possible, we expand the entropy of state f in a Taylor series of powers of ΔU_1 and ΔV_1,

$$S_f^{\text{tot}} = S_0^{\text{tot}} + \left[\left(\frac{\partial S_1}{\partial U_1}\right)_{V_1} - \left(\frac{\partial S_2}{\partial U_2}\right)_{V_2}\right]_0 \Delta U_1 + \left[\left(\frac{\partial S_1}{\partial V_1}\right)_{U_1} - \left(\frac{\partial S_2}{\partial V_2}\right)_{U_2}\right]_0 \Delta V_1 + \text{higher order terms.} \tag{11.64}$$

The coefficients in brackets are evaluated in state 0. We now assume that state 0 is an equilibrium state of the system. Since the entropy of an adiabatically isolated system will increase if it can, the final entropy must be less than or equal to the initial entropy ($S_f^{\text{tot}} \leq S_0^{\text{tot}}$), otherwise the system would spontaneously change, which would imply that state 0 was not an equilibrium state. Since S_0^{tot} is the maximum value of the entropy, the derivatives of S_f^{tot} with respect to ΔU_1 and ΔV_1 are zero. Since these derivatives are the coefficients of ΔU_1 and ΔV_1 in (11.64), setting them equal to zero implies that

$$\left(\frac{\partial S_1}{\partial U_1}\right)_{V_1} = \left(\frac{\partial S_2}{\partial U_2}\right)_{V_2} \quad \text{and} \quad \left(\frac{\partial S_1}{\partial V_1}\right)_{U_1} = \left(\frac{\partial S_2}{\partial V_2}\right)_{U_2}. \tag{11.65}$$

These are the basic forms of the equilibrium conditions.

By combining $dS = đQ/T$ and $đW = PdV$ with the first law $đQ = dU + đW$, we find that

$$dS = \left(\frac{1}{T}\right) dU + \left(\frac{P}{T}\right) dV. \tag{11.66}$$

Since dS is exact, it follows that

$$\left(\frac{\partial S}{\partial U}\right)_V = \frac{1}{T} \quad \text{and} \quad \left(\frac{\partial S}{\partial V}\right)_U = \frac{P}{T}. \tag{11.67}$$

By labeling the values of T and P in the subsystems with subscripts 1 and 2, the conditions in (11.65) become $1/T_1 = 1/T_2$ and $P_1/T_1 = P_2/T_2$, which imply that

$$\boxed{T_1 = T_2 \quad \text{and} \quad P_1 = P_2.} \tag{11.68}$$

These are the practical forms of the *equilibrium conditions* for *PVT* systems.

Since subsystem 1 is of arbitrary size and shape and can be located anywhere within the combined system, the equilibrium conditions imply that all parts of the system have the same temperature and pressure. In other words, *temperature and pressure are uniform throughout a system in equilibrium.* Implicit in this conclusion is the assumption that the system is a fluid and has no adiabatically isolated parts. Solids are excluded because solids can sustain an uneven distribution of pressure (stress) indefinitely. In contrast, fluids flow to conform to the shape of their container.

The above derivation treated the total entropy as a function of U_1, U_2, V_1, and V_2, all of which are extensive. When deriving equilibrium conditions for other types of systems, V_1 and V_2 are replaced by other variables. Those variables need to be extensive so that equations similar to (11.62) are satisfied.

Stability Conditions

Satisfying equilibrium conditions does not guarantee stability. The concept of stability is nicely illustrated by an example from mechanics. Imagine balancing a cube on one corner so that its center of gravity is directly above it. Since there are no unbalanced forces acting on it, the cube is in mechanical equilibrium. However, it is not stable, as any small disturbance will cause it to fall on one of its faces. In contrast, a cube that is resting face down is stable, since any small disturbance introduces forces that return it to its original face down configuration.

The stability conditions are obtained from the second order terms in the expansion in (11.64). For simplicity, we assume that the volume is constant ($\Delta V = 0$), and only consider changes that result from an exchange of heat between the subsystems. We assume that state 0 is a stable equilibrium state and expand the final entropy in powers of ΔU_1, where $\Delta U_2 = -\Delta U_1$. The expansion through second order terms is

$$S_f^{\text{tot}} = S_0^{\text{tot}} + \left[\frac{1}{T_1} - \frac{1}{T_2}\right]_0 \Delta U_1 + \frac{1}{2}\left[\left(\frac{\partial^2 S_1}{\partial {U_1}^2}\right)_{V_1} + \left(\frac{\partial^2 S_2}{\partial {U_2}^2}\right)_{V_2}\right]_0 (\Delta U_1)^2 + \cdots. \tag{11.69}$$

where $(\partial U/\partial V)_V = 1/T$ has been used. The first order term vanishes because of the equilibrium condition $T_1 = T_2$. The third and higher order terms can be neglected when ΔU_1 is small. To simplify the coefficient of $(\Delta U_1)^2$, we consider subsystems of the same size so that the two derivatives in the second order term are equal. By adding them and omitting the subscripts, we obtain

$$S_f^{\text{tot}} = S_0^{\text{tot}} + \left(\frac{\partial^2 S}{\partial U^2}\right)_V (\Delta U)^2. \tag{11.70}$$

For state 0 to be stable, transitions to a hypothetical state f must be prohibited, which requires that the entropy in state f is less than the entropy in state 0 ($S_f^{\text{tot}} < S_0^{\text{tot}}$). Since $(\Delta U)^2 > 0$, this implies that $(\partial^2 S/\partial U^2)_V < 0$. By using the relationships $C_V = (\partial U/\partial T)_V$ and $(\partial S/\partial U)_V = 1/T$, we find that

$$\left(\frac{\partial^2 S}{\partial U^2}\right)_V = \frac{\partial}{\partial U}\frac{1}{T(U,V)} = \frac{-1}{T^2}\left(\frac{\partial T}{\partial U}\right)_V = \frac{-1}{T^2(\partial U/\partial T)_V} = \frac{-1}{T^2 C_V} < 0. \tag{11.71}$$

Thus, stability requires that the constant-volume heat capacity be positive,

$$\boxed{C_V > 0.} \tag{11.72}$$

This is one of the stability conditions. Other conditions are obtained in Section 12.5.

The above stability condition, obtained as a consequence of the second law, is also intuitively reasonable. Since the heat capacity C_V equals $(\partial U/\partial T)_V$, a heat capacity that is contrary to (11.72) would have $(\partial T/\partial U)_V < 0$. However, if this was the case, any small accidental increase in the internal energy would cause the temperature to decrease, which in turn would cause additional heat to be transferred from the surroundings, which would further increase the internal energy, which would cause the temperature to become even lower, and so forth. In contrast, an accidental increase in the internal energy of a stable system causes the temperature to increase, which causes heat to be transferred from the system, which tends to return the internal energy to its original value.

11.6 Mixing

Systems that are not in equilibrium evolve irreversibly to equilibrium. Since an increase in entropy is the quantitative manifestation that irreversible behavior has occurred, we should expect the total amount of entropy to increase whenever there is irreversibility. We now investigate what happens when two gases are mixed, such as when pure oxygen is released and mixes with the air in the room. We know that the process is irreversible because the reversed process, in which pure oxygen spontaneously separates from the atmosphere, never occurs. To calculate the resulting increase in entropy, we need the concept of partial pressures.

Partial Pressures

In earlier discussions, we have assumed that the pressure of an ideal gas depends on temperature, volume, and number of moles, independent of the type of molecule present. *Dalton's law of partial pressures*, named after John Dalton, is needed for the analysis of gases containing more than one type of molecule. It states that: *The pressure exerted by a gas is the sum of the partial pressures of the individual components*. The partial pressure of a component is the pressure it would exerts if it alone occupied the volume, where different *components* are the different types of molecule, such as O_2, N_2, CO_2, etc. The law is summarized by

$$P = \sum_i P_i, \tag{11.73}$$

where P is the total pressure exerted by the gas and P_i is the partial pressure of component i. Dalton's law is only exact for ideal gases, but is an accurate approximation in many applications. The partial pressure of component i of an ideal gas is $P_i = n_i RT/V$ where n_i is the number of moles of component i. Combining this with (11.73) indicates that the total pressure is

$$P = \sum_i \frac{n_i RT}{V} = \frac{n_{\text{tot}} RT}{V}, \tag{11.74}$$

where $n_{\text{tot}} = \sum_i n_i$ is the total number of moles present. Thus, as assumed earlier, the pressure of an ideal gas is determined by the total number of moles present.

Mixing of Gases

We consider an experiment with ideal gases of type A and type B, such as pure carbon dioxide and pure helium, and assume their specific heats are constant. As diagrammed in **Figure 11.3**, the gases are initially separated by a partition with equal amounts of each type in the same volume and at the same temperature. A hole is made in the partition so that the gases mix and come to equilibrium with no work done and no exchange of heat with the container, which has rigid insulating walls.

Since no work is done and no heat is exchanged, the internal energy of our system – the two gases – is unchanged. Since the internal energy of an ideal gas is a function of temperature only, the initial and final temperatures are the same. The only significant change is in the volume occupied by each gas, which increases from $\frac{1}{2}V_0$ to V_0, where V_0 is the volume of final mixture. According to Dalton's law, the pressure of the mixture is

$$P_0 = P_A + P_B = \frac{n_o RT}{V_0} + \frac{n_o RT}{V_0} = \frac{2n_o RT}{V_0}, \tag{11.75}$$

where n_o is the number of moles of each type. Since its initial volume is $\frac{1}{2}V_0$, the initial pressure of each gas is $2n_o RT/V_0$, which is the same as the final pressure of the mixture.

The entropy of an ideal gas as given in (11.23) is

$$S(T, V) = nc_V \ln T + nR \ln V + k. \tag{11.76}$$

Since the initial and final temperatures are the same, the change in the entropy of the gas of type A is

$$\Delta S_A = S_A(T, V_0) - S_A\left(T, \frac{1}{2}V_0\right) = n_0 R\left[\ln V_0 - \ln\left(\frac{1}{2}V_0\right)\right] = n_o R \ln 2. \tag{11.77}$$

The change in entropy of the gas of type B is the same, so that the total change in entropy is

$$\Delta S_{\text{mix}} = \Delta S_A + \Delta S_B = 2n_o R \ln 2. \tag{11.78}$$

Since this is greater than zero and the system is adiabatically isolated, the second law specifies that the process is possible and that the reversed process – the spontaneous separation of the two types of gas – never occurs.

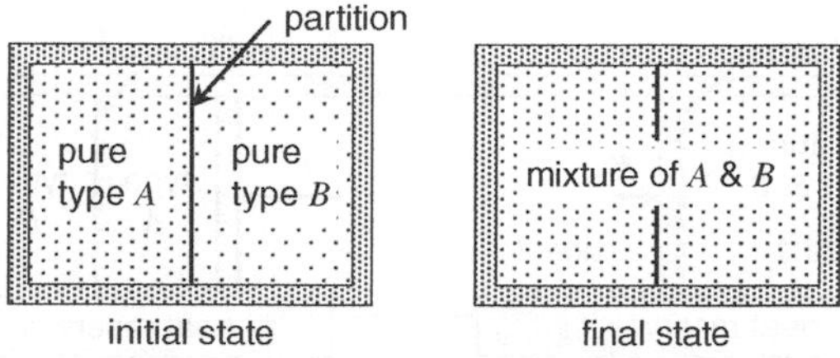

Figure 11.3 *Mixing of Two Gases*

Reversible Mixing

The result in (11.78) is correct, but as we used an entropy state function derived without considering the possibility of mixing, there is a conceptual problem with its derivation. To correct this problem, we calculate the entropy change ΔS_{mix} by applying the fundamental expression of entropy changes from (10.7), which asserts for a reversible process represented by line $L_{0\to f}$ that

$$\Delta S = S_f - S_0 = \int_{L_{0\to f}} \frac{đQ}{T}. \tag{11.79}$$

To utilize this expression, we need to find a reversible process that takes the system from an initial separated state to a final mixed state, or vice versa. The process we consider uses a *semipermeable membrane*. These membranes allow one type of molecule to pass through while preventing the passage of another type. One can imagine a barrier with microscopically small holes large enough for small molecules to pass through but too small for large molecules. The membranes that enclose biological cells are semipermeable membranes. Other examples are the membranes used in the purification of water and in the desalination of seawater in the process called reverse osmosis. Since the passage of molecules through membranes is typically slow, experiments that use them can only be considered reversible if performed very slowly.

To illustrate how the entropy principle can be applied to mixing, we consider an experiment (process) that starts with a mixture of two types of gas diagrammed in **Figure 11.4**. In the initial state (state a) we have a mixture of gases of types A and B confined in a cylinder by pistons held in place by external forces F_1 and F_2. The cylinder is divided by a semipermeable membrane that allows type B molecules to pass but is impermeable to type A molecules. Initially there is a mixture of n_o moles of gas of type A and n_o moles of gas of type B on the left side of the membrane (side 1), and the piston on the right side (side 2) is adjacent to the membrane. The pressure of the mixture is P_0, and the temperature is held constant by a heat reservoir at temperature T. To separate the two types of gas, we move the piston on side 2 so that the type B gas can pass through the membrane. The pressure on side 1 is maintained at P_0, while the piston on side 2 is moved until all of the type B gas is on side 2, which requires a very low pressure. After this state (state b), we prevent any further passage of gas by replacing the membrane with an impermeable partition. The piston on side 2 is then moved so that the final pressure on both sides is P_0. In this final state (state c) the two types of gas are fully separated (unmixed).

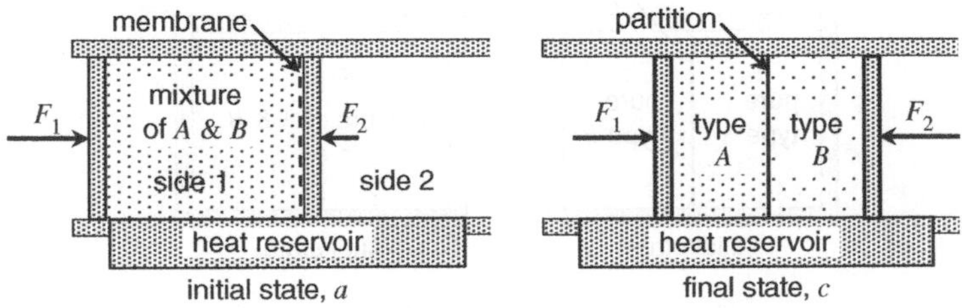

Figure 11.4 An Experiment for Separating Two Gases

If the experiment is performed at low pressure, we can treat the gas as ideal. Since the temperature is held constant, the internal energy of the gas does not change ($\Delta U_{gas} = 0$). The first law $Q = \Delta U + W$ then requires that $Q_{gas} = W_{gas}$. As shown below, work is done *on* the gas, which implies that the work done *by* the gas W_{gas} is negative, so that Q_{gas} is negative. This means that the heat transferred *to* the reservoir is positive and its entropy increases ($\Delta S_{res} = Q_{res}/T > 0$). Since the process is reversible and the gas and reservoir are adiabatically isolated, the entropy of the combination remains unchanged. Since the entropy of the combination is unchanged, the entropy of the gas has decreased, which tells us that the entropy of the initial mixture is higher than the entropy of the gases when separated.

The entropy change of the gas is found by calculating the work done by the pistons. In initial state a, the amount of gas on side 1 is $2n_o$ and the pressure is P_0, so that

$$P_0 V_0 = 2n_o RT, \tag{11.80}$$

where V_0 is the initial volume of side 1. Since the pressure is maintained at P_0 and the amount of gas on side 1 goes from $2n_o$ to n_o, the volume goes from V_0 to $\frac{1}{2}V_0$. Since work is done *on* the gas by the piston, the work done *by* the gas is negative as we go from state a to state b,

$$W_1 = \int_{V_0}^{V_0/2} P_0 dV_1 = P_0\left(\frac{1}{2}V_0 - V_0\right) = -\frac{1}{2}P_0V_0. \tag{11.81}$$

If the membrane is permeable to a gas, the equilibrium partial pressure is the same on both sides of the membrane. We can use this property to find the work done by the piston on side 2. When applied to our experiment, this property specifies that $P_{B,1} = P_{B,2}$, where $P_{B,1}$ and $P_{B,2}$ are the partial pressures of gas B on sides 1 and 2. Since the only gas on side 2 is of type B, the pressure there is

$$P_{B,2} = X_2 n_o \frac{RT}{V_2}, \tag{11.82}$$

where X_2 is the fraction of gas B on side 2. The fraction goes from $X_2 = 0$ in state a to $X_2 = 1$ in states b and c.

Since the volume V_2 of side 2 is zero in state a, all of the type B gas is in initial volume V_0. Hence, the partial pressure on side 2 is $P_{B,2} = n_oRT/V_0$. Since the pressure on side 2 tends to zero in state b, the volume and pressure tend to $V_{2,b} = \infty$ and $P_{B,2} = 0$, a condition that can only be approached. After the impermeable partition is inserted and the gas is compressed to pressure P_0 in state c, the volume of side 2 becomes $V_2 = n_oRT/P_0$, which equals $\frac{1}{2}V_0$. Since the volume of side 2 in state a is $V_{2,a} = 0$, the work done *by* the gas on side 2 during process $a \to b \to c$ is

$$W_2 = \int_0^{\infty} \frac{(X_2 n_o)RT}{V_2} dV_2 + \int_{\infty}^{V_0/2} \frac{n_oRT}{V_2} dV_2. \tag{11.83}$$

By using $P_0V_0 = 2n_oRT$ and introducing the ratio $r_2 = V_2/V_0$, this simplifies to

$$W_2 = \frac{1}{2}P_0V_0 \int_0^{\alpha} \frac{X_2}{r_2} dr_2 + \frac{1}{2}P_0V_0 \int_{\alpha}^{1/2} \frac{1}{r_2} dr_2, \tag{11.84}$$

where $\alpha \to \infty$. Before we can evaluate this, we need to know how the fraction X_2 depends on V_2, which is somewhat involved. As shown in the next paragraph, the result is $W_2 = \frac{1}{2}P_0V_0(1 - 2\ln 2) < 0$.

By using the equality of the partial pressures, $P_{B,1} = P_{B,2}$, we find that

$$\frac{(n_o - X_2 n_o)RT}{V_1} = \frac{(X_2 n_o)RT}{V_2}, \tag{11.85}$$

which simplifies to $(1 - X_2)V_2 = X_2V_1$. During process $a \to b$, the amount of gas on side 1 is

$$n_{\text{tot}} = n_o + n_o(1 - X_2) = (2 - X_2)n_o. \tag{11.86}$$

Since $P_0V_0 = 2n_oRT$ and the pressure is maintained at P_0, the pressure on side 1 is

$$P_0 = \frac{(2 - X_2)n_oRT}{V_1} = \frac{(2 - X_2)}{V_1}\frac{1}{2}P_0V_0 = \left(1 - \frac{1}{2}X_2\right)\frac{P_0V_0}{V_1}, \tag{11.87}$$

which implies $V_1/V_0 = 1 - \frac{1}{2}X_2$. Substituting this into $(1 - X_2)V_2 = X_2V_1$ gives

$$(1 - X_2)r_2 = X_2\left(1 - \frac{1}{2}X_2\right), \tag{11.88}$$

where $r_2 = V_2/V_0$. By rearranging terms, we obtain the quadratic equation

$$\frac{1}{2}{X_2}^2 - (r_2 + 1)X_2 + r_2 = 0. \tag{11.89}$$

Since $X_2 = 0$ when $r_2 = 0$, the appropriate solution is $X_2 = (r_2 + 1) - \sqrt{{r_2}^2 + 1}$, so that

$$\frac{X_2}{r_2} = 1 + \frac{1}{r_2}(1 - \sqrt{{r_2}^2 + 1}). \tag{11.90}$$

To integrate X_2/r_2 as required in (11.84), we need the integral formula

$$\int \frac{1}{r}(1 - \sqrt{r^2 + 1})dr = -\sqrt{r^2 + 1} + \ln(1 + \sqrt{r^2 + 1}). \tag{11.91}$$

(This can be verified by differentiating the expression on the right.) By using this formula and taking the limit $\alpha \to \infty$, the expression for W_2 in (11.84) becomes

$$\begin{aligned} W_2 &= \frac{1}{2}P_0V_0\left[r_2 - \sqrt{{r_2}^2 + 1} + \ln\left(1 + \sqrt{{r_2}^2 + 1}\right)\right]\Bigg|_0^{\alpha} + \frac{1}{2}P_0V_0 \ln r_2\Big|_{\alpha}^{1/2} \\ &= \frac{1}{2}P_0V_0\left[\left(\alpha - \sqrt{\alpha^2 + 1} + \ln\left(1 + \sqrt{\alpha^2 + 1}\right)\right) - (-1 + \ln 2) + \left(\ln\frac{1}{2} - \ln\alpha\right)\right] \\ &= \frac{1}{2}P_0V_0(1 - 2\ln 2). \end{aligned} \tag{11.92}$$

The total work done by the gas as both pistons are moved is

$$W_{\text{gas}} = W_1 + W_2 = -\frac{1}{2}P_0V_0 + \frac{1}{2}P_0V_0[1 - 2\ln 2] = -P_0V_0\ln 2. \tag{11.93}$$

By using $W_{\text{gas}} = Q_{\text{gas}}$ and $P_0 V_0 = 2n_o RT$, we find that $Q_{\text{gas}} = -2n_o RT \ln 2$. Since T is constant, the entropy change of the gas in process $a \to b \to c$ is

$$\Delta S_{\text{gas}} = S_c - S_a = \frac{Q_{\text{gas}}}{T} = -2n_o RT. \tag{11.94}$$

The entropy of mixing is obtained by subtracting the entropy of the unmixed gas in state c from the entropy of mixed gas in state a, which is the negative of the entropy change in (11.94). Thus,

$$\Delta S_{\text{mix}} = S_a - S_c = 2n_o RT. \tag{11.95}$$

This is the same as the value of ΔS_{mix} in (11.78) which was obtained by using the entropy function in (11.76). The agreement of two values for ΔS_{mix} indicates that the mixing of two gases can be described by employing the expression for entropy in (11.76), even though it was obtained without considering the possibility of mixing.

Entropy of Mixing

Although we have only considered the mixing of equal amounts of the two gases, the expression for the entropy of mixing is easily generalized for unequal amounts. Let n_A and n_B represent the number of moles of each gas, which are initially in separate containers of volumes V_A and V_B at the same temperature and pressure. After allowing them to mix, the volume occupied by each gas is $V_A + V_B$ and their temperature is unchanged. The total change in entropy can be found by using (11.76) to find the initial and final entropies and adding,

$$\begin{aligned}\Delta S_{\text{mix}} &= [S_A(T, V_A + V_B) - S_A(T, V_A)] + [S_B(T, V_A + V_B) - S_B(T, V_B)] \\ &= n_A R \ln\left(\frac{V_A + V_B}{V_A}\right) + n_B R \ln\left(\frac{V_A + V_B}{V_B}\right).\end{aligned} \tag{11.96}$$

The initial pressures of the gases are $P = n_A RT/V_A$ and $P = n_B RT/V_B$, which are the same. The pressure of the mixture is $P = (n_A + n_B)RT/(V_A + V_B)$. Since the mixing does not change the pressure, the entropy of mixing can be expressed as

$$\Delta S_{\text{mix}} = n_A R \ln\left(\frac{n_A + n_B}{n_A}\right) + n_B R \ln\left(\frac{n_A + n_B}{n_B}\right). \tag{11.97}$$

This result is for ideal gases. Since most materials behave like ideal gases at high temperatures and low densities, the result is useful for finding the entropy of mixing in other situations. For example, to find the entropy of mixing for two fluids at temperature T_0 and pressure P_0, we could first reversibly heat the fluids individually until they behave like ideal gases, then allow the two gases to mix, and finally reversibly cool the mixture back to T_0 and P_0. The total entropy change would be the sum of three entropy changes. The entropy changes in the first and final steps would be calculated by integrating $đQ/T$ along a reversible path, and the entropy change caused by the mixing would be given by ΔS_{mix}.

Problems

11.1 Starting with the differential for work for a system with coordinates (T, P), show that $C_P = T\left(\frac{\partial S}{\partial T}\right)_P$.

11.2 Consider a simple solid with $P = a_o T - b_o \ln(V/V_o)$ and heat capacity $C_V = M c_o$.

(a) Find dU and integrate it to determine the internal energy $U(T, V)$.
(b) Find dS and integrate it to determine the entropy $S(T, V)$.
(c) Find the adiabatic bulk modulus $B_S(T, V)$.

11.3 Consider a gas with constant specific heat c_V and the van der Waals equation of state $(P + a/v^2)(v - b) = RT$, where $v = V/n$.

(a) Find du and the specific internal energy $u = U/n$.
(b) Find ds and the specific entropy $s = S/n$.

11.4 The equation of state and heat capacity of a gas are

$$P = \frac{nRT}{V} + b_o T^2 \quad \text{and} \quad C_V = \left(\frac{\partial U}{\partial T}\right)_V = nc_o + 2b_o n^k T^m V^j$$

where j, k, and m are integers and b_o and c_o are intensive constants.

(a) Find the values of j, k, and m.
(b) Find the differentials of internal energy dU and entropy dS.
(c) Find the internal energy $U(V, T)$ and the entropy $S(V, T)$.
(d) Find the differential for heat $đQ$ and show that it is not exact.

11.5 Use Clapeyron's equation to estimate the boiling temperature of water in Denver, Colorado, when atmospheric pressure there is 83.6 kPa. (Remember: 100 °C is the boiling temperature of water at one standard atmosphere.)

12

Thermodynamic Potentials

A full description of a system's equilibrium behavior is contained in its thermodynamic potentials. All of its equilibrium properties can be obtained from any one of the potentials by differentiations, substitutions, and algebraic manipulations. Internal energy is a thermodynamic potential when expressed as a function of entropy and volume. As temperature is often a more useful variable than entropy, it is desirable to have potentials that are functions of temperature. These potentials are called free energies and are obtained by transformations that introduce new potentials in the process of changing the independent variables. When expressed as functions of specific variables, entropy and enthalpy are also thermodynamic potentials.

12.1 Internal Energy

Internal energy U is of special importance because of its role in the first law of thermodynamics, which is summarized by

$$dU = đQ - đW. \tag{12.1}$$

This relationship is more useful when the differentials for the process dependent quantities Q and W are expressed as changes in the system's state functions. It is a consequence of the mechanical description of PVT systems that the differential for work is $đW = PdV$ and a consequence of the second law that the differential for heat is $đQ = TdS$. By substituting these relationships into (12.1), the first law as applied to equilibrium processes becomes a relationship between exact differentials,

$$\boxed{dU = T\,dS - P\,dV.} \tag{12.2}$$

By using (12.2) instead of (12.1), we can avoid having to deal with inexact differentials.

The pairs of variables appearing together in the differential dU are called *conjugate variables*. Temperature T and entropy S are conjugate variables. The other pair of conjugate

Thermodynamics and Statistical Mechanics: An Integrated Approach, First Edition.
Robert J Hardy and Christian Binek.

variable for *PVT* systems are pressure P and volume V. The conjugate variables for describing electrical and magnetic interactions are introduced in Chapter 14.

Equation (12.2) suggests that we use S and V as the system's thermodynamic coordinates. The differential of the resulting function for the internal energy is

$$dU = \left(\frac{\partial U}{\partial S}\right)_V dS + \left(\frac{\partial U}{\partial V}\right)_S dV. \tag{12.3}$$

Comparing this with (12.2) indicates that

$$T = \left(\frac{\partial U}{\partial S}\right)_V \quad \text{and} \quad P = -\left(\frac{\partial U}{\partial V}\right)_S. \tag{12.4}$$

The function $U(S, V)$ is a *thermodynamic potential*, and entropy and volume are its *natural coordinates*. A function is only a thermodynamic potential when expressed as a function of its natural coordinates. Specifically, internal energy is only a thermodynamic potential when expressed as a function of entropy and volume.

12.2 Free Energies

Legendre Transformations[1]

Thermodynamic potentials with natural coordinates other than entropy and volume are obtained by Legendre transformations, which replace an independent variable (natural coordinate) by its conjugate variable and introduce a new potential. As an illustration, consider a function (thermodynamic potential) Φ with independent variables (X, Z) and differential $d\Phi = W\,dX + Y\,dZ$. The roles of W and X are exchanged by introducing a new function $\Psi = \Phi - WX$. By combining the differential $d\Psi = d\Phi - WdX - XdW$ with $d\Phi$, we obtain $d\Psi = -X\,dW + Y\,dZ$, which makes it natural to treat Ψ as a function of (W, Z).

Helmholtz Free Energy

The transformation that changes the natural coordinates from (S, V) to (T, V) introduces the *Helmholtz free energy* F named after Hermann von Helmholtz,

$$\boxed{F = U - TS.} \tag{12.5}$$

By using the relationship $d(TS) = TdS + SdT$, the differential of F becomes

$$dF = dU - T\,dS - S\,dT. \tag{12.6}$$

Combining this with $dU = TdS - P\,dV$ from (12.2) gives

$$dF = (T\,dS - P\,dV) - T\,dS - S\,dT, \tag{12.7}$$

[1] An alternate approach for exchanging the independent variables (natural coordinates) is outlined in Appendix 12.A.

which simplifies to

$$\boxed{dF = -S\,dT - P\,dV.} \tag{12.8}$$

Since dF is exact and thus is the differential of a function, it follows that

$$S = -\left(\frac{\partial F}{\partial T}\right)_V \quad \text{and} \quad P = -\left(\frac{\partial F}{\partial V}\right)_T. \tag{12.9}$$

$F(T, V)$ is the thermodynamic potential with temperature and volume as natural coordinates.

Gibbs Free Energy

An additional Legendre transformation changes the natural coordinates from (T, V) to (T, P) and introduces the *Gibbs free energy* G named after Willard Gibbs,

$$\boxed{G = F + PV.} \tag{12.10}$$

The differential of G is

$$dG = dF + d(PV) = dF + P\,dV + V\,dP. \tag{12.11}$$

Substituting $dF = -SdT - PdV$ into this gives

$$dG = (-S\,dT - P\,dV) + P\,dV + V\,dP, \tag{12.12}$$

which simplifies to

$$\boxed{dG = -S\,dT + V\,dP.} \tag{12.13}$$

As dG is the differential of a function, it follows that

$$S = -\left(\frac{\partial G}{\partial T}\right)_P \quad \text{and} \quad V = \left(\frac{\partial G}{\partial P}\right)_T. \tag{12.14}$$

$G(T, P)$ is the thermodynamic potential with temperature and pressure as natural coordinates.

Enthalpy

The Legendre transformation that changes the natural coordinates from (S, V) to (S, P) introduces the *enthalpy*,

$$\boxed{H = U + PV.} \tag{12.15}$$

$H(S, P)$ is the thermodynamic potential with entropy and pressure as natural coordinates. The differential of enthalpy is

$$dH = T\,dS + V\,dP, \tag{12.16}$$

and its derivatives are

$$T = \left(\frac{\partial H}{\partial S}\right)_V \quad \text{and} \quad V = \left(\frac{\partial H}{\partial P}\right)_S. \tag{12.17}$$

As indicated in Chapter 3, enthalpy includes both the internal energy and the energy PV required to insert the system into a constant-pressure environment. It follows from the above expressions for F, G, and H that the Gibbs free energy and enthalpy are related by

$$G = H - TS. \tag{12.18}$$

Comparing this with $F = U - TS$ indicates that the relationship between enthalpy and Gibbs free energy is analogous to the relationship between internal energy and the Helmholtz free energy. The expressions for the constant-pressure and constant-pressure heat capacities are

$$C_P = \left(\frac{\partial H}{\partial T}\right)_P \quad \text{and} \quad C_V = \left(\frac{\partial U}{\partial T}\right)_V, \tag{12.19}$$

which indicate that the role of enthalpy in the description of constant-pressure processes is analogous to the role of internal energy in the description of constant-volume processes.

12.3 Properties From Potentials

If we had no theory of thermal phenomena, the full description of the equilibrium properties of systems would require the specification of many unrelated functions. For example, a PVT system is described by an equation of state, internal energy, entropy, heat capacity, etc. and the associated functions $P(T, V)$, $U(T, V)$, $S(T, V)$, $C_V(T, V)$, etc., which might be unrelated to each other. We found in Chapter 3 that some of these functions are related by partial differentiation, for example, $C_V(T, V) = (\partial U/\partial T)_V$. We found in Chapter 11 that the internal energy and the entropy can be determined from $P(T, V)$ and $C_V(T, V)$. Now that we have the thermodynamic potentials, the equilibrium properties of a system can be determined from a single function.

Maxwell Relations

As discussed in Section 9.3, the cross derivatives of exact differentials are equal. For example, since $dF = -SdT - PdV$ is exact, it follows that

$$\frac{\partial^2 F}{\partial V \partial T} = \left(\frac{\partial S}{\partial V}\right)_T = \left(\frac{\partial P}{\partial T}\right)_V. \tag{12.20}$$

Equalities like this, which follow from the independence of the order of differentiation, are called *Maxwell relations*. The characteristics of the thermodynamic potentials U, F, G, and H are summarized in Table 12.1.

Helmholtz Free Energy

As indicated in the table, entropy and pressure are determined by the first derivatives of the Helmholtz free energy $F(T, V)$. In addition, the second derivatives determine the heat

Table 12.1 *Thermodynamic potentials for PVT systems*

Internal energy	Helmholtz free energy	Gibbs free energy	Enthalpy
$U(S, V)$	$F(T, V)$	$G(T, P)$	$H(S, P)$ $H = U + PV$
	$F = U - TS$	$G = U - TS + PV$	
$dU = TdS - PdV$	$dF = SdT - PdV$	$dG = -SdT + VdP$	$dH = TdS + VdP$
$T = \left(\frac{\partial U}{\partial S}\right)_V$	$S = \left(\frac{\partial F}{\partial T}\right)_V$	$S = \left(\frac{\partial G}{\partial T}\right)_P$	$T = \left(\frac{\partial H}{\partial S}\right)_P$
$P = \left(\frac{\partial U}{\partial S}\right)_S$	$P = \left(\frac{\partial F}{\partial V}\right)_T$	$V = \left(\frac{\partial G}{\partial P}\right)_T$	$V = \left(\frac{\partial H}{\partial P}\right)_S$
$\left(\frac{\partial T}{\partial V}\right)_S = \left(\frac{\partial P}{\partial S}\right)_V$	$\left(\frac{\partial S}{\partial V}\right)_T = \left(\frac{\partial P}{\partial T}\right)_V$	$-\left(\frac{\partial S}{\partial P}\right)_T = \left(\frac{\partial V}{\partial T}\right)_P$	$-\left(\frac{\partial T}{\partial P}\right)_S = \left(\frac{\partial V}{\partial S}\right)_P$

capacity C_V, bulk modulus B_T, and volume coefficient of thermal expansion α_V. By using (11.12) it follows that the constant-volume heat capacity is

$$C_V = T\left(\frac{\partial S}{\partial T}\right)_V = -T\left(\frac{\partial^2 F}{\partial T^2}\right)_V. \tag{12.21}$$

It follows from (3.21) that the isothermal bulk modulus is

$$B_T = -V\left(\frac{\partial P}{\partial V}\right)_T = V\left(\frac{\partial^2 F}{\partial V^2}\right)_T. \tag{12.22}$$

It follows from (3.55) that

$$\alpha_V B_T = \left(\frac{\partial P}{\partial T}\right)_V = -\frac{\partial^2 F}{\partial T \partial V}. \tag{12.23}$$

Before the above results can be used, we need the function $F(T, V)$ that gives the system's free energy. The function can be found from $F = U - TS$ and the functions $U(T, V)$ and $S(T, V)$, which can be determined from measurable properties by the methods outlined in the previous chapter. The free energy function can also be found from the atomic-level description of the system by the methods of statistical mechanics. Once $F(T, V)$ is known, it can be used to find properties of the system that were not utilized in its derivation. Finding $F(T, V)$ might not be easy and could involve extensive numerical analysis. To illustrate how equilibrium properties are obtained from the Helmholtz free energy, we consider an ideal gas.

Example 12.1 An ideal gas

Find the Helmholtz free energy of an ideal gas with constant c_V.

Solution. The Helmholtz free energy is found by combining its definition $F = U - TS$ with the expressions for U and S. According to (3.8), the internal energy of an ideal gas is $U = n(u_o + c_V T)$. Setting the constant u_o equal to zero gives

$$U = nc_V T.$$

According to (11.28), the entropy of an ideal gas is

$$S = nc_V \ln \frac{T}{T_o} + nR \ln \frac{V}{V_o} + ns_o,$$

where $S = ns_o$ at $T = T_o$ and $V = V_o$. By combining the expressions for U and S with $F = U - TS$ and rearranging terms, the Helmholtz free energy becomes

$$F(T, V) = nc_V T - nc_V T \ln \frac{T}{T_o} - nRT \ln \frac{V}{V_o} - ns_o T. \tag{12.24}$$

Example 12.2 Properties of an ideal gas

Starting with the Helmholtz free energy for an ideal gas, find S, P, U, C_V, B_T, α_V, C_P, and B_S as functions of T and V.

Solution. The Helmholtz free energy of an ideal gas with constant specific heats is given in (12.24). The entropy and equation of state are obtained by using Table 12.1,

$$S = -\left(\frac{\partial F}{\partial T}\right)_V = nc_V \ln \frac{T}{T_o} + nR \ln \frac{V}{V_o} + ns_o$$

$$P = -\left(\frac{\partial F}{\partial V}\right)_T = -\left(\frac{-nRT}{V}\right) = \frac{nRT}{V}.$$

The internal energy is obtained by rewriting $F = U - TS$ as $U = F + TS$ and using the expressions for F and S,

$$U = + nc_V T - nc_V T \ln \frac{T}{T_o} - nRT \ln \frac{V}{V_o} - ns_o T$$
$$+ T\left[nc_V \ln \frac{T}{T_o} + nR \ln \frac{V}{V_o} + ns_o\right].$$

This simplifies to

$$U = nc_V T.$$

The constant-volume heat capacity can be obtained from (12.21) or from the internal energy,

$$C_V = \left(\frac{\partial U}{\partial T}\right)_V = nc_V.$$

The isothermal bulk modulus can be obtained from (12.22) or from the equation of state,

$$B_T = -V\left(\frac{\partial P}{\partial V}\right)_T = -V\left(\frac{-nRT}{V^2}\right) = \frac{nRT}{V}.$$

The coefficient of thermal expansion can be obtained from (12.23),

$$\alpha_V = \frac{1}{B_T}\left(\frac{\partial P}{\partial T}\right)_V = \frac{V}{nRT}\frac{nR}{V} = \frac{1}{T}.$$

The constant-pressure heat capacity and the adiabatic bulk modulus can be found from the above results by using the relationships between C_V, C_P, B_T, and B_S derived in Chapter 11. According to (11.38) and (11.43),

$$C_P = C_V + TVB_T\alpha_V^2 \quad \text{and} \quad \frac{C_P}{C_V} = \frac{B_S}{B_T}.$$

Using the above results for C_V, B_T, and α_V gives

$$C_P = nc_V + TV\frac{1}{T^2}\frac{nRT}{V} = n(c_V + R).$$

Using the results for C_V, C_P, and B_T in $C_P/C_V = B_S/B_T$ and rearranging terms gives

$$B_S = \frac{C_P}{C_V}B_T = \frac{c_P}{c_V}B_T = \gamma\frac{nRT}{V},$$

where γ is the ratio of specific heats. These results all agree with the properties of ideal gases found earlier.

Gibbs Free Energy

As indicated in Table 12.1, entropy and volume are given by the first derivatives of the Gibbs free energy. Its second derivatives determine C_P, K_T, and α_V. The constant-pressure heat capacity can be found from the entropy (see Table 11.2),

$$C_P = T\left(\frac{\partial S}{\partial T}\right)_P = -T\left(\frac{\partial^2 G}{\partial T^2}\right)_P. \tag{12.25}$$

It follows from (3.46) that the isothermal compressibility is

$$K_T = -\frac{1}{V}\left(\frac{\partial V}{\partial P}\right)_T = -\frac{1}{V}\left(\frac{\partial^2 G}{\partial P^2}\right)_T. \tag{12.26}$$

It follows from (3.21) that the volume coefficient of thermal expansion is

$$\alpha_V = -\frac{1}{V}\left(\frac{\partial V}{\partial T}\right)_P = -\frac{1}{V}\frac{\partial}{\partial T}\left(\frac{\partial G}{\partial P}\right)_T = -\frac{1}{V}\frac{\partial G}{\partial T\partial P}. \tag{12.27}$$

Entropy

When expressed as a function of internal energy and volume, entropy is a thermodynamic potential. Unlike the free energies, it is not obtained from the potential $U(S, V)$ by a Legendre transformation but instead is found by inverting the function $U(S, V)$, *i.e.*, by re-expressing the relationship $U = U(S, V)$ as $S = S(U, V)$, where U and V are the natural coordinates. The differential dS can be obtained from the differential of energy by considering a state space in which the thermodynamics coordinates are U and V. Solving $dU = TdS - PdV$ for dS gives

$$dS = \frac{1}{T}dU + \frac{P}{T}dV. \tag{12.28}$$

As dS is the differential of a function, it follows that

$$\frac{1}{T} = \left(\frac{\partial S}{\partial U}\right)_V \quad \text{and} \quad \frac{P}{T} = \left(\frac{\partial S}{\partial V}\right)_U. \tag{12.29}$$

The equality of cross derivatives implies that

$$\left(\frac{\partial}{\partial V}\frac{1}{T}\right)_U = \left(\frac{\partial}{\partial U}\frac{P}{T}\right)_V. \tag{12.30}$$

After some rearrangement this becomes

$$\left(\frac{\partial T}{\partial V}\right)_U = P\left(\frac{\partial T}{\partial U}\right)_V - T\left(\frac{\partial P}{\partial U}\right)_V. \tag{12.31}$$

Finding expressions for the thermodynamic potentials $S(U, V)$ and $U(S, V)$ can be somewhat involved. One way to proceed is to find $T(U, V)$ by inverting the function $U(T, V)$ and substituting the result into $S(T, V)$. The resulting composite function $S(T(U, V), V)$ is the thermodynamic potential $S(U, V)$. The potential $U(S, V)$ is then found by inverting $S(U, V)$. The following example illustrates the procedure.

Example 12.3 S and U for an ideal gas

Find the thermodynamic potentials $S(U, V)$ and $U(S, V)$ for an ideal gas with constant c_V.

Solution. Because $U = U(T)$ in an ideal gas, inverting $U(T, V)$ to find $T(U, V)$ is quite easy, especially when c_V is constant. According to (3.8), the internal energy of an ideal gas with constant c_V is $U = n(u_o + c_V T)$. Setting $u_o = 0$ and solving for T gives $T = U/(nc_V)$. Substituting this into the expression for $S(T, V)$ in (11.28) gives

$$S(U, V) = nc_V \ln \frac{U}{nc_V T_o} + nR \ln \frac{V}{V_o} + ns_o,$$

where s_o is the entropy in the state with temperature T_o and volume V_o. By introducing the constant $a_o = s_o - c_V \ln(c_V T_o) - R \ln(V_o/n)$, this becomes

$$S(U, V) = nc_V \ln \frac{U}{n} + nR \ln \frac{V}{n} + na_o, \tag{12.32}$$

which is the desired function $S(U, V)$. Since n, U, and V are extensive and a_o is intensive, this expression for entropy is clearly extensive. Rearranging terms in (12.32) gives

$$\ln \frac{U}{n} = -\frac{a_o}{c_V} + \frac{S}{nc_V} - \frac{R}{c_V} \ln \frac{V}{n}.$$

Since $c_P - c_V = R$ in an ideal gas, it follows that $R/c_V = (c_P - c_V)/c_V = \gamma - 1$ where $\gamma = c_P/c_V$. By using the properties of logarithms and exponentials, the thermodynamic potential $U(S, V)$ for an ideal gas becomes

$$U(S, V) = n^{\gamma} e^{-a_o/c_V} e^{S/(nc_V)} V^{1-\gamma}. \tag{12.33}$$

We have emphasized that internal energy is only a thermodynamic potential when expressed as a function of its natural coordinates. Now that we have an expression for the thermodynamic potential $U(S, V)$, we can contrast it with the function $U(T, V)$ which is not a thermodynamic potential. Specifically, it can be shown that the equation of state of an ideal gas is not determined by $U(T, V)$ but is determined by $U(S, V)$. As shown in Section 11.1, internal energy and pressure are related by

$$\left(\frac{\partial U}{\partial V}\right)_T = T\left(\frac{\partial P}{\partial T}\right)_V - P. \tag{12.34}$$

Since the internal energy of an ideal gas is independent of volume, it follows that $(\partial U/\partial V)_T = 0$, in which case (12.34) becomes $P = T(\partial P/\partial T)_V$. This equation is satisfied by all equations of state of the form $P = Tf(V)$, where $f(V)$ is any differentiable function of V. As a result, the system's equation of state is not fully determined.

In contrast, the complete equation of state can be found from $U(S, V)$. Since $dU = TdS - PdV$ is an exact differential, it follows that $T = (\partial U/\partial S)_V$ and $P = -(\partial U/\partial V)_S$. By differentiating the expression for $U(S, V)$ in (12.33), we find that

$$T = n^\gamma e^{-a_o/c_V} \frac{e^{S/(nc_V)}}{nc_V} V^{1-\gamma} \quad \text{and} \quad P = -n^\gamma e^{-a_o/c_V} e^{S/(nc_V)} V^{-\gamma}(1-\gamma). \tag{12.35}$$

Combining these results and using $\gamma - 1 = R/c_V$ indicates that

$$PV = -n^\gamma e^{-a_o/c_V} e^{S/(nc_V)} V^{1-\gamma} \left(\frac{-R}{c_V}\right) = nRT. \tag{12.36}$$

Thus, the system's equation of state ($PV = nRT$) is fully determined by $U(S, V)$.

12.4 Systems in Contact with a Heat Reservoir

To apply the entropy principle, we need an adiabatically isolated system, which is sometimes obtained by combining the system of interest with an external source of energy. The systems we study are often maintained at constant temperature through contact with a heat reservoir, such as the atmosphere or a body of water. To facilitate the analysis of these systems, we apply the entropy principle to the adiabatically isolated composite system made up of the system of interest plus the reservoir. As the entropy of an adiabatically system never decreases, it follows that

$$(S_f - S_0) + \Delta S_R \geq 0, \tag{12.37}$$

where S_0 and S_f are the initial and final entropies of the system and ΔS_R is the entropy change of the reservoir. Since a heat reservoir has a fixed temperature T, its entropy change is $\Delta S_R = Q_R/T$, where Q_R is the heat transferred *to* it. Combining this with (12.37) indicates that

$$T(S_f - S_0) + Q_R \geq 0. \tag{12.38}$$

The amount of work done W is obtained by applying the first law to the system exclusive of the reservoir, which gives

$$Q_{\text{sys}} = (U_f - U_0) + W, \tag{12.39}$$

where U_0 and U_f are the initial and final values of its internal energy. As the only heat exchanges are between the system and heat reservoir, we have $Q_R = -Q_{sys}$. Hence,

$$T(S_f - S_0) \geq Q_{sys} = (U_f - U_0) + W. \tag{12.40}$$

Rearranging terms gives

$$W \leq -(U_f - TS_f) + (U_0 - TS_0) = -F_f + F_0, \tag{12.41}$$

where $F = U - TS$ is the Helmholtz free energy. This result implies that an upper bound exists on the work done, so that $W \leq W_{max}$ where

$$\boxed{W_{max} = F_0 - F_f.} \tag{12.42}$$

This result tells us that the energy that is free is the energy available (free) for obtaining work from a system maintained at constant temperature through contact with a heat reservoir. *The maximum amount of work obtainable from a system in contact with a heat reservoir equals the decrease in its Helmholtz free energy.*

12.5 Minimum Free Energy

To apply the principle that the entropy never decreases in adiabatically isolated systems to systems maintained at a constant temperature, we use the result $W \leq W_{max}$ to derive similar principles for systems in contact with heat reservoirs. The resulting principles impose limits on an increase instead of a decrease. Specifically, the Helmholtz free energy never increases in systems held at constant temperature and volume, and the Gibbs free energy never increases in systems held at constant temperature and pressure.

Helmholtz Free Energy

No work W is done by a PVT system when its volume V is constant. When $W = 0$, the relationship $W \leq F_0 - F_f$ in (12.42) implies that

$$\boxed{F_f \leq F_0.} \tag{12.43}$$

The Helmholtz free energy of a system in equilibrium at constant temperature and volume has the minimum value consistent with the constraints on it. Just as the entropy of an adiabatically isolated system will increase if it can, this result indicates that:
The Helmholtz free energy of a system at constant temperature and volume will decrease if it can.

Stability Conditions

By using the minimum principle for free energy, we can show that the bulk modulus B_T is positive by using reasoning similar to that employed to show that C_V is positive in

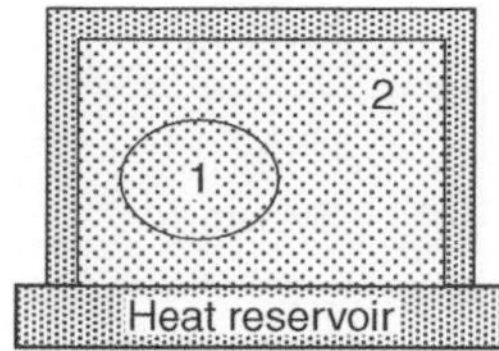

Figure 12.1 *A Two-Part System*

Section 11.5. As indicated in **Figure 12.1**, the system we consider is divided into two subsystems, 1 and 2, which are maintained at a constant temperature. The total volume $V^{\text{tot}} = V_1 + V_2$ is held fixed so that $\Delta V_1 = -\Delta V_2$.

The initial state 0 is assumed to be a stable equilibrium state. We now find the change in the Helmholtz free energy that would result from a transition to a final hypothetical state f. The free energies in states 0 and f are

$$F_0^{\text{tot}} = F^{\text{tot}}(T, V_{1,0}, V_{2,0}) \quad \text{and} \quad F_f^{\text{tot}} = F^{\text{tot}}(T, V_{1,0} + \Delta V_1, V_{2,0} - \Delta V_1). \tag{12.44}$$

ΔV_1 is the free variable that is determined by the minimum principle, while $V_{1,0}$ $V_{2,0}$, and T are fixed variables. As free energy is extensive, the free energy of the combined system is $F^{\text{tot}} = F_1 + F_2$, so that expanding F_f^{tot} in a Taylor series gives

$$F_f^{\text{tot}} = F_0^{\text{tot}} + [-P_1 + P_2]_0 \Delta V_1 + \frac{1}{2}\left[\left(\frac{\partial^2 F_1}{\partial {V_1}^2}\right)_{T_1} + \left(\frac{\partial^2 F_2}{\partial {V_2}^2}\right)_{T_2}\right]_0 \Delta {V_1}^2 + \cdots. \tag{12.45}$$

where $P = -(\partial F/\partial V)_T$ has been used. The coefficients in brackets are evaluated in state 0. The first order term vanishes because of the equilibrium condition $P_1 = P_2$. The third and higher order terms can be neglected when ΔV_1 is small. To simplify the coefficient of $\Delta {V_1}^2$, we consider subsystems of the same size, so that the two derivatives in the coefficient are equal. By adding them and omitting the subscripts, we obtain

$$F_f^{\text{tot}} = F_0^{\text{tot}} + \left(\frac{\partial^2 F}{\partial V^2}\right)_T \Delta {V_1}^2 = F_0^{\text{tot}} - \left(\frac{\partial P}{\partial V}\right)_T \Delta V^2. \tag{12.46}$$

For 0 to be stable, the Helmholtz free energy must be at a minimum in state 0. This means that the free energy in all hypothetical final states must be greater than the free energy in state 0 otherwise the system would spontaneously change to state f, which would imply that state 0 is not an equilibrium state. As $\Delta V^2 > 0$, the requirement $F_f^{\text{tot}} > F_0^{\text{tot}}$ implies that the derivative in (12.46) is negative, *i.e.*, $(\partial P/\partial V)_T < 0$. By using $B_T = -V(\partial P/\partial V)_T$, it follows that the isothermal bulk modulus is positive,

$$\boxed{B_T > 0.} \tag{12.47}$$

Equations that relate C_V, C_P, B_T, B_S, and α_V were found in Chapter 11. It follows from (11.38) and (11.44) and the stability conditions $C_V > 0$ and $B_T > 0$ that $C_P \geq C_V$ and $B_S \geq B_T$. The equal signs apply when $\alpha_V = 0$. This gives us two more stability conditions,

$$\boxed{C_P > 0 \quad \text{and} \quad B_S > 0.} \tag{12.48}$$

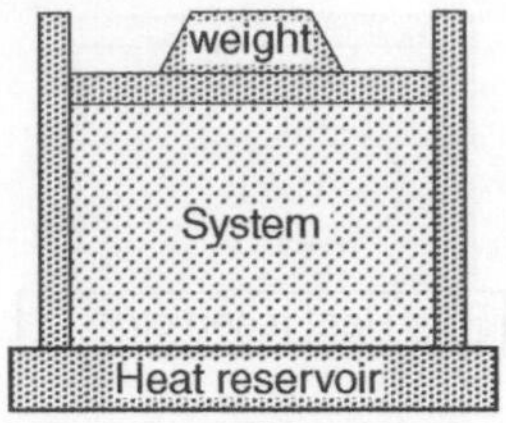

Figure 12.2 *System Maintained at Constant Temperature and Pressure*

The requirement that both bulk moduli are positive is intuitively reasonable: If either B_T or B_S was negative, a small random reduction in volume would reduce the pressure exerted by the system on its surroundings. That would cause the external forces on the system to exceed the forces opposing them, which would cause the volume to decrease even more, which would further decrease the pressure, and so forth. In contrast, small reductions in volume in stable systems introduce forces that return the volume to its original value. The special case in which $B_T = 0$ occurs at a critical point.

Gibbs Free Energy

Experiments are often performed on systems in contact with the atmosphere, which effectively maintains both the pressure and temperature at constant values. To facilitate the analysis of such experiments, consider the arrangement diagrammed in **Figure 12.2**, where the pressure is held fixed by a piston with a weight on it. The system can do work on its environment by moving the piston. The combination of the system and reservoir is adiabatically isolated.

When the pressure P is constant pressure, the work done *by* the system on the piston is $P(V_f - V_0)$, where V_0 and V_f are its initial and final volumes. The energy available for doing work *other than on the piston* is obtained by subtracting $P(V_f - V_0)$ from the upper bound on W given in (12.42), which gives

$$W = W - P(V_f - V_0) \leq -(F_f - F_0) - P(V_f - V_0) = -(G_f - G_0), \tag{12.49}$$

where $G = F + PV$ is the Gibbs free energy defined in (12.10). Thus, the upper limit on the amount of work – other than $P\Delta V$ work – that can be obtained from a system held at constant temperature and pressure equals the decrease in its Gibbs free energy.

When no work is done other than $P\Delta V$ work, the above result becomes $0 \leq -(G_f - G_0)$, which implies that

$$\boxed{G_f \leq G_0.} \tag{12.50}$$

The Gibbs free energy of a system in equilibrium at constant temperature and pressure has the minimum value consistent with the constraints on it. The Gibbs free energy of a system at constant temperature and pressure will decrease if it can. The next chapter uses the minimum principles to investigate the nature of phase transitions and the behavior of open systems.

Problems

12.1 Derive the relationship $\left(\frac{\partial T}{\partial V}\right)_U = P\left(\frac{\partial T}{\partial U}\right)_V - T\left(\frac{\partial P}{\partial U}\right)_V$.

12.2 **(a)** Find Helmholtz free energy $F(V,T)$ of a simple solid.
(b) Use the result of part (a) to verify that $(\partial F/\partial T)_V$ and $(\partial F/\partial V)_T$ are consistent with $S(T,V)$ from problem 11.2 and $P(T,V)$ from equation (1.15).

12.3 Find the Gibbs free energy $G(T,P)$ for an ideal gas with constant c_V.

12.4 Verify the Maxwell relations associated with the Helmholtz free energy F and the Gibbs free energy G by using the equation of state an ideal gas and the expressions for $S(T,V)$ and $S(P,V)$ when c_V is constant.

Appendix 12.A

Derivatives of Potentials

The derivatives of the free energies were found in Section 12.2 by using the properties of differentials. In this appendix we present an alternate derivation of these derivatives by focusing on the composite functions implicit in the definitions of the thermodynamic potentials and using the derivatives of internal energy and entropy found in earlier chapters. As given in (3.5) and (11.6), the derivatives of the internal energy $U(T,V)$ are

$$\left(\frac{\partial U}{\partial T}\right)_V = C_V \quad \text{and} \quad \left(\frac{\partial U}{\partial V}\right)_T = T\left(\frac{\partial P}{\partial T}\right)_V - P. \tag{12.A.1}$$

As given in (11.11), the derivatives of the entropy $S(T,V)$ are

$$\left(\frac{\partial S}{\partial T}\right)_V = \frac{C_V}{T} \quad \text{and} \quad \left(\frac{\partial S}{\partial V}\right)_T = \left(\frac{\partial P}{\partial T}\right)_V. \tag{12.A.2}$$

Since the Helmholtz free energy is defined as $F = U - TS$, the state function $F(T,V)$ is

$$F(T,V) = U(T,V) - TS(T,V), \tag{12.A.3}$$

so that its temperature derivative is

$$\left(\frac{\partial F}{\partial T}\right)_V = \left(\frac{\partial U}{\partial T}\right)_V - T\left(\frac{\partial S}{\partial T}\right)_V - S(T,V). \tag{12.A.4}$$

By using the expressions for $(\partial U/\partial T)_V$ and $(\partial S/\partial T)_V$ in (3.5) and (11.11), this becomes

$$\boxed{S = -\left(\frac{\partial F}{\partial T}\right)_V.} \tag{12.A.5}$$

The volume derivatives of $F(T,V)$ is

$$\left(\frac{\partial F}{\partial V}\right)_T = \left(\frac{\partial U}{\partial V}\right)_T - T\left(\frac{\partial S}{\partial V}\right)_T, \tag{12.A.6}$$

which becomes

$$\left(\frac{\partial F}{\partial V}\right)_T = \left[T\left(\frac{\partial P}{\partial T}\right)_V - P\right] - T\left(\frac{\partial P}{\partial T}\right)_V = -P. \tag{12.A.7}$$

Thus,

$$\boxed{P = -\left(\frac{\partial F}{\partial V}\right)_T.} \tag{12.A.8}$$

To find the thermodynamic potential $U(S, V)$, we invert the function $S(T, V)$ to obtain $T(S, V)$ and introduce the composite function

$$U(S, V) = U(T(S, V), V). \tag{12.A.9}$$

The definition $F = U - TS$ implies that $U = F + TS$, so that

$$U(S, V) = F(T(S, V), V) + T(S, V)S. \tag{12.A.10}$$

The derivative of this with respect to entropy is

$$\left(\frac{\partial U}{\partial S}\right)_V = \left(\frac{\partial F}{\partial T}\right)_V\left(\frac{\partial T}{\partial S}\right)_V + \left(\frac{\partial T}{\partial S}\right)_V S + T(S, V). \tag{12.A.11}$$

By using the result $(\partial F/\partial T)_V = -S$ in (12.A.5), we obtain

$$\boxed{T = \left(\frac{\partial U}{\partial S}\right)_V.} \tag{12.A.12}$$

The other derivative of $U(S, V)$ is

$$\left(\frac{\partial U}{\partial V}\right)_S = \left(\frac{\partial F}{\partial T}\right)_V\left(\frac{\partial T}{\partial V}\right)_S + \left(\frac{\partial F}{\partial V}\right)_T + \left(\frac{\partial T}{\partial V}\right)_S S. \tag{12.A.13}$$

By using the derivatives of F, this becomes

$$\left(\frac{\partial U}{\partial V}\right)_S = -S\left(\frac{\partial T}{\partial V}\right)_S - P + \left(\frac{\partial T}{\partial V}\right)_S S = -P. \tag{12.A.14}$$

Thus,

$$\boxed{P = P(S, V) = -\left(\frac{\partial U}{\partial V}\right)_S.} \tag{12.A.15}$$

Since enthalpy is $H = U + VP$, the thermodynamic potential $H(S, P)$ is determined by $U(S, P)$ and $V(S, P)$. The function $V(S, P)$ is obtained by inverting the function $P(S, V)$, so that the enthalpy is given by the composite function

$$H(S, P) = U(S, V(S, P)) + V(S, P)P. \tag{12.A.16}$$

The entropy derivative of this is

$$\left(\frac{\partial H}{\partial S}\right)_P = \left(\frac{\partial U}{\partial S}\right)_V + \left(\frac{\partial U}{\partial V}\right)_S\left(\frac{\partial V}{\partial S}\right)_P + \left(\frac{\partial V}{\partial S}\right)_P P. \tag{12.A.17}$$

By using $(\partial U/\partial S)_V = T$ and $(\partial U/\partial V)_S = -P$, this becomes

$$\left(\frac{\partial H}{\partial S}\right)_P = T - P\left(\frac{\partial V}{\partial S}\right)_P + \left(\frac{\partial V}{\partial S}\right)_P P = T, \tag{12.A.18}$$

so that

$$\boxed{T = \left(\frac{\partial H}{\partial S}\right)_P.} \tag{12.A.19}$$

The pressure derivatives of $H(S, P)$ is

$$\left(\frac{\partial H}{\partial P}\right)_S = \left(\frac{\partial U}{\partial V}\right)_S\left(\frac{\partial V}{\partial P}\right)_S + \left(\frac{\partial V}{\partial P}\right)_S P + V(S, P). \tag{12.A.20}$$

Since $(\partial U/\partial V)_S = -P$, it follows that

$$\boxed{V = \left(\frac{\partial H}{\partial P}\right)_S.} \tag{12.A.21}$$

As defined in (12.10), the Gibbs free energy is $G = F + PV$. Combining this with $F = U - TS$ and $H = U + VP$ gives $G = H - TS$, so that

$$G(T, P) = H(S(T, P), P) - TS(T, P), \tag{12.A.22}$$

where $S(T, P)$ can be obtained either by inverting $T(S, P)$ from (12.A.19) or by creating the composite function $S(T, V(T, P))$. The temperature derivative of $G(T, P)$ is

$$\left(\frac{\partial G}{\partial T}\right)_P = \left(\frac{\partial H}{\partial S}\right)_P\left(\frac{\partial S}{\partial T}\right)_P - S(T, V) - T\left(\frac{\partial S}{\partial T}\right)_P. \tag{12.A.23}$$

Since $(\partial H/\partial S)_P = T$, this becomes

$$\boxed{S = -\left(\frac{\partial G}{\partial T}\right)_P.} \tag{12.A.24}$$

The pressure derivative of $G(T, P)$ is

$$\left(\frac{\partial G}{\partial P}\right)_T = \left(\frac{\partial H}{\partial S}\right)_P\left(\frac{\partial S}{\partial P}\right)_T + \left(\frac{\partial H}{\partial P}\right)_S - T\left(\frac{\partial S}{\partial P}\right)_T, \tag{12.A.25}$$

By using $(\partial H/\partial S)_P = T$, this simplifies to $(\partial G/\partial P)_T = (\partial H/\partial P)_S$. It then follows from $(\partial H/\partial P)_S = V$ that

$$\boxed{V = \left(\frac{\partial G}{\partial P}\right)_T.} \tag{12.A.26}$$

The boxed equations give the derivatives of the thermodynamic potentials $F(T, V)$, $U(S, V)$, $H(S, P)$, and $G(T, P)$ summarized in Table 12.1.

13

Phase Transitions and Open Systems

The free energies introduced in the previous chapter are especially useful in the study of phase transitions. Because of the exchange of matter (molecules) between different phases, each phase can be treated as an open system, *i.e.*, a system that can exchange matter with other systems. Just as systems that can exchange heat have the same temperature when in equilibrium, systems that can exchange matter have the same chemical potential. Many systems contain several chemical constituents, called components, such as H_2O, CO_2, O_2, and N_2, and each component has a chemical potential. In an open system, the matter is distributed so that the chemical potential of each component is the same in every phase. We begin by using the Helmholtz and Gibbs free energies to explain why systems spontaneously separate into more than one phase.

13.1 Two-Phase Equilibrium

To illustrate the reason for phase separations, we analyze a single component system with the properties of the van der Waals fluid described in Section 5.3. When the specific volume $v = V/n$ is expressed in terms of volume V and number of moles n, the van der Waals equation of state in (5.9) becomes

$$P = \frac{nRT}{(V - bn)} - \frac{an^2}{V^2}. \tag{13.1}$$

We use this equation to investigate the situation diagrammed in **Figure 13.1**, where a fluid at constant temperature has separated into its liquid and vapor phases. Although the volume of the container is fixed, the volume of each phase can vary through an exchange of molecules.

The PV-diagram in **Figure 13.2** shows typical isotherms for temperatures above, below, and at the critical temperature T_c as determined by the equation of state in (13.1). (The plotted isotherms are at $0.875\,T_c$, $1.125\,T_c$, and T_c.) The curve through the points labeled

Thermodynamics and Statistical Mechanics: An Integrated Approach, First Edition.
Robert J Hardy and Christian Binek.

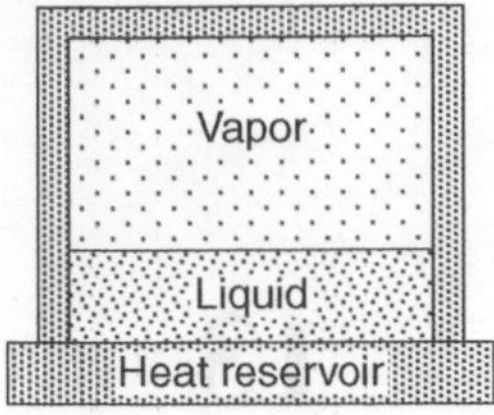

Figure 13.1 *Liquid and Vapor Phases in Equilibrium*

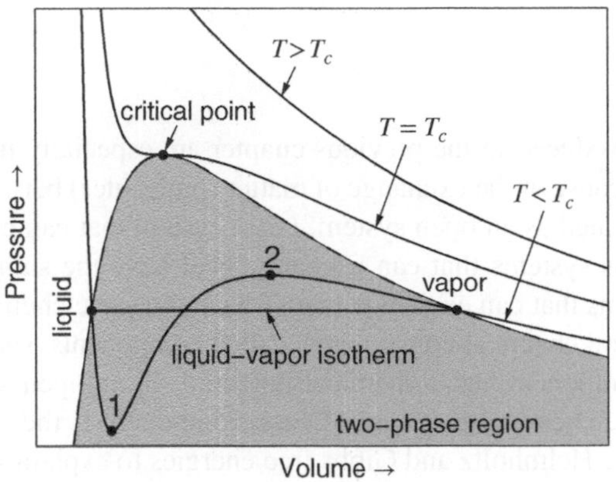

Figure 13.2 *Isotherms on a PV-Diagram*

1 and 2 on the isotherm for $T < T_c$ represents a fluid in a single phase, while the straight line represents it after it has separated into two phases. The shaded region identifies the pressures and volumes at which the system spontaneously separates into two phases.

Helmholtz Free Energy

Since both the equation of state and the heat capacity are needed to determine the free energy, we consider a system in which the constant-volume heat capacity $C_V = nc_V$ is a constant. A typical value for diatomic gases is $c_V = 2.5R$ (see Section 15.3). It is left as a problem to show that the free energy of a system described by the van der Waals equation of state and constant heat capacity C_V is

$$F(T, V) = ncRT(1 - \ln T) - nRT\ln(V - nb) - \frac{an^2}{V} - kT, \tag{13.2}$$

where c and k are a constants. It is easily verified by using $P = -(\partial F/\partial V)_T$ and $C_V = T(\partial^2 F/\partial T^2)_V$ that this yields the correct equation of state and heat capacity. The free energy for $T < T_c$ is plotted in **Figure 13.3**, where the line predicted by (13.2) is the single-phase isotherm. (The figure is plotted for $T = 0.875\,T_c$ and $c = 2.5$.) The location of the straight line that represents the free energy of a liquid-vapor system is determined by the principle that the free energy will decrease if it can.

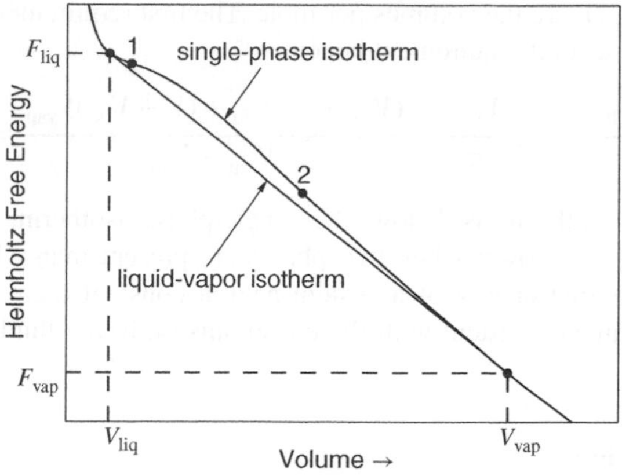

Figure 13.3 *Isotherm of the Helmholtz Free Energy in a Two-Phase Region*

The straight line in the figure is drawn tangent to the single-phase isotherm at the two points (V_{liq}, F_{liq}) and (V_{vap}, F_{vap}). These points represent properties of the single-phase system at the edges of the two-phase region in Figure 13.2. The equation of the straight line in Figure 13.3 is

$$F(V) = F_{liq} + \frac{F_{vap} - F_{liq}}{V_{vap} - V_{liq}}(V - V_{liq}), \tag{13.3}$$

where $(F_{vap} - F_{liq})/(V_{vap} - V_{liq})$ is its slope. By rearranging terms this becomes

$$F(V) = n_{liq}(V)F_{liq} + n_{vap}(V)F_{vap}, \tag{13.4}$$

where

$$n_{liq}(V) = \frac{V_{vap} - V}{V_{vap} - V_{liq}}n \quad \text{and} \quad n_{vap}(V) = \frac{V - V_{liq}}{V_{vap} - V_{liq}}n, \tag{13.5}$$

and n is the total number of moles in the system. $(\partial F/\partial V)_T$ is the slope of a line in the figure. Because the single-phase isotherm is tangent to the straight line at both V_{liq} and V_{vap}, the pressure $P = -(\partial F/\partial V)_T$ of the phases at both points is the same. Thus, the straight line represents a system with two phases present, where the system has $n_{liq}(V)$ moles of the phase at point (V_{liq}, F_{liq}) and $n_{vap}(V)$ moles of the phase at point (V_{vap}, F_{vap}). As the density n/V_{liq} at the first point is greater than the density n/V_{vap} at the second point, the two points represent respectively the liquid and vapor phases of the system. Since the total number of moles n in the system and total volume V are fixed, we need to check that

$$n_{liq} + n_{vap} = n \quad \text{and} \quad n_{liq}\frac{V_{liq}}{n} + n_{vap}\frac{V_{vap}}{n} = V. \tag{13.6}$$

where V_{liq}/n and V_{liq}/n are the volumes per mole. The first requirement is easily verified. By using (13.5), the second requirement becomes

$$n_{\text{liq}}\frac{V_{\text{liq}}}{n} + n_{\text{vap}}\frac{V_{\text{vap}}}{n} = \frac{(V_{\text{vap}} - V)V_{\text{liq}} + (V - V_{\text{liq}})V_{\text{vap}}}{V_{\text{vap}} - V_{\text{liq}}} = V. \tag{13.7}$$

The liquid-vapor isotherm is below the single-phase isotherm, which means the Helmholtz free energy is lower when two phases are present than when one is present. Since the Helmholtz free energy of a system held at constant temperature and volume has the minimum value consistent with the constraints on it, the fluid separates into two phases.

Maxwell Construction

The region of a *PV*-diagram in which two phases are present can be found by employing the Maxwell equal area construction. The *PV*-diagram in Figure 13.2 is redrawn in **Figure 13.4**. The idea of the Maxwell construction is to draw the liquid-vapor isotherm so that the areas of the two shaded regions are equal, *i.e.*, so that $A_1 = A_2$.

To show that the Maxwell construction locates the liquid-vapor isotherm correctly, we integrate the pressure from V_{liq} to V_{vap} and use the relationship $P = -(\partial F/\partial V)_T$,

$$\int_{V_{\text{liq}}}^{V_{\text{vap}}} P(V)\,dV = -\int_{V_{\text{liq}}}^{V_{\text{vap}}} \left(\frac{\partial F}{\partial V}\right)_T dV = F_{\text{liq}} - F_{\text{vap}}. \tag{13.8}$$

This result is valid whether the pressure considered is the pressure $P_{l\text{-}v}(V)$ along the liquid-vapor isotherm or the pressure $P_{s\text{-}p}(V)$ along the single-phase isotherm,

$$\int_{V_{\text{liq}}}^{V_{\text{vap}}} P_{l\text{-}v}(V)\,dV = \int_{V_{\text{liq}}}^{V_{\text{vap}}} P_{s\text{-}p}(V)\,dV. \tag{13.9}$$

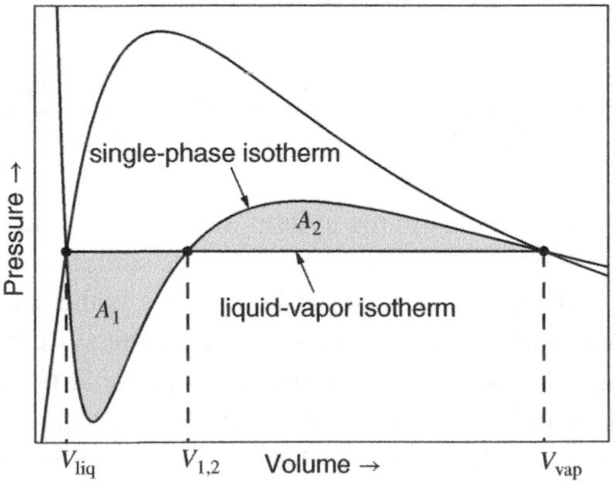

Figure 13.4 Maxwell Construction

As the two integrals are equal, it follows that

$$\int_{V_{\text{liq}}}^{V_{\text{vap}}} \left[P_{l\text{-}v}(V) - P_{s\text{-}p}(V)\right] dV = 0. \tag{13.10}$$

Let $V_{1,2}$ represent the volume at which the liquid-vapor isotherm crosses the single-phase isotherm, and let I_1 and I_2 represent integrations from V_{liq} to $V_{1,2}$ and from $V_{1,2}$ to V_{vap},

$$I_1 = \int_{V_{\text{liq}}}^{V_{1,2}} \left[P_{l\text{-}v}(V) - P_{s\text{-}p}(V)\right] dV \quad \text{and} \quad I_2 = \int_{V_{1,2}}^{V_{\text{vap}}} \left[P_{l\text{-}v}(V) - P_{s\text{-}p}(V)\right] dV. \tag{13.11}$$

It follows from (13.10) that $I_1 + I_2 = 0$. It can be seen from the figure that I_1 is negative and I_2 is positive. Consequently, the areas of the regions are $A_1 = -I_1$ and $A_2 = I_2$, which indicates that $-A_1 + A_2 = 0$, or equivalently that $A_1 = A_2$. Thus, the two-phase region can be found by adjusting the pressure at each temperature so that the two areas are equal.

Although the van der Waals model is useful for explaining the separation of a fluid into two phases, it possesses an unphysical aspect which should be mentioned. According to the stability condition in (12.47), a system is only stable in response to small changes in volume if the isothermal bulk modulus B_T is positive ($B_T > 0$). Since $B_T = -V(\partial P/\partial V)_T$, this requires that $(\partial P/\partial V)_T < 0$, which implies that the slopes of the isotherms on the PV-diagram in Figure 13.2 should be negative. Since the slope of the single-phase isotherm between points 1 and 2 is positive, that part of the isotherm represents states that are unstable, *i.e.*, states that are not physically realizable.

There are points within the two-phase region in which the slope of the single-phase isotherm is negative and thus represents states in which the system is stable. Such states are physically realizable, although they are not the most thermodynamically stable states. They are *meta-stable states* which can only be observed in delicate experiments on very pure substances. The states in the two-phase region to the left of point 1 in Figures 13.2 and 13.3 represent a *superheated liquid* and the states to the right of point 2 represent a *supercooled vapor*.

Gibbs Free Energy

The Gibbs free energy $G(T, P)$ is especially useful for describing systems with two phases present because both phases have the same temperature and pressure. The function $G(T, P)$ is obtained by combining the Helmholtz free energy $F(T, V)$ with the equation of state $V = V(T, P)$ to form the composite function

$$G(T, P) = F(T, V(T, P)) + PV(T, P). \tag{13.12}$$

The values of the Gibbs free energy for the van der Waals model are shown in **Figure 13.5**. The temperature is constant in the graph at the left, while the pressure is constant in the graph at the right. The curves in the figure represent the system when a single phase is present.

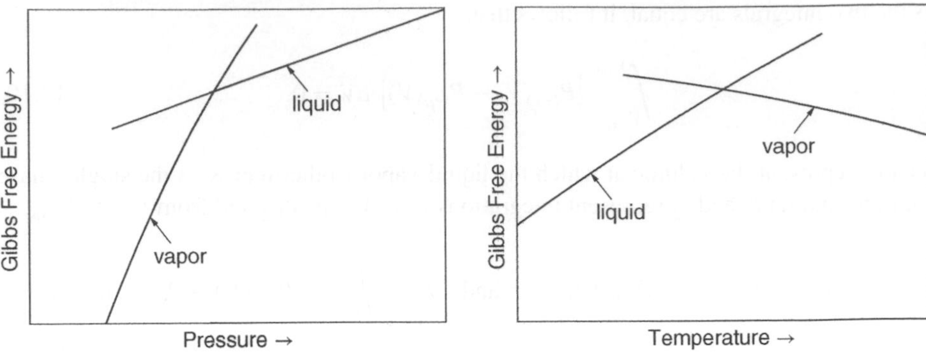

Figure 13.5 *Gibbs Free Energy near a Phase Transition*

The temperature and pressure at which two phases can coexist in equilibrium are determined by the principle that the Gibbs free energy will decrease if it can. The graph at the left indicates that the vapor phase is more stable than the liquid phase at low pressures, and the graph at the right indicates that the liquid phase is more stable than the vapor phase at low temperatures. The temperature and pressure at which the phases can coexist in equilibrium are given by the values for T and P where the curves cross. The parts of the curves where the free energy of one phase is greater than the other represent meta-stable states.

Additional information about the system is given by the slope of the curve in Figure 13.5. According to (12.14), a system's volume and its Gibbs free energy are related by $V = (\partial G/\partial P)_T$. Since volume is never negative, this implies that the slopes of the isotherms must be positive, which they are.

The stability conditions $B_T > 0$ and $C_P > 0$ from Section 11.5 give us information about the curvatures of the lines in the figure. According to (12.26), the isothermal compressibility is $K_T = -V^{-1}(\partial^2 G/\partial P^2)_V$. It then follows from $K_T = 1/B_T$ and $B_T > 0$ that $K_T > 0$, which implies that $(\partial^2 G/\partial V^2)_T < 0$. Thus, the second derivatives of the curves in the graph at the left should be negative, which they are. According to (12.25), the constant-pressure heat capacity is $C_P = -T(\partial^2 G/\partial T^2)_P$, so that condition $C_P > 0$ implies that $(\partial^2 G/\partial T^2)_P < 0$. Thus, the second derivatives of the curves in the graph at the right should be negative, which they are.

13.2 Chemical Potential

The chemical potential is the intensive property obtained by dividing the Gibbs free energy by the number of particles. It is valuable for describing the interactions of the components of open systems. Before considering its uses in multi-component systems, we investigate its significance in single component systems.

In addition to the two thermodynamic coordinates that specify the states of a closed *PVT* system, we need a variable to describe the amount of matter present when the system is open. The convenient choice is the number of particles N, which is related to the number of moles n by

$$N = nN_A, \tag{13.13}$$

where *Avogadro's number* is the number of molecules per mole, $N_A = 6.022 \times 10^{23}$. With this choice, the Gibbs free energy is the thermodynamic potential $G(T, P, N)$ with temperature, pressure, and number of particles as natural coordinates. The differential of $G(T, P, N)$ is

$$dG = \left(\frac{\partial G}{\partial T}\right)_{PN} dT + \left(\frac{\partial G}{\partial P}\right)_{TN} dP + \left(\frac{\partial G}{\partial N}\right)_{TP} dN. \tag{13.14}$$

The partial derivatives with N fixed are the derivatives for closed systems, which are given in Table 12.1, which by including a subscript to indicate that N is constant become

$$S = -\left(\frac{\partial G}{\partial T}\right)_{PN} \quad \text{and} \quad V = \left(\frac{\partial G}{\partial P}\right)_{TN}, \tag{13.15}$$

where S is entropy and V is volume. The Gibbs free energy per particle – the *chemical potential* – is[1]

$$\boxed{\mu = \left(\frac{\partial G}{\partial N}\right)_{TP}.} \tag{13.16}$$

Combining these definitions gives

$$\boxed{dG = -SdT + VdP + \mu dN.} \tag{13.17}$$

The relationships between the thermodynamic potentials G, F, U, and H in Table 12.1 are unchanged for open systems. Since $G = F + PV$, or equivalently $F = G - PV$, the differential of the Helmholtz free energy is

$$dF = dG - PdV - VdP. \tag{13.18}$$

Combining this with (13.17) gives[2]

$$dF = -SdT - PdV + \mu dN. \tag{13.19}$$

Since $F = U - TS$, the internal energy is $U = F + TS$ and its differential is

$$dU = TdS - PdV + \mu dN. \tag{13.20}$$

The differential of enthalpy $H = U + PV$ is

$$dH = TdS + VdP + \mu dN. \tag{13.21}$$

The entropy $S(U, V, N)$ is found by inverting the internal energy function $U(S, V, N)$. Its differential is obtained by rearranging terms in (13.20),

$$dS = \frac{1}{T}dU + \frac{P}{T}dV - \frac{\mu}{T}dN. \tag{13.22}$$

---

[1] Some authors define the chemical potential as the Gibbs free energy *per mole* instead of *per particle*.

[2] For multi-component systems the term μdN in the differentials dG, dF, dU and dH is replaced by $\sum_j \mu_j dN_j$.

The chemical potential μ is related to the thermodynamic potentials by

$$\mu = \left(\frac{\partial G}{\partial N}\right)_{TP} = \left(\frac{\partial F}{\partial N}\right)_{TV} = \left(\frac{\partial U}{\partial N}\right)_{SV} = \left(\frac{\partial H}{\partial N}\right)_{SP} = -T\left(\frac{\partial S}{\partial N}\right)_{UV}. \tag{13.23}$$

The microscopic significance of the chemical potential is indicated by the expression for the differential of internal energy in (13.20): *The chemical potential is the energy needed to introduce one molecule into a system when no heat is exchanged* ($dS = 0$) *and no work is done* ($dV = 0$).

Equilibrium Conditions for Open Systems

The equilibrium conditions $T_1 = T_2$ and $P_1 = P_2$ were derived in Chapter 11 without allowing for an exchange of matter. They were found by investigating the change in entropy that results from an exchange of energy and a change in volume. The additional condition can be obtained by generalizing the arguments in Section 11.5.

The additional condition ($\mu_1 = \mu_2$) can also be obtained by using the principle that *the Gibbs free energy is at a minimum when T and P are held constant*, derived in Section 12.5. The principle is applied to a total system that is divided into subsystems 1 and 2 which are allowed to exchange matter while maintained at constant temperature and pressure. **Figure 13.6** diagrams one way to achieve this situation. Subsystem 1 is a small part of the total system, and subsystem 2 is the remainder. The number of particles $N^{tot} = N_1 + N_2$ in the combined system is fixed, so that $\Delta N_1 = -\Delta N_2$, where ΔN_1 and ΔN_2 are the changes in the numbers of particles.

We consider a process in which the subsystems are individually in equilibrium at the start and finish of the process and focus our attention on the possibility that they are not in equilibrium with each other. As the Gibbs free energy is extensive, its value for the combined system is

$$G^{tot} = G_1(T, P, N_1) + G_2(T, P, N_2). \tag{13.24}$$

By identifying the initial state by subscript 0 and the final start by *f*, the initial free energy is

$$G_0^{tot} = G_1(T, P, N_{1,0}) + G_2(T, P, N_{2,0}) \tag{13.25}$$

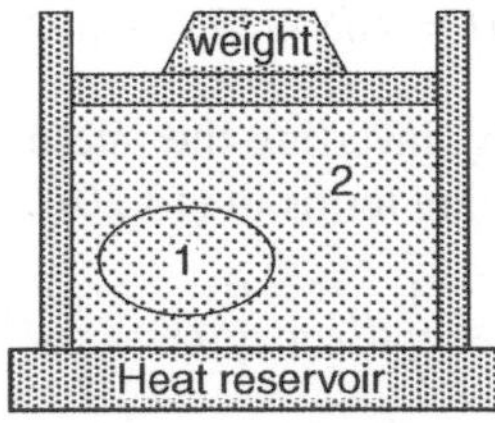

Figure 13.6 *System Maintained at Constant Temperature and Pressure*

and the final free energy is

$$G_f^{\text{tot}} = G_1(T, P, N_{1,0} + \Delta N_1) + G_2(T, P, N_{2,0} - \Delta N_1). \tag{13.26}$$

By expanding G_f^{tot} in powers of ΔN_1 and using $\mu = (\partial G/\partial N)_{TP}$, it follows that

$$G_f^{\text{tot}} = G_0^{\text{tot}} + \left[\left(\frac{\partial G_1}{\partial N_1}\right)_{TP} - \left(\frac{\partial G_2}{\partial N_2}\right)_{TP}\right]_0 \Delta N_1 + \cdots = G_0^{\text{tot}} + [\mu_1 - \mu_2]_0 \Delta N_1 + \cdots. \tag{13.27}$$

The higher order terms are negligible when ΔN_1 is small. If the subsystems are in equilibrium with each other in the initial state, the Gibbs free energy will be at a minimum, which implies that the free energy G_f^{tot} in any other state must be greater than the free energy G_0^{tot} independent of whether ΔN_1 is positive or negative. Hence, the coefficient of ΔN_1 in (13.27) is zero, which implies that $\mu_1 = \mu_2$. By combining this with the conditions derived in Chapter 11 (see (11.68)), the equilibrium conditions for open systems are

$$\boxed{\mu_1 = \mu_2, \qquad T_1 = T_2, \qquad P_1 = P_2.} \tag{13.28}$$

Since subsystem 1 can be anywhere within the total system, the equality $\mu_1 = \mu_2$ indicates that: *The chemical potential has the same value everywhere in a system in thermal equilibrium.*

Homogeneous Systems

As discussed in Section 1.5, the size dependence of a homogenous system can be characterized by a continuous paramcter λ with the system's extensive properties proportional to λ and the intensive properties are independent of λ. Since G and N are extensive while T and P are intensive, it follows that[3]

$$\lambda G(T, P, N) = G(T, P, \lambda N). \tag{13.29}$$

Differentiating with respect to λ and setting $\lambda = 1$ gives

$$G(T, P, N) = N \frac{\partial G(T, P, N)}{\partial N}. \tag{13.30}$$

This can be treated as a collection of differential equations $G(N) = N(\partial G/\partial N)_{TP}$, where T and P identify different equations. The solution of the differential equations is $G(N) = KN$, where K is a constant that depends on T and P. When this dependence is made explicit, we obtain

$$G(T, P, N) = K(T, P)N. \tag{13.31}$$

Combining this with $G(N) = N(\partial G/\partial N)_{TP}$ implies that $K(T, P) = (\partial G/\partial N)_{TP}$, so that the chemical potential is defined as $\mu = (\partial G/\partial N)_{TP}$ and

$$\boxed{G(T, P, N) = N\mu(T, P).} \tag{13.32}$$

[3] This is a special case of a homogeneous function, which are functions such that $\lambda^n f(x, y, \ldots) = f(\lambda x, \lambda y, \ldots)$. Euler's theorem states that such functions satisfy $nf(x, y, \ldots) = x\partial f(x, y, \ldots)/\partial x + y\partial f(x, y, \ldots)/\partial y + \ldots$.

This equation summarizes two important results: (i) The free energy is proportional to N, and (ii) *the chemical potential is independent of N.*

We now check that the discussion of phase transitions in Chapter 12 is consistent with the equality $\mu_1 = \mu_2$. To apply this equilibrium condition to a two-phase system, we treat each phase as a subsystem. Subsystem 1 is the liquid phase and subsystem 2 is the vapor phase. In Figure 13.5 the Gibbs free energy is plotted for both the liquid and vapor phases and the curves cross at the temperature and pressure where the phases coexist in equilibrium. As the free energy is proportional to the number of particles, the curves can be interpreted as scaled versions of the chemical potential $\mu = G/N$. Since the curves for the two phases cross at the point where their values are equal, we have $\mu_{\text{liq}} = \mu_{\text{vap}}$.

Clapeyron's Equation

The equilibrium condition $\mu_1(T,P) = \mu_2(T,P)$ can be used to derive Clapeyron's equation for the slopes of the two-phase coexistence curves on a phase diagrams, a result found in Section 11.4 by using the properties of entropy. Along the coexistence curve the two phases (subsystems) have the same temperature and pressure. Since the chemical potentials of the phases depend on T and P in different ways, the equilibrium condition implies there is one value of P for each value of T, which implies that $P = P_{2\text{ph}}(T)$ where $P_{2\text{ph}}(T)$ defines a curve on a phase diagram. Along this coexistence curve, the equilibrium conditions requires that

$$\mu_1\left(T, P_{2\text{ph}}(T)\right) - \mu_2\left(T, P_{2\text{ph}}(T)\right) = 0. \tag{13.33}$$

Differentiating this with respect to T gives

$$\left[\left(\frac{\partial\mu_1}{\partial T}\right)_P - \left(\frac{\partial\mu_2}{\partial T}\right)_P\right] + \left[\left(\frac{\partial\mu_1}{\partial P}\right)_T - \left(\frac{\partial\mu_2}{\partial P}\right)_T\right]\frac{dP_{2\text{ph}}(T)}{dT} = 0. \tag{13.34}$$

The derivatives of the chemical potential can be found by using the expressions for S and V in (13.15). By using $\mu = G/N$, the derivatives of the chemical potential become

$$S = -\left(\frac{\partial G}{\partial T}\right)_{PN} = -\frac{1}{N}\left(\frac{\partial\mu}{\partial T}\right)_{PN} = -\frac{S}{N} \tag{13.35}$$

and

$$V = \left(\frac{\partial G}{\partial P}\right)_{TN} = \frac{1}{N}\left(\frac{\partial\mu}{\partial P}\right)_{TN} = \frac{V}{N}. \tag{13.36}$$

Equation (13.34) then becomes

$$-\left[\frac{S_1}{N_1} - \frac{S_2}{N_2}\right] + \left[\frac{V_1}{N_1} - \frac{V_2}{N_2}\right]\frac{dP_{2\text{ph}}(T)}{dT} = 0. \tag{13.37}$$

Now, consider a process in which a system starts entirely in phase 1 and ends entirely in phase 2, while its temperature is held constant. In the initial state $N_1 = N_{\text{tot}}$ and $N_2 = 0$, while in the final state $N_1 = 0$ and $N_2 = N_{\text{tot}}$. With these values of N_1 and N_2

equation (13.37) can be rewritten as

$$\frac{dP_{2ph}(T)}{dT} = \frac{S_2 - S_1}{V_2 - V_1}, \tag{13.38}$$

where S_1 and V_1 are the initial entropy and volume and S_2 and V_2 are their final values. Since the temperature is constant, the change in entropy is $S_2 - S_1 = Q_{1\to2}/T$, where $Q_{1\to2}$ is the heat transferred to the system. The heat required to transform mass M of a substance from phase 1 to phase 2 is $Q_{1\to2} = ML_{1\to2}$, where $L_{1\to2}$ is the latent heat of transformation (see (5.1)). Thus, the change in entropy is $S_2 - S_1 = ML_{1\to2}/T$, so that

$$\boxed{\frac{dP_{2ph}}{dT} = \frac{1}{T}\frac{L_{1\to2}}{v_2 - v_1},} \tag{13.39}$$

where $v_1 = V_1/M$ and $v_2 = V_2/M$ are the specific volumes of the two phases. This is *Clapeyron's equation.*

For example, when phase 1 is liquid and phase 2 is vapor, v_2 is greater than v_1 and $L_{1\to2}$ is the heat of vaporization. Since $v_2 - v_1 > 0$ and $L_{1\to2} > 0$, Clapeyron's equation indicates that the slope of the vaporization curve on a phase diagram is positive.

13.3 Multi-Component Systems

PVT systems with several components (chemical constituent) have a different chemical potential for each component. We start by investigating a single-phase system and identify different components by index j. In addition to temperature and pressure, the Gibbs free energy is a function of the number of particles N_j in each component,

$$G = G(N_1, \ldots, N_j, \ldots, T, P). \tag{13.40}$$

The chemical potential of component j is

$$\mu_j = (\partial G/\partial N_j)_{T,P,N\ldots}, \tag{13.41}$$

where $N\cdots$ indicates that all but one of the N_j's is fixed. As G and the N_j's are extensive while T and P are intensive, changing the size of a system by a factor of λ implies that

$$\lambda G(N_1, \ldots, N_j, \ldots, T, P) = G(\lambda N_1, \ldots, \lambda N_j, \ldots, T, P). \tag{13.42}$$

Differentiating this with respect to λ and setting $\lambda = 1$ gives

$$G(N_1, \ldots, N_j, \ldots, T, P) = \sum_j \left(\frac{\partial G}{\partial N_j}\right)_{T,P,N\ldots} N_j. \tag{13.43}$$

Thus,

$$G = \sum_j \mu_j N_j. \tag{13.44}$$

To show that the chemical potentials of a multi-component system only depend on intensive variables, we consider a systems with J components ($j = 1, 2, \ldots, J$) and treat $\mu_j(N_1, \ldots, N_J, T, P)$ as a function of the concentrations c_j of $(J-1)$ of the component and the number of particles in the remaining component. The concentrations $c_j = N_j/N_J$ are measured relative to component J, so that

$$\mu_j = \mu_j\left(\frac{N_1}{N_J}, \ldots, \frac{N_{J-1}}{N_J}, N_J, T, P\right) = \mu_j(c_1, \ldots, c_{J-1}, N_J, T, P). \tag{13.45}$$

To eliminate the remaining extensive variable N_J, we multiply the size of the system by λ. Since all of the variables are intensive except N_J, it follows that

$$\mu_j(c_1, \ldots, c_{J-1}, N_J, T, P) = \mu_j(c_1, \ldots, c_{J-1}, \lambda N_J, T, P). \tag{13.46}$$

Setting $\lambda = 1/N_J$ gives

$$\mu_j = \mu_j(c_1, \ldots, c_{J-1}, 1, T, P) = \mu_j(c_1, \ldots, c_{J-1}, T, P), \tag{13.47}$$

where each chemical potential is a function of temperature, pressure, and $(J-1)$ concentrations, all of which are intensive.

Multi-Phase Systems

The Gibbs free energy of a multi-phase system is the sum of the free energies of the different phases. By identifying different phases by index k and applying the result in (13.44) to each phase, the Gibbs free energy of a multi-component multi-phase system becomes

$$\boxed{G = \sum_{j,k} \mu_j^k N_j^k,} \tag{13.48}$$

where N_j^k is the number of particles of component j in phase k. Differentiating G with respect to N_j^k indicates that the associated chemical potential is

$$\boxed{\mu_j^k = \left(\frac{\partial G}{\partial N_j^k}\right)_{T,P,N\ldots}.} \tag{13.49}$$

A relationship between the chemical potentials μ_j^k is found by making arguments similar to those used to find the equilibrium conditions for a single component system. Consider a system that is held in equilibrium at constant temperature and pressure, where the Gibbs free energy in initial state 0 is

$$G_0 = G(\ldots, N_{j,0}^k, \ldots, T, P) \tag{13.50}$$

and the free energy in a nearby state f is

$$G_f = G(\ldots, N_{j,0}^k + \Delta N_j^k, \ldots + \Delta N_J, T, P). \tag{13.51}$$

Expanding G_f in a Taylor series of powers of the ΔN_j^k gives

$$G_f = G_0 + \sum_j \mu_{j,0}^k \Delta N_j^k + \text{higher order terms}, \tag{13.52}$$

where $\mu_{j,0}^k$ is the initial value of the chemical potential. Since the Gibbs free energy at constant T and P has the minimum value consistent with the constraints on the system, it follows that $G_f \geq G_0$. When the linear terms in (13.52) dominate, replacing each ΔN_j^k by its negative $-\Delta N_j^k$ changes $G_f \geq G_0$ to $G_f \leq G_0$. Since both the original ΔN_j^k and their negatives are allowed, it follows that $G_f = G_0$. The changes ΔN_j^k are restricted to those that cause the sum in (13.52) to vanish, so that

$$\boxed{\sum_{j,k} \mu_j^k \Delta N_j^k = 0.} \tag{13.53}$$

Let $c_j^{(k\to l)}$ represent the number of particles of component j that is transferred *from* phase k *to* phase l. When all possible transfers are included, the total change in the number of particles N_j^k is

$$\Delta N_j^k = \sum_l c_j^{(k\to l)}. \tag{13.54}$$

Since the matter that leaves one phase enters the other, it follows that $c_j^{(k\to l)} = -c_j^{(l\to k)}$, which implies that $c_j^{(k\to k)} = 0$. Substituting (13.54) into (13.53) gives

$$\sum_{j,k,l} \mu_j^k c_j^{(k\to l)} = 0. \tag{13.55}$$

By exchanging indices k and l, adding the result to the original sum, and using $c_j^{(k\to l)} = -c_j^{(l\to k)}$, we find that

$$\sum_{j,k,l} \left(\mu_j^k c_j^{(k\to l)} + \mu_j^l c_j^{(l\to k)} \right) = \sum_{j,k,l} (\mu_j^k - \mu_j^l) c_j^{(k\to l)} = 0. \tag{13.56}$$

As this is true for all possible transfers, it is true when only one $c_j^{(k\to l)}$ is non-zero. In this case, equation (13.56) simplifies to $(\mu_j^k - \mu_j^l)c_j^{(k\to l)} = 0$, which implies that

$$\boxed{\mu_j^k = \mu_j^l.} \tag{13.57}$$

Thus, *the chemical potential of each component has the same value in all phases.*

13.4 Gibbs Phase Rule

The above equalities limit the number of way in which multi-component multi-phase systems can change. Consider a system of J components ($j = 1, 2, \ldots, J$) and K phases ($k = 1, 2, \ldots, K$). A different equality $\mu_j^k = \mu_j^l$ exists for each j and each pair (k, l), and there are $K(K-1)/2$ such pairs. The equalities $\mu_j^k = \mu_j^l$ are not independent, since they can all be obtained from the following $(K-1)$ equalities:

$$\mu_j^1 = \mu_j^2 = \cdots = \mu_j^{K-1} = \mu_j^K. \tag{13.58}$$

As there are J components and $(K-1)$ independent equalities for each component, these equalities impose $J(K-1)$ constraints on the values of the chemical potentials.

We found that the chemical potentials of the various components of a single-phase system are functions of T and P and $(J-1)$ particle concentrations. By treating each phase as a different system, the result can be applied to each phase, so that

$$\mu_j^k = \mu_j^k(c_j^k, \ldots, c_{J-1}^k, T, P), \tag{13.59}$$

where $c_j^k = N_j^k / N_J^k$ is the concentration of component j in phase k relative to component J. As the two variables T and P have the same value in every phase, the chemical potentials for a system with J components in K phases is determined by $[2 + (J-1)K]$ variables. However, these variables are not independent. Because of the $J(K-1)$ constraints found in the previous paragraph, the number of variables that can be chosen independently is

$$D = [2 + (J-1)K] - J(K-1), \tag{13.60}$$

which simplifies to

$$\boxed{D = 2 + J - K.} \tag{13.61}$$

D is called the *degree of variability* or *the number of degrees of freedom*. Equation (13.61) summarizes *Gibbs' phase rule*.

The implications of Gibbs' phase rule are illustrated by the behavior of PVT systems. Consider a pure substance in a single phase (gas, liquid, or solid). Since it has one component ($J = 1$) and one phase ($K = 1$), Gibbs' phase rule tells us that it has two degrees of freedom ($D = 2$). Temperature and pressure are the two degrees of freedom.

Now, consider a liquid in equilibrium with its vapor, which is a single component system ($J = 1$) with two phases present ($K = 2$). According to Gibbs' phase rule, the system has one degree of freedom ($D = 1$). This conclusion is consistent with the fact that the system's state is represented by a point on the vaporization curve, which implies that its pressure is determined by its temperature.

Next, consider a system at a triple point. The number of components is one ($J = 1$) and the number of phases is three ($K = 3$). According to Gibbs' phase rule, the number of degrees of freedom is zero ($D = 0$). This lack of variability is reflected in the unique values of temperature and pressure at triple points.

Finally, consider a system of water and nitrogen at room temperature and atmospheric pressure. It has two components (H_2O and N_2) and two phases (liquid and gas). According to Gibbs' phase rule, there are two degrees of freedom ($D = 2$). This reflects the fact that the concentration of water vapor in the gas and the concentration of dissolved nitrogen in the liquid are determined by the temperature and pressure.

13.5 Chemical Reactions

Chemical reactions can alter the amount of matter in the different components. For example, consider a system that consists of O_2, CO, and CO_2 gases. At sufficiently high temperatures, these compounds react as described by the chemical equation

$$O_2 + 2CO \rightleftharpoons 2CO_2, \tag{13.62}$$

where $\rightleftharpoons$ indicates that the reaction can proceed in either direction. When active, such reactions impose restrictions on the chemical potentials of systems in equilibrium. A chemical equation specifies the amounts by which different components increase or decrease when a reaction occurs and thereby imposes a restriction on the changes ΔN_j to the numbers of particles in the different components. To find the restriction, we rewrite the chemical equation as

$$O_2 + 2CO - 2CO_2 = 0. \tag{13.63}$$

We then represent a small change in the amount of O_2 present by α, so that $\Delta N_{O_2} = \alpha$. The changes in the amounts of CO and CO_2 are then $\Delta N_{CO} = 2\alpha$ and $\Delta N_{CO_2} = -2\alpha$.

The notation in the example can be generalized by letting (A_1, A_2, A_3) represent (O_2, CO, CO_2), so that (13.63) becomes

$$A_1 + 2A_2 - 2A_3 = 0. \tag{13.64}$$

This generalizes to

$$\sum_j I_j A_j = 0, \tag{13.65}$$

where the values of the integers I_j describe the chemical equation being considered. In the above example, $I_1 = 1$, $I_2 = 2$, and $I_3 = -2$. Furthermore, the changes in the numbers of particles of components A_1, A_2, and A_3 are $\Delta N_1 = I_1\alpha$, $\Delta N_2 = I_2\alpha$, and $\Delta N_3 = I_3\alpha$, which are of the form

$$\Delta N_j = I_j\alpha. \tag{13.66}$$

This result is valid for any reaction with a chemical equation that can be expressed as in (13.64).

For single-phase systems the restriction in (13.53) becomes

$$\sum_j \mu_j \Delta N_j = 0. \tag{13.67}$$

Substituting (13.66) into this gives

$$\sum_j \mu_j I_j \alpha = 0, \tag{13.68}$$

where α determines the extent of the chemical reaction. For (13.68) to be independent of the value of α, the chemical potentials must be such that

$$\boxed{\sum_j I_j \mu_j = 0.} \tag{13.69}$$

For example, for the reaction in (13.62) we have

$$\mu_{O_2} + 2\mu_{CO} = 2\mu_{CO_2}. \tag{13.70}$$

A restriction like (13.70) is satisfied for every chemical equation that represents an active reaction in a system in thermal, mechanical, and chemical equilibrium. Whether a particular chemical reaction is active or not depends on the temperature and pressure.

Example 13.1 Gibbs–Duhem equation

Find an equation that relates the differentials of the chemical potentials to the differentials of temperature and pressure.

Solution. As given in (13.44), the free energy of a single-phase system is $G = \sum_j \mu_j N_j$. The differential of this is

$$dG = \sum_j (\mu_j dN_j + N_j d\mu_j) = \sum_j \mu_j dN_j + \sum_j N_j d\mu_j,$$

where $\mu_j = (\partial G/\partial N_j)_{T,P,N\cdots}$. The Gibbs free energy is given by a function $G(N_1, \ldots, N_j, \ldots, T, P)$ whose differential is

$$dG = \sum_j \left(\frac{\partial G}{\partial N_j}\right)_{T,P,N\cdots} dn_j + \left(\frac{\partial G}{\partial T}\right)_{P,N\cdots} dT + \left(\frac{\partial G}{\partial P}\right)_{T,N\cdots} dP.$$

When expressed in terms of entropy, volume, and chemical potential, this becomes

$$dG = \sum_j \mu_j dN_j - S\,dT + V dP.$$

Subtracting this from the first expression for dG gives

$$S\,dT - VdP + \sum_j N_j d\mu_j = 0.$$

This is the *Gibbs–Duhem equation*. It tells us that, when temperature and pressure are held constant, changes in the chemical potentials are related by

$$\sum_j N_j d\mu_j = 0.$$

Problems

13.1 **(a)** Derive the free energy $F(T, V)$ of a van der Waals gas given in equation (13.2). Use the results $u = c_V T - a/\text{v} + u_o$ and $s = c_V \ln T + R\ln(\text{v} - b) + s_o$ of Problem 11.3,

(b) Find the equation of state, entropy, internal energy, constant-volume heat capacity, and isothermal bulk modulus of the gas from the free energy $F(T, V)$.

13.2 The function $\Phi(T, V, \mu)$ defined by $\Phi = G - PV - \mu N$ is called the grand potential.

(a) Show that $d\Phi = -SdT - PdV - Nd\mu$.

(b) The grand potential for a gas (found in Chapter 25) is

$$\Phi(T, V, \mu) = -K_o e^{\mu/k_B T}(k_B T)^{5/2} V,$$

where K_o is a constant and $k_B = R/N_A$ is the Boltzmann constant.
Find the equation of state of the gas by using the partial derivatives of Φ implied by the differential $d\Phi$.

13.3 Deduce the equilibrium condition $\mu_1 = \mu_2$ by generalizing the derivation of the equilibrium condition in Chapter 11 to allow for the exchange of matter.

14

Dielectric and Magnetic Systems

The thermodynamic properties of matter depend on the strength of the electric and magnetic fields, as well as on temperature and pressure. To analyze the effects of the fields, we need expressions for the work done by their interactions with matter. Just as the differential for work for *PVT* systems was obtained from mechanics, the differentials for work for electric and magnetic systems are obtained from electromagnetic theory. We consider uniform fields and focus on the exchanges of energy that occur as the fields are varied. In the final section, magnetic systems are used to investigate the unusual behavior of thermodynamic systems in the vicinity of critical points. Critical behavior occurs at second order phase transitions, including the paramagnetic-ferromagnetic transition, the liquid-gas transition in fluids, and the λ-transition in liquid helium. SI units are used throughout.

14.1 Dielectrics

We start by analyzing the response of dielectrics to an external electric field. The convenient device to consider is the parallel plate capacitor diagrammed in **Figure 14.1**, where a homogeneous dielectric material is sandwiched between two charged conducting plates of area A. The separation of the plates is a, and the volume of material is $V = aA$. The separation is sufficiently small that the fringing fields at the edges of the plates can be ignored, so that the electric field $\mathcal{E}$, displacement $\mathcal{D}$, and polarization $\mathcal{P}$ are uniform.

The magnitude of the electric field in the capacitor is

$$\mathcal{E} = \frac{V_{\mathrm{el}}}{a}, \tag{14.1}$$

where V_{el} is the externally controlled potential difference (voltage) between the plates. (The subscript distinguishes voltage V_{el} from volume V.) The displacement $\mathcal{D}$ is determined by Maxwell's equation

$$\nabla \cdot \mathcal{D} = \rho, \tag{14.2}$$

Thermodynamics and Statistical Mechanics: An Integrated Approach, First Edition.
Robert J Hardy and Christian Binek.

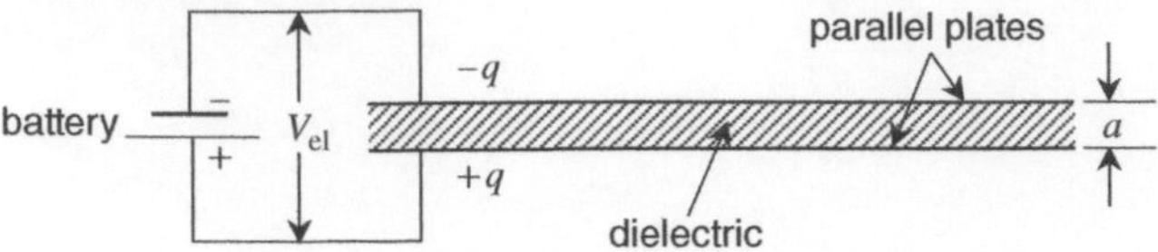

Figure 14.1 *Parallel Plate Capacitor*

where ρ is the density of free electric charge. In a capacitor, the charges on the plates are free charges while the charges within the dielectric material are bound charges.

The relationship of the $\mathcal{D}$-field to the charge of the capacitor is found by introducing a surface that encloses the positive plate. Integrating (14.2) over the volume enclosed by this surface gives

$$\int \nabla \cdot \mathcal{D} \, dr^3 = \int \rho \, dr^3. \tag{14.3}$$

By using the divergence theorem, the volume integral of $\nabla \cdot \mathcal{D}$ becomes the surface integral $\int \mathcal{D} \cdot d\mathbf{A}$. As the field is only non-zero between the plates, the surface integral equals $A\mathcal{D}$, where $\mathcal{D}$ is the field in the dielectric. As ρ is charge per unit volume, the volume integral on the right of (14.3) equals the total charge q enclosed within the surface. Hence, the charge on the plate is $q = A\mathcal{D}$.

When the value of q is reduced, work is done by the capacitor on the device that maintains the voltage V_{el}, which in the figure is a battery. The work done *by* the capacitor is

$$W_{cap} = -\int V_{el} \, dq, \tag{14.4}$$

where the minus sign indicates that the work done *by* the capacitor is positive when the charge on it decreases. By assuming that the area A of the plates and their separation a are constant, it follows from $\mathcal{E} = V_{el}/a$ and $q = A\mathcal{D}$ that $V_{el}\, dq = \mathcal{E} V d\mathcal{D}$, where $V = aA$. By using time t to parameterize the process and changing the integration variable from $\mathcal{D}$ to t, it follows that

$$W_{cap} = -V \int_{t_0}^{t_f} \mathcal{E}(t) \frac{d\mathcal{D}(t)}{dt} \, dt. \tag{14.5}$$

Since we are interested in the work done by the dielectric material, we need to subtract from W_{cap} the work that would be done in its absence. In the absence of the material the $\mathcal{D}$-field is $\mathcal{D} = \varepsilon_0 \mathcal{E}$, where ε_0 is the *electric constant* (also called *vacuum permittivity* and *permittivity of free space*), so that the work done by the material is

$$W_{sys} = -V \int_{t_0}^{t_f} \mathcal{E}(t) \left(\frac{d\mathcal{D}(t)}{dt} - \varepsilon_0 \frac{d\mathcal{E}(t)}{dt} \right) dt. \tag{14.6}$$

In electromagnetic theory the vector fields $\mathcal{E}$ and $\mathcal{D}$ are related to the *polarization* $\mathcal{P}$ by

$$\mathcal{D} = \varepsilon_0 \mathcal{E} + \mathcal{P}, \tag{14.7}$$

so that their magnitudes are related by $\mathcal{D} - \varepsilon_0 \mathcal{E} = \mathcal{P}$ when the vectors are parallel, as they are in isotropic materials. The resulting prediction of electromagnetic theory for the work done *by* the dielectric system is

$$W_{\text{sys}} = -V \int_{t_0}^{t_f} \mathcal{E}(t) \frac{d\mathcal{P}(t)}{dt} dt. \tag{14.8}$$

We limit our investigation to equilibrium processes, *i.e.*, to experiments performed slowly enough that the system is effectively in equilibrium at all times. The polarization is then given by a function $\mathcal{P}(T, \mathcal{E})$ of temperature and electric field, so that

$$\frac{\partial \mathcal{P}(t)}{\partial t} = \frac{\partial \mathcal{P}}{\partial T}\frac{dT(t)}{dt} + \frac{\partial \mathcal{P}}{\partial \mathcal{E}}\frac{d\mathcal{E}(t)}{dt}, \tag{14.9}$$

where the functions $T(t)$ and $\mathcal{E}(t)$ describe a line in a space with coordinates $(T, \mathcal{E})$. Substituting this into (14.8) gives

$$W_{sys} = -V \int_{t_0}^{t_f} \mathcal{E}(t) \left[\frac{\partial \mathcal{P}}{\partial t}\frac{dT\,(t)}{dt} + \frac{\partial \mathcal{P}}{\partial \mathcal{E}}\frac{d\mathcal{E}(t)}{dt} \right] dt, \tag{14.10}$$

which has the form of a line integral. When written in terms of differentials, the line integral becomes

$$W_{\text{sys}} = \int_L đW, \tag{14.11}$$

where $đW = -\mathcal{E}\, Vd\mathcal{P}$ is the differential for work and $d\mathcal{P}$ is the differential of polarization. Since polarization is the dipole moment per unit volume, which is uniform throughout the system, the *dipole moment* of the system is

$$P_{\text{el}} = V\mathcal{P}, \tag{14.12}$$

where V is volume. When V is constant, the differential of the dipole moment is $dP_{\text{el}} = Vd\mathcal{P}$, and the differential for the work done *by* the dielectric system is

$$\boxed{đW = -\mathcal{E}\, dP_{\text{el}}.} \tag{14.13}$$

Thermodynamic Potentials

The first law is summarized by $dU = đQ - đW$. By using the above expression for $đW$, the differential dU for dielectric systems is

$$dU = TdS + \mathcal{E}\, dP_{\text{el}}, \tag{14.14}$$

where $\mathcal{E}$ and P_{el} are conjugate variables. Comparing this with the corresponding result for PVT systems ($dU = TdS - PdV$) indicates that $-\mathcal{E}$ is analogous to pressure P and dipole moment P_{el} is analogous to volume V,

$$-\mathcal{E} \leftrightarrow P \quad \text{and} \quad P_{\text{el}} \leftrightarrow V. \tag{14.15}$$

Note that $\mathcal{E}$ and P are both intensive and that P_{el} and V are both extensive. Internal energy U is a thermodynamic potential when expressed as a function of its natural coordinates, which

***Table 14.1** Thermodynamic potentials for dielectric systems*

Internal energy	Helmholtz free energy	Gibbs free energy	Enthalpy
$U(S, P_{el})$	$F(T, P_{el})$	$G(T, \mathcal{E})$	$E_{th}(S, \mathcal{E})$
	$F = U - TS$	$G = U - TS - \mathcal{E}P_{el}$	$E_{th} = U - \mathcal{E}P_{el}$
$dU = TdS + \mathcal{E}dP_{el}$	$dF = -SdT + \mathcal{E}dP_{el}$	$dG = -SdT - P_{el}d\mathcal{E}$	$dE_{th} = TdS - P_{el}d\mathcal{E}$
$T = \left(\frac{\partial U}{\partial S}\right)_{P_{el}}$	$S = -\left(\frac{\partial F}{\partial T}\right)_{P_{el}}$	$S = -\left(\frac{\partial G}{\partial T}\right)_{\mathcal{E}}$	$T = \left(\frac{\partial E_{th}}{\partial S}\right)_{\mathcal{E}}$
$\mathcal{E} = \left(\frac{\partial U}{\partial P_{el}}\right)_S$	$\mathcal{E} = \left(\frac{\partial F}{\partial P_{el}}\right)_T$	$P_{el} = -\left(\frac{\partial G}{\partial \mathcal{E}}\right)_T$	$P_{th} = -\left(\frac{\partial E_{th}}{\partial \mathcal{E}}\right)_S$
$\left(\frac{\partial T}{\partial P_{el}}\right)_S = \left(\frac{\partial \mathcal{E}}{\partial S}\right)_{P_{el}}$	$-\left(\frac{\partial S}{\partial P_{el}}\right)_T = \left(\frac{\partial \mathcal{E}}{\partial T}\right)_{P_{el}}$	$\left(\frac{\partial S}{\partial \mathcal{E}}\right)_T = \left(\frac{\partial P_{el}}{\partial T}\right)_{\mathcal{E}}$	$\left(\frac{\partial T}{\partial \mathcal{E}}\right)_S = -\left(\frac{\partial P_{el}}{\partial S}\right)_{\mathcal{E}}$

for dielectrics are S and P_{el}. A full description of the equilibrium properties of a dielectric system is contained in any one of its thermodynamic potentials. As described in Chapter 12, internal energy, enthalpy, and the free energies are related by Legendre transformations. The thermodynamic potentials, their differentials, and their derivatives are summarized in Table 14.1.[1] (For dielectric and magnetic systems, enthalpy is represented by E_{th}.) The entropy $S(U, P_{el})$ is also a thermodynamic potential.

Since the $\mathcal{E}$-field in a capacitor can be varied by adjusting the voltage between the plates, $\mathcal{E}$ is often a more convenient thermodynamic coordinate than the dipole moment P_{el}. As a result, the Gibbs free energy and enthalpy are often more convenient properties for describing dielectrics than the Helmholtz free energy and internal energy. Just as the enthalpy of a PVT system includes the internal energy U and the energy PV, the enthalpy E_{th} of a dielectric system includes both the internal energy and the energy $-\mathcal{E}P_{el}$. As the work done *by* a system is the negative of the work done *on* it, $\mathcal{E}P_{el}$ is the energy that would need to be transferred *to* the system to take its dipole moment from zero to P_{el} in a constant field environment.

Heat Capacity

The empirical definition of the heat capacity with the electric field constant is

$$C_{\mathcal{E}} = \lim_{Q \to 0} \left. \frac{Q}{\Delta T} \right|_{\mathcal{E}=\text{const}}, \tag{14.16}$$

[1] The names Helmholtz and Gibbs have been assigned to the free energies following the analogy with PVT systems in (14.15). The analogy for magnetic systems is given in (14.38). When there is only one independent variable in addition to T, the additional variable (V, P_{el}, or M) is extensive in the Helmholtz free energy, while the variable (P, $\mathcal{E}$, or $\mathcal{H}$) is intensive in the Gibbs free energy. In situations requiring two or more additional variables, such as V and $\mathcal{E}$, the term "free energy" should be used without a modifier.

where the limiting process is indicated by $Q \to 0$, because Q is the quantity controlled experimentally. When the heat Q can be transferred reversibly, the differential of enthalpy is $dE_{th} = TdS - P_{el}d\mathcal{E}$. Since $d\mathcal{E} = 0$ when the field is constant, the *heat capacity at constant electric field* is

$$\boxed{C_{\mathcal{E}} = \left(\frac{đQ}{dT}\right)_{\mathcal{E}} = T\left(\frac{\partial S}{\partial T}\right)_{\mathcal{E}} = \left(\frac{\partial E_{th}}{\partial T}\right)_{\mathcal{E}}.} \tag{14.17}$$

It follows from $S = -(\partial G/\partial T)_{\mathcal{E}}$ that the heat capacity is related to the Gibbs free energy by $C_{\mathcal{E}} = -T(\partial^2 G/\partial T^2)_{\mathcal{E}}$.

Electric Susceptibility

In many applications the polarization $\mathcal{P}$ is proportional to the electric field $\mathcal{E}$, so that

$$\mathcal{P} = \varepsilon_0 \chi_{el} \mathcal{E}, \tag{14.18}$$

where the *electric susceptibility* χ_{el} is independent of $\mathcal{E}$. From this and $\mathcal{D} = \varepsilon_0\mathcal{E} + \mathcal{P}$, it follows that

$$\mathcal{D} = \varepsilon_0(1 + \chi_{el})\mathcal{E} = \varepsilon\mathcal{E} = \kappa_{el}\,\varepsilon_0\,\mathcal{E}, \tag{14.19}$$

where ε is the *permittivity* and $\kappa_{el} = 1 + \chi_{el}$ is the *dielectric constant* (also called *relative permittivity*), which is dimensionless. The parameters χ_{el}, ε, and κ_{el} characterize the dielectric properties of materials, which in general are functions of temperature. The dielectric constants of a few materials are given in Table 14.2.

The *isothermal electric susceptibility* and *adiabatic electric susceptibility* are

$$\chi_T = \frac{1}{\varepsilon_0}\left(\frac{\partial \mathcal{P}}{\partial \mathcal{E}}\right)_T \quad \text{and} \quad \chi_S = \frac{1}{\varepsilon_0}\left(\frac{\partial \mathcal{P}}{\partial \mathcal{E}}\right)_S. \tag{14.20}$$

In experiments performed at constant temperature, the susceptibility is χ_T. In experiments performed adiabatically the susceptibility is χ_S, where the adiabatic value is appropriate for experiments performed sufficiently rapidly that any transfer of heat is negligible. When changes in volume are negligible, it follows from $P_{el} = V\mathcal{P}$ and the expression for P_{el} in Table 14.1 that

$$\chi_T = \frac{-1}{\varepsilon_0 V}\left(\frac{\partial^2 G}{\partial \mathcal{E}^2}\right)_T. \tag{14.21}$$

The following example helps relate the above concepts to the use of capacitors as devices for storing electrical energy.

Table 14.2 *Dielectric constants κ_{el} at room temperature*

Material	Diamond, C	Table salt, NaCl	Cupric oxide, CuO	Lead oxide, PbO	Strontium titanate, $SiTiO_3$
κ_{el}	5.5	5.9	18.1	25.9	332

Example 14.1 Stored energy

Find the energy stored in a capacitor. Assume the temperature and dimensions of the capacitor are fixed.

Solution. The electrical energy stored in a capacitor is the energy transferred to it as work as the charge on the plates is taken from zero to its final value. As the work done *on* the capacitor by the external source of energy is the negative of the work done *by* the capacitor, it follows from (14.4) that the energy stored is

$$E_{\text{cap}} = \int V_{\text{el}}\, dq.$$

Since charge q and voltage V_{el} are related to the fields $\mathcal{D}$ and $\mathcal{E}$ by $q = \mathcal{D}A$ and $\mathcal{E} = V_{\text{el}}/a$, it follows from $\mathcal{D} = \kappa_{\text{el}}\,\varepsilon_0\,\mathcal{E}$ that $q = \kappa_{\text{el}}\varepsilon_0 A V_{\text{el}}/a$, which indicates that the charge is proportional to the voltage. When area A and separation a are fixed and κ_{el} is constant, the energy stored is

$$E_{\text{cap}} = \left(\frac{\kappa_{\text{el}}\varepsilon_0 A}{a}\right)\int V_{\text{el}}\, dV_{\text{el}}.$$

Since $\int V_{\text{el}}\, dV_{\text{el}} = \frac{1}{2}V_{\text{cap}}^2$ in processes that take the voltage from zero to V_{cap}, it follows that

$$E_{\text{cap}} = \tfrac{1}{2}\kappa_{\text{el}} C_0\, V_{\text{cap}}^2 = (1+\chi_T)\tfrac{1}{2}C_0\, V_{\text{cap}}^2,$$

where $C_0 = \varepsilon_0 A/a$ is the *capacitance* in the absence of a dielectric. When the final voltage V_{cap} is maintained by a battery or other source, the energy stored in a capacitor with a dielectric is $\frac{1}{2}\kappa_{\text{el}} C_0\, V_{\text{cap}}^2$.

The example tells us that including the dielectric multiplies the energy stored by the dielectric constant κ_{el}. The energy stored in the dielectric system is $\frac{1}{2}\chi_T C_0\, V_{\text{cap}}^2$, which is the difference between the energy stored with and without the dielectric. This energy results from the interaction of the dielectric with the electric field created by the charges on the capacitor plates.

14.2 Magnetic Materials

The most familiar magnetically ordered phase is the ferromagnetic phase. Materials also exhibit paramagnetic and diamagnetic behavior. Paramagnetism enhances the applied field and is caused by the response of molecules with permanent magnetic moments which are free to change their orientation. Diamagnetism creates a weak field in opposition to the applied field and results from the effect of the external field on the orbital motion of the electrons. Ferromagnetism occurs in materials in which the microscopic moments are aligned by interactions with each other. Ferromagnetic materials go through second phase transitions and become paramagnetic at high temperature.

To simplify the analysis of the response of matter to magnetic fields, we investigate a device in which the $\mathcal{B}$ and $\mathcal{H}$ fields are parallel. A device with this property is the Roland

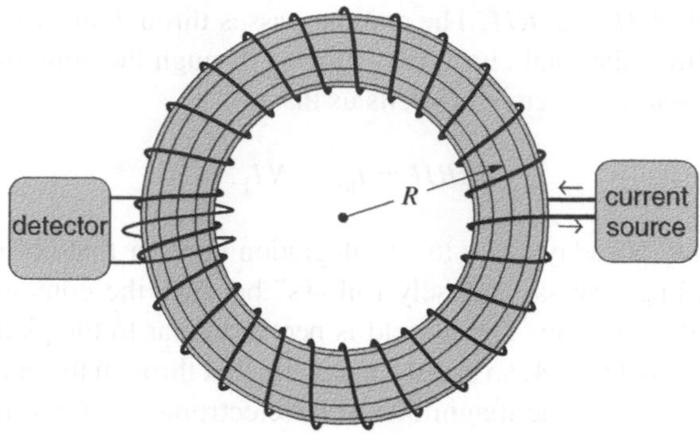

Figure 14.2 Rowland Ring

ring diagrammed in **Figure 14.2**. The device consists of a toroid of homogeneous magnetic material in the $\mathcal{H}$-field created by a current in the primary winding which is wound tightly around the ring. Changes in the $\mathcal{B}$-field are detected by a secondary winding of a few turns (at the left of the diagram). The magnetic induction $\mathcal{B}$, the magnetic intensity $\mathcal{H}$, and the magnetization $\mathcal{M}$ are parallel and can be represented by field lines (not shown) which are circles that loop around inside the material. When the current is in the direction indicated in the figure, the fields are counterclockwise. The thickness of the ring is sufficiently small compared to its mean radius R that the magnitudes $\mathcal{B}$, $\mathcal{H}$, and $\mathcal{M}$ are essentially uniform. The volume of the ring is $V_{\text{ring}} = 2\pi RA$ and can be treated as a constant, where the circumference of the ring is $2\pi R$ and A is its cross-sectional area.

The $\mathcal{H}$-field created by the current in the primary winding and the $\mathcal{B}$-field detected by the secondary winding are related to the magnetization $\mathcal{M}$ of the material by

$$\mathcal{B} = \mu_0(\mathcal{H} + \mathcal{M}), \tag{14.22}$$

where μ_0 is the *magnetic constant* (also called *vacuum permeability* and *permeability of free space*). The magnetic and electric fields satisfy Ampere's law

$$\oint \mathcal{H} \cdot d\boldsymbol{r} = I_{\text{tot}} \tag{14.23}$$

and Faraday's law

$$\oint \mathcal{E}(\boldsymbol{r}) \cdot d\boldsymbol{r} = -\frac{d\Phi_B}{dt}, \tag{14.24}$$

which are the integral forms of Maxwell's equations

$$\nabla \times \mathcal{H} = \boldsymbol{J} + \frac{\partial \mathcal{D}}{\partial t} \quad \text{and} \quad \nabla \times \mathcal{E} = -\frac{\partial \mathcal{B}}{\partial t}. \tag{14.25}$$

Materials with no free charges are considered, so that $\mathcal{D}$ and $\partial \mathcal{D}/\partial t$ are zero.

To find $\mathcal{H}$, we apply Ampere's law and use an integration contour in the direction of the fields, which is a circle of radius R within the material. Since the $\mathcal{H}$-field is parallel to this contour, the integral of $\mathcal{H}$ is $2\pi R\mathcal{H}$. The contour passes through all N turns of wire in the primary winding. Since the total current I_{tot} flowing through the contour is NI_1, where I_1 is the current in the wire, Ampere's law tells us that

$$2\pi R\mathcal{H} = I_{tot} = NI_1. \tag{14.26}$$

To find $\mathcal{B}$, we apply Faraday's law to an integration contour that closely follows a turn of the primary winding. We say "closely follows" because the contour closes on itself, unlike the turns of the winding. The $\mathcal{B}$-field is perpendicular to the plane of this contour, which has cross-sectional area A, so that the magnetic flux through the contour is $\Phi_{\mathcal{B}} = \mathcal{B}A$. According to Faraday's law, the magnitude of the electromotive force in each turn of the primary winding is

$$\text{emf}_1 = \oint \mathcal{E}(\boldsymbol{r}) \cdot d\boldsymbol{r} = -A\frac{\partial \mathcal{B}}{\partial t}, \tag{14.27}$$

where the minus sign is a manifestation of Lenz's law, which asserts that the direction of an induced current opposes the change that produced it.

When the externally controlled current in the primary winding is reduced, the magnetic material does work on the external current source. The rate at which work is done (power) *by* the ring on the source is

$$\frac{dW_{\text{ring}}}{dt} = -N\text{emf}_1 I_1, \tag{14.28}$$

where $N\text{emf}_1$ is the total electromotive force (back emf) in the N turns of the primary winding and I_1 is the current in the winding. It follows from (14.26) through (14.28) that

$$\frac{dW_{\text{ring}}}{dt} = -\left(NA\frac{\partial \mathcal{B}}{\partial t}\right)\left(\frac{2\pi R\mathcal{H}}{N}\right) = -V_{\text{ring}}\mathcal{H}\frac{\partial \mathcal{B}}{\partial t}, \tag{14.29}$$

where $V_{\text{ring}} = 2\pi RA$. The minus sign indicates that dW_{ring}/dt is positive when $\partial\mathcal{B}/\partial t$ is negative.

We are interested in the properties of the magnetic material and need to correct for the work done when no material is present. Since the $\mathcal{B}$ and $\mathcal{H}$ fields are related by $\mathcal{B} = \mu_0\mathcal{H}$ in the absence of material, the rate at which work is done *by* the magnetic material is

$$\frac{dW}{dt} = -V_{\text{ring}}\mathcal{H}\left[\frac{d\mathcal{B}}{dt} - \mu_0\frac{d\mathcal{H}}{dt}\right]. \tag{14.30}$$

The magnetization $\mathcal{M}$ and the vector fields $\mathcal{B}$ and $\mathcal{H}$ are related by $\mathcal{B} = \mu_0(\mathcal{H} + \mathcal{M})$, so that their magnitudes are related by $\mu_0\mathcal{M} = \mathcal{B} - \mu_0\mathcal{H}$ when the fields are parallel. Hence,

$$\frac{dW}{dt} = -\mu_0\mathcal{H}\,V_{\text{ring}}\frac{d\mathcal{M}}{dt}. \tag{14.31}$$

We now consider a small sample of the material over which the direction of $\mathcal{M}$ is essentially constant. Since magnetization is magnetic moment per unit volume, the *magnetic moment* of the small system is

$$M = V\mathcal{M}, \tag{14.32}$$

where V is its volume. The work done by the small system is reduced from the work done by the entire ring by a factor of V/V_{ring}, so that

$$\frac{dW_{\text{sys}}}{dt} = \frac{V}{V_{\text{ring}}}\frac{dW}{dt} = -\mu_0 \mathcal{H} V \frac{d\mathcal{M}}{dt} = -\mu_0 \mathcal{H}\frac{dM}{dt}. \tag{14.33}$$

The work done between times t_0 and t_f is

$$W_{\text{sys}} = -\int_{t_0}^{t_f} \mu_0 \mathcal{H}\frac{dM}{dt}dt. \tag{14.34}$$

When the system is in equilibrium, the magnetic field can be treated as a function of magnetic moment and temperature, in which case (14.34) becomes a line integral in a space with coordinates (T, M). Thus,

$$W_{\text{sys}} = -\int_L \mu_0 \mathcal{H} dM = \int_L đW, \tag{14.35}$$

where the differential for work for a magnetic system is

$$\boxed{đW = -\mu_0 \mathcal{H}\, dM.} \tag{14.36}$$

Thermodynamic Properties

Substituting $đQ = TdS$ and $đW = -\mu_0 \mathcal{H}\, dM$ into $dU = đQ - đW$ gives

$$dU = TdS + \mu_0 \mathcal{H}\, dM, \tag{14.37}$$

where $\mu_0 \mathcal{H}$ and M are conjugate variables. Comparing this with the relationship for PVT systems, *i.e.*, $dU = TdS - PdV$, indicates that $-\mu_0 \mathcal{H}$ is analogous to pressure P and magnetic moment M is analogous to volume V,

$$-\mu_0 \mathcal{H} \leftrightarrow P \quad \text{and} \quad M \leftrightarrow V, \tag{14.38}$$

where M and V are extensive and $\mu_0 \mathcal{H}$ and P are intensive. The internal energy $U(S, M)$ is a thermodynamic potentials for magnetic systems. The thermodynamic potentials related to it by Legendre transformations are summarized in Table 14.3. Also included are their differentials and the properties obtainable from them. $S(U, M)$ is also a thermodynamic potential.

The $\mathcal{H}$-field, which can be varied by adjusting the current in the Roland ring, is often preferable to the moment M as an independent variable. As a result, the Gibbs free energy is often more useful for describing magnetic systems than the Helmholtz free energy. The enthalpy $E_{\text{th}} = U - \mu_0 \mathcal{H} M$ of a magnetic system includes both the internal energy U and the energy $-\mu_0 \mathcal{H} M$, which describes the interaction of the system with the externally established field. Just as $\mathcal{E} P_{\text{el}}$ is the energy that would need to be transferred *to* the system to take its dipole moment from zero to P_{el} in a constant field environment, $\mu_0 \mathcal{H} M$ is the energy that is transferred *to* a magnetic system when its moment is taken from zero to M in a constant $\mathcal{H}$-field, *i.e.*, $\mu_0 \mathcal{H} M$ is the energy transferred *to* the system *from* some external device.

Table 14.3 Thermodynamic potentials for magnetic systems

Internal energy	Helmholtz free energy	Gibbs free energy	Enthalpy
$U(S, M)$	$F(T, M)$	$G(T, \mathcal{H})$	$E_{\text{th}}(S, \mathcal{H})$
	$F = U - TS$	$G = U - TS - \mu_0 \mathcal{H} M$	$E_{\text{th}} = U - \mu_0 \mathcal{H} M$
$dU = TdS + \mu_0 \mathcal{H} dM$	$dF = -SdT + \mu_0 \mathcal{H}\, dM$	$dG = -SdT - M\mu_0 d\mathcal{H}$	$dE_{\text{th}} = TdS - M\mu_0 d\mathcal{H}$
$T = \left(\frac{\partial U}{\partial S}\right)_M$	$S = -\left(\frac{\partial F}{\partial T}\right)_M$	$S = -\left(\frac{\partial G}{\partial T}\right)_{\mathcal{H}}$	$T = \left(\frac{\partial E_{\text{th}}}{\partial S}\right)_{\mathcal{H}}$
$\mu_0 \mathcal{H} = \left(\frac{\partial U}{\partial M}\right)_S$	$\mu_0 \mathcal{H} = \left(\frac{\partial F}{\partial M}\right)_T$	$M = -\frac{1}{\mu_0}\left(\frac{\partial G}{\partial \mathcal{H}}\right)_T$	$M = -\frac{1}{\mu_0}\left(\frac{\partial E_{th}}{\partial \mathcal{H}}\right)_S$
$\left(\frac{\partial T}{\partial M}\right)_S = \mu_0 \left(\frac{\partial \mathcal{H}}{\partial S}\right)_M$	$-\left(\frac{\partial S}{\partial M}\right)_T = \mu_0 \left(\frac{\partial \mathcal{H}}{\partial T}\right)_M$	$\frac{1}{\mu_0}\left(\frac{\partial S}{\partial \mathcal{H}}\right)_T = \left(\frac{\partial M}{\partial T}\right)_{\mathcal{H}}$	$\frac{1}{\mu_0}\left(\frac{\partial T}{\partial \mathcal{H}}\right)_S = -\left(\frac{\partial M}{\partial S}\right)_{\mathcal{H}}$

Heat Capacity

The definition of heat capacity at constant magnetic field $C_{\mathcal{H}}$ is similar to the definition of $C_{\mathcal{E}}$ in (14.16) and can be expressed as $C_{\mathcal{H}} = (đQ/dT)_{\mathcal{H}}$. By using the differential $dE_{\text{th}} = TdS - M\mu_0 d\mathcal{H}$ and using similar reasoning to that employed to derive (14.17), we find that *heat capacity at constant magnetic field* is

$$C_{\mathcal{H}} = \left(\frac{đQ}{dT}\right)_{\mathcal{H}} = T\left(\frac{\partial S}{\partial T}\right)_{\mathcal{H}} = \left(\frac{\partial E_{\text{th}}}{\partial T}\right)_{\mathcal{H}}. \tag{14.39}$$

The heat capacity is also given by $C_{\mathcal{H}} = -T(\partial^2 G/\partial T^2)_{\mathcal{H}}$.

Magnetic Susceptibility

The *magnetic susceptibility* χ_m is the dimensionless proportional constant that relates the magnetization to the applied field,

$$\mathcal{M} = \chi_m \mathcal{H}. \tag{14.40}$$

By combining this with $B = \mu_0(\mathcal{H} + \mathcal{M})$, we obtain

$$B = \mu_0(1 + \chi_m)\mathcal{H} = \mu \mathcal{H} = \kappa_m \mu_0 \mathcal{H}, \tag{14.41}$$

where μ is the *permeability* and κ_m is the *relative permeability*. When the magnetization $\mathcal{M}$ is proportional to $\mathcal{H}$, the material properties χ_m, μ, and κ_m are functions of temperature and independent of the field strength. The magnetic susceptibilities of a few diamagnetic ($\chi_m < 0$) and paramagnetic ($\chi_m > 0$) materials are given in Table 14.4. Because of hysteresis effects, the susceptibilities of ferromagnetic materials cannot be represented by a single number but typically have large values of the order of 10^{+3} and greater.

Table 14.4 *Magnetic susceptibilities at room temperature*

Material	Water	Diamond	Magnesium	Tungsten
χ_m	-9.035×10^{-6}	-2.2×10^{-5}	1.2×10^{-5}	6.8×10^{-5}

The *isothermal magnetic susceptibility* χ_T and the *adiabatic magnetic susceptibility* χ_S are given by

$$\chi_T = \left(\frac{\partial \mathcal{M}}{\partial \mathcal{H}}\right)_T \quad \text{and} \quad \chi_S = \left(\frac{\partial \mathcal{M}}{\partial \mathcal{H}}\right)_S. \tag{14.42}$$

In experiments performed at constant temperature χ_m is χ_T, provided the susceptibility is constant. In experiments performed adiabatically χ_m is equal to χ_S. It follows from $M = V\mathcal{M}$ and $\mu_0 M = -(\partial G/\partial \mathcal{H})_T$ that

$$\chi_T = \frac{-1}{\mu_0 V}\left(\frac{\partial^2 G}{\partial \mathcal{H}^2}\right)_T. \tag{14.43}$$

It is left as a problem to show that the energy stored in a Roland ring is $\frac{1}{2}LI^2$, where L is inductance.

14.3 Critical Phenomena

Critical behavior occurs in many different systems and has characteristics that are independent of the type of system involved. The intrinsic instability of systems at a critical point is associated with the divergence of the compressibility K_T in fluids and the divergence of the susceptibility χ_T in magnetic systems. The paramagnetic-ferromagnetic transition is a good illustration of what happens in the vicinity of a critical point.

The critical point in magnetic systems – called the *Curie point* – occurs at zero magnetic field ($\mathcal{H} = 0$) and zero magnetic moment ($M = 0$), as shown at the left in **Figure 14.3**. The Curie point occurs at the critical temperature, which for iron is at 770 °C. At zero-field

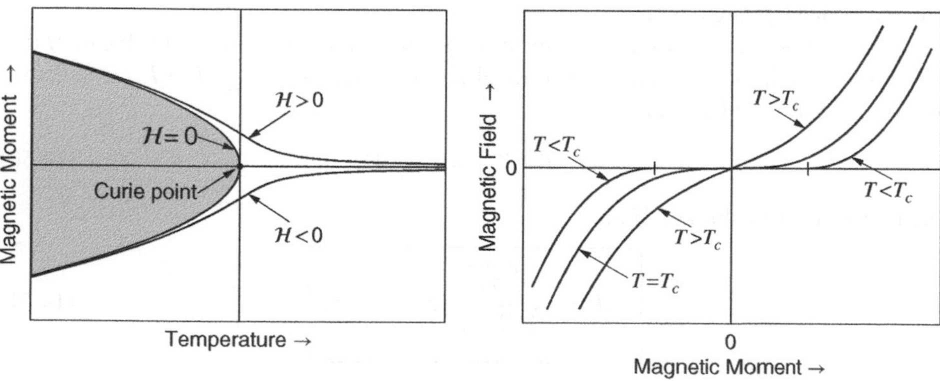

Figure 14.3 *Magnetic Moment M and Field $\mathcal{H}$ in Ferromagnetic Systems*

($\mathcal{H} = 0$) the magnetic moment vanishes in the paramagnetic phase above T_c and is non-zero in the ferromagnetic phase below T_c. The graph at the right indicates that the magnetic field is an odd function of the magnetic moment, *i.e.*, $\mathcal{H}(M) = -\mathcal{H}(-M)$. It also shows that the inverse susceptibility χ_T^{-1} vanishes at the critical point,

$$\chi_T^{-1} = \left(\frac{\partial \mathcal{H}}{\partial \mathcal{M}}\right)_T = 0, \tag{14.44}$$

which implies that the susceptibility $\chi_T = (\partial \mathcal{M}/\partial \mathcal{H})_T$ diverges at the critical point.

To find the dependence of the magnetic moment on the field strength at the critical point, we express the Helmholtz free energy $F(T, M)$ in a series of powers of the deviations of T and M from their values at the critical point. Since $\mathcal{H}$ is an odd function of M and $\mu_0\mathcal{H} = (\partial F/\partial M)_T$, the free energy is an even function of M, so that

$$F(T, M) = A_0(T) + A_2(T)M^2 + A_4(T)M^4 + \cdots \tag{14.45}$$

and

$$\mu_0\mathcal{H} = \left(\frac{\partial F}{\partial M}\right)_T = 2A_2(T)M + 4A_4(T)M^3 + \cdots. \tag{14.46}$$

Expanding the coefficients $A_2(T)$ and $A_4(T)$ in powers of $T - T_c$ gives

$$\mu_0\mathcal{H} = 2\left[a_{2,0} + a_{2,1}\left(T - T_c\right) + \cdots\right]M + 4\left[a_{4,0} + a_{4,1}\left(T - T_c\right) + \cdots\right]M^3 + \cdots. \tag{14.47}$$

As $\mathcal{M} = M/V$, it follows that $\chi_T^{-1} = (\partial \mathcal{H}/\partial \mathcal{M})_T = V(\partial \mathcal{H}/\partial M)_T$, so that

$$\left(\frac{V}{\mu_0}\right)\chi_T^{-1} = 2\left[a_{2,0} + a_{2,1}\left(T - T_c\right) + \cdots\right] + 12\left[a_{4,0} + a_{4,1}\left(T - T_c\right) + \cdots\right]M^2 + \cdots. \tag{14.48}$$

Since $\chi_T^{-1} = 0$ at the critical point, evaluating (14.48) at $T = T_c$ and $M = 0$ implies that $a_{2,0} = 0$. Then, by considering $T > T_c$ and $M = 0$, it follows from $\chi_T^{-1} \geq 0$ that $a_{2,1} > 0$. We can assume that $a_{4,0} > 0$, otherwise χ_T^{-1} and χ_T could be negative below T_c when M is non-zero. As $\mathcal{H} = 0$ at the critical point, it follows from (14.47) that

$$2\left[a_{2,1}\left(T - T_c\right) + \cdots\right]M + 4\left[a_{4,0} + a_{4,1}\left(T - T_c\right) + \cdots\right]M^3 = 0, \tag{14.49}$$

where $a_{2,1}$ and $a_{4,0}$ are positive.

Since $a_{2,1}(T - T_c)$ is positive above the critical temperature, the only real value of M that satisfies (14.49) is $M = 0$. Below the critical temperature, where $a_{2,1}(T - T_c)$ is negative, equation (14.49) implies that

$$a_{2,1}|T - T_c| = 2a_{4,0}M^2. \tag{14.50}$$

The two non-zero values of M are

$$\boxed{M = \pm\sqrt{\frac{a_{2,1}}{2a_{4,0}}}(T_c - T)^{1/2}.} \tag{14.51}$$

Below T_c the solution with $M = 0$ is unphysical, because χ_T^{-1} would then be negative according to (14.48).

Critical Exponents

The critical exponents specify the functional form of the system's properties as they approach a critical point. For example, Equation (14.51) indicates that the magnetic moment approaches its critical value from below as $(T - T_c)^{1/2}$, where the 1/2 is a *critical exponent*. Another example is the behavior of the inverse susceptibility χ_T^{-1}. After setting $M = 0$ and $a_{2,0} = 0$, equation (14.48) indicates that

$$\chi_T^{-1} = 2a_{2,1}\left(\frac{\mu_0}{V}\right)(T - T_c), \tag{14.52}$$

as the critical point is approached from above. Since χ_T^{-1} approaches its critical value of zero as $(T - T_c)^{+1}$, the *critical exponent* for the inverse susceptibility above the critical point is +1. To find the critical exponent for χ_T^{-1} below the critical point, we need to account for the non-zero value of M by substituting (14.51) into (14.48). The resulting expression for $T < T_c$ is

$$\chi_T^{-1} = \frac{\mu_0}{V}\left[-2a_{2,1}\left(T_c - T\right) + 12a_{4,0}\left(\frac{a_{2,1}}{2a_{4,0}}\left(T_c - T\right)\right)\right] = 4a_{2,1}\left(\frac{\mu_0}{V}\right)(T_c - T). \tag{14.53}$$

Thus, the critical exponent for the inverse susceptibility is +1 both above and below the critical temperature. Critical exponents are defined for other properties including specific heats.

The critical exponents obtained above are referred to as classical values. What is surprising is that they are only approximate. For example, depending on their "universality class" the critical exponent corresponding to the 1/2 in (14.51) range from 0.3 to 0.4. The theory which yields correct values, *renormalization group theory*, is outside the scope of this book.

Landau Theory[2]

The theory of critical phenomena developed by Lev D. Landau predicts the classical values of the critical exponents. It is presented here because of the insights it gives into the nature of critical behavior. An essential concept of Landau theory is the *order parameter*, which is a property of the system that is non-zero below the critical temperature and vanishes above it. In ferromagnetic systems, the order parameter is the magnetic moment, which describes the ordering of the microscopic magnetic moments. The microscopic moments are disordered at high temperatures and tend to align in the same direction at low temperatures. Another example of an order parameter is the electric dipole moment in ferroelectric materials. The order parameter in *PVT* systems is the difference in the densities of the liquid and vapor phases. The order parameter in binary alloys that undergo an order-disorder transition is the fraction of one of the atomic species in a particular sublattice.

Landau theory treats the order parameter η as a free variable whose value is determined by minimizing the free energy $F_g(T, \eta)$. The subscript on $F_g(T, \eta)$ distinguishes it from the free energy $F(T, M)$ considered earlier. $F(T, M)$ is a function of the equilibrium magnetic moment M, while $F_g(T, \eta)$ is a generalized free energy that is defined for both equilibrium

[2] Landau, L. D. and Lifshitz, E. M. (1969) *Statistical Physics*, 2nd edn, Addison-Wesley, p. 429.

and non-equilibrium values of the order parameter. The theory is illustrated by the behavior of ferromagnetic systems in zero field ($\mathcal{H} = 0$).

The theory is illustrated by analyzing the free energy $F_g(T, \eta)$ above and below the critical temperature. As shown in **Figure 14.4**, there is a single minimum at $\eta = 0$ for $T > T_c$ which becomes quite broad at $T = T_c$. Below T_c, there are two minima with the same free energy. Since the equilibrium value of the order parameter is at the minimum (or minimums) of the free energy, the order parameter vanishes above the critical temperature and is non-zero below it. Notice that below T_c the second derivative $\partial^2 F_g/\partial\eta^2$ at $\eta = 0$ is negative below T_c and is positive above T_c and that it vanishes at $T = T_c$.

Figure 14.4 represents a ferromagnetic system when the $\mathcal{H}$-field is zero. When the field is non-zero, the two minima below T_c cease to have the same free energy, as indicated in **Figure 14.5**. In this situation there is single value of the order parameter at which the free energy is minimized, so that there is single value of the magnetic moment which is positive when $\mathcal{H}$ is positive and is negative when $\mathcal{H}$ is negative.

Since η and $T - T_c$ are small in the neighborhood of critical point, the free energy $F_g(T, \eta)$ can be represented by a series of powers of η and $T - T_c$. (The representation is not a convergent expansion.) The free energy is then expressed as

$$F_g(T, \eta) = \left[f_{0,0} + \cdots\right] + \left[f_{2,0} + f_{2,1}\left(T - T_c\right) + \cdots\right]\eta^2 + \left[f_{4,0} + \cdots\right]\eta^4 + \cdots, \quad (14.54)$$

where it is assumed that $F_g(T, \eta)$ is an even function of η. The appearance of (14.54) is similar to (14.45), but its significance is very different. The variable M in (14.45) represents the equilibrium value of the magnetic moment, and the derivative $(\partial F/\partial M)_T$ is equal to

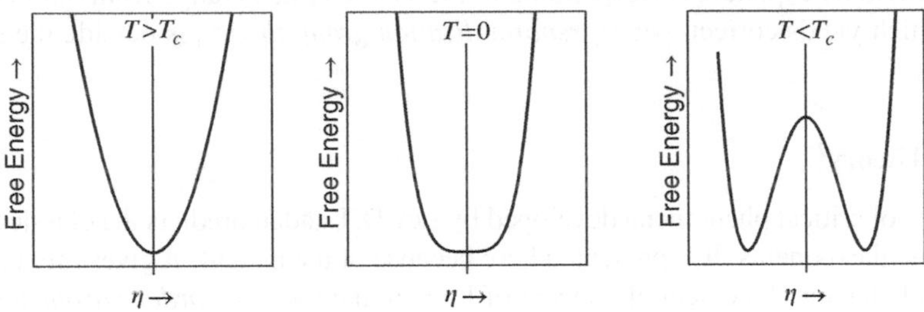

Figure 14.4 *Free Energy Versus Order Parameter in Zero Field*

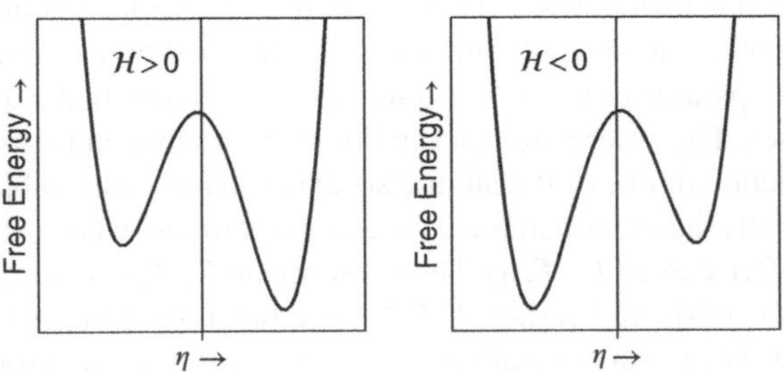

Figure 14.5 *Free Energy Versus Order Parameter at* $T < T_c$

$\mu_0\mathcal{H}$. In contrast, the variable η represents both equilibrium and non-equilibrium values of the magnetic moment. The first two derivatives of $F_g(T,\eta)$ are

$$\left(\frac{\partial F_g}{\partial \eta}\right)_T = 2\left[f_{2,0} + f_{2,1}\left(T - T_c\right) + \cdots\right]\eta + 4\left[f_{4,0} + \cdots\right]\eta^3 + \cdots \tag{14.55}$$

and

$$(\partial^2 F_g/\partial\eta^2)_T = 2\left[f_{2,0} + f_{2,1}\left(T - T_c\right) + \cdots\right] + 12\left[f_{4,0} + \cdots\right]\eta^2 + \cdots. \tag{14.56}$$

Since $\partial^2 F_g/\partial\eta^2 = 0$ at $\eta = 0$ and $T = T_c$, it follows that $f_{2,0} = 0$. Then, for $\partial^2 F_g/\partial\eta^2$ to be positive above T_c requires that $f_{2,1} > 0$. For the free energy to increase away from the minima requires that $f_{4,0} > 0$. The equilibrium value of η is obtained by minimizing the free energy, which requires that $(\partial F_g/\partial\eta)_T = 0$, so that it follows from (14.55) that

$$f_{2,1}(T - T_c)\eta + 2f_{4,0}\eta^3 = 0. \tag{14.57}$$

When $T > T_c$, the only real value of η that satisfies this is $\eta = 0$. When $T < T_c$, the two values of η at which the free energy is minimized are

$$\boxed{\eta = \pm\sqrt{\frac{f_{2,1}}{2f_{4,0}}}(T_c - T)^{1/2}.} \tag{14.58}$$

The above result indicates that the order parameter approaches its critical value from below as $(T_c - T)^{1/2}$, where 1/2 is the critical exponent. For ferromagnetic systems this is the result given in (14.51). The significance of (14.58) is that its derivation via Landau theory is based on general characteristics of the free energy that exist in the neighborhood of critical points in all kinds of systems.

Why does Landau theory yield incorrect results? The reason is that thermodynamics does not account for fluctuations in the properties of systems that result from the random motions of the constituent particles. As discussed in Chapter 22, fluctuations are usually so small that they can be ignored, but this is not the case in the vicinity of a critical point. Renormalization theory is needed to obtain correct results because of the divergent size of the fluctuations.

Problems

14.1 Consider a dielectric with polarization $\mathcal{P} = (a + b/T)\mathcal{E}$ and heat capacity $C_{\mathcal{E}} = A + B\mathcal{E}$, where a and b are constants. Show that $B = 2bV/T^2$ and find the isothermal dielectric constant κ_{el}. (Use the properties of $G(T,\mathcal{E})$ in Table 14.1.)

14.2 Show that the energy stored in a Roland ring with no magnetic field present is $\frac{1}{2}LI^2$. Express L as function of the cross-sectional area A and circumference $2\pi R$.

14.3 Show that if a paramagnetic system obeys Curie's law, $\mathcal{M} = C\mathcal{H}/T$ ($\mathcal{M}$ is magnetization and C is the Curie constant), then there is a temperature dependent contribution to the heat capacity $C_{\mathcal{H}}$ proportional to $\mathcal{H}$. Find this contribution. (Use a Maxwell relation.)

Part III
Statistical Thermodynamics

The science of thermodynamics gives us the tools for describing the properties of matter but gives no indication of their microscopic origins. As viewed macroscopically, systems only change in response to changes in their environments, while at the atomic level they are very dynamic, with molecules, atoms, ions, and electrons in continuous random motion. Microscopic models of thermal behavior were introduced in the nineteenth century at the same time as the foundations of thermodynamics. Although created before the atomic structure of matter was fully accepted, these models give valuable insights into the nature of thermal phenomena. Part III discusses some of these models and insights. The origins of thermal behavior are most easily seen in systems of weakly interacting particles. Since gas molecules are independent except when colliding with other particles or the walls of their container, we focus on the properties of gases and investigate the origin of pressure, the distribution of molecular velocities, and the microscopic significance of entropy.

Thermodynamics and Statistical Mechanics: An Integrated Approach, First Edition.
Robert J Hardy and Christian Binek.

15
Molecular Models

In this chapter we show how the pressure of gases results from the collisions of molecules with their container and how their specific heats result from the distribution of the available energy among the microscopic degrees of freedom. We also investigate the atomic level distinction between heat and work. Although the explanations given are qualitatively correct, the quantitative estimates are sometimes not precise because of the simplifications involved. We start by reviewing a few of the terms employed.

15.1 Microscopic Descriptions

The size of a system is conveniently indicated at the atomic level by the number of particles N, instead of the number of moles n. The two numbers are related by

$$N = nN_A, \tag{15.1}$$

where N_A is given by *Avogadro's number*

$$N_A = 6.022 \times 10^{23}\ \text{mole}^{-1}, \tag{15.2}$$

which is the number of particles per mole. To determine the effects of the atomic level motions, we need to know the masses of the particles. The mass in grams of a mole of particles is called *molecular weight* (or molecular mass). Since there are N_A molecules in a mole, the mass per molecule is

$$m = \frac{(\text{molecular weight})(1\text{g})}{N_A} = (\text{molecular weight})\, m_u, \tag{15.3}$$

where m_u is the *atomic mass unit*

$$m_u = \frac{1\text{g}}{N_A} = 1.6605 \times 10^{-27}\ \text{kg}.$$

Thermodynamics and Statistical Mechanics: An Integrated Approach, First Edition.
Robert J Hardy and Christian Binek.

m_u is sometimes called an *amu* (atomic mass unit). As an illustration, the atomic weight of oxygen is 15.9994. As oxygen normally occurs in diatomic O_2 molecules, its molecular weight to three significant figures is $2 \times 16.0 = 32.0$. Hence, the average mass of an O_2 molecule is

$$m = 32.0\, m_u = 5.31 \times 10^{-26} \text{ kg}. \tag{15.4}$$

The *Boltzmann constant* k_B is essential to the microscopic description of thermal phenomena. It is effectively the universal gas constant R expressed in terms of particles instead of moles

$$k_B = \frac{R}{N_A} = 1.381 \times 10^{-23} \text{J K}^{-1}. \tag{15.5}$$

From this and (15.1) it follows that

$$N k_B = nR, \tag{15.6}$$

so that the ideal gas equation of state becomes

$$PV = N k_B T. \tag{15.7}$$

15.2 Gas Pressure

A gas is a collection of molecules moving in random directions at close to the speed of sound with frequent collisions with other molecules and with the walls of their container. To understand the microscopic origin of pressure and the equation of state, we investigate a system of N molecules in a cubic container of volume $V = L^3$. A two dimensional representation of a container is shown in **Figure 15.1**. The forces needed to deflect the molecules when they collide with the walls are the reactions (in the sense of Newton's third law) to the force associated with pressure.

Let $\boldsymbol{F}(t)$ represent the force exerted on the molecule as it collides with the wall at $x = 0$. The time integral of $\boldsymbol{F}(t)$, called impulse, equals the change in momentum. Applying Newton's second law to the motion of the molecule indicates that the x-component of

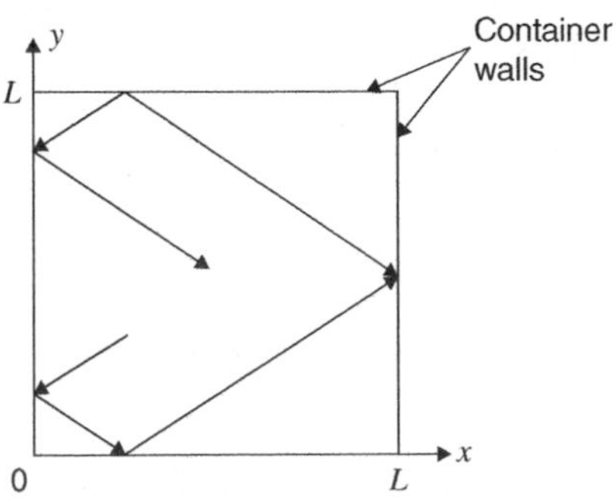

Figure 15.1 Motion of a Gas Molecule Without Intermolecular Collisions

the impulse for the time interval from t to $t + \Delta t$ is

$$\int_{t}^{t+\Delta t} F_x(t')\,dt' = \int_{t}^{t+\Delta t} m\,a(t')\,dt' = m\,v_x(t+\Delta t) - m\,v_x(t), \tag{15.8}$$

where $a(t)$ is the x-component of the acceleration and $m\,v_x(t)$ and $m\,v_x(t+\Delta t)$ are the initial and final components of the momentum. Impulse is conveniently written as $\boldsymbol{F}^{\text{av}}\Delta t$. The x-component of the average force is

$$F_x^{\text{av}} = \frac{1}{\Delta t}\int_{t}^{t+\Delta t} F_x(t')\,dt'. \tag{15.9}$$

For simplicity, assume that the walls are flat and rigid and the collisions are elastic. The magnitude of the momentum is then unchanged by the collision, its component perpendicular to the wall is reversed, and the components parallel to the wall are unchanged. Thus,

$$m\,v_x(t+\Delta t) - m\,v_x(t) = m\,|v_x| - m(-|v_x|) = 2m\,|v_x|. \tag{15.10}$$

Combining the above results indicates that

$$F_x^{\text{av}}\,\Delta t = 2m\,|v_x|. \tag{15.11}$$

A value for Δt is needed. We want F_x^{av} to represent a typical value of the force exerted on the molecule averaged over the time Δt between successive collisions. Since the force $\boldsymbol{F}(t)$ is only non-zero for the very short time during which the molecule is in contact with the wall, the integral of $\boldsymbol{F}(t)$ is independent of Δt for collisions that occur during the interval Δt. Intermolecular collisions tend to keep individual molecules localized, so that the time between successive collisions with a wall will depend on where the molecule is located. Since intermolecular collisions tend to direct some molecules away from a wall while directing others toward it, we can estimate Δt by simply ignoring intermolecular collisions. The trajectory of a molecule in the absence of collisions is shown in **Figure 15.1**. With flat rigid walls the rate at which a molecule moves in the $+x$ and $-x$ directions is unaffected by its motion in the y and z directions. As a result, the distance traveled from the time the molecule leaves the wall at $x = L$, collides with the wall at $x = 0$, and then returns to the wall at $x = L$ is twice the width of the container. The time needed to travel a distance $2L$ in the $\pm x$ directions is

$$\Delta t = \frac{2L}{|v_x|}, \tag{15.12}$$

where $|v_x|$ is the magnitude of the x-component of velocity. Combining this with (15.11) indicates that

$$F_x^{\text{av}} = \frac{m\,|v_x|^2}{L} = \frac{m\,{v_x}^2}{L}. \tag{15.13}$$

The total force exerted by the wall on all N molecules is obtained by adding up the results for the different molecules. To facilitate this, the molecules are labeled by subscript i with integer values from 1 through N. Introducing the subscript into (15.13) gives

$$F_{x,i}^{\text{av}} = \frac{m_i\,{v_{x,i}}^2}{L}. \tag{15.14}$$

The total force exerted by the wall on the N molecules is

$$F_x^{\rm tot} = F_{x,1}^{\rm av} + F_{x,2}^{\rm av} + \cdots + F_{x,N}^{\rm av} = \frac{m_1 {v_{x,1}}^2}{L} + \frac{m_2 {v_{x,2}}^2}{L} + \cdots + \frac{m_N {v_{x,N}}^2}{L}. \quad (15.15)$$

When the standard notation for sums is used, this becomes

$$F_x^{\rm tot} = \sum_{i=1}^{N} F_{x,i}^{\rm av} = \sum_{i=1}^{N} \frac{m_i {v_{x,i}}^2}{L}. \quad (15.16)$$

The pressure on a surface is the perpendicular component of the force per unit of area. For a cubic container the area of a wall is L^2. As the force exerted *on* the molecules *by* the wall has the same magnitude as the force exerted *on* the wall *by* the molecules, the pressure on the wall at $x = 0$ is

$$P = \frac{\text{force}}{\text{area}} = \frac{F_x^{\rm tot}}{L^2} = \frac{1}{L^2}\sum_{i=1}^{N} \frac{m_i {v_{x,i}}^2}{L} = \frac{1}{V}\sum_{i=1}^{N} m_i {v_{x,i}}^2, \quad (15.17)$$

where $V = L^3$ is the volume of the container. Similar expressions with x replaced by y or by z can be obtained by considering the walls at $y = 0$ or at $z = 0$. Since the pressure exerted by a gas is the same in all directions, the results for the walls at $x = 0$, $y = 0$, and $z = 0$ can be combined to give

$$\begin{aligned} P &= \frac{1}{3}\left[\frac{1}{V}\sum_{i=1}^{N} m_i {v_{x,i}}^2 + \frac{1}{V}\sum_{i=1}^{N} m_i {v_{y,i}}^2 + \frac{1}{V}\sum_{i=1}^{N} m_i {v_{z,i}}^2\right] \\ &= \frac{1}{3V}\sum_{i=1}^{N} m_i \left({v_{x,i}}^2 + {v_{y,i}}^2 + {v_{z,i}}^2\right), \end{aligned} \quad (15.18)$$

where $(v_{x,i}, v_{y,i}, v_{z,i})$ are the components of velocity $\boldsymbol{v}_i$ of particle i. By multiplying and dividing by N, the pressure becomes

$$P = \frac{N}{3V}\left(\frac{1}{N}\sum_{i=1}^{N} m_i {v_i}^2\right), \quad (15.19)$$

where ${v_i}^2 = {v_{x,i}}^2 + {v_{y,i}}^2 + {v_{z,i}}^2$.

Averages

The *average* of a property X_i of the molecules in a system of N molecules is represented by $\overline{X}$, where

$$\boxed{\overline{X} = \frac{X_1 + X_2 + \cdots + X_N}{N} = \frac{1}{N}\sum_{i=1}^{N} X_i.} \quad (15.20)$$

This type of average is an *arithmetic mean*. Since the summation eliminates the dependence on i, index i is a *dummy index*.

The *average kinetic energy per molecule* is

$$\overline{\frac{1}{2}mv^2} = \frac{\frac{1}{2}m_1{v_1}^2 + \frac{1}{2}m_2{v_2}^2 + \cdots + \frac{1}{2}m_N{v_N}^2}{N} = \frac{1}{N}\sum_{i=1}^{N}\left(\frac{1}{2}m_i{v_i}^2\right). \tag{15.21}$$

The $\frac{1}{2}$ can be factored out of each term, so that

$$\overline{\frac{1}{2}mv^2} = \frac{1}{2}\overline{mv^2}. \tag{15.22}$$

When the masses of the molecules are the same, they can be factored out of the average, so that $\frac{1}{2}\overline{mv^2}$ becomes $\frac{1}{2}m\overline{v^2}$, where $\overline{v^2}$ is the *mean-square velocity*. Combining (15.19) and (15.21) indicates that the pressure of the gas is

$$PV = N\frac{2}{3}\left(\frac{1}{2}\overline{mv^2}\right). \tag{15.23}$$

Temperature

The relationship between P, V, and N is similar to the relationship in the ideal gas equation of state, $PV = Nk_BT$. This suggests that we equate the value of PV/N implied by (15.23) to the ideal gas value, which implies that

$$\frac{PV}{N} = \frac{2}{3}\left(\frac{1}{2}\overline{mv^2}\right) = k_BT. \tag{15.24}$$

It should be emphasized that we have not derived the ideal gas equation of state but only found a microscopic reason for the proportionality of P to N/V. Nevertheless, equation (15.24) does imply a useful interpretation of absolute temperature: *Absolute temperature is proportional to the average kinetic energy of the molecules*. Specifically, it indicates that the average kinetic energy of each gas molecule is

$$\frac{1}{2}\overline{mv^2} = \frac{3}{2}k_BT. \tag{15.25}$$

These indications are correct to the extent that classical mechanics is valid, but fail at low temperatures when quantum effects become significant.

A useful estimate for the typical speed of a molecule is given by the *root-mean-square velocity* v_{rms}, which for molecules of the same mass is

$$v_{\text{rms}} = \sqrt{\overline{v^2}}. \tag{15.26}$$

For molecules with different masses, the root-mean-square velocity is

$$v_{\text{rms}} = \sqrt{\frac{\overline{mv^2}}{\overline{m}}}, \tag{15.27}$$

where $\overline{m}$ is the average mass. It then follows from (15.25) that

$$v_{\text{rms}} = \sqrt{\frac{3k_BT}{\overline{m}}}. \tag{15.28}$$

Example 15.1 Root-mean-square velocity

Find the root-mean-square velocity of molecules in oxygen at room temperature.

Solution. We can use the mass of an O_2 molecule in (15.4). Since naturally occurring oxygen contains several isotopes, this is essentially $\overline{m}$. By considering room temperature to be 293 K (20 °C), we obtain

$$v_{\text{rms}} = \sqrt{\frac{3k_BT}{\overline{m}}} = \sqrt{\frac{3\,(1.381\times10^{-23}\text{J K}^{-1})(273+20)\text{K}}{5.31\times10^{-26}\text{ kg}}} = 484 \text{ m s}^{-1}.$$

Example 15.2 Frequency of wall collisions

Estimate the number of collisions per second with 1 mm^2 of a solid surface. Consider a gas of oxygen at atmospheric pressure and room temperature.

Solution. Consider a 1.0 mm^2 section of the wall at $x = L$ in the container in Figure 15.1. The change in the momentum of a molecule as it collides with the wall is $2m\,|v_x|$. The total change in the momentum of the molecules after N_c collisions is $N_c 2m\,\overline{|v_x|}$, where $\overline{|v_x|}$ is the average of $|v_x|$. Since impulse $F_x^{\text{av}}\Delta t$ equals change in momentum, the average force exerted by the surface on the gas molecules is

$$F_x^{\text{av}} = \frac{N_c 2m\,\overline{|v_x|}}{\Delta t}.$$

This equals PA, where P is the pressure of the gas and A is 1.0 mm^2. Replacing F_x^{av} with PA in the preceding equation and solving for the number of collisions per unit time gives

$$\frac{N_c}{\Delta t} = \frac{PA}{2m\,\overline{|v_x|}}.$$

By using the root-mean-square velocity v_{rms} to estimate $\overline{|v_x|}$, this becomes

$$\frac{N_c}{\Delta t} \approx \frac{PA}{2m\,v_{\text{rms}}}.$$

By taking atmospheric pressure to be 100 kPa and using the mass in (15.4) and the value of v_{rms} from the previous example, our estimate for the number of collisions per second is

$$\frac{N_c}{\Delta t} \approx \frac{(100\times10^3\text{ Pa})(0.00100\text{ m})^2}{2(5.31\times10^{-26}\text{ kg})(484\text{ m s}^{-1})} = 1.95\times10^{21}\text{ s}^{-1}.$$

The above estimate for the number of collisions with 1 mm^2 of a solid surface is truly huge! Such very large numbers are typical of discussions of the microscopic origin of thermal behavior. Because of them, some of the intuitions we have formed by dealing with numbers of ordinary size may need to be modified.

Independent Particles

Since intermolecular collisions have been neglected, the ideal gas model is an *independent particle model*. When determining the pressure of gases at low densities, the neglect of interactions is justified. Nevertheless, independent particle models are necessarily approximate. Although small, some interaction between particles is necessary, otherwise there would be no mechanism for the exchange of energy needed for creating the randomness characteristic of thermal equilibrium at the atomic level.

15.3 Equipartition of Energy

Degrees of Freedom

The term "degrees of freedom" is used differently in thermal physics than in mechanics. In mechanics the number of degrees of freedom is the number of independent coordinates needed to specify the configuration of a system. For example, three coordinates are needed to specify the position of an unconstrained particle in three dimensions. Hence, each particle has three mechanical degrees of freedom. *In thermal physics the number of degrees of freedom is the number of ways in which a system can take up energy*. For example, since kinetic energy is associated with each of the three components of velocity, three degrees of freedom are associated with the velocity of the center of mass of each molecule, but no degrees of freedom are associated with the positions of molecules in an ideal gas, because no energy is associated with their positions.

Equipartition Hypothesis

The energy associated with a degree of freedom is continually changing. Nevertheless, its value is well defined when averaged over a large number of particles. The equipartition hypothesis asserts that *on the average there is* $\frac{1}{2}k_BT$ *of energy associated with each degree of freedom.*

As three degrees of freedom are associated with the velocity of the center of mass, the average energy associated with each molecule in a gas is

$$\frac{1}{2}m\overline{v^2} = 3\left(\frac{1}{2}k_BT\right). \tag{15.29}$$

This is the same as the value in (15.25) which was found by investigating the microscopic origin of the ideal gas equation of state.

According to the equipartition hypothesis, the internal energy of a gas of N molecules with f degrees of freedom per molecule, is

$$U(T) = Nf\left(\frac{1}{2}k_BT\right) = nf\frac{1}{2}RT, \tag{15.30}$$

where $Nk_B = nR$ has been used. This is the internal energy function of an ideal gas with constant c_V (with the constant u_0 in (3.8) equal to zero). The constant-volume specific heat is

$$c_V = \frac{1}{n}\left(\frac{\partial U}{\partial T}\right)_V = f\frac{1}{2}R. \tag{15.31}$$

Table 15.1 *Properties of monatomic gases. Specific heats in* kcal kmol K^{-1}

	c_V	c_P	γ
Theory	2.98	4.97	1.67
He	2.99	4.97	1.66
Ar	2.98	4.98	1.67
Kr	2.95	4.99	1.69
Xe	3.01	5.02	1.67

The value of R in terms of kilocalories is 1.987 kcal (kmol K)$^{-1}$. Hence, to three significant figures, the value of R is two. From this and (15.31) it follows that the value of c_V in units of kcal (kmol K)$^{-1}$ should equal the number of degrees of freedom per molecule. Since $c_P = c_V + R$ in ideal gases, the constant-pressure specific heat is

$$c_P = \left(1 + \frac{f}{2}\right) R, \tag{15.32}$$

and the ratio of specific heats is

$$\gamma = \frac{c_P}{c_V} = 1 + \frac{2}{f}. \tag{15.33}$$

As different types of molecules have different numbers of degrees of freedom, they have different specific heats. The rare gases are helium, neon, argon, krypton, and xenon (He, Ne, Ar, Kr, and Xe). Since they are monatomic gases (one atom per molecule), the only significant degrees of freedom are the three associated with the velocity of the center of mass. Hence,

$$f = 3, \; c_V = 1.5R, \; c_P = 2.5R, \; \gamma = \frac{5}{3} = 1.67. \tag{15.34}$$

These theoretical values are compared with experimental values in Table 15.1. The agreement between theory and experiment is excellent.

Some diatomic molecules can be treated as *rigid dumbbells, i.e.*, as two point masses a fixed distance apart. Two variables are needed to specify the orientation of a molecule. For example, it could be done by placing one of the atoms at the origin of coordinates and giving the spherical coordinates θ and ϕ of the location of the other atom. Since rotational kinetic energy is associated with the rates of change of these angles, there are two rotational degrees of freedom per molecule. Adding these to the three translational degrees of freedom associated with the center of mass motion gives five degrees of freedom per molecule. Hence,

$$f = 5, \; c_V = 2.5R, \; c_P = 3.5R, \; \gamma = 1.4. \tag{15.35}$$

These theoretical values are compared with experimental values in Table 15.2. The agreement between theory and experiment is very good with the exception of chlorine (Cl_2).

Molecules with more than two atoms require three variables to specify their orientation. Angles θ and ϕ could be used to specify the orientation of the line drawn between two of the atoms with a third angle specifying the orientation (rotation) of the other atoms around

Table 15.2 Properties of diatomic gases. Specific heats in kcal kmol K^{-1}

	c_V	c_P	γ
Theory	4.97	6.95	1.40
H_2	4.87	6.89	1.41
N_2	4.97	6.96	1.40
O_2	5.03	7.02	1.40
Cl_2	5.99	8.15	1.36

Table 15.3 Properties of polyatomic gases. Specific heats in kcal kmol K^{-1}

	Theory	CO_2	NH_3	SO_2	C_2H_6
c_V	5.96	6.75	6.80	7.43	10.35

that line. Kinetic energy is associated with each of these three angles. Adding the three rotational degrees of freedom to the three translational degrees of freedom gives a total of six. Hence,

$$f = 6,\ c_V = 3R,\ c_P = 4R,\ \gamma = \left(\frac{4}{3}\right) = 1.33. \tag{15.36}$$

This theoretical value for c_V is compared with experimental values in Table 15.3. The agreement between theory and experiment is not good.

The above examples assume that the molecules are rigid. What happens if the atoms can vibrate? The energy associated with the vibrational motion would then need to be accounted for. A vibrating diatomic molecule is modeled as two masses connected by a spring. In addition to the angles θ and ϕ specifying the orientation of the masses, the distance between them can change. Let Δr represent the deviation of this distance from the value that minimizes the potential energy. Since both kinetic energy $\frac{1}{2}m(d\Delta r/dt)^2$ and potential energy $\frac{1}{2}K(\Delta r)^2$ are associated with vibrations, where K is an effective spring constant, there are two degrees of freedom associated with the motion. Adding two vibrational degrees of freedom to the two rotational degrees and the three translational degrees gives a total of seven degrees of freedom. Hence,

$$f = 7,\ c_V = 3.5R,\ c_P = 4.5R,\ \gamma = \left(\frac{9}{7}\right) = 1.29. \tag{15.37}$$

Note that the vibrating model overestimates the specific heat of Cl_2 in Table 15.2 by the same amount that the rigid dumbbell model underestimates it.

The number of degrees of freedom and the type of energy for different kinds of gas molecules are summarized in Table 15.4. Although the equipartition hypothesis gives accurate predictions for the simpler molecules, it is of limited value for more complex systems. In a later chapter we prove an *equipartition theorem* based on the use of classical mechanics to describe atomic level behavior. The limitations of the equipartition idea are primarily a consequence of the use of classical mechanics. Quantum mechanics is needed to obtain predictions that are accurate at all temperatures.

Table 15.4 Degrees of freedom per molecule

	Translational (kinetic)	Rotational (kinetic)	Vibrational (kinetic)	Vibrational (potential)	f (total)
Monatomic	3	0	0	0	3
Rigid diatomic	3	2	0	0	5
Rigid polyatomic	3	3	0	0	6
Vibrating diatomic	3	2	1	1	7

15.4 Internal Energy of Solids

The atoms in solids vibrate about average positions, and these positions form a regular lattice of points in space. As a first approximation, the atoms can be modeled as point masses connected to neighboring atoms by springs. When pressure is applied, these springs are compressed and energy is stored in them. This is the static energy $\Phi(V)$ included in the simple solid model in Section 1.1. Its contribution to the internal energy of a solid is independent of temperature. The temperature-dependent contribution arises from the vibrations of the atoms and can be estimated with the equipartition hypothesis.

Law of Dulong and Petit

As discussed in the previous section, both potential and kinetic energies are associated with vibrational motion. Potential energy is associated with the three components of the vector that specify the position of each atom, and kinetic energy is associated with the three components of its velocity. Hence, there are six degrees of freedom per atom ($f = 6$). It then follows from the equipartition hypothesis that the average energy per atom is $6\left(\frac{1}{2}k_BT\right) = 3k_BT$. This result is summarized by the Law of Dulong and Petit (which is called a law for historical reasons). It states that: *On the average there is* $3\,k_BT$ *of vibrational energy associated with every atom in a solid.*

By combining the static and vibrational contributions, the internal energy of a solid becomes

$$U(V,T) = \Phi(V) + 3Nk_BT = \Phi(V) + 3nRT, \tag{15.38}$$

so that the constant-volume specific heat is

$$c_V = \frac{1}{n}\left(\frac{\partial U}{\partial T}\right)_V = 3R. \tag{15.39}$$

This result can be used to compare the Law of Dulong and Petit with experimental data. The specific heat $3R$ is a heat capacity per mole. The experimental data in Table 1.2 is for specific heats that are heat capacities per kilogram. It is left as a problem to convert $3R$ to estimates for the specific heats of a few materials in terms of kilograms. The agreement between theory and experiment is only accurate to about two significant figures.

In solids and liquids, as in gases, pressure is caused by the forces exerted by the system's particles on whatever is confining them. In solids, these forces are primarily associated with the static energy $\Phi(V)$. Because of this, the constant-pressure and constant-volume specific heats of solids are approximately equal ($c_P \approx c_V$), as was found in Example 3.4.

15.5 Inactive Degrees of Freedom

The molecules in gases move about at high speeds with occasional collisions with each other and with the walls of their container. The atoms in solids vibrate about average positions. When in thermal contact, small quantities of energy are exchanged between systems. If this microscopic activity could be slowed down, magnified, and observed with the eye, it would appear extremely random and chaotic. The instantaneous energy associated with each degree of freedom fluctuates and is usually quite different from the average energy. Nevertheless, when averaged over a long time, or when the same type of energy is averaged over many particles, the average is well defined and is shared in a predictable way among the different degrees of freedom. The equipartition hypothesis indicates that the kinetic energy of each particle contributes $\frac{3}{2}k_BT$ to the internal energy. For the purpose of order-of-magnitude estimates, *k_BT of thermal energy is available to each degree of freedom*. (Multiplying k_BT by $\frac{1}{2}$, as suggested by the equipartition hypothesis, would imply a greater level of significance to this statement than is appropriate.)

Quantum Effects

When discussing diatomic gases, it would be reasonable to ask why some molecules can be treated as rigid dumbbells. Why are they not vibrating? We might also wonder why the energy associated with the motions of the electrons in atoms have not been considered. Answering these questions requires quantum mechanics. In quantum mechanics, the energy associated with a particular type of motion is restricted to discrete values referred to as energy levels. Let $\Delta\varepsilon$ represent the energy difference between the lowest energy level for some particular type of motion. *If the thermal energy k_BT is much less than the energy level spacing $\Delta\varepsilon$, the type of motion will be inactive – or frozen out*. Since the energy available decreases as temperature decreases, the different types of motion tend to become inactive at low temperatures, where the meaning of low is determined by $\Delta\varepsilon$. Also, when k_BT is significantly greater than $\Delta\varepsilon$, quantum effects become insignificant and the motion can then be treated classically.

As an illustration, consider the vibrational motion of diatomic molecules, which can be modeled by treating them as harmonic oscillators. The energy levels of a harmonic oscillator are $\varepsilon(n) = \left(n + \frac{1}{2}\right)\hbar\omega$, where n is an integer ($n \geq 0$), ω is the classical angular frequency, and $\hbar$ is Planck's constant divided by 2π. Since the energy difference between levels is $\Delta\varepsilon = \hbar\omega$, the motion is frozen out when $k_BT \ll \hbar\omega$, and the molecule can be treated as a rigid dumbbell.

The temperature dependence of the constant-volume specific heat of hydrogen gas is plotted in **Figure 15.2**. Note the logarithmic scale of the temperature axis. It is clear from the figure that the specific heat is only constant over restricted ranges of temperature. When interpreted in terms of the equipartition hypothesis, the figure indicates that H_2 molecules

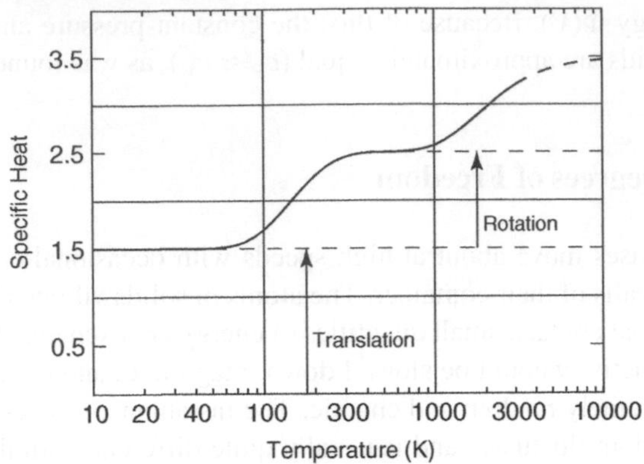

Figure 15.2 *Constant-Volume Specific Heat of Hydrogen Gas*

behave like monatomic molecules ($f = 3$) below about 70 K and behave like rigid dumbbells ($f = 5$) between about 300 and 600 K. The effect of vibrational motion begins to appear at about 1000 K. However, classical vibrational motion does develop because the molecules begin to dissociate above 3000 K.

A useful relationship for comparing thermal behavior with atomic and molecular properties is obtained by expressing the inverse of the Boltzmann constant in units of kelvins per electron-volt,

$$\frac{1}{k_B} = \frac{1}{1.381 \times 10^{-23} \text{J K}^{-1}} \left(\frac{1.602 \times 10^{-19} \text{ J}}{\text{eV}} \right) = 11\,600 \text{ K eV}^{-1}. \tag{15.40}$$

This result associates one electron-volt with a temperature of 11 600 K. The energy difference between the lowest energy levels for the rotational motion of H_2 molecules is $\Delta\varepsilon = 0.015$ eV, which suggests that rotational motion will begin to become inactive at about

$$T \approx \frac{\Delta\varepsilon}{k_B} = (0.015 \text{ eV}) (11\,600 \text{ K eV}^{-1}) = 174 \text{ K}. \tag{15.41}$$

This is consistent with the figure. The value of $\Delta\varepsilon$ for the electronic excitation of H_2 molecules is 11.4 eV. This indicates that the approximate temperature above which the electronic degrees of freedom are active is (11 eV) (10^4 K eV^{-1}) $\approx$ 10^5K, which is well above the temperature at which the molecules dissociate. As the accurate treatment of quantum effects requires knowledge of more than just the lowest energy levels, the temperatures calculated here are only rough estimates. Nevertheless, such estimates give useful insights into the types of atomic level motion that are important.

15.6 Microscopic Significance of Heat

The distinction between the two ways of transferring energy – as heat Q or as work W – is central to thermodynamics. To illustrate the microscopic significance of the distinction, we investigate the interaction of the gas in a cylinder with the piston that confines it. The system

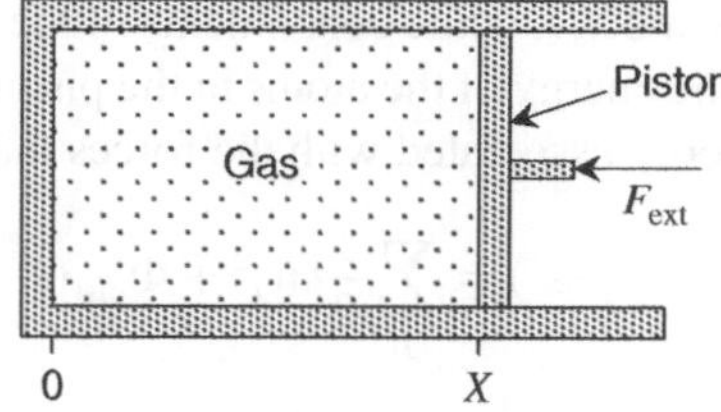

Figure 15.3 Gas Confined in a Cylinder

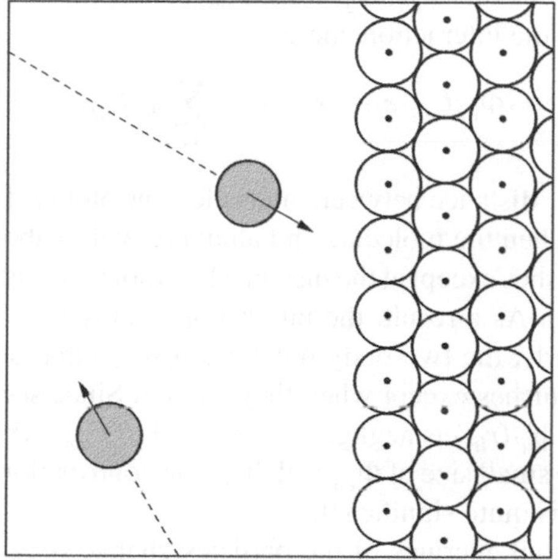

Figure 15.4 Microscopic View of the Piston Face

is diagrammed in **Figure 15.3**, where the piston is held in place by an external force $\boldsymbol{F}_{ext}$ that balances the force exerted by the gas.

Our attention will be focused on the interaction between the molecules in the gas and the atoms in the piston. A small section of the face of the piston is sketched in **Figure 15.4**. The shaded circles on the left represent gas molecules, while the open circles at the right represent atoms in the piston. The atoms in solids vibrate about average positions, which are indicated by the dots inside the open circles. The instantaneous displacements from the average positions of a few atoms are visible in the figure. The dimensions of the microscopic view in Figure 15.4 are measured in nanometers, while the dimensions of the macroscopic view in Figure 15.3 would typically be in centimeters.

The energy of the molecules in the gas is the sum of their kinetic energies plus a small amount of potential energy associated with the forces that arise during intermolecular collisions

$$E_{gas}(\cdots \boldsymbol{v}_i \cdots \boldsymbol{r}_i \cdots) = \sum_i \frac{1}{2} m_i v_i^2 + \Phi_{gas}(\cdots \boldsymbol{r}_i \cdots), \tag{15.42}$$

where m_i is the mass of a molecule, $v_i = |\boldsymbol{v}_i|$ is the magnitude of its velocity, and $\boldsymbol{r}_i = (x_i, y_i, z_i)$ is its position. For simplicity we will consider a monatomic gas. To distinguish the particles in the gas from those in the piston, the constituents of the gas will be called molecules and be identified by index i, while constituents of the piston will be called atoms and be identified by index j. The energy of the atoms in the piston is the sum of their kinetic energies plus the potential energy associated with the forces that bind them together

$$E_{\text{pist}}(\cdots \boldsymbol{v}_j \cdots \boldsymbol{r}_j \cdots) = \sum_j \frac{1}{2} m v_j^2 + \Phi_{\text{pist}}(\cdots \boldsymbol{r}_j \cdots). \tag{15.43}$$

The forces that arise when the gas molecules collide with the piston are the interactions that transfer energy between the gas and piston. The potential energy associated with these forces is represented by Φ_{int}.

The potential energy functions Φ_{int}, Φ_{gas}, and Φ_{pist} can be expressed as sums of two-body potentials. For example, the interaction energy is

$$\Phi_{\text{int}}(\cdots \boldsymbol{r}_i \cdots \boldsymbol{r}_j \cdots) = \sum_{ij} \phi_{ij}(r_{ij}), \tag{15.44}$$

where $r_{ij} = |\boldsymbol{r}_i - \boldsymbol{r}_j|$ is the distance between molecule i and atom j. The two-body potentials $\phi_{ij}(r_{ij})$ are zero except when the molecule and atom are within about a nanometer of each other, so that $\phi_{ij}(r_{ij})$ vanishes except at the instant when molecule i is bouncing off the piston at the location of atom j. As a result, the interaction energy Φ_{int} is negligible compared to E_{gas} and E_{pist}. Similarly, the two-body potential $\phi_{ii'}(r_{ii'})$ that describes the interaction of molecules i and i' vanishes except when they collide. Since specific molecules collide infrequently, the energy $\phi_{ii'}(r_{ii'})$ is negligible compared to E_{gas}, which is almost all kinetic energy. In contrast to the small size of Φ_{int} and Φ_{gas}, the contribution of the potential energy Φ_{pist} to the energy E_{pist} is quite significant.

The kinetic and potential energies of the particles change because their velocities and positions change. It follows from the expression for E_{gas} in (15.42) that its time rate of change is

$$\frac{dE_{\text{gas}}}{dt} = \sum_i m_i \boldsymbol{a}_i \cdot \boldsymbol{v}_i + \sum_i \boldsymbol{v}_i \cdot \frac{\partial}{\partial \boldsymbol{r}_i} \Phi_{\text{gas}}, \tag{15.45}$$

where $\boldsymbol{a}_i = d\boldsymbol{v}_i/dt$ is the acceleration of particle i and $\boldsymbol{v}_i = d\boldsymbol{r}_i/dt$ is its velocity. The symbol $\partial/\partial \boldsymbol{r}_i$ indicates the components of the vector obtained by the differentiations $(\partial/\partial x_i, \partial/\partial y_i, \partial/\partial z_i)$. The time dependence of the velocities and positions are determined by Newton's second law, which asserts that $m_i \boldsymbol{a}_i = \boldsymbol{F}_i$, where $\boldsymbol{F}_i$ is the total force *on* particle i exerted *by* all other particles. Forces are derived from potential energies by differentiating. Specifically, the force on particle i is $\boldsymbol{F}_i = -\partial\Phi/\partial \boldsymbol{r}_i$, where $\Phi = \Phi_{\text{gas}} + \Phi_{\text{int}} + \Phi_{\text{pist}}$. Since Φ_{pist} is not a function of the positions $\boldsymbol{r}_i$ of the gas molecules, we have

$$m_i \boldsymbol{a}_i = \boldsymbol{F}_i = -\frac{\partial}{\partial \boldsymbol{r}_i}(\Phi_{\text{gas}} + \Phi_{\text{int}}). \tag{15.46}$$

Substituting this into (15.45) gives

$$\frac{dE_{\text{gas}}}{dt} = -\sum_i \frac{\partial}{\partial \boldsymbol{r}_i}(\Phi_{\text{gas}} + \Phi_{\text{int}}) \cdot \boldsymbol{v}_i + \sum_i \boldsymbol{v}_i \cdot \frac{\partial}{\partial \boldsymbol{r}_i} \Phi_{\text{gas}} = -\sum_i \boldsymbol{v}_i \cdot \frac{\partial}{\partial \boldsymbol{r}_i} \Phi_{\text{int}}. \tag{15.47}$$

The forces derived from Φ_{gas} leave the value of the gas unchanged, because they simply transfer energy from one molecule to another. Since the interaction energy Φ_{int} depends on the positions of particles in both the gas and the piston, its time rate of change is

$$\frac{d\Phi_{\text{int}}}{dt} = \sum_i \boldsymbol{v}_i \cdot \frac{\partial}{\partial \boldsymbol{r}_i}\Phi_{\text{int}} + \sum_j \boldsymbol{v}_j \cdot \frac{\partial}{\partial \boldsymbol{r}_j}\Phi_{\text{int}}. \tag{15.48}$$

Adding this to (14.47) gives

$$\frac{d}{dt}(E_{\text{gas}} + \Phi_{\text{int}}) = \sum_j \boldsymbol{v}_j \cdot \frac{\partial}{\partial \boldsymbol{r}_j}\Phi_{\text{int}} = -\sum_j \boldsymbol{v}_j \cdot \boldsymbol{F}_j^{\text{int}}, \tag{15.49}$$

where $\boldsymbol{F}_j^{\text{int}} = -\partial\Phi_{\text{int}}/\partial\boldsymbol{r}_j$ is the force *on* molecule *j* in the piston caused by its interaction with the gas. We now use (15.44) to express Φ_{int} as a sum of two-body potentials, so that the force *on* molecule *j* becomes

$$\boldsymbol{F}_j^{\text{int}} = -\frac{\partial}{\partial \boldsymbol{r}_j}\Phi_{\text{int}} = -\frac{\partial}{\partial \boldsymbol{r}_j}\sum_{ij}\phi_{ij}(r_{ij}) = -\sum_i \frac{\partial}{\partial \boldsymbol{r}_j}\phi_{ij}(r_{ij}) = \sum_i \boldsymbol{F}_{j,i}. \tag{15.50}$$

$\boldsymbol{F}_{j,i}$ equals $-\partial\phi_{ij}(r_{ij})/\partial\boldsymbol{r}_j$ which is the force exerted *on* atom *j by* molecule *i*. Thus,

$$\frac{d}{dt}(E_{\text{gas}} + \Phi_{\text{int}}) = -\sum_{ij} \boldsymbol{v}_j \cdot \boldsymbol{F}_{j,i}. \tag{15.51}$$

This tells us that the time rate of change of $(E_{\text{gas}} + \Phi_{\text{int}})$ is the negative of the rate at which work is done *on* the atoms in the piston *by* the molecules in the gas.

The atoms in the piston vibrate about average positions, called lattice sites, and the amplitudes of vibration are small compared to the distance between adjacent sites. To separate the vibrational motion from the macroscopic motion, the displacement of atom *j* from its lattice site is represented by $\boldsymbol{q}_j$ and the location of its lattice site is represented by $\boldsymbol{R}_j$, so that

$$\boldsymbol{r}_j = \boldsymbol{R}_j + \boldsymbol{q}_j. \tag{15.52}$$

This is diagrammed in **Figure 15.5**. Differentiating $\boldsymbol{r}_j$ with respect to time gives

$$\boldsymbol{v}_j = \boldsymbol{V}_j + \boldsymbol{u}_j. \tag{15.53}$$

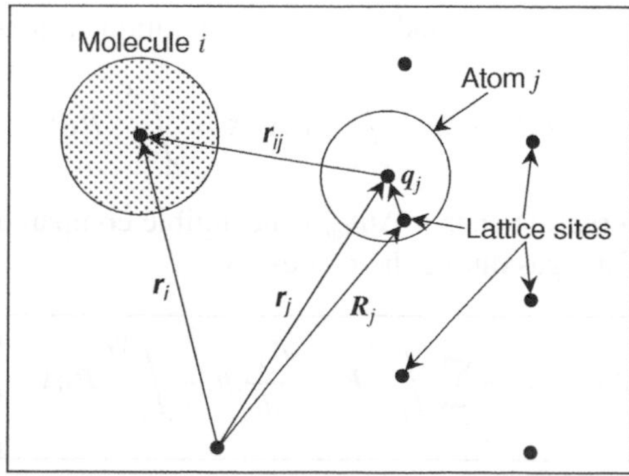

Figure 15.5 *Particles near the Face of the Piston*

$\boldsymbol{v}_j = d\boldsymbol{r}_j/dt$ is the velocity of the atom in a coordinate system fixed in space, $\boldsymbol{u}_j = d\boldsymbol{q}_j/dt$ is its velocity *relative* to its lattice site, and $\boldsymbol{V}_j = d\boldsymbol{R}_j/dt$ is the velocity of the site. As indicated in Figure 15.3, the distance from the left end of the cylinder to the face of the piston is X so that the velocity of the piston is $(dX/dt, 0, 0)$. The lattice sites have fixed positions within the piston, and thus move with the same velocity, which is $\boldsymbol{V}_j = (dX/dt, 0, 0)$. The time rate of change of energy in (15.51) now becomes

$$\frac{d}{dt}(E_{\text{gas}} + \Phi_{\text{int}}) = -\sum_{i,j} \boldsymbol{u}_j \cdot \boldsymbol{F}_{j,i} - \frac{dX}{dt}\sum_{i,j} F^x_{j,i}, \tag{15.54}$$

where $F^x_{j,i}$ is the x-component of $\boldsymbol{F}_{j,i}$.

This result can be simplified by introducing the pressure P of the gas. *Pressure* is the normal component of force on a surface divided by its area. The pressure of the gas is the sum of forces exerted *on* all of the atoms in the piston *by* all of the molecules in the gas,

$$P = \frac{1}{A}\sum_{ij} F^x_{j,i}, \tag{15.55}$$

where A is the area of the piston. Combining this with (15.54) result gives

$$\frac{d}{dt}(E_{\text{gas}} + \Phi_{\text{int}}) = -\sum_{i,j} \boldsymbol{u}_j \cdot \boldsymbol{F}_{j,i} - PA\frac{dX}{dt}. \tag{15.56}$$

The macroscopic velocity of the piston dX/dt is controlled by the external force that holds the piston in place, while the relative velocities $\boldsymbol{u}_j$ are random and chaotic.

The change in energy during a process that begins at initial time t_0 and ends at time t_f is obtained by integrating the time rate of change,

$$\begin{aligned}\Delta(E_{\text{gas}} + \Phi_{\text{int}}) &= \int_{t_0}^{t_f} \frac{d}{dt}(E_{\text{gas}} + \Phi_{\text{int}})dt \\ &= -\sum_{i,j}\int_{t_0}^{t_f} \boldsymbol{F}_{j,i}\cdot \boldsymbol{u}_j dt - \int_{t_0}^{t_f} PA\frac{dX}{dt}dt.\end{aligned} \tag{15.57}$$

We can find the work done by the gas by finding the time rate of change of volume, which is $dV/dt = A(dX/dt)$. By using this and changing integration variables, we find that

$$\int_{t_0}^{t_f} PA\frac{dX}{dt}dt = \int_{X_0}^{X_f} PA\,dX = \int_{V_0}^{V_f} P\,dV. \tag{15.58}$$

The change in the interaction energy $\Delta\Phi_{\text{int}}$ is negligible compared to ΔE_{gas}, so that the change in the energy of the gas during the process is

$$\boxed{\Delta E_{\text{gas}} = -\sum_{i,j}\int_{t_0}^{t_f} \boldsymbol{F}_{j,i}\cdot\frac{d\boldsymbol{q}_j}{dt}dt - \int_{V_0}^{V_f} P\,dV.} \tag{15.59}$$

This result establishes the distinction between the two ways of transferring energy. The work done *by* the gas as a consequence of the motion of the piston is

$$\boxed{W = \int_{V_0}^{V_f} P dV.} \tag{15.60}$$

The term $\boldsymbol{F}_{j,i} \cdot (d\boldsymbol{q}_j/dt)$ is the contribution to the work done *on* atoms j in the piston *by* molecule i in the gas which results from the random chaotic motions of the atoms relative to their lattice sites. When summed over all gas molecules and all atoms in the piston, these terms determine the heat Q_{pist} transferred *to* the piston *from* the gas. Since the heat Q_{pist} is the negative of the heat Q transferred *to* the gas *from* the piston ($Q_{\text{pist}} = -Q$), it follows that the heat transferred to the gas is

$$\boxed{Q = -\sum_{i,j} \int_{t_0}^{t_f} \boldsymbol{F}_{j,i} \cdot \frac{d\boldsymbol{q}_j}{dt} dt.} \tag{15.61}$$

Combining (15.59) through (15.61) gives

$$\boxed{\Delta E_{\text{gas}} = Q - W.} \tag{15.62}$$

Since internal energy U is an equilibrium state function, the change ΔE_{gas} can only be equated to the change in the internal energy when the gas is in equilibrium at the beginning and end of the process. In which case

$$\Delta U = Q - W. \tag{15.63}$$

This is not a full restatement of the first law of thermodynamics, because the first law asserts that internal energy is a state function, which implies that U is a function of the system's thermodynamic coordinates, such as T and V. In contrast, the energy E_{gas} in (15.42) is a function of the velocities and positions $(\cdots \boldsymbol{v}_i \cdots \boldsymbol{r}_i \cdots)$ of the system's constituents. Nevertheless, as a practical matter, the value of E_{gas} is determined by the thermodynamic coordinates when in equilibrium. *Thus, (i) heat is the energy transferred between systems as a result of the random microscopic motions of the constituent particles, and (ii) work is the energy transferred between systems as a result of the macroscopic motion of the interface between them.*

Problems

15.1 In the laboratory 10^{-10} torr is a high vacuum. Use the ideal gas equation of state to estimate the number of molecules in one cubic millimeter of a gas at this pressure when the temperature is 20 °C?

15.2 Air is approximately 80% N_2 and 20% O_2.

(a) Estimate the average molecular weight of air.
(b) Estimate the root-mean-square speed $v_{\rm rms}$ of air molecules at 30 °C.

15.3 **(a)** Starting with the velocity of sound $v_{\rm snd} = \sqrt{\gamma P/\rho}$, where $\rho = M/V$ is mass density, show that $v_{\rm snd} = \sqrt{\gamma k_B T/m}$.
(b) Find a simple expression for the ratio $v_{\rm rms}/v_{\rm snd}$ of the root-mean-square speed to the speed of sound. Calculate its value for air.

15.4 Use the law of Dulong and Petit to estimate the specific heats c_V^M of aluminum, copper, and lead in units of kcal kg^{-1} K^{-1}. Compare with the experimental values in Table 1.2. (The difference between c_P and c_V can be neglected.)

16

Kinetic Theory of Gases

The previous chapter established the dependence of the average speed of gas molecules on temperature but gave no indication about how those speeds are distributed or about the distance the molecules travel between collisions. These questions are answered by the kinetic theory of gases. The distribution of speeds is found by assuming that the speeds of the molecules are randomized by intermolecular collisions and then transforming the physical problem into a combinatorial one. The resulting distribution was predicted by James C. Maxwell and verified many years later. To estimate how far the molecules travel between collisions, we treat them as tiny spheres that move in straight lines except when colliding with another molecule.

16.1 Velocity Distribution

Since speed is the magnitude of velocity, the distribution of molecular speeds is determined by the distribution of velocities. To describe a velocity distribution, we introduce *velocity space* whose coordinates are the components (v_x, v_y, v_z) of velocity. The space is divided into cells of equal volume and the number of molecules with velocities in each cell is specified. For example, small cells in velocity space of volume $(\Delta v)^3$ can be created by introducing planes at $v_x = \left(I_x + \frac{1}{2}\right) \Delta v$, $v_y = \left(I_y + \frac{1}{2}\right) \Delta v$, and $v_z = \left(I_z + \frac{1}{2}\right) \Delta v$, where I_x, I_y, and I_z are integers, and the cells are the small cubic regions between adjacent planes. We will use index j to identify different cells and let n_j represent the number of molecules with velocities in cell j. The *distribution of velocities* is specified by the numbers $n_1, \ldots, n_j, \ldots$

The number of molecules in the gas is fixed, so that

$$N = \sum_j n_j, \tag{16.1}$$

where N is the total number of molecules. The total kinetic energy of N molecules of mass m is

$$E = \sum_j n_j \epsilon_j, \tag{16.2}$$

Thermodynamics and Statistical Mechanics: An Integrated Approach, First Edition.
Robert J Hardy and Christian Binek.

where $\epsilon_j = \frac{1}{2}mv_j^2$ is the kinetic energy of a molecule in cell j and $v_j = |\mathbf{v}_j|$ is its speed. The association of a single speed with all of the molecules in a cell is justified by making the cells limitingly small. When in equilibrium, the value of E is given by the equipartition hypothesis, which associates an average energy of $\frac{1}{2}k_BT$ with each component of velocity, so that $E = \frac{3}{2}Nk_BT$.

The distribution $n_1, \dots, n_j, \dots$ does not tell us which molecules have velocities in which cell, and there are many different ways that the velocities can be assigned to the molecules without altering the distribution. Let $W(n_1 \dots n_j \dots)$ represent the number of ways in which velocities can be assigned so that there are n_1 molecules with velocities in cell 1, n_2 molecules with velocities in cell 2, and so forth. The distribution of velocities in thermal equilibrium is found by using the following assumption:

The velocity distribution of a gas in equilibrium is given by the numbers $n_1, \dots, n_j, \dots$ that maximize the number of arrangements $W(n_1 \dots n_j \dots)$ consistent with the specified values of N and E.

Although this assumption is plausible, it is *ad hoc i.e.*, not a consequence of any previously established principles. Its validity is established by verifying that the predictions obtained agree with experimental results.

16.2 Combinatorics

Finding $W(n_1 \dots n_j \dots)$ is equivalent to the combinatorial problem of finding the number of ways of arranging N objects (molecules) in J boxes (cells). Each arrangement corresponds to a different way of associating velocities with molecules.

Multinomial Distribution

The number of ways of arranging N objects in J boxes with n_j objects in box j is given by the multinomial coefficient

$$W(n_1 \dots n_j \dots) = \frac{N!}{n_1! \dots n_j! \dots n_J!}, \tag{16.3}$$

where $j = 1, 2, \dots, J$. The W's are called multinomial coefficients because they are the coefficients in the expansion

$$(x_1 + \cdots + x_j + \cdots + x_J)^N = \sum W(n_1 \dots n_j \dots) x_1^{n_1} \dots x_j^{n_j} \dots x_J^{n_J}, \tag{16.4}$$

where the sum includes all non-negative values of the n_j such that $\sum_j n_j = N$.

We can understand (16.3) by imagining a collection of boxes like those in **Figure 16.1**. The objects are distinct (distinguishable) in the sense that they can be identified by indices. Initially, we ignore the division into boxes and consider the indices (objects) appearing in the positions at the bottom of the figure. There are N different ways of choosing the object to put in the first position. (In the figure object 21 is there.) As one object has been used, there are $N - 1$ ways of choosing the object to put in the second position. (In the figure object 83 is there.) As two objects have now been used, there are $N - 2$ ways of

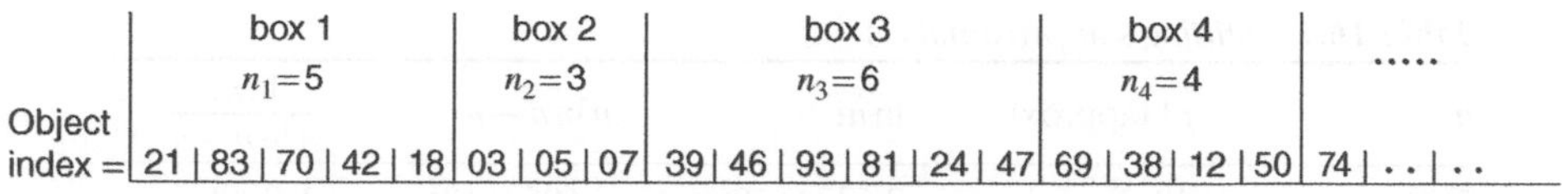

	box 1	box 2	box 3	box 4	.....
	$n_1=5$	$n_2=3$	$n_3=6$	$n_4=4$	
Object index =	21 83 70 42 18	03 05 07	39 46 93 81 24 47	69 38 12 50	74 .. ..

Figure 16.1 *An Arrangement of Objects in Boxes*

choosing the object to put in the third position, and so forth. After choosing the object for the next-to-last position, only one object is left for the Nth position. Thus:

The number of ways that N distinct objects can be arranged, or ordered, is

$$N(N-1)(N-2)\cdots(2)(1) = N! \tag{16.5}$$

We now consider the division into boxes and focus on the objects in each box while ignoring how they are arranged within the box. There are $n_j!$ different arrangements of the objects in box j. In the figure there are $n_2! = 3! = 6$ different arrangements of the three objects in box 2, which are [3 | 5 | 7], [7 | 5 | 3], [5 | 7 | 3], [3 | 7 | 5], [7 | 3 | 5], and [5 | 3 | 7]. In the previous paragraph the objects in box 1 were counted $n_1!$ times, the objects in box 2 were counted $n_2!$ times, and so forth. Furthermore, the number of different ways of arranging all N of the objects is $N!$. To ignore the arrangements within the boxes, we must divide $N!$ by the number of arrangements of the objects within each box. The result is $N!/(n_1!\cdots n_j!\cdots)$, which is the number given in (16.3). Since $0! = 1$, the possibility $n_j = 0$ does not lead to a division by zero.

Binomial Distribution

An important special case of the multinomial distribution occurs when N objects are arranged in two boxes. The binomial distribution is obtained by setting $J = 2$ and $n_1 = M$ in (16.3). Since $n_1 + n_2 = N$, it follows that $n_2 = N - M$. The distribution is commonly described as follows:

The number of ways of choosing M objects from a collection of N objects without regard to order is given by the binomial coefficient

$$\binom{N}{M} = \frac{N!}{M!(N-M)!}. \tag{16.6}$$

It is called the binomial coefficient because $(x+y)^N = \sum_{M=0}^{N} \binom{N}{M} x^{N-M} y^M$.

Stirling's Approximation

Because of the large numbers involved, Stirling's approximation for the factorial function is used. It states that $n! \approx n^n e^{-n}$.[1] The natural logarithm of $n!$ is

$$\boxed{\ln n! \approx n \ln n - n.} \tag{16.7}$$

[1] The approximation $\ln n! \approx n \ln n - n + \ln \sqrt{2\pi n}$ is more accurate, but the contribution of $\ln \sqrt{2\pi n}$ is negligible for the values of n of interest.

Table 16.1 Stirling's approximation

n	$n!$ (approx)	$\ln n!$	$n \ln n - n$	$\frac{\ln n!}{n \ln n - n}$
100	10^{258}	3.637×10^2	3.605×10^2	1.0089
1 000	$10^{2\,568}$	5.912×10^3	5.908×10^3	1.00074
10 000	$10^{35\,659}$	8.211×10^4	8.210×10^4	1.000067
100 000	$10^{456\,573}$	1.051×10^6	1.051×10^6	1.0000063
1 000 000	$10^{5\,567\,708}$	1.282×10^7	1.282×10^7	1.00000061

To illustrate the very large values that $n!$ can have, the orders of magnitude of $n!$ are given for a few values of n in the second column of Table 16.1. The exact values of $\ln n!$ divided by the approximate value are given in the last column. We see that Stirling's approximation is accurate to better than 0.1% when n is equal to 1000 or larger.

Although only integer values of n_j are physically meaningful, the factorial function can be generalized to include non-integer values, so that $n!$ can be treated as a differentiable function of n. The maximum value of $W(n_1 \dots n_j \dots)$ can then be found by setting its derivatives with respect to the n_j's equal to zero. The maximum value of W can be found by maximizing $\ln W$, since

$$\frac{\partial \ln W}{\partial n_j} = \frac{1}{W}\frac{\partial W}{\partial n_j}. \tag{16.8}$$

By using Stirling's approximation, the logarithm of the multinomial coefficient in (16.3) becomes

$$\ln W(n_1 \dots n_j \dots) = (N \ln N - N) - \sum_j (n_j \ln n_j - n_j). \tag{16.9}$$

16.3 Method of Undetermined Multipliers

The equilibrium velocity distribution is given by the values of $n_1 \cdots n_j \cdots$ that maximize $\ln W(n_1 \cdots n_j \cdots)$ subject to the conditions on N and E. These conditions could be used to eliminate two of the arguments of $\ln W$, so that the remaining variables are independent, but it is much easier to use *Lagrange's method of undetermined multipliers*.

To illustrate the method, consider the problem of finding the extrema (maxima and minima) of a function of two variables $f(x, y)$ when x and y are subject to the condition

$$g(x, y) = 0. \tag{16.10}$$

Lagrange's method proceeds by multiplying $g(x, y)$ by an undetermined multiplier α and adding it to the function $f(x, y)$. This yields the *auxiliary function* of three variables

$$F_{aux}(x, y; \alpha) = f(x, y) + \alpha g(x, y). \tag{16.11}$$

The constrained extrema of $f(x, y)$ are found by setting the partial derivatives of $F_{ex}(x, y, \alpha)$ with respect to x and y equal to zero,

$$\frac{\partial F_{aux}(x, y, \alpha)}{\partial x} = 0 \quad \text{and} \quad \frac{\partial F_{aux}(x, y, \alpha)}{\partial y} = 0, \tag{16.12}$$

and then selecting the value of α that satisfies the given condition. The extrema are found by finding the values of (x, y, α) that simultaneously satisfy equations (16.10) and (16.12). Let $(\widehat{x}, \widehat{y}, \widehat{\alpha})$ represent those values. Since $g(\widehat{x}, \widehat{y}) = 0$, it follows that $F_{ex}(\widehat{x}, \widehat{y}, \widehat{\alpha}) = f(\widehat{x}, \widehat{y})$. The extrema of $F(x, y, \alpha)$ are also the extrema of $f(x, y)$. The following example illustrates the method.

Example 16.1 Lagrange's method

Find the extrema of $f(x, y) = x^2y$ subject to the condition $y = e^{-x^2/a}$.

Solution. When expressed in the form $g(x, y) = 0$, the condition $y = e^{-x^2/a}$ becomes $y - e^{-ax^2} = 0$. The auxiliary function for the problem is

$$F_{\text{aux}}(x, y; \alpha) = x^2y + \alpha\left(y - e^{-x^2/a}\right),$$

where α is an undetermined multiplier. The three equations whose simultaneous solution is needed are

$$\frac{\partial F_{\text{aux}}(x, y; \alpha)}{\partial x} = 2xy + \alpha(2x/a)e^{-x^2/a} = 0,$$

$$\frac{\partial F_{\text{aux}}(x, y; \alpha)}{\partial y} = x^2 + \alpha = 0, \quad \text{and} \quad y - e^{-x^2/a} = 0.$$

It is straightforward to verify that the two sets of values of (x, y, α) that satisfy these equations are

$$\left(\widehat{x} = +\sqrt{a},\ \widehat{y} = e^{-1},\ \widehat{\alpha} = -a\right) \quad \text{and} \quad \left(\widehat{x} = -\sqrt{a},\ \widehat{y} = e^{-1},\ \widehat{\alpha} = -a\right).$$

The value of $f(x, y)$ at both extrema (which are maxima) is

$$f(\widehat{x}, \widehat{y}) = \widehat{x}^2\widehat{y} = \left(\pm\sqrt{a}\right)^2 e^{-1} = a e^{-1}.$$

Because of the simplicity of this example, the extrema could easily be found by using the condition $y(x) = e^{-x^2/a}$ to eliminate the second argument of $f(x, y)$, which yields the composite function

$$\overline{f}(x) = f(x, y(x)) = x^2 e^{-x^2/a}.$$

The extrema of $\overline{f}(x)$ are found by solving

$$\frac{d\overline{f}(x)}{dx} = 2x e^{-x^2/a} - \frac{2x^3}{a} e^{-x^2/a} = 0,$$

which gives $\widehat{x} = \pm\sqrt{a}$. Thus, the value of $\overline{f}(x) = f(x, y(x))$ at both extrema is $f(\widehat{x}, \widehat{y}) = a e^{-1}$, which is the same as the result found by Lagrange's method.

16.4 Maxwell Distribution

The number of undetermined multipliers equals the number of conditions. Since the values of $n_1, \ldots, n_j, \ldots$ that maximize $\ln W(n_1 \cdots n_j \cdots)$ are subject to conditions on N and on E, the auxiliary function is

$$F_{\text{aux}}(n_1 \ldots n_j \ldots; \alpha, \beta) = \ln W(n_1 \ldots n_j \ldots) + \alpha\left(N - \sum_j n_j\right) + \beta\left(E - \sum_j n_j \epsilon_j\right), \tag{16.13}$$

where α and β are undetermined multipliers. The maximum of $\ln W$ occurs at the values of the n_j that satisfy the equations

$$\frac{\partial F_{\text{aux}}(n_1 \ldots n_j \ldots; \alpha, \beta)}{\partial n_j} = \frac{\partial \ln W(n_1 \ldots n_j \ldots)}{\partial n_j} - \alpha - \beta\epsilon_j = 0. \tag{16.14}$$

It follows from the expression for $\ln W$ in (16.9) that

$$\frac{\partial \ln W}{\partial n_j} = -\frac{\partial}{\partial n_j}(n_j \ln n_j - n_j) = -\ln n_j. \tag{16.15}$$

Using this in (16.14) gives $-\ln n_j - \alpha - \beta\epsilon_j = 0$. Hence, the values of the n_j are

$$\hat{n}_j = e^{-\alpha} e^{-\beta\epsilon_j}. \tag{16.16}$$

The equations that determine α and β are obtained by substituting $\hat{n}_j$ into the conditions on N and E in (16.1) and (16.2), which gives

$$N = \sum_j e^{-\alpha} e^{-\beta\epsilon_j} \quad \text{and} \quad E = \sum_j e^{-\alpha} e^{-\beta\epsilon_j} \epsilon_j. \tag{16.17}$$

The quantity used to describe the distribution of velocities is the *number of molecules per unit volume of velocity space*, $n_{\text{vel}}(\boldsymbol{v})$. It is obtained by dividing the number of molecules with velocities in each cell by the volume of the cell,

$$n_{\text{vel}}(\boldsymbol{v}_j) = \frac{\hat{n}_j}{\Delta v^3} = \left(\frac{e^{-\alpha}}{\Delta v^3}\right) e^{-\beta\epsilon_j}. \tag{16.18}$$

$\boldsymbol{v}_j = (v_{x,j}, v_{y,j}, v_{z,j})$ is the velocity of the molecules in cell j. By using (16.18) and multiplying and dividing by $(\Delta v)^3 = \Delta v_x \Delta v_y \Delta v_z$, the conditions in (16.17) become

$$N = \sum_j \frac{e^{-\alpha}}{(\Delta v)^3} e^{-\beta\epsilon_j} \Delta v_x \Delta v_y \Delta v_z = \sum_j n_{vel}(\boldsymbol{v}_j) \Delta v_x \Delta v_y \Delta v_z \tag{16.19}$$

and

$$E = \sum_j \frac{e^{-\alpha}}{(\Delta v)^3} e^{-\beta\epsilon_j} \epsilon_j \Delta v_x \Delta v_y \Delta v_z = \sum_j n_{vel}(\boldsymbol{v}_j) \left(\frac{1}{2} m v_j^2\right) \Delta v_x \Delta v_y \Delta v_z, \tag{16.20}$$

where

$$\epsilon_j = \frac{1}{2}mv_j^2 = \frac{1}{2}m(v_{x,j}^2 + v_{y,j}^2 + v_{z,j}^2). \tag{16.21}$$

When the cells are sufficiently small, the sums over j become the integrals, so that

$$N = \int_{-\infty}^{+\infty} dv_x \int_{-\infty}^{+\infty} dv_y \int_{-\infty}^{+\infty} dv_z\, n_{\text{vel}}(\boldsymbol{v}) \tag{16.22}$$

and

$$E = \int_{-\infty}^{+\infty} dv_x \int_{-\infty}^{+\infty} dv_y \int_{-\infty}^{+\infty} dv_z\, n_{\text{vel}}(\boldsymbol{v}) \left(\frac{1}{2}mv^2\right). \tag{16.23}$$

Index j has been omitted because the cells are now identified by the value of $\boldsymbol{v}$. By using $\epsilon = \frac{1}{2}mv^2$ the number of molecules per unit volume becomes

$$n_{\text{vel}}(\boldsymbol{v}) = \left(\frac{e^{-\alpha}}{\Delta v^3}\right) e^{-\beta\frac{1}{2}mv^2}. \tag{16.24}$$

It follows from this that the integral in (16.23) equals minus the derivative with respect to β of the integral in (16.22), so that

$$N = \left(\frac{e^{-\alpha}}{(\Delta v)^3}\right) I(\beta) \quad \text{and} \quad E = -\left(\frac{e^{-\alpha}}{(\Delta v)^3}\right)\frac{dI(\beta)}{d\beta}, \tag{16.25}$$

where

$$I(\beta) = \int_{-\infty}^{+\infty} dv_x \int_{-\infty}^{+\infty} dv_y \int_{-\infty}^{+\infty} dv_z\, e^{-\beta\frac{1}{2}mv^2}. \tag{16.26}$$

As $v^2 = v_x^2 + v_y^2 + v_z^2$, we have $e^{-\beta\frac{1}{2}mv^2} = e^{-\beta\frac{1}{2}mv_x^2}e^{-\beta\frac{1}{2}mv_y^2}e^{-\beta\frac{1}{2}mv_z^2}$. From this and $\int_{-\infty}^{+\infty} e^{-u^2/a}du = \sqrt{\pi a}$ it follows that

$$I(\beta) = \int_{-\infty}^{+\infty} e^{-\beta\frac{1}{2}mv_x^2}dv_x \int_{-\infty}^{+\infty} e^{-\beta\frac{1}{2}mv_y^2}dv_y \int_{-\infty}^{+\infty} e^{-\beta\frac{1}{2}mv_z^2}dv_z = \left(\frac{2\pi}{m\beta}\right)^{\frac{3}{2}}, \tag{16.27}$$

and

$$\frac{dI(\beta)}{d\beta} = -I(\beta)\frac{3}{2}\frac{1}{\beta}. \tag{16.28}$$

Substituting these into (16.25) gives

$$N = \frac{e^{-\alpha}}{(\Delta v)^3}\left(\frac{2\pi}{m\beta}\right)^{\frac{3}{2}} \quad \text{and} \quad E = \frac{e^{-\alpha}}{\Delta v^3}\left(\frac{2\pi}{m\beta}\right)^{\frac{3}{2}}\frac{3}{2}\frac{1}{\beta}. \tag{16.29}$$

Rearranging terms and combining results gives

$$\frac{e^{-\alpha}}{(\Delta v)^3} = N\left(\frac{m\beta}{2\pi}\right)^{\frac{3}{2}} \quad \text{and} \quad E = \frac{3}{2}\frac{N}{\beta}. \tag{16.30}$$

As the contribution of the translational degrees of freedom to the internal energy is $E = \frac{3}{2}Nk_BT$, it follows from $E = \frac{3}{2}N/\beta$ that

$$\boxed{\beta = 1/k_BT.} \tag{16.31}$$

By using this and the value for $e^{-\alpha}/(\Delta v)^3$ from (16.30), equation (16.24) becomes

$$\boxed{n_{\text{vel}}(\boldsymbol{v}) = N\left(\frac{m}{2\pi k_BT}\right)^{\frac{3}{2}} e^{-m|\boldsymbol{v}|^2/(2k_BT)}.} \tag{16.32}$$

This is the *Maxwell distribution of velocities*. Its significance is summarized by stating that $n_{\text{vel}}(\boldsymbol{v})\,dv^3$ is the number of molecules with velocities between $\boldsymbol{v}$ and $\boldsymbol{v} + d\boldsymbol{v}$. To emphasize that the volume element is small, the symbol Δ has been replaced by d, so that $(\Delta v)^3$ becomes dv^3. The phrase "molecules with velocities between $\boldsymbol{v}$ and $\boldsymbol{v} + d\boldsymbol{v}$" is an abbreviated way of saying "molecules whose velocities are in the cell of volume $dv^3 = dv_x dv_y dv_z$ that includes velocity $\boldsymbol{v}$".

Statistical Interpretation

Implicit in the derivation of the Maxwell velocity distribution is the assumption that each of the different assignments of velocities to the N molecules is equally probable. With this assumption, the probability of distribution $n_1, \ldots, n_j, \ldots$ equals the probable p_o of each arrangement times the number of assignments that yield the specified distribution, so that the probability of a distribution equals $p_o W(n_1 \ldots n_j \ldots)$. This probability can be maximized by maximizing $W(n_1 \ldots n_j \ldots)$. As a result, our derivation of the Maxwell distribution can be interpreted as an application of the following principle:
When in thermal equilibrium, the distribution of molecular velocities in gases is the most probable distribution.

Distribution of Speeds

The distribution of speeds $n(v)$ is obtained from the distribution of velocities $n_{\text{vel}}(\boldsymbol{v})$ by integrating over all the directions. To do this, we express the velocity in spherical coordinates, so that $\boldsymbol{v} = (v, \theta, \varphi)$ where speed v is the magnitude of $\boldsymbol{v}$. The volume of velocity space between v and $v + dv$ is $dv^3 = v^2 \sin\theta\, dv\, d\theta\, d\varphi$. Since the Maxwell distribution is independent of θ and φ, it follows that *the number of molecules with speeds between v and $v + dv$* is

$$n(v)\,dv = \int_0^{\pi} \sin\theta\, d\theta \int_0^{2\pi} d\phi\, v^2 n_{\text{vel}}(\boldsymbol{v})\,dv = 4\pi v^2 n_{\text{vel}}(v)\,dv. \tag{16.33}$$

Combining this with (16.32) gives

$$\boxed{n(v)dv = 4\pi N\left(\frac{m}{2\pi k_B T}\right)^{\frac{3}{2}} v^2 e^{-mv^2/(2k_B T)}dv.} \tag{16.34}$$

$n(v)$ is the *Maxwell distribution of speeds.*

When the above result is describing a real system, there is a lower limit on the size of the interval dv. Since N is finite, the number of molecules with speeds between v and $v + dv$ may be zero, or some small number, if the interval dv is too small. When analyzing experimental data, the number of particles with speeds in the different intervals will typically vary smoothly and change only slightly as the size of dv is varied – provided N is sufficiently large and dv is not too small. Such practical considerations are sometimes indicated by stating that the intervals must be "macroscopically small and microscopically large".

Averages

It is often convenient to describe the properties of gases in terms of the speed of the molecules. By changing the conditions on N and E in (16.22) and (16.23) from Cartesian coordinates to spherical coordinates, integrating over θ and φ, and using $n = 4\pi v^2 n_{\text{vel}}$, it follows that

$$N = \int_0^\infty n(v)\,dv \tag{16.35}$$

and

$$E = \int_0^\infty n(v)\left(\frac{1}{2}mv^2\right)dv. \tag{16.36}$$

By using $\int_0^{+\infty} u^2 e^{-u^2/a}du = \frac{1}{4}\pi^{\frac{1}{2}}a^{\frac{3}{2}}$ and $\int_0^{+\infty} u^4 e^{-u^2/a}du = \frac{3}{8}\pi^{\frac{1}{2}}a^{\frac{5}{2}}$ the above equations simplify to $N = N$ and $E = \frac{3}{2}Nk_BT$, as they should. *Average speed* is

$$\bar{v} = \frac{1}{N}\int_0^\infty n(v)\,v\,dv, \tag{16.37}$$

and the *mean-square velocity* is

$$\overline{v^2} = \frac{1}{N}\int_0^\infty n(v)\,v^2\,dv. \tag{16.38}$$

The average energy is related to the mean-square velocity by

$$\frac{1}{2}m\overline{v^2} = \frac{1}{N}\int_0^\infty n(v)\left(\frac{1}{2}mv^2\right)dv = \frac{E}{N}. \tag{16.39}$$

Since $E = \frac{3}{2}Nk_BT$, it follows that

$$\overline{v^2} = \frac{3k_BT}{m}. \tag{16.40}$$

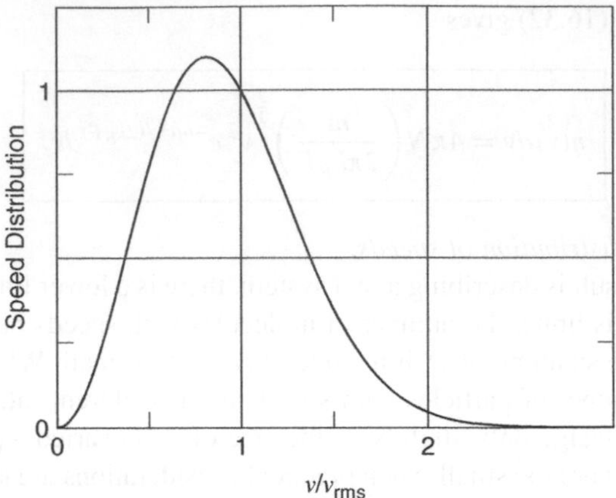

Figure 16.2 *Maxwell Distribution of Speeds*

The *root-mean-square velocity* is

$$v_{\rm rms} = \sqrt{\overline{v^2}} = \sqrt{\frac{3k_BT}{m}}. \tag{16.41}$$

This is the same value obtained in Section 15.2. The Maxwell distribution of speeds is plotted versus $v/v_{\rm rms}$ in **Figure 16.2**. The speed at the maximum is the *most probable speed* $\widehat{v}$, which is obtained by solving $\partial n(\widehat{v})/\partial v = 0$, which gives

$$\widehat{v} = \sqrt{\frac{2k_BT}{m}}. \tag{16.42}$$

It is left as a problem to find the average speed $\overline{v}$. Notice that the most probable speed is less than the average speed which is less than $v_{\rm rms}$,

$$\widehat{v} < \overline{v} < v_{\rm rms}. \tag{16.43}$$

Although derived for monatomic ideal gases, the validity of the Maxwell distribution is not limited to those systems. In fact, it describes the distribution of the translational velocities of the molecules in gases of all kinds as well as the velocity distribution associated with the vibrational motion of the atoms in solids. The principal restriction on its validity is that the temperature is sufficiently high that quantum effects are negligible.

Verification

The first direct measurement of molecular velocities was made in 1920 by O. Stern.[2] This original work was followed by more accurate experiments. In 1930, Zartman used an apparatus like that diagrammed in **Figure 16.3** to measure the velocity distribution in a vapor of bismuth.[3] Although bismuth is a solid at ordinary temperatures, it forms a vapor of

[2] Stern, O. (1920) *Zeitschrift fur Physik*, **2**, 49 and **3**, 415.

[3] Zartman, I. F. *Physical Review*, **37**, 383–391. Also Ko, C. C. (1934) *Journal of the Franklin Institute*, **217**, 173–199.

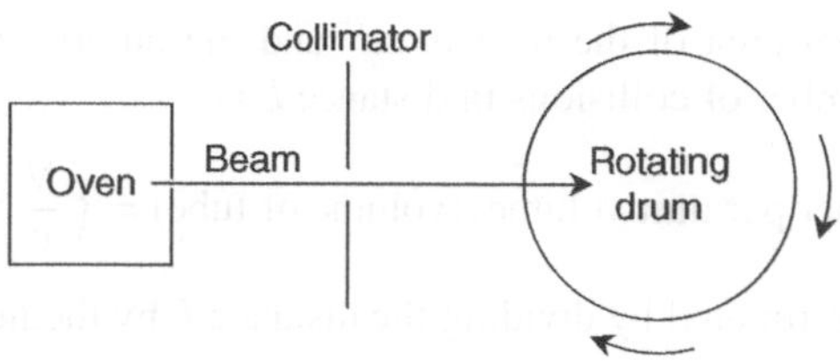

Figure 16.3 Apparatus for Measuring Velocity Distributions

monatomic and diatomic molecules at high temperatures. As indicated in the diagram, the bismuth is heated in an oven, a few of the particles pass through a hole in the wall of the oven, are collimated, and continue towards a rapidly rotating drum. When a narrow slit in the wall of the drum crosses the path of the beam, a few molecules enter the drum, cross to the far side, and stick to the inside surface. The faster molecules reach the far side before the slower ones. Because the drum is rotating, the place where a molecule adheres to the surface is determined by its speed, the diameter of the drum, and the rotational speed. This allows the distribution of speeds to be determined from the profile of the deposit on the inside surface. Since there are two types of molecule in the vapor, the profile is a sum of two distributions. Good agreement with the Maxwell distribution was obtained.

16.5 Mean-Free-Path

Interactions between particles are needed for systems to come to equilibrium. In gases, the interactions are the collisions of molecules with each other. Between collisions, the trajectory of a molecule is a straight line. The average distance that a molecule travels between collisions is called the *mean-free-path* and is designated by $\bar{l}$.

We can estimate the mean-free-path by treating the molecules as small spheres of diameter d. We then focus on one molecule and ignore the motion of the others. As indicated in **Figure 16.4**, the moving molecule collides with other molecules whose centers are a distance d or less from its trajectory. It misses molecule 1 in the figure but collides with molecule 2. We can imagine that the moving molecule traces out a tube of radius d as it travels and collides with the molecules with centers within the tube. The tube has a bend everywhere that a collision occurs. Thus, the number of collisions experienced by the molecule as it moves a distance L along the tube equals the number of other molecules within length L.

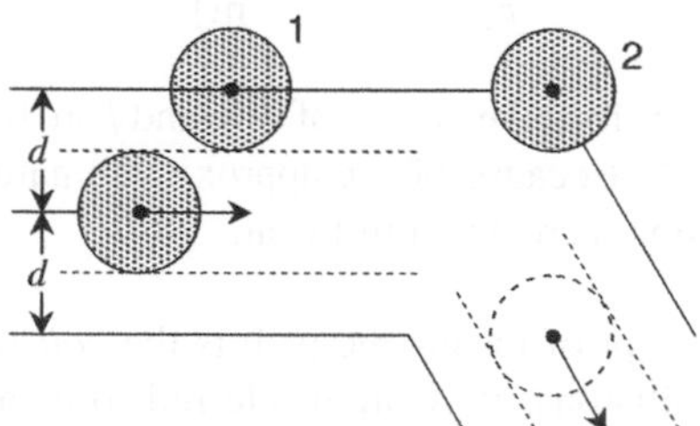

Figure 16.4 An Intermolecular Collision

Since the cross-sectional area of the tube is πd^2 and the number of molecules per unit volume is N/V, the number of collisions in distance L is

$$N_{\text{collisons}} = (\text{number per unit volume})(\text{volume of tube}) = \left(\frac{N}{V}\right)(\pi d^2 L). \tag{16.44}$$

The mean-free-path is estimated by dividing the distance L by the number of collisions,

$$\bar{l} = \frac{L}{N_{\text{collisons}}} = \frac{1}{(N/V)\pi d^2}. \tag{16.45}$$

Example 16.2 Microscopic distances

Estimate the typical distance between molecules and the mean-free-path in argon gas at 0.0 °C and 100 kPa. The effective diameter of an atom (molecule) of argon is $d \approx$ 0.30 nm.

Solution. The distance between molecules is continuously changing, but a rough estimate of the intermolecular separation can be obtained by placing the molecules in a cubic array and finding the distance between adjacent molecules. The Cartesian coordinates of the molecules in a cubic array are $(n_x\bar{s}, n_y\bar{s}, n_z\bar{s})$, where n_x, n_y, and n_z are integers and $\bar{s}$ is the distance between adjacent molecules. The volume per molecule is $\bar{s}^3$, so that the volume of a system of N molecules is $V \approx N\bar{s}^3$. Solving for $\bar{s}$ gives

$$\bar{s} \approx \left(\frac{V}{N}\right)^{\frac{1}{3}}.$$

The value of V/N can be obtained from the ideal gas equation of state ($PV = Nk_BT$),

$$\frac{V}{N} = \frac{k_BT}{P} = \frac{(1.38 \times 10^{-23}\text{J K}^{-1})(273\text{ K})}{1.00 \times 10^5\text{ Pa}} = 3.77 \times 10^{-26}\text{ m}^3.$$

Thus, the typical distance between atoms is of the order of

$$\bar{s} \approx \left(\frac{V}{N}\right)^{\frac{1}{3}} = 3.4 \times 10^{-9}\text{ m} = 3.4\text{ nm}.$$

The mean-free-path is estimated with (16.45). By using the above value for V/N, we obtain

$$\bar{l} = \frac{1}{(N/V)\pi d^2} = \left(\frac{V}{N}\right)\frac{1}{\pi d^2} = \frac{3.77 \times 10^{-26}\text{ m}^3}{\pi(3.0 \times 10^{-10}\text{ m})^2} = 1.33 \times 10^{-7}\text{ m} = 130\text{ nm}.$$

Hence, under ordinary conditions the values of d, $\bar{s}$, and $\bar{l}$ are 0.30, 3.4, and 130 nanometers, or 3.0, 34, and 1300 Å. Because of the approximate nature of these estimates, the second digits in these numbers are not significant.

Closely related to the concept of mean-free-path is the *collision time* τ, which characterizes the time between collisions. It is often referred to as a *relaxation time* and is a convenient estimate for the time needed for the velocities of the particles in systems not in equilibrium to acquire the Maxwell distribution. It is estimated by dividing the average distance between collisions by a typical value for the speed of molecules. By using the root-mean-square velocity v_{rms}, we obtain

$$\tau = \frac{\bar{l}}{v_{rms}}. \tag{16.46}$$

Example 16.3 Collision time

Estimate the collision time for argon gas at 0.0 °C and 100 kPa.

Solution. The molecular weight of an argon atom is 39.95, and its mass is $39.95\, m_u = 6.634 \times 10^{-26}$ kg. Its root-mean-square velocity is

$$v_{rms} = \sqrt{\frac{3k_BT}{\overline{m}}} = \sqrt{\frac{3(1.381 \times 10^{-23}\ \mathrm{J\,K^{-1}})(273\,\mathrm{K})}{6.634 \times 10^{-26}\ \mathrm{kg}}} = 413\ \mathrm{m\,s^{-1}}.$$

By combining this with the mean-free path found in the previous example, the estimate for the collision time in (16.46) becomes

$$\tau = \frac{\bar{l}}{v_{rms}} = \frac{1.33 \times 10^{-7}\ \mathrm{m}}{413\ \mathrm{m\,s^{-1}}} = 3.2 \times 10^{-10}\ \mathrm{s}.$$

Problems

16.1 Find the average speed $\bar{v}$ of the Maxwell distribution and verify that $\hat{v} < \bar{v} < v_{rms}$.

16.2 A two-level system is a collection of N independent particles with two possible energies. Let n_1 and n_2 represent the number of particles with energy ϵ_1 and ϵ_2, respectively. Find the values of n_1 and n_2 that maximize the number of arrangements $W(n_1, n_2) = N!/(n_1!n_2!)$ consistent with the specified values of N and energy $E = n_1\epsilon_1 + n_2\epsilon_2$. Show that $E = N(e^{-\beta\epsilon_1}\epsilon_1 + e^{-\beta\epsilon_2}\epsilon_2)(e^{-\beta\epsilon_1} + e^{-\beta\epsilon_2})^{-1}$.

16.3 A typical value for the effective diameter of a neon atom is 0.33 nm. Find the average distance between atoms and the mean-free-path for neon gas:

(a) at 100 kPa pressure and 293 K.
(b) at 0.10 kPa pressure and 293 K.

Appendix 16.A

Quantum Distributions

The methods used to derive the Maxwell distribution of velocities can also be used to find the Fermi-Dirac and Bose-Einstein distributions. The derivations presented in this appendix are alternatives to those based on the quantum mechanical principle of indistinguishability discussed in Chapter 26. They are made available here for students interested in the origins of these quantum distributions before the methods of statistical mechanics are developed in later chapters. The Fermi-Dirac distribution describes systems of identical *fermions*, such as electrons and ^{3}He atoms, while the Bose-Einstein distribution describes systems of identical *bosons*, such as ^{4}He atoms.

Fermions obey the *Pauli exclusion principle*, which states that no more than one fermion can occupy any single-particle state. Bosons have the property that any number of particles can occupy any single-particle state. These restrictions are applied by dividing the possible energies into intervals of width $\delta\epsilon$, which have a role similar to the cells in velocity space used in deriving the Maxwell distribution. The single-particle energy states are arranged in these intervals, which are identified by index j. Since there are large numbers of closely spaced energy levels,[4] we can have many states in each interval and still have the width $\delta\epsilon$ sufficiently small that the states in each interval have essentially the same energy, which we designate by ϵ_j. The intervals are labeled so that $\epsilon_{j'} < \epsilon_j$ when $j' < j$. Because of the indistinguishability of identical particles, the only way to create new arrangements of particles is to change the number of particles in the different intervals.

Systems of N identical particles with total energy E are considered. Our problem is to find the number of ways that the particles can be arranged among the intervals. The distribution of particles is described by the numbers $n_1, \ldots, n_j, \ldots$, where n_j is the number of particles in the interval j. The distribution is constrained by the conditions that

$$N = \sum_j n_j \quad \text{and} \quad E = \sum_j n_j \epsilon_j. \tag{16.A.1}$$

The equilibrium distributions for both fermions and bosons are obtained by assuming that: *The distribution of particles in thermal equilibrium is given by the numbers $n_1, \ldots, n_j, \ldots$ that maximize the number of arrangements consistent with the values of N and E.*

Fermi-Dirac distribution

Let K_j represent the number of single-particle states in interval j. To find the Fermi-Dirac distribution, we need to determine the number of ways that n_j particles can be arranged among K_j states with no more than one particle in each single-particle state. Represent the number of ways by $w_f(n_j)$. Finding $w_f(n_j)$ is equivalent to the combinatorial problem of finding the number of ways of choosing n_j objects (fermions) from a collection of K_j objects (states) without regard to order. According to (16.6), the number of ways of choosing M objects from a collection of N distinct objects without regard to order is given by the binomial coefficient $N!/(M!(N-M)!)$. By replacing the N in this by K_j and the M by n_j, we obtain the number of ways that n_j identical particles can be arranged in K_j single-particle states,

$$w_f(n_j) = \frac{K_j!}{n_j!(K_j - n_j)!}. \tag{16.A.2}$$

[4] The single-particle energies of an ideal gas are derived in Section 17.2.

When the particles are indistinguishable, the number of ways of arranging n_j particles among the states in interval j is independent of the arrangements of the particles in the other intervals. Consequently, the number of arrangements W_f of particles for the entire system is the product of the numbers for all of the intervals,

$$W_f(n_1 \ldots n_j \ldots) = w_f(n_1)w_f(n_2) \cdots w_f(n_j) \cdots. \tag{16.A.3}$$

The distribution $n_1 \cdots n_j \cdots$ that maximizes W_f can be found by maximizing its logarithm, which is

$$\ln W_f(n_1 \cdots n_j \cdots) = \ln w_f(n_1) + \ln w_f(n_2) + \cdots + \ln w_f(n_j) + \cdots. \tag{16.A.4}$$

By using Stirling's approximation $\ln n! = n \ln n - n$ and the above expression for $w_f(n_j)$, we find that

$$\begin{aligned} \ln W_f(n_1 \ldots n_j \ldots) &= \sum_j \ln w_f(n_j) \\ &= \sum_j [K_j(\ln K_j - 1) - n_j(\ln n_j - 1) - (K_j - n_j)(\ln(K_j - n_j) - 1)]. \end{aligned} \tag{16.A.5}$$

The distribution $n_1, \ldots, n_j, \ldots$ that maximizes this is found by using Lagrange's method of undetermined multipliers. By using α and β as the undetermined multipliers for the conditions on N and E, the auxiliary function is

$$F_{\text{aux}}(n_1 \ldots n_j \ldots ; \alpha, \beta) = \ln W_f(n_1 \ldots n_j \ldots) + \alpha\left(N - \sum_j n_j\right) + \beta\left(E - \sum_j n_j \epsilon_j\right). \tag{16.A.6}$$

By using the expression for $\ln W_f$ in (16.A.5), the derivative of this with respect to n_j becomes

$$\begin{aligned} \frac{\partial F_{\text{aux}}(n_1 \ldots n_j \ldots ; \alpha, \beta)}{\partial n_j} &= - \ln n_j + \ln(K_j - n_j) - \alpha - \beta\epsilon_j \\ &= \ln(K_j/n_j - 1) - \alpha - \beta\epsilon_j = 0. \end{aligned} \tag{16.A.7}$$

Let $\hat{n}_j$ represent the value of n_j that maximizes the number of arrangements. It then follows that $\ln(K_j/\hat{n}_j - 1) = \alpha + \beta\epsilon_j$, so that

$$\hat{n}_j = \frac{K_j}{e^{\alpha+\beta\epsilon_j} + 1}. \tag{16.A.8}$$

The values of the multipliers α and β are determined by the conditions on N and E, which are conveniently expressed in terms of the number of states per unit of energy, which for interval j is

$$D(\epsilon_j) = \frac{K_j}{\delta\epsilon}, \tag{16.A.9}$$

where $D(\epsilon)$ is the single-particle density of states.[5] By substituting the values of $\widehat{n}_j$ into the conditions on N and E in (16.A.1), we obtain

$$N = \sum_j \frac{D(\epsilon_j)}{e^{\alpha+\beta\epsilon_j}+1}\delta\epsilon \quad \text{and} \quad E = \sum_j \frac{D(\epsilon_j)\epsilon_j}{e^{\alpha+\beta\epsilon_j}+1}\delta\epsilon. \tag{16.A.10}$$

We state without proof that α and β are related to the absolute temperature T and the chemical potential μ by

$$\beta = \frac{1}{k_BT} \quad \text{and} \quad \alpha = \frac{-\mu}{k_BT}. \tag{16.A.11}$$

By making $\delta\epsilon$ limitingly small, the sums in (16.A.10) become integrals, so that

$$N = \int \frac{D(\epsilon)}{e^{(\epsilon-\mu)/k_BT}+1}d\epsilon \quad \text{and} \quad E = \int \frac{D(\epsilon)\epsilon}{e^{(\epsilon-\mu)/k_BT}+1}d\epsilon. \tag{16.A.12}$$

These relationships can be used to determine the energy E and chemical potential μ from the number of particles N and the temperature T. At absolute zero ($T = 0$) all single-particle states with energies less than a maximum value are occupied and all states with energies greater than the maximum are unoccupied. The maximum energy is called the *Fermi energy*, and the chemical potential equals the Fermi energy when $T = 0$.

The average number of particles in each state is more useful than the number $\widehat{n}_j$. By using index k to identify individual single-particle states, the average number of particles in state k in interval j is

$$\overline{N}_k = \frac{\widehat{n}_j}{K_j}. \tag{16.A.13}$$

From this and (16.A.8) it follows that the average occupation number of state k is

$$\boxed{\overline{N}_k = \frac{1}{e^{(\epsilon_k-\mu)/k_BT}+1},} \tag{16.A.14}$$

where ϵ_k is the energy of the state. This is the *Fermi-Dirac distribution*, which is also called the *Fermi distribution*. A graph of $\overline{N}_k$ versus ϵ_k is shown in Figure 27.1.

Bose–Einstein Distribution

To find the Bose-Einstein distribution, we need the number of ways that n_j particles can be arranged among K_j single-particle states, when any number of bosons are allowed in each state. We call this number $w_f(n_j)$. Finding it is equivalent to the combinatorial problem of finding the number of ways that N white objects (bosons) can be placed in M boxes (states) without distinguishing the objects from each other.

The combinatorial problem is solved by imagining that we have N distinct white objects and $M - 1$ distinct black objects arranged along a line, as diagrammed in **Figure 16.A.1**.

[5] A density of states is a number of things per unit of some property, but the things are not always single-particle states and the property is not always energy.

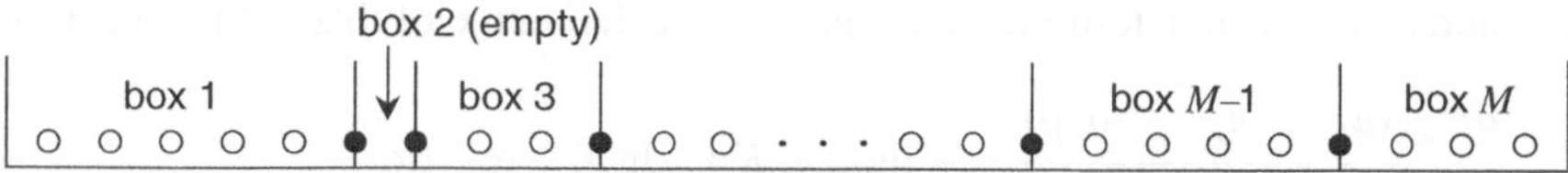

Figure 16.A.1 *A Sequence of White and Black Objects*

The black objects are considered to be walls that separate adjacent boxes. A box can be empty, like box number 2 in the figure. The $M-1$ walls create M boxes. Since the number of ways that n distinct objects can be arranged is $n!$ (see (16.5)), the number of ways that the $N+M-1$ black and white objects can be arranged along the line is $(N+M-1)!$. As the boxes can be identified by their locations along the line, rearranging the $M-1$ black objects has no effect on our ability to identify them. As there are $(M-1)!$ ways in which they can be arranged, we divide by $(M-1)!$. Since the white objects are distinct, there are $N!$ ways of arranging them. To obtain the number of arrangements when the objects are not distinguished from each other, we divide by $N!$. Thus:
The number of ways that N white objects can be placed in M boxes without distinguishing the object from each other is

$$(N+M-1)!/(N!(M-1)!). \tag{16.A.15}$$

When this combinatorial result is applied to bosons, the N white objects correspond to the n_j particles in interval j and the M boxes correspond to the K_j states in the interval. The resulting value for the number of ways of assigning n_j identical bosons to the K_j single-particle states is

$$w_b(n_j) = \frac{(n_j+K_j-1)!}{n_j!(K_j-1)!}. \tag{16.A.16}$$

Since the particles are indistinguishable, the number of arrangements in any interval is unaffected by the arrangements in the other intervals, so that the number of arrangements for the entire system is

$$W_b(n_1\ \dots\ n_j\ \dots) = w_b(n_1)w_b(n_2)\cdots w_b(n_j)\cdots. \tag{16.A.17}$$

By using Stirling's approximation and the expression for $w_b(n_j)$, the logarithm of W_b becomes

$$\begin{aligned} \ln W_b(n_1\ \dots\ n_j\ \dots) &= \sum_j \ln w_b(n_j) \\ &= \sum_j [(n_j+K_j-1)(\ln(n_j+K_j-1)-1) - n_j(\ln n_j - 1) - K_j(\ln(K_j-1)-1)]. \end{aligned} \tag{16.A.18}$$

The maximum number of arrangements is found by using Lagrange's method to maximize $\ln W_b$. The needed auxiliary function is

$$F_{\text{aux}}(n_1\ \dots\ n_j\ \dots\ ;\alpha,\beta) = \ln W_b(n_1\ \dots\ n_j\ \dots) + \alpha\left(N - \sum_j n_j\right) + \beta\left(E - \sum_j n_j\epsilon_j\right), \tag{16.A.19}$$

where α and β are the undetermined multipliers. The derivative of this with respect to n_j is

$$\frac{\partial F_{\text{aux}}(n_1 \ldots n_j \ldots ; \alpha, \beta)}{\partial n_j} = +\ln(n_j + K_j) - \ln n_j - \alpha - \beta\epsilon_j$$

$$= \ln\left(1 + \frac{(K_j - 1)}{n_j}\right) - \alpha - \beta\epsilon_j = 0. \tag{16.A.20}$$

Since K_j is typically much larger than one, the -1 can be omitted. Then, by letting $\hat{n}_j$ represent the value of n_j that satisfies (16.A.20), we find that $\ln(K_j/\hat{n}_j + 1) = \alpha + \beta\epsilon_j$. Solving for $\hat{n}_j$ gives

$$\hat{n}_j = \frac{K_j}{e^{\alpha+\beta\epsilon_j} - 1}. \tag{16.A.21}$$

By substituting $\hat{n}_j$ into the conditions on N and E, introducing the density of states $D(\epsilon_j) = K_j/\delta\epsilon$, and making $\delta\epsilon$ limitingly small, the conditions in (16.A.1) become

$$N = \int \frac{D(\epsilon)}{e^{\alpha+\beta\epsilon} - 1} d\epsilon \quad \text{and} \quad E = \int \frac{D(\epsilon)}{e^{\alpha+\beta\epsilon} - 1} d\epsilon. \tag{16.A.22}$$

As for fermions, the multipliers α and β are related to absolute temperature T and the chemical potential μ by

$$\beta = \frac{1}{k_B T} \quad \text{and} \quad \alpha = -\frac{\mu}{k_B T}, \tag{16.A.23}$$

so that

$$N = \int \frac{D(\epsilon)}{e^{(\epsilon-\mu)/k_B T} - 1} d\epsilon \quad \text{and} \quad E = \int \frac{D(\epsilon)\epsilon}{e^{(\epsilon-\mu)/k_B T} + 1} d\epsilon. \tag{16.A.24}$$

These relationships can be used to determine the energy E and chemical potential μ from the number of particles N and the temperature T. By introducing $\overline{N}_k = \hat{n}_j/K_j$, which gives the average number of particles in each state in interval j, the average occupation number for a single-particle state of energy ϵ_k in a system of identical bosons in equilibrium is

$$\boxed{\overline{N}_k = \frac{1}{e^{(\epsilon_k-\mu)/k_B T} - 1}.} \tag{16.A.25}$$

This is *Bose–Einstein distribution*, which is also called the *Bose distribution.*

17
Microscopic Significance of Entropy

The fundamental concept of entropy was originally deduced by employing macroscopic arguments about the nature of heat. Since the macroscopic characteristics of systems reflect the behavior of their atomic-level constituents, it is natural to ask what microscopic aspect of a system is reflected in its entropy. That question is answered by Boltzmann's insight that entropy is a measure of the number of microscopic configurations accessible to a system once its macroscopic state has been specified. To illustrate the implications of the insight, it is used to predict the properties of an ideal gas.

17.1 Boltzmann Entropy

Microstates

Entropy is a measure of the amount by which a system's microscopic configuration – its *microstate* – can be varied without changing its thermodynamic state – its *macrostate*. When described quantum mechanically, microstates are identified with the system's energy eigenstates, and the variability of a system is measured by the *number of accessible states* $\Omega(E)$, where

$$\boxed{\Omega(E) = \text{Number of microstates with energy between } E - \delta E \text{ and } E.} \tag{17.1}$$

The sum of the kinetic and potential energies of the system's constituents is represented by E. The interval of energy δE, which is negligible compared to E, is needed because of the discrete nature of energy eigenvalues.

The connection between the microscopic and macroscopic properties of systems is made by identifying *Boltzmann entropy* S_B with thermodynamic entropy S and identifying E with

Thermodynamics and Statistical Mechanics: An Integrated Approach, First Edition.
Robert J Hardy and Christian Binek.

the internal energy U. Boltzmann's formula for entropy is[1]

$$\boxed{S_B = k_B \ln \Omega(E).} \tag{17.2}$$

The need for the logarithm of $\Omega(E)$ will become clear when a specific system is considered. Boltzmann's formula tells us that entropy is determined by the number of microstates accessible to a system with the specified value of energy.

Two useful concepts related to $\Omega(E)$ are the *number of microstates with energy less than or equal to E*

$$\Sigma(E) = \text{Number of microstates with energy less than or equal to } E, \tag{17.3}$$

and the *number of microstates per unit of energy* is

$$\omega(E) = \frac{\partial \Sigma}{\partial E}. \tag{17.4}$$

It follows from these definitions that $\Omega(E) = \Sigma(E) - \Sigma(E - \delta E)$, which for small values of δE becomes

$$\Omega(E) = \frac{\partial \Sigma}{\partial E}\delta E = \omega(E)\delta E. \tag{17.5}$$

17.2 Ideal Gas

The significance of Boltzmann entropy S_B can be seen by predicting the properties of a monatomic ideal gas. To verify that Boltzmann's formula yields the known properties of the gas, we need to find $\Omega(E)$, which is done by calculating $\Sigma(E)$ and using (17.5). We begin by finding the energies of the system's microstates. In the ideal gas model, the energy E of a monatomic gas is the sum of the kinetic energies of the individual molecules (which are atoms).

The kinetic energy of an atom (particle) is determined by its momentum. In quantum mechanics the magnitude of momentum is $p = h/\lambda$, where λ is the de Broglie wavelength and h is Planck's constant. When a traveling wave moves between parallel walls, a standing wave pattern with an integer number of half wavelengths results, so that $\lambda = 2L/n$. Thus $p = nh/(2L)$ where n is an integer and L is the distance between the walls. As a result, the kinetic energy of the particle of mass m is $\varepsilon = p^2/2m$, where $\varepsilon = n^2(h^2/8mL^2)$. By applying this result to each coordinate direction, the energies of a particle in a cubic container become

$$\varepsilon(n_x, n_y, n_z) = (n_x{}^2 + n_y{}^2 + n_z{}^2)\frac{h^2}{8mL^2}. \tag{17.6}$$

These *single-particle energies* are identified by the *quantum numbers* (n_x, n_y, n_z), which are positive integers. Note that many different combinations of n_x, n_y, and n_z yield the same value of $\varepsilon(n_x, n_y, n_z)$.

[1] This famous formula is engraved on Boltzmann's tombstone in Vienna, but with the symbol Ω replaced by W.

For simplicity, we consider a gas of N identical particles (atoms) in a cubic container of volume $V = L^3$. Since the particles in the ideal gas model are independent, the system's microstates are determined by the states of every particle. We identify the particles by index i, where $i = 1, 2, \cdots, N$, so that the state of particle i is specified by $(n_{x,i}, n_{y,i}, n_{z,i})$. The system's microstates are then specified by the $3N$ quantum numbers $(n_{x,1}, n_{y,1}, n_{z,1}, n_{x,2}, \ldots, n_{y,N}, n_{z,N})$, and the energy of a microstate (eigenvalue) is the sum of the energies of the particles,[2]

$$E_N(n_{x,1}, n_{y,1}, n_{z,1}, \ldots, n_{z,N}) = \sum_{i=1}^{N} (n_{x,i}^2 + n_{y,i}^2 + n_{z,i}^2) \frac{h^2}{8mL^2}. \tag{17.7}$$

It is convenient to label the quantum numbers by a single subscript j, so that $(n_{x,1}, n_{y,1}, n_{z,1}, \cdots, n_{z,N})$ becomes $(n_1, n_2, n_3, \ldots, n_{3N})$, and the energies of the microstates become

$$E_N(n_1, n_2, \ldots, n_{3N}) = \sum_{j=1}^{3N} \frac{1}{2m}(n_j \Delta p)^2, \tag{17.8}$$

where $\Delta p = h/(2L)$. The quantum numbers n_j are positive integers ($n_j = 1, 2, \cdots, \infty$), and Δp has the dimensions of momentum.

To find the number of microstates with energy less than or equal to E, we need the number of combinations of numbers $(n_1 \cdots n_{3N})$ associated with energies $E_N(n_1 \cdots n_{3N})$ that are less than or equal to E. The formal expression for this number is

$$\Sigma(E) = \sum_{n_1=1}^{+\infty} \cdots \sum_{n_{3N}=1}^{+\infty} \begin{pmatrix} 1, & \text{if } \sum_j \frac{1}{2m}(n_j \Delta p)^2 \leq E \\ 0, & \text{otherwise} \end{pmatrix}. \tag{17.9}$$

where $j = 1, 2, \cdots, 3N$. The value of $\Sigma(E)$ are estimated by replacing the sums by integrations. By introducing $p_j = n_j \Delta p$, summing over p_j, and multiplying and dividing by $\Delta p = h/(2L)$, we obtain

$$\Sigma(E) = L^{3N} \left(\frac{2}{h}\right)^{3N} \sum_{p_1} \Delta p \cdots \sum_{p_{3N}} \Delta p \cdots \begin{pmatrix} 1, & \text{if } \sum_j \frac{1}{2m} p_j^2 \leq E \\ 0, & \text{otherwise} \end{pmatrix}. \tag{17.10}$$

By using $V = L^3$ and taking the limit $\Delta p \to 0$, this becomes

$$\Sigma(E) = V^N \left(\frac{2}{h}\right)^{3N} \int_0^{+\infty} dp_1 \cdots \int_0^{+\infty} dp_{3N} \begin{pmatrix} 1, & \text{if } \sum_j \frac{1}{2m} p_j^2 \leq E \\ 0, & \text{otherwise} \end{pmatrix}. \tag{17.11}$$

Changing the $3N$ integration variables p_j to the dimensionless variables u_j, where $u_j = p_j/\sqrt{2mE}$, gives

$$\Sigma(E) = V^N \left(\frac{2}{h}\right)^{3N} (2mE)^{\frac{3}{2}N} \int_0^{+\infty} du_1 \cdots \int_0^{+\infty} du_{3N} \begin{pmatrix} 1, & \text{if } \sum_j u_j^2 \leq 1 \\ 0, & \text{otherwise} \end{pmatrix}. \tag{17.12}$$

[2] These eigenvalues are derived from the Schrödinger equation in Appendix 20.A.

Since $u_j^2 = (-u_j)^2$, we can change the lower integration limit zero to minus infinity ($-\infty$) and compensate by dividing by two. Hence,

$$\Sigma(E) = V^N \left(\frac{2mE}{h^2}\right)^{\frac{3}{2}N} I_{3N}, \tag{17.13}$$

where

$$I_{3N} = \int_{-\infty}^{+\infty} du_1 \cdots \int_{-\infty}^{+\infty} du_{3N} \begin{pmatrix} 1, & \text{if } \sum_j u_j^2 \leq 1 \\ 0, & \text{otherwise} \end{pmatrix}. \tag{17.14}$$

As shown in Appendix 17.A, the value of I_{3N} is $\pi^{\frac{3}{2}N} / \left(\frac{3}{2}N\right)!$, so that[3]

$$\boxed{\Sigma(E) = V^N \frac{E^{\frac{3}{2}N}}{(3N/2)!} \left(\frac{2\pi m}{h^2}\right)^{\frac{3}{2}N}.} \tag{17.15}$$

The value of I_{3N} contributes to the entropy of the gas, but does not affect the equation of state or internal energy of the gas.

It follows from this (17.15) that

$$\omega(E) = \frac{\partial \Sigma(E)}{\partial E} = V^N \frac{E^{\frac{3}{2}N-1}}{(3N/2-1)!} \left(\frac{2\pi m}{h^2}\right)^{\frac{3}{2}N}. \tag{17.16}$$

By multiplying by $\frac{3}{2}N/E$ and using $\Omega(E) = \omega(E)\delta E$, the number of accessible states becomes

$$\Omega(E) = V^N E^{\frac{3}{2}N} \frac{1}{(3N/2)!} \left(\frac{2\pi m}{h^2}\right)^{\frac{3}{2}N} \frac{3N}{2} \frac{\delta E}{E}. \tag{17.17}$$

By using Stirling's approximation $\ln n! = n \ln n - n$, we find that

$$\begin{aligned} \ln \Omega(E) = & N \ln V + \frac{3}{2} N \ln E + \frac{3}{2} N \ln \left(\frac{2\pi m}{h^2}\right) \\ & - \frac{3}{2}\left[\ln\left(\frac{3N}{2}\right) - 1\right] + \ln\left(\frac{3}{2}\frac{N\delta E}{E}\right). \end{aligned} \tag{17.18}$$

With the exception of the final term, the terms on the right hand side of this are proportional to N, which is of the order of Avogadro's number. It is always possible to choose δE so that the contribution of the final term is negligible. Thus, for a monatomic ideal gas Boltzmann entropy as defined in (17.2) is[4]

$$\boxed{S_B = N k_B \left[\ln V + \frac{3}{2} \ln(E/N) + \frac{3}{2} \ln\left(\frac{2\pi m}{h^2}\right) + \frac{3}{2}\right].} \tag{17.19}$$

[3] Estimating the sums over quantum numbers by integrations has given us the classical value of $\Sigma(E)$. This is discussed in Section 22.4.

[4] The expression for S_B is modified in Section 22.3 by dividing $\Omega(E)$ by $N!$, where $N!$ is the number of indistinguishable arrangements of N identical particles. Dividing by $N!$ changes the $\ln V$ term in (17.19) to $\ln(V/N)$, which makes the expression for entropy fully extensive.

Thermodynamic Properties

The thermodynamic property identified with energy E is the internal energy U. By replacing E by U, the Boltzmann entropy in (17.19) becomes

$$S(U, V) = Nk_B \ln V + \frac{3}{2} Nk_B \ln U + S_o, \tag{17.20}$$

where $S_o = \frac{3}{2} Nk_B \left[-\ln N + \ln \left(\frac{4\pi m}{3h^2} \right) + 1 \right]$. As discussed in Chapter 12, the function $S(U, V)$ that gives the dependence of the entropy on U and V is a thermodynamic potential, which means the properties of the system can be obtained from $S(U, V)$ by differentiation and arithmetic manipulations. By using the expressions for $1/T$ and P/T in (12.29), we find that

$$\frac{1}{T} = \left(\frac{\partial S}{\partial U} \right)_V = \frac{3}{2} Nk_B \frac{1}{U} \quad \text{and} \quad \frac{P}{T} = \left(\frac{\partial S}{\partial V} \right)_U = \frac{Nk_B}{V}. \tag{17.21}$$

It then follows that

$$U = \frac{3}{2} Nk_B T \quad \text{and} \quad PV = Nk_B T. \tag{17.22}$$

The constant-volume heat capacity is $C_V = (\partial U/\partial T)_V = \frac{3}{2} Nk_B$. These are the familiar expressions for the internal energy, equation of state, and heat capacity of a monatomic ideal gas. The above calculation indicates the need for taking the logarithm of $\Omega(E)$. It also illustrates the incredible number of accessible states that exist in macroscopic systems.

Example 17.1 Accessible states

Estimate the number of accessible states $\Omega(E)$ in a monatomic ideal gas of $N = 10^6$ helium atoms in a cubic box of dimension $L = 1.0$mm at temperature $T = 10$K.

Solution. Because of the very large numbers involved, we calculate $\Omega(E)$ by first finding its logarithm and then using the relationship $\Omega = 10^{\log \Omega}$, where $\log \Omega = \ln \Omega / \ln 10$, $\ln x$ is the natural logarithm of x, and $\log x$ is its logarithm base 10. Since helium is a monatomic gas, internal energy is $E = \frac{3}{2} Nk_B T$. By using $V = L^3$, the expression for $\ln \Omega(E)$ in (17.18) becomes

$$\ln \Omega(E) = N \left[3 \ln L + \frac{3}{2} \ln k_B T + \frac{3}{2} \ln \left(\frac{2\pi m}{h^2} \right) + \frac{3}{2} \right].$$

Since $k_B T = 1.38 \times 10^{-22}$ J, $m = 4m_u = 6.64 \times 10^{-27}$ kg, and $h = 6.63 \times 10^{-34}$ J s, we obtain

$$\ln \Omega(E) = 4.681 \times 10^7 \quad \text{and} \quad \Omega(E) = 10^{20\,329\,000}.$$

Very large numbers

The value of $\Omega(E)$ found above is *not* an "ordinary" number like 20 million, which would be written as a 2 followed by 7 zeros. Instead, it would be written as a one followed by 20 million zeros or other digits! Such incredibly large numbers are not often encountered.

As the value $N = 10^6$ used in the example is small, even this value for $\Omega(E)$ is quite small: The equation of state $PV = Nk_BT$ can be used to predict the value of N. For a gas in volume $V = (1.0\,\text{mm})^3$ at room temperature and atmospheric pressure (300 K and 100 kPa) we obtain $N = 2.4 \times 10^{16}$. Because $\ln\Omega(E)$ in the example is proportional to N, the value of $\ln\Omega(E)$ at this temperature and pressure is 2.4×10^{10} times larger than the value in the example! Such large values of $\ln\Omega(E)$ and the associated *very large* values of $\Omega(E)$ are characteristic of macroscopic systems.

17.3 Statistical Interpretation

The statistical nature of thermodynamic behavior is implicit in Boltzmann's formula for entropy, which is a measure of the number of microscopic configurations available to a system. Since energy is conserved, the microstates accessible to an isolated system all have essentially the same energy. The microstates of a monatomic ideal gas were identified by the values of the quantum numbers $(n_1, n_2, \ldots, n_{3N})$. It is convenient to identify the microstates (eigenstates) by a single index α instead of $3N$ indices, where

$$\alpha = (n_1, n_2, \ldots, n_{3N}). \tag{17.23}$$

We could assign integer values 1, 2, 3, etc. to index α and create a procedure for putting the values of α into one to one correspondence with different sets of quantum numbers, or we can think of index α as simply an abbreviated notation for collections of quantum numbers. In terms of α, the number of accessible states of a system with energy E is

$$\Omega(E) = \sum_{\alpha} \begin{pmatrix} 1, & \text{if } E - \delta E < E_{N,\alpha} \leq E \\ 0, & \text{otherwise} \end{pmatrix}, \tag{17.24}$$

where E_α is the energy of microstate α. Implicit in Boltzmann's formula for entropy is the assignment of equal *a priori* probability to every accessible microstate:[5] *Every accessible microstate of an isolated system in thermal equilibrium is equally probable.*

Probabilities

Statistics is essentially applied probability theory. The probabilities assigned to the outcomes of experiments are required to be non-negative and normalized, so that

$$\rho_n \geq 0 \tag{17.25}$$

and

$$\sum_n \rho_n = 1, \tag{17.26}$$

[5] "*a priori*" means based on a hypothesis rather than experiment or experience.

where all possible outcomes n are included in the sum. In the analysis of thermodynamic behavior, the terms "outcome" and "experiment" become "microstate" and "system," respectively. Since every accessible state is equally probable, the normalization condition $\sum_\alpha \rho_\alpha = 1$ requires that the probability of each accessible state is $1/\Omega(E)$, so that the probability assigned to microstate α for a system with energy E is

$$\rho_\alpha(E) = \begin{cases} 1/\Omega(E), & \text{if } E - \delta E < E_{N,\alpha} \leq E. \\ 0, & \text{otherwise.} \end{cases} \tag{17.27}$$

Because $\Omega(E)$ is so large, the probability of any particular microstate is extremely small. Because of the random chaotic motion of the constituent atoms and molecules, the microstate that describes the configuration of a system is continually changing.[6]

In the statistical interpretation of thermodynamic behavior, the macroscopic properties of systems are averages obtained by averaging over the values of the properties in different microstates. Properties such as pressure and energy fluctuate about average values, and the size of the fluctuations can be predicted statistically. In most situations, the fluctuations are of the order of $1/\sqrt{N}$ times the average value, where N is the number of particles in the system. Because N is typically of the order of Avogadro's number ($N_A = 6 \times 10^{23}$), the fluctuations are of the order of $1/\sqrt{N_A} = 1.3 \times 10^{-12}$ times the average value. Since this is much smaller than the typical uncertainty in measured values, fluctuations can often be ignored.[7]

The entropy of a system indicates its disorder, or randomness, at the atomic level. For example, when solids melt and become liquid, the system's structural disorder increases. This is the disorder we would see if the instantaneous arrangement of the particle could be magnified and seen with the naked eye. This increase in structural disorder upon melting is not directly measured by the increase in the value of $\Omega(E)$, which describes the number of *different* configurations accessible to a system. Nevertheless, at every instant the system in just one of those configurations. The reason why liquids are more disordered than solids is because the number of accessible microstates with liquid-like structural disorder is very much greater than the number with solid-like the order.

17.4 Thermodynamic Properties

In our use of Boltzmann's formula to predict the properties of an ideal gas, the number of accessible states $\Omega(E)$ was a function of energy E. Using a specific value of E implies the system is isolated, because the energy of a system that can exchange energy with other systems is not constant. To describe systems that are free to exchange energy with other systems we apply the following principle: *When in thermal equilibrium with other systems, a system's (or subsystem's) internal energy U is identified with the most probable energy $\widehat{E}$ and its entropy S is identified with $k_B \ln \Omega(\widehat{E})$.*

[6] A system's microstates have been identified with energy eigenstates, which are time-independent states, while the actual microscopic configuration evolves over time.

[7] The averaging process is developed in Chapters 18 through 21. Fluctuations are discussed in Chapter 22.

Composite Systems

To verify that Boltzmann entropy and the above principle are consistent with the properties of systems established by thermodynamics, we investigate the behavior of composite systems. We consider systems made up of two parts, called subsystem 1 and subsystem 2. Results for composite systems with more than two parts are obtained by successively considering each part to be subsystem 1 while treating all other parts as subsystem 2. The combined (total) system is assumed to be isolated, so that its energy E_{tot} is fixed. Because of the exchange of energy as heat results from the interactions of the particles at the interface where the subsystems are in contact, the energy of these interactions is negligible compared to the energies E_1 and E_2 of the subsystems, so that

$$E_{tot} = E_1 + E_2. \tag{17.28}$$

For simplicity we assume the minimum possible values of E_1 and E_2 are zero. Since E_{tot} is fixed, we can treat E_1 as an independent variable by letting E_2 depend on E_1 through the relationship $E_2 = E_{tot} - E_1$.

In a *PVT* system the number of accessible states depends on its energy E, volume V, and number of particles N. So that we can focus on the distribution of energy which results from the exchange of energy as heat, the values of V and N are held fixed. When volumes V_1 and V_2 are fixed (or when the analogous variables for other types of systems are fixed), no work is done by the subsystems.

Most Probable Energies

Each microstate α_{tot} of the combined system consists of a microstate α_1 of subsystem 1 and a microstate α_2 of subsystem 2, that is, $\alpha_{tot} = (\alpha_1, \alpha_2)$. For each microstate α_1 with energy E_1 there are $\Omega_2\,(E_{tot} - E_1)$ microstates α_2 accessible to the subsystem 2. As there are $\Omega_1\,(E_1)$ microstates α_1 of subsystem 1 with energy E_1, the number of microstates of the combined system consistent with the specified values of E_{tot} and E_1 is

$$\Omega_{tot}(E_{tot}, E_1) = \Omega_1(E_1)\Omega_2(E_{tot} - E_1). \tag{17.29}$$

The principle that every accessible microstate of an isolated system is equally probable can be used to find the most probable subsystem energies $\hat{E}_1$ and $\hat{E}_2$. Because the probability of a set of microstates is the sum of the probabilities for the individual states, the probability $P_{tot}(E_1)$ that the energy of the combined system when subsystem 1 has energy E_1 is proportional to $\Omega_{tot}(E_{tot}, E_1)$, so that

$$P_{tot}(E_1) = K_0\,\Omega_{tot}(E_{tot}, E_1) = K_0\,\Omega_1(E_1)\Omega_2(E_{tot} - E_1), \tag{17.30}$$

where K_0 is a proportionality constant.

The most probable value of E_1 is found by treating E_1 as a continuous variable and setting the derivative of $P_{tot}(E_1)$ equal to zero,

$$\frac{\partial P_{tot}(\hat{E}_1)}{\partial E_1} = K_0 \left[\frac{\partial \Omega_1\left(\hat{E}_1\right)}{\partial E_1} \Omega_2(E_{tot} - \hat{E}_1) + \Omega_1(\hat{E}_1) \frac{\partial \Omega_2(E_{tot} - \hat{E}_1)}{\partial E_1} \right] = 0. \tag{17.31}$$

Since $E_2 = E_{\text{tot}} - E_1$, the derivative $\partial\Omega_2/\partial E_1$ equals $-\partial\Omega_2/\partial E_2$, so that (17.31) implies that $(\partial\Omega_1/\partial E_1)\Omega_2 - \Omega_1(\partial\Omega_2/\partial E_2) = 0$ which when divided by $\Omega_1\,\Omega_2$ gives

$$\frac{1}{\Omega_1(\hat{E}_1)}\frac{\partial\Omega_1(\hat{E}_1)}{\partial E_1} = \frac{1}{\Omega_2(\hat{E}_2)}\frac{\partial\Omega_2(\hat{E}_2)}{\partial E_2}, \tag{17.32}$$

where the most probable energy of subsystem 2 is

$$\hat{E}_2 = E_{\text{tot}} - \hat{E}_1. \tag{17.33}$$

By using $d\ln x/dx = 1/x$, equation (17.32) can be expressed as

$$\frac{\partial \ln\Omega_1(\hat{E}_1)}{\partial E_1} = \frac{\partial \ln\Omega_2(\hat{E}_2)}{\partial E_2}. \tag{17.34}$$

Temperature

The principle that a subsystem's internal energy is identified with its most probable energy specifies that

$$U_1 = \hat{E}_1 \quad \text{and} \quad U_2 = \hat{E}_2, \tag{17.35}$$

which has important implications for the relationship between entropy and temperature. When the relationship $S = k_B \ln\Omega$ is used and the internal energies are identified with $\hat{E}_1$ and $\hat{E}_2$, equation (17.34) specifies that

$$\frac{\partial S_1(U_1)}{\partial U_1} = \frac{\partial S_2(U_2)}{\partial U_2}. \tag{17.36}$$

The essential characteristic of temperature is its equality among interacting systems that are in equilibrium with each other. Equation (17.36) indicates that their $\partial S/\partial U$ has the characteristics of a temperature. When *PVT* systems are considered and we explicitly indicate that their volume are held fixed, the equation becomes $(\partial S_1/\partial U_1)_{V_1} = (\partial S_2/\partial U_2)_{V_2}$, which implies that $(\partial U_1/\partial S_1)_{V_1} = (\partial U_2/\partial S_2)_{V_2}$. Since the thermodynamic relationship between entropy and absolute temperatures from (12.4) is $T = (\partial U/\partial S)_V$, the energy derivative of Boltzmann entropy gives a microscopic interpretation of absolute temperature T.

Internal Energy

Since the energies E_1 and E_2 are not fixed in the composite system being considered, we need to verify that these values for U_1 and U_2 are consistent with thermodynamics, in which internal energies are well defined. We do this by showing that the probability that E_1 and E_2 differ significantly from their most probable values is extremely small.

We found in (17.17) that the number of accessible states $\Omega(E)$ in a monatomic ideal gas is proportional to E^{cN} with $c = 3/2$. This kind of behavior with various values of c is typical of many systems.[8] To illustrate the distribution of the values of E_1, we consider

[8] A situation in which $\Omega(E)$ is not an increasing function of E is discussed in Appendix 21.A.

subsystems in which the dependence of the numbers of accessible states on their energies is

$$\Omega_1(E_1) = K_1 {E_1}^{c_1 N_1} \quad \text{and} \quad \Omega_2(E_2) = K_2 {E_2}^{c_2 N_2}, \tag{17.37}$$

where K_1 and K_2 are proportionality constants and N_1 and N_2 are the numbers of particles in the two subsystems. By substituting these into the expressions for $P_{tot}(E_1)$ in (17.30), the probability that the energy E_{tot} is distributed with energy E_1 in subsystem 1 becomes

$$P_{tot}(E_1) = K_0 K_1 K_2 {E_1}^{c_1 N_1} (E_{tot} - E_1)^{c_2 N_2}. \tag{17.38}$$

The most probable energy $\widehat{E}_1$ of subsystem 1 is determined by requiring that $\partial P_{tot}(\widehat{E}_1)/\partial E_1 = 0$, which implies that

$$\frac{\widehat{E}_1}{c_1 N_1} = \frac{\widehat{E}_2}{c_2 N_2}, \tag{17.39}$$

where $\widehat{E}_2 = \widehat{E}_{tot} - \widehat{E}_1$. The probability that subsystem 1 has energy E_1 relative to the maximum probability $P_{tot}(\widehat{E}_1)$ is plotted in **Figure 17.1** for the special case with $c_1 = c_2 = 3/2$ and $N_1 = N_2$.

Since the subsystems are of the same type and size, the most probable energies $\widehat{E}_1$ and $\widehat{E}_2$ are both equal to $\frac{1}{2}E_{tot}$. The results in the figure for $N_1 = 10^2$ and $N_1 = 10^4$ indicate that the distribution of energy between subsystems is strongly localized about the most probable distribution and that the localization increases as the number of particles increases.[9] If the result for $N_1 = 10^6$ had been plotted, it would appear as a vertical line at $E_1/E_{tot} = \frac{1}{2}$. Since the number of particles in macroscopic systems is of the order of Avogadro's number, the probability that E_1 and E_2 are different from $\widehat{E}_1$ and $\widehat{E}_2$ is extremely small. Because of this, we are justified in identifying the internal energies U_1 and U_2 with the

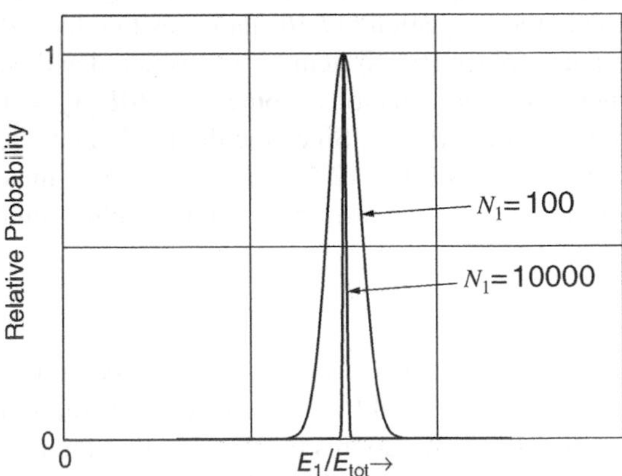

Figure 17.1 *Relative Probability $P_{tot}(E_1)/P_{tot}(\widehat{E}_1)$*

[9] It is left as a problem to show that the widths of the curves in the figure are proportional to $1/\sqrt{N}$.

most probable energies $\widehat{E}_1$ and $\widehat{E}_2$. Then, $E_{tot} = \widehat{E}_1 + \widehat{E}_2$ implies that $U_{tot} = U_1 + U_2$, so that internal energy has the additive property of an extensive state function.

The consistency of the microscopic description of thermal phenomena with the thermodynamic description depend on the high probability that energy is distributed in the most probable way. Since the existence of the high probability depends on the large value of the number of constituent particles, the identification of internal energy and entropy with the corresponding thermodynamic properties requires systems of macroscopic size.

Entropy

To verify that Boltzmann entropy $S_B = k_B \ln \Omega$ has the characteristics of the corresponding state function of thermodynamics, we need to show that it is consistent with the entropy statement of the second law (see Section 10.3). Specifically, we need to show (i) that Boltzmann entropy is extensive, (ii) that it increases as internal energy increases when no work is done, and (iii) that it never decreases in an adiabatically isolated system.

That Boltzmann entropy has the second characteristic is indicated by the behavior of $\Omega(E)$ in ideal gases. Although the dependence of Ω on E in (17.17) is specific to a monatomic gas, it is typical that $\Omega(E)$ is a rapidly increasing function of E when volume V is fixed. Because no work is done when V is fixed, the value of $k_B \ln \Omega(E)$ increases as internal energy increases when no work is done, and this is true whether internal energy is identified with E or with $\widehat{E}$.

Entropy is Extensive

When subsystems 1 and 2 are isolated, the energies E_1 and E_2 are fixed and are the subsystem internal energies. According to (17.29), the number of microstates accessible to the combined system is

$$\Omega_{tot}(E_{tot}, E_1) = \Omega_1(E_1)\Omega_2(E_2), \tag{17.40}$$

where $E_2 = E_{tot} - E_1$. Taking the logarithm of this and multiplying by k_B gives

$$k_B \ln \Omega_{tot}(E_{tot}, E_1) = k_B \ln \Omega_1(E_1) + k_B \ln \Omega_2(E_2). \tag{17.41}$$

It then follows from Boltzmann's formula that

$$S_{tot}(E_{tot}) = S_1(E_1) + S_2(E_2). \tag{17.42}$$

Thus, entropy has the additive property of an extensive state function when the subsystems are isolated.

When a system is in equilibrium, it is a matter of choice in thermodynamics whether an isolated system is treated as a single entity or as a combination of subsystems. When treated as a single entity, its Boltzmann entropy is $S_{tot} = k_B \ln \Omega_{tot}$. When treated as two subsystems, the subsystem entropies are $S_1 = S_1(\widehat{E}_1)$ and $S_2 = S_2(\widehat{E}_2)$, and the extensive characteristic of entropy requires that $S_{tot} = S_1 + S_2$. Evaluating (17.40) at $E_1 = \widehat{E}_1$ and $E_2 = \widehat{E}_2$ gives

$$\Omega_{tot}(E_{tot}, \widehat{E}_1) = \Omega_1(\widehat{E}_1)\Omega_2(\widehat{E}_2). \tag{17.43}$$

Taking the logarithm of this, multiplying by k_B, and using Boltzmann's formula gives

$$S_{\text{tot}}(E_{\text{tot}}) = S_1(\widehat{E}_1) + S_2(\widehat{E}_2). \tag{17.44}$$

Thus, Boltzmann entropy has the additive characteristic of an extensive state function.

Entropy Never Deceases

Since entropy is an equilibrium property, we consider processes that begin and end in equilibrium. We start with a system in equilibrium, disturb it by removing a constraint or constraints, and then allow it to come to equilibrium. The isolated two-part systems discussed above can be used to illustrate why entropy tends to increase. When the subsystems are isolated, the number of microstates accessible to the system is given by (17.40), so that the initial entropy of the combined systems is

$$S_0 = k_B \ln \Omega_{\text{tot}}(E_{\text{tot},} E_1) = S_1(E_1) + S_2(E_2). \tag{17.45}$$

We then disturb the system by removing the constraint of adiabatic isolation. After the two-part system comes to equilibrium, the number of accessible microstates is given by (17.43), so that the final entropy of the combined systems is

$$S_f = k_B \ln \Omega_{\text{tot}}(E_{\text{tot},} \widehat{E}_1) = S_1(\widehat{E}_1) + S_2(\widehat{E}_2). \tag{17.46}$$

Since $\Omega_{\text{tot}}(E_{\text{tot}}, \widehat{E}_1)$ is the maximum value of $\Omega_{\text{tot}}(E_{\text{tot}}, E_1)$, the final entropy S_f can only be greater than or equal to the initial entropy S_0.

Another illustration of the origin of the increase in entropy is its increase in a free expansion experiment. These experiments use two containers with rigid insulating walls and start with one container filled with gas and the other evacuated. A constraint is removed by opening a valve so that the gas can flow freely between containers. Since opening the valve does not prevent the gas from remaining where it was, the microstates that were initially accessible remain accessible. However, after the valve is opened, there are additional microstates in which there is gas in both containers. The resulting increase in the number of accessible states Ω causes $S_B = k_B \ln \Omega$ to increase. We could try to decrease S_B by reversing the effect of the expansion by reducing the total volume of the two containers to its original value, which could be done by moving a piston in some appropriately designed apparatus. The volume reduction would indeed tend to decrease the value of Ω, but the work done by the piston would increases the internal energy of the gas, which in turn tends to increase the value of Ω. The end result (which we have not proved in general) is that effects that tend to decrease the value of Ω are always compensated by other effects which tend to increase it in such a way that the number of accessible states never decreases.

17.5 Boltzmann Factors

The probability that a system in equilibrium with other systems is in microstate α is proportional to $e^{-E_{N,\alpha}/k_BT}$, where T is its temperature. Expressions of the form e^{-E/k_BT} are called *Boltzmann factors*. To demonstrate this proportionality, we consider a system made up of two interacting subsystems and focus our attention on subsystem 1. The combined (total) system is isolated with microstates specified by (α', α''), where α' and α'' are the microstates of the subsystems 1 and 2, respectively. We are seeking the probability $\rho_{\alpha'}^{(1)}$ that the subsystem 1 is in microstate α'.

When the energy of subsystem 1 is $E_{\alpha'}^{(1)}$ and the energy of the combined system is E_{tot}, the number of microstates accessible to subsystem 2 is $\Omega_2(E_2)$, where $E_2 = E_{\text{tot}} - E_{\alpha'}^{(1)}$. Since microstates of the combined system have been assumed to be equally probable, the probability that subsystem 1 is in microstate α' is proportional to $\Omega_2(E_{\text{tot}} - E_{1,\alpha'})$, so that

$$\rho_{\alpha'}^{(1)} = K\Omega_2(E_{tot} - E_{\alpha'}^{(1)}), \tag{17.47}$$

where K is a proportionality constant. Because the energy of system 1 is highly localized about the most probable value, we can expand the dependence of $\rho_{\alpha'}^{(1)}$ on $E_{\alpha'}^{(1)}$ in powers of the deviation from the most probable energy $\widehat{E}_1$ and neglect higher order terms. Since the number of accessible states Ω_2 is so large, we first use the relationship $x = e^{\ln x}$ to express $\rho_{\alpha'}^{(1)}$ as

$$\rho_{\alpha'}^{(1)} = K\, e^{\ln \Omega_2(E_{\text{tot}} - E_{\alpha'}^{(1)})} \tag{17.48}$$

and then expand the exponent. The first terms in the Taylor series expansion in powers of the deviation from $\widehat{E}_1$ is

$$\ln \Omega_2(E_{\text{tot}} - E_{\alpha'}^{(1)}) = \ln \Omega_2(\widehat{E}_2) - \frac{\partial \ln \Omega_2(\widehat{E}_2)}{\partial E_2}(E_{\alpha'}^{(1)} - \widehat{E}_1) + \cdots, \tag{17.49}$$

where $\widehat{E}_2 = E_{\text{tot}} - \widehat{E}_1$. Since the entropy of subsystem 2 is $S_2 = k_B \ln \Omega_2(\widehat{E}_2)$, it follows from $(\partial S/\partial U)_V = 1/T$ that $\partial \ln \Omega_2(\widehat{E}_2)/\partial E_2 = 1/k_B T_2$. When the combined system is in equilibrium, the temperatures T_2 and T_1 are equal, so that the probability that subsystem 1 is in microstate α' becomes

$$\rho_{\alpha'}^{(1)} = K\, e^{\ln \Omega_2(\widehat{E}_2)} e^{-(E_{\alpha'}^{(1)} - \widehat{E}_1)/k_B T_1} = \frac{e^{-E_{\alpha'}^{(1)}/k_B T_1}}{Z_N}, \tag{17.50}$$

where $1/Z_N = K\, e^{\ln \Omega_2(\widehat{E}_2)} e^{\widehat{E}_1/k_B T_1}$. Notice that Z_N does not depend on the microstate being considered.

Canonical probabilities

Since our interest is in subsystem 1, the prime on α' can be omitted. After adding subscript N to $E_{N,\alpha}$ to indicate that it is an eigenvalue of an N particle system, the resulting probabilities become proportional to the Boltzmann factor $e^{-E_{N,\alpha}/k_B T}$,

$$\boxed{\rho_{\alpha}^{c}(T) = \frac{e^{-E_{N,\alpha}/k_B T}}{Z_N},} \tag{17.51}$$

where superscript "c" indicates that these are the *canonical probabilities* which describe a system in equilibrium at temperature T. The normalization condition determines the value of Z_N, which is called the *partition function*. Substituting $\rho^c_\alpha(T)$ into $\sum_\alpha \rho^c_\alpha(T) = 1$ and solving for Z_N gives

$$Z_N = \sum_\alpha e^{-E_{N,\alpha}/k_BT}. \tag{17.52}$$

As discussed in the next chapter, the probabilities $\rho^c_\alpha(T)$ describe what is known as the *canonical ensemble*, while the probabilities $\rho_\alpha(E)$ given in (17.27) describe the *microcanonical ensemble*.

Problems

17.1 Find the number of accessible states, the entropy, and the internal energy of a classical solid of N atoms with energy $E = \sum_j x_j^2$, where the components of momentum $(p_1^x, \ldots, p_{3N}^z)$ divided by $\sqrt{2m}$ are represented by $(x_1, \ldots, x_{3N})$ and the components of the atomic displacements $(q_1^x, \ldots, q_{3N}^z)$ are proportional to $(x_{3N+1}, \ldots, x_{6N})$. (This is a classical version of the Einstein model.) Start with

$$\Sigma(E) = \frac{1}{h^{3N}} \int_{-\infty}^{\infty} dx_1 \cdots \int_{-\infty}^{\infty} dx_{6N} \begin{pmatrix} 1, & \text{if } \sum_j x_j^2 \le E \\ 0, & \text{otherwise} \end{pmatrix}.$$

Compare your result with the law of Dulong and Petit.

17.2 Show that the widths $|E_1 - \hat{E}_1|$ of the curves in Figure 17.1 are proportional to $1/\sqrt{N}$ when the relative probability $P_{\text{tot}}(E_1)/P_{\text{tot}}(\hat{E}_1)$ is $1/2$. Assume that $N_1 = N_2$ and $c_1 = c_2$. (Hint: Expand $P_{\text{tot}}(E_1)$ in a Taylor series about $E_1 = \hat{E}_1$ and neglect higher powers of $E_1 - \hat{E}_1$.)

Appendix 17.A

Evaluation of I_{3N}

The value of the integral I_{3N} needed for determining the number of accessible states $\Omega(E)$ for an ideal gas is derived in this appendix. We start by defining the function

$$J_n(X) = \int_{-\infty}^{+\infty} dx_1 \cdots \int_{-\infty}^{+\infty} dx_{2n} F(X, x_1 \ldots x_{2n}), \tag{17.A.1}$$

where

$$F(X, x_1 \ldots x_{2n}) = \begin{pmatrix} 1, & \text{if } \sum_j x_j^2 \le X \\ 0, & \text{otherwise} \end{pmatrix}. \tag{17.A.2}$$

The equation $X = \sum_j x_j^2$ defines a "surface" in a $2n$ dimensional space with coordinates $(x_1, x_2, \ldots, x_{2n})$, and $F(X, x_1 \ldots x_{2n})$ is the "volume" inside the surface. Changing the integration variables from x_j to u_j, where $x_j = u_j\sqrt{X}$ and $dx_j = du_j\sqrt{X}$, gives

$$J_n(X) = X^n \int_{-\infty}^{+\infty} du_1 \cdots \int_{-\infty}^{+\infty} du_{2n} F(1, \ldots u_j \ldots). \tag{17.A.3}$$

The requirement that $\sum_j x_j^2 \leq X$ becomes $\sum_j u_j^2 \leq 1$, so that

$$J_n(X) = X^n I_{2n}, \tag{17.A.4}$$

where

$$I_{2n} = \int_{-\infty}^{+\infty} du_1 \cdots \int_{-\infty}^{+\infty} du_{2n} \begin{pmatrix} 1, & \text{if } \sum_j u_j^2 \leq 1 \\ 0, & \text{otherwise} \end{pmatrix}. \tag{17.A.5}$$

Setting $2n = 3N$ in this yields the quantity I_{3N} defined in (17.14).

The value of I_{2n} can be found by evaluating the following integral in two ways and combining the results. Define

$$K_n = \int_{-\infty}^{+\infty} dx_1 \cdots \int_{-\infty}^{+\infty} dx_{2n}\, e^{-\sum_j x_j^2}, \tag{17.A.6}$$

where $j = 1, 2, \cdots, 2n$. The first way uses $e^{a+b+c+\cdots} = e^a e^b e^c \cdots$ to rewrite K_n as a product of $2n$ integrals. It then follows from $\int_{-\infty}^{+\infty} e^{-x^2} dx = \sqrt{\pi}$ that

$$K_n = \left(\int_{-\infty}^{+\infty} e^{-x_1^2} dx_1\right) \cdots \left(\int_{-\infty}^{+\infty} e^{-x_{2n}^2} dx_{2n}\right) = \left(\int_{-\infty}^{+\infty} e^{-x^2} dx\right)^{2n} = \pi^n. \tag{17.A.7}$$

The second way to express K_n represents it as a sum of integrals over regions in which $X = \sum_j x_j^2$ is essentially constant and then adding up the integrals over the different regions. To do this, we define $X_k = k\,\Delta X$, where ΔX is infinitesimal and k is a non-negative integer $(k = 0, 1, 2, \cdots)$. Since $F(X, x_1 \cdots x_{2n})$ equals one inside the surface at X and is zero outside, the difference $F(X_{k+1}, x_1 \cdots x_{2n}) - F(X_k, x_1 \cdots x_{2n})$ vanishes except between the surfaces at X_k and X_{k+1}. Since the sum over k includes all of x- space, it follows from (17.A.2) that

$$\sum_{k=0}^{\infty} [F(X_{k+1}, x_1 \ldots x_{2n}) - F(X_k, x_1 \ldots x_{2n})] = 1. \tag{17.A.8}$$

By multiplying the integrand of K_n in (17.A.6) by (17.A.8) and exchanging the sum and the integrations, we obtain

$$K_n = \sum_{k=0}^{\infty} \int_{-\infty}^{+\infty} dx_1 \cdots \int_{-\infty}^{+\infty} dx_{2n}\, e^{-\sum_j x_j^2} [F(X_{k+1}, x_1 \ldots x_{2n}) - F(X_k, x_1 \ldots x_{2n})]. \tag{17.A.9}$$

As $\Delta X \to 0$, the exponential $e^{-\sum_j x_j^2} = e^{-X}$ approaches e^{-X_k}. Then, by taking e^{-X_k} outside the integrations and multiplying and dividing by ΔX, it follows from (17.A.1) that

$$\begin{aligned} K_n &= \sum_{k=0}^{\infty} e^{-X_k} \int_{-\infty}^{+\infty} dx_1 \cdots \int_{-\infty}^{+\infty} dx_{2n} [F(X_{k+1}, x_1 \ldots x_{2n}) - F(X_k, x_1 \ldots x_{2n})] \\ &= \sum_{k=0}^{\infty} e^{-X_k} \frac{J_n(X_k + \delta X) - J_n(X_k)}{\Delta X} \Delta X. \end{aligned} \tag{17.A.10}$$

In the limit $\Delta X \to 0$ this becomes

$$K_n = \int_0^{\infty} e^{-X} \frac{dJ_n(X)}{dX} dX. \tag{17.A.11}$$

It follows from (17.A.4) that $\partial J_n/\partial X = nX^{n-1}I_{2n}$, so that

$$K_n = nI_{2n}\int_0^\infty e^{-X}X^{n-1}dX = n\Gamma(n)I_{2n}. \tag{17.A.12}$$

$\Gamma(n) = \int_0^\infty e^{-X}X^{n-1}dX$ is the well known gamma function, which has the values $\Gamma(1) = 1$ and $\Gamma(n) = (n-1)\Gamma(n-1)$, as can verified by integrating the parts. Since $\Gamma(n) = (n-1)!$, multiplying by n gives $n\Gamma(n) = n!$. Thus,

$$K_n = I_{2n}\,n!. \tag{17.A.13}$$

Combining this with the result $K_n = \pi^n$ from (17.A.7) implies that $I_{2n} = \pi^n/n!$. Setting $2n = 3N$ gives

$$\boxed{I_{3N} = \frac{\pi^{\frac{3}{2}N}}{\left(\frac{3}{2}N\right)!}} \tag{17.A.14}$$

Part IV
Statistical Mechanics I

Statistical methods are needed because of the large number of microscopic degrees of freedom in thermodynamic systems. The probabilities of statistical physics are interpreted in terms of ensembles, which are collections of imaginary copies of the real system being studied. The three basic ensembles – microcanonical, canonical, and grand canonical – can be found by using concepts from information theory to obtain unbiased assignments of probabilities to a system's microscopic configurations – called microstates. Part IV focuses on the canonical ensemble, which assigns a probability proportional to e^{-E/k_BT} to a microstate of energy E. An important consequence of this assignment is the partition function formula, which determines a system's macroscopic properties from its microscopic structure. Atomic-level behavior in both classical and quantum mechanics is determined by its Hamiltonian. The Hamiltonian determines a system's microstate energies, which determines its partition function, which in turn determines the Helmholtz free energy. Since the free energy is a thermodynamic potential and every potential contains a full description of a system's equilibrium properties, the partition function formula establishes a relationship between the macroscopic and microscopic properties of general validity. Thermodynamic properties can also be found by calculating ensemble averages.

The practical use of ensemble averages and the partition function requires some idealization of the physical system being studied. In many cases, useful results can be obtained by employing model systems consisting of noninteracting particles or independent modes of oscillation. The properties of systems in equilibrium appear to be static with well defined values while in fact they experience small random fluctuations. Macroscopic properties appear to be static because the deviations from their average values are very small (except near a critical point). Statistical mechanics gives us the tools for finding both the averages and the fluctuations. We start by deriving the three basic ensembles.

Thermodynamics and Statistical Mechanics: An Integrated Approach, First Edition.
Robert J Hardy and Christian Binek.

18

Ensembles[1]

At the microscopic level, the constituent particles of thermodynamic systems are in continual chaotic motion and pass through many different microscopic configurations (microstates), all of which correspond to the same macroscopic state. This situation allows us to employ a statistical analysis which assigns probabilities to the system's microstates. The probabilities characteristic of the microcanonical, canonical, and grand canonical ensembles are obtained by describing the condition of thermal equilibrium in different ways.

18.1 Probabilities and Averages

We begin by reviewing some terms and notations by investigating the distribution of energy in gases. The kinetic energy of a gas particle i is

$$K_i = \frac{1}{2}mv_i^2, \tag{18.1}$$

where v_i is its speed (magnitude of velocity). The kinetic energy per particle in a gas of N particles is given by the *arithmetic mean,*

$$\overline{K} = \frac{K_1 + K_2 + \cdots + K_N}{N} = \frac{1}{N}\sum_{i=1}^{N}\left(\frac{1}{2}mv_i^2\right). \tag{18.2}$$

The average $\overline{K}$ can also be described by introducing intervals of speed of width Δv and letting v_n represent the speed associated with interval n, where $n = 1, 2, 3, \ldots$ and $n\Delta v \leq v_n < (n+1)\Delta v$. By representing the number of particles with speeds in interval n by N_n,

[1] This chapter utilizes concepts discussed in Chapter 16 but is independent of Chapter 17.

Thermodynamics and Statistical Mechanics: An Integrated Approach, First Edition.
Robert J Hardy and Christian Binek.

the average becomes

$$\overline{K} = \frac{1}{N}\sum_{n=0}^{\infty} N_n \left(\frac{1}{2}m{v_n}^2\right) = \sum_{n=0}^{\infty} \frac{N_n}{N} K_n, \tag{18.3}$$

where $K_n = \frac{1}{2}m{v_n}^2$. The ratio N_n/N is the *relative frequency* of speed v_n.

A relative frequency can be interpreted as a *probability* ρ_n and vice versa, where

$$\rho_n = \frac{N_n}{N}. \tag{18.4}$$

Since probabilities are *non-negative* and *normalized, i.e.*, since they satisfy the conditions $\rho_n \geq 0$ and $\sum_n \rho_n = 1$, we need to verify that relative frequencies satisfy these conditions. It follows from the meaning of N_n as a "number of things" that N_n/N is non-negative. When there are N "things" and each one is associated with a value of index n, it follows that $N = \sum_n N_n$. Dividing this by N gives $\sum_n N_n/N = 1$, which indicates that the normalization condition is satisfied. By representing the relative frequency of speed v_n as ρ_n, the average kinetic energy in (18.3) then becomes

$$\langle K \rangle = \sum_n \rho_n K_n, \tag{18.5}$$

where the pointed bracket notation indicates that the average is given by a sum weighted by probabilities. The general expression for the *average value* of any property X is

$$\langle X \rangle = \sum_n \rho_n X_n, \tag{18.6}$$

where X_n is the value of the property identified by index n.

Ensembles

Since kinetic energy is a property of individual particles, $\langle K \rangle$ is a *single-particle average*. The procedure for finding single-particle averages is straightforward: Add up the contributions of different particles and divide by the number of particles. The resulting average is an arithmetic mean. However, this procedure does not work for properties such as potential energy that are intrinsically dependent on the interactions of two or more particles. Nevertheless, it is helpful to be able to interpret the averages of such properties as arithmetic means. The ensemble concept does that. One of the founders of statistical mechanics, J. Willard Gibbs,[2] wrote: "Let us imagine a great number of independent systems, identical in nature, but differing in their condition with respect to configuration and velocity. The forces are supposed to be determined for every system by the same law". This imaginary collection of independent systems is called an *ensemble*. An ensemble is a collection of things that work together, like the members of a musical ensemble. The average values of the properties of physical systems are interpreted as arithmetic means of their values in an ensemble. In the statistical interpretation of thermodynamics, the properties of systems are given by averages calculated with an appropriate ensemble.

---

[2] Gibbs, J.W. (1902) *Elementary Principles of Statistical Mechanics,* Yale University Press, reprinted by Dover Publications, 1960. See p. 5 in the Dover edition.

Microstates

The members of an ensemble are copies of the same physical system in which each copy is in a different microscopic configuration, *i.e.*, a different *microstate*. Because of the importance of energy and because quantum mechanical wave functions can be expanded in a series of energy eigenfunctions, the microstates of systems in thermal equilibrium are identified with *energy eigenstates*, which are solutions of the time-independent Schrödinger equation. Different eigenstates are identified by index α and the energy *eigenvalue* of an N-particle is represented by $E_{N,\alpha}$. In practice, α is an abbreviation for a collection of many indices. For example, index α in the previous chapter was used as a short-hand notation for the $3N$ quantum numbers $(n_{x,1}, n_{y,1}, n_{z,1}, \cdots, n_{z,N})$ that identify the eigenstates of a monatomic ideal gas. In the two-level system considered below, α is an abbreviation for the N indices $(n_1, n_2, \cdots, n_N)$.

The relative frequency of the members of an ensemble that are in microstate α is given by probability ρ_α, so that the *ensemble average* of property X is

$$\boxed{\langle X\rangle = \sum_\alpha \rho_\alpha X_\alpha,} \tag{18.7}$$

where X_α is the value of the property X when the system is in microstate α. Specifically, the average energy is

$$\langle E_N\rangle = \sum_\alpha \rho_\alpha E_{N,\alpha}, \tag{18.8}$$

where $E_{N,\alpha}$ is the energy eigenvalue of microstate α. The *normalization condition* is

$$\sum_\alpha \rho_\alpha = 1, \tag{18.9}$$

where the sum is over all microstates. The incredibly large number of microstates that exist in many particle systems can be seen by investigating systems in which each particle has only two states.

18.2 Two-Level Systems

The statistical description of thermodynamic systems is nicely illustrated by a collection of independent particles each of which has two possible energies. It is called a *two-level system*. One such system is the collection of spin-$\frac{1}{2}$ particles in paramagnetic systems. Each particle has two spin states, *up* and *down*, with lower energy in the state in which the magnetic moment is parallel to the external magnetic field. Another example is the collection of molecules in ammonia gas. The nitrogen atom in each ammonia molecule (NH_3) can be oriented either *up* or *down* relative to the plane of the three hydrogen atoms. In an external electric field the energy of a molecule depends on the orientation of the nitrogen atom. In many systems, the two-level behavior is sufficiently uncoupled from the system's other characteristics that the two-level behavior can be analyzed separately. Let ε_o represent the difference in the energy of the two states of a particle, and identify each particle by index i. Then, the energy of a system of N particles is

$$E_N(n_1, n_2, \cdots, n_N) = \sum_{i=1}^{N} n_i \varepsilon_o, \tag{18.10}$$

where $n_i = 1$ when particle i is in the higher energy state and $n_i = 0$ when in the lower energy state. $E_N(n_1 \cdots n_N)$ is the energy *eigenvalue* of the *eigenstate* (microstate) specified by the *quantum numbers* $(n_1, n_2, \cdots, n_N)$. For two-level systems, index α is an abbreviated notation for $(n_1, n_2, \cdots, n_N)$, where

$$\alpha = (n_1, n_2, \cdots, n_N). \tag{18.11}$$

The energy of microstate α in this concise notation is

$$E_{N,\alpha} = E_N(n_1, n_2, \cdots, n_N) = \varepsilon_o \sum_i n_i. \tag{18.12}$$

The sums over microstates in (18.7) through (18.9) represent sums over all values of the N quantum numbers, so that

$$\sum_\alpha = \sum_{n_1=0,1} \sum_{n_2=0,1} \cdots \sum_{n_N=0,1}. \tag{18.13}$$

Accessible Microstates

The number of microstates in thermodynamic systems is very large. We demonstrated this in Example 17.1 where we found the number of accessible states $\Omega(E)$ in an ideal gas. To illustrate that this is not simply a characteristic of ideal gases, the value of $\Omega(E)$ for a two-level system is found. The definition of *number of accessible states* $\Omega(E)$ from (17.1) is

$$\Omega(E) = \text{Number of microstates with energy between } E - \delta E \text{ and } E, \tag{18.14}$$

where the interval δE is small compared to the system's energy E.

In a two-level system of N particles with K particles in the higher energy state, the energy is $K\varepsilon_o$, where $K = \sum_i n_i$. Then, the number of particles in the zero energy state is $N - K$. Finding the number of microstates $E_{N,\alpha}$ with energy $K\varepsilon_o$ is equivalent to the combinatorial problem of finding the number of ways of choosing K objects (particles) from a collection of N objects (particles) without regard to order. This number is given by the binomial coefficient (see (16.6)), so that the number of microstates with energy $K\varepsilon_o$ is

$$\Omega_K = \frac{N!}{K!(N-K)!}. \tag{18.15}$$

Since the total energy E of a two-level system increases in steps of size ε_o, a reasonable choice for the interval δE is $\delta E = \varepsilon_o$. With this choice $\Omega(E)$ is equal to Ω_K. By using Stirling's approximation, $\ln n! = n \ln n - n$, the logarithm of $\Omega(E)$ for energies between $E \geq K\varepsilon_o$ and $E < (K+1)\varepsilon_o$ is

$$\ln \Omega(E) = N \ln N - K \ln K - (N-K) \ln(N-K). \tag{18.16}$$

It is convenient to express this result in terms of the fraction $f = K/N$ of particles in the higher energy state, so that

$$\ln \Omega(E) = N[-f \ln f - (1-f)\ln(1-f)], \tag{18.17}$$

where $E = Nf\varepsilon_o$. So that we can differentiate it, $\ln \Omega(E)$ will be treated as smooth function of E.

For a system of a million particles ($N = 10^6$) with 25% of them in the higher energy state and 75% with zero energy, so that $f = \frac{1}{4}$, the value of $\ln \Omega(E)$ is 5.62×10^5. By using $\Omega = 10^{\log \Omega}$ and $\log \Omega = \ln \Omega / \ln 10$, it follows that

$$\Omega(E) = 10^{244\,000}. \tag{18.18}$$

This value for the number of accessible states is appreciably less than the value for a monatomic ideal gas found in Example 17.1. Nevertheless, a number $10,000,\cdots$ with 244 thousand zeros is still a very large number!

18.3 Information Theory

Many situations exist in which the information available about an experiment is insufficient to determine the probabilities of the outcomes. Although our interest is in assigning probabilities to the microstates of thermodynamic systems, the terms outcomes and experiment are used in this section because of the general applicability of the concepts discussed. We would like to know the least biased assignment of probability consistent with the available information about the experiment. We do this by using techniques from *information theory*, which is a method of analysis based on the work on communication theory by C. E. Shannon[3] which was developed for use in statistical mechanics by E. T. Jaynes.[4]

Information-Entropy

An assignment of probability that is unbiased can be found by interpreting a hypothetical experiment that is closely related to the real experiment. The hypothetical experiment considered is patterned after the idea used to predict the Maxwell distribution of velocities in Chapter 16, where we assumed that the distribution of the molecules among cells in velocity space maximizes the number of arrangements consistent with the number of molecules N and their energy E. In the hypothetical experiment, we imagine arranging N_0 objects in J boxes with n_j objects in box j, where N_0 is an arbitrary number and $j = 1, 2, \cdots, J$. The numbers $n_1 \cdots n_J$ specify the distribution of objects, and the relative frequency n_j/N_0 can be interpreted as the probability ρ_j of outcome j. The relationship between the hypothetical and real experiments is established by setting J equal to the number of different possible outcomes of the real experiment and by requiring that the numbers $n_1 \cdots n_J$ are consistent with the available information about the real experiment expressed in terms of probabilities. The least biased assignment of probability is identified with the distribution

[3] Shannon, C.E. (1948) *Bell Telephone Technical Journal*, **27**, 379–423 & 623–656.

[4] Jaynes, E.T. (1956) Phys. Rev. **106**, 620; Jaynes, E.T. (1957) Phys. Rev. **108**, 171.

$n_1 \cdots n_J$ that maximizes the number of ways $W(n_1 \cdots n_J)$ of arranging the objects in the boxes: *The least biased assignment of probability is given by the relative frequencies* $\rho_j = n_j/N_o$, *where the numbers* $n_1 \cdots n_J$ *are chosen so that the number of arrangements of* N_0 *objects in J boxes is maximized consistent with the available information.*

The number of ways of arranging N_0 objects in J boxes with n_j in box j is given by the multinomial coefficient (see (16.3))

$$W(n_1 \cdots n_j \cdots n_J) = \frac{N_0!}{n_1! \cdots n_j! \cdots n_J!}, \tag{18.19}$$

where $\sum_j n_j = N_0$. We can maximize $W(n_1 \cdots n_J)$ by maximizing its logarithm $\ln W(n_1 \cdots n_J)$. By assuming that N_0 is large and using Stirling's approximation $\ln n! = n \ln n - n$, it follows that

$$\ln W(n_1 \cdots n_j \cdots n_J) = (N_0 \ln N_0 - N_0) - \sum_{j=1}^{J} (n_j \ln n_j - n_j)$$

$$= - N_0 \sum_{j=1}^{J} \frac{n_j}{N_0} \ln \frac{n_j}{N_0}. \tag{18.20}$$

We now use the relative frequencies $\rho_j = n_j/N_o$ to express $\ln W$ as a function of $\rho_1 \cdots \rho_J$. After multiplying the function obtained by an arbitrary positive constant k/N_o and designating the result by $S_I(\rho_1 \cdots \rho_J)$, we obtain

$$\boxed{S_I(\rho_1 \cdots \rho_J) = -k \sum_{j=1}^{J} \rho_j \ln \rho_j.} \tag{18.21}$$

$S_I(\rho_1 \cdots \rho_J)$ is called *information-entropy*.[5] The probabilities $\rho_1 \cdots \rho_J$ that maximize the information-entropy $S_I(\rho_1 \cdots \rho_J)$ also maximize the number of arrangements $W(n_1 \cdots n_J)$. Combining the above results yields the following principle:

> ***Principle of Maximum Entropy***
> *The least biased description of an experiment is given by the probabilities that maximize the information-entropy* S_I *consistent with the available information.*

Information-entropy S_I has the intuitive properties of a measure of uncertainty: (i) It is zero when there is no uncertainty about the outcome of the experiment. (ii) Its value is additive for systems with statistically independent parts. (iii) Its value increases as our uncertainty about the outcome increases. Note that, because of the bounds $0 \leq \rho_j$ and $\rho_j \leq 1$ on probabilities, every term $-\rho_j \ln \rho_j$ is non-negative which implies that information-entropy is non-negative.

[5] The uniqueness of S_I as a measure of information is deduced from three reasonable assumptions in Appendix 18.A.

The absence of uncertainty about the outcome of an experiment implies there is only one possible outcome, which means that the probabilities of all other outcomes are zero. Let j_o identify the one possible outcome. The normalization condition then implies that $\rho_{j_o} = 1$, which implies that $\rho_{j_o} \ln \rho_{j_o} = 0$. Since $\rho_j = 0$ for the other outcomes, the fact that $x \ln x \to 0$ in the limit $x \to 0$ implies that $\rho_j \ln \rho_j = 0$ when $j \neq j_o$. Hence, information-entropy is zero when there is no uncertainty about the outcome.

Statistical Independence

The concept of statistical independence is illustrated by the behavior of the dice used in games of chance. When two dice are thrown, the outcome is determined by the number of dots showing on each die. If we let m represent the number showing on one die and n represent the number showing on the other, the outcomes will be identified by (m, n). Since each die is unaffected by the behavior of the other, their outcomes are described by separate probabilities ρ_m and ρ_n. In this case, the probability $\rho_{m,n}$ of outcome (m, n) is

$$\rho_{m,n} = \rho_m \rho_n. \tag{18.22}$$

This kind of factorization is characteristic of the probabilities that describe systems with statistically independent parts. The information-entropy for the combined (total) system is obtained by substituting $\rho_{m,n} = \rho_m \rho_n$ into (18.21), which becomes

$$S_I = -k \sum_{m,n} \rho_{m,n} \ln \rho_{m,n} = -k \sum_{m,n} \rho_m \rho_n \ln(\rho_m \rho_n). \tag{18.23}$$

By using $\ln xy = \ln x + \ln y$ and the normalization conditions $\sum_m \rho_m = 1$ and $\sum_n \rho_n = 1$, we find that

$$S_I^{\text{tot}} = -k \sum_{n,m} \rho_n \rho_m \ln \rho_m - k \sum_{m,n} \rho_m \rho_n \ln \rho_n = S_{I,1} + S_{I,2}, \tag{18.24}$$

where $S_{I,1}$ and $S_{I,2}$ are the information-entropies of the two parts.

The outcomes of systems with three independent parts can be identified by three indices (a, b, c) and the probabilities can be factorized as $\rho_a \rho_b \rho_c$. By treating two of the three parts as a system, identifying outcome (m, n) in (18.24) with (b, c), and representing it by $\rho_{b,c} = \rho_b \rho_c$, the information-entropy of the system becomes $S_{I,B} + S_{I,C}$. Then, by identifying outcome m in (18.24) with a and outcome n with (b, c), the information-entropy of the three-part system becomes $S_{I,A} + S_{I,B} + S_{I,C}$. This result can be generalized to show that, when the probabilities factorize as $\rho_a \cdots \rho_k$, the information-entropy of the combined system is $S_{I,A} + \cdots + S_{I,K}$. That is, the information-entropy of systems with statistically independent parts is the sum of the information-entropies of the parts.

The number of outcomes with nonzero probabilities can be used as a measure of the uncertainty of the outcome of an experiment. Let M represent this number and identify the outcomes with nonzero probabilities by indices $j = 1, 2, \cdots, M$. As the only information we have is the value of M, there is no reason for assigning different values to the different nonzero probabilities. If M probabilities have the same value and all other probabilities are zero, the normalization condition requires that $\rho_j = 1/M$. Then, since $\ln(1/M) = -\ln M$, the information-entropy becomes

$$S_I = -k\sum_{j=1}^{M} \frac{1}{M} \ln \frac{1}{M} = k \ln M. \tag{18.25}$$

Since $\ln M$ is a monotonically increasing function of M and M is our measure of uncertainty, the information-entropy of the experiment increases as our uncertainty about the outcome increases.

The assertion that "there is no reason for assigning different values to the different nonzero probabilities" is an example of the idea known as the *principle of insufficient reason* or the *principle of indifference*. Unfortunately, the usefulness of this principle is limited to situations in which the only information available is the number of outcomes with nonzero probabilities. However, we can show that the principle is a special case of the *principle of maximum entropy*, which is applicable to a much wider range of situations. To show this, we use Lagrange's method of undetermined multipliers discussed in Section 16.3. By including the normalization condition $\sum_j \rho_j = 1$ with multiplier λ and employing the information that $\rho_j \neq 0$ for $1 \leq j \leq M$ and $\rho_j = 0$ for $j > M$, the auxiliary function becomes

$$F_{\text{aux}}(\rho_1 \cdots \rho_M; \lambda) = -k\sum\nolimits_j \rho_j \ln \rho_j + \lambda\left(\sum\nolimits_j \rho_j - 1\right), \tag{18.26}$$

where only values of j from 1 through M are included in the sums. The auxiliary function is maximized by the ρ_j that satisfy $\partial F_{\text{ex}}/\partial \rho_j = 0$. Differentiating with respect to ρ_j and setting the result equal to zero specifies that $\ln \rho_j + 1 - \lambda/k = 0$. Since λ and k are independent of j, this result implies that probabilities ρ_j are independent of j for $1 \leq j \leq M$, which is the result implied by the principle of insufficient reason. The normalization condition again requires that $\rho_j = 1/M$, which again yields the result $S_I = k \ln M$.

18.4 Equilibrium Ensembles

The ensembles that describe systems in thermal equilibrium can be found by applying the principle of maximum entropy. When describing thermodynamic systems, the terms "microstate" and "system" are used instead of "outcome" and "experiment." The following principle is fundamental: *A thermodynamic system in equilibrium is described by assigning to its microstates the probabilities that maximizes the information-entropy consistent with its macroscopic state.*

Microcanonical Ensemble[6]

The microcanonical ensemble is obtained by assuming that the only information available is the number of microstates with nonzero probabilities. Because of the discrete nature of energy eigenvalues, a small interval of energy δE is introduced so that the microstates with nonzero probabilities are those with energies between $E - \delta E$ and E. The resulting number is the *number of accessible states* $\Omega(E)$ defined in (18.14). This is the kind of situation in which the principle of insufficient reason can be used. Since there is no reason

[6] The microcanonical ensemble and Boltzmann entropy are discussed in more detail in Chapters 17 and 22.

for assigning different values to the different nonzero probabilities, the normalization condition requires that

$$\rho_\alpha^{\text{mc}}(E) = \begin{cases} 1/\Omega\,(E), & \text{if } E - \delta E < E_{N,\alpha} \leq E. \\ 0, & \text{otherwise.} \end{cases} \tag{18.27}$$

The maximum value of information-entropy is given by (18.25) with M replaced by $\Omega(E)$, so that

$$S_I = k_B \ln \Omega(E), \tag{18.28}$$

which is *Boltzmann entropy*, as defined in Chapter 17. To obtain the correct units for entropy, the Boltzmann constant k_B has been used for the constant k in (18.21). The connection of the microcanonical ensemble to thermodynamics is made by identifying S_B with thermodynamic entropy and identifying E with the internal energy U. Since Boltzmann entropy corresponds to the maximum information-entropy, it follows that *entropy is a measure of our uncertainty about a system's microstate.*

Canonical Ensemble

The equilibrium characteristics of systems in equilibrium at specific temperatures are described by the *canonical ensemble*. As used here, "canonical" means "standard" or "according to a general rule." Since measuring a system's temperature requires interacting with another system – a thermometer – its energy does not have a unique value. In this case, the internal energy U is identified with the average energy $\langle E_N \rangle$, and the system is described by the probabilities that maximize the information-entropy consistent with $U = \langle E_N \rangle$.

By replacing index j in (18.21) with α, information-entropy becomes

$$S_I = -k \sum_\alpha \rho_\alpha \ln \rho_\alpha. \tag{18.29}$$

We are interested in the probabilities ρ_α that maximize S_I consistent with the normalization condition $\sum_\alpha \rho_\alpha = 1$ and the assumption that $U = \langle E_N \rangle$, which imposes the requirement that

$$U = \sum_\alpha \rho_\alpha E_{N,\alpha}, \tag{18.30}$$

where $E_{N,\alpha}$ is the energy of microstate α. The auxiliary function needed to apply the method of undetermined multipliers is (see Section 16.2)

$$F_{\text{aux}}(\cdots \rho_\alpha \cdots ; \lambda, \gamma) = -k \sum_\alpha \rho_\alpha \ln \rho_\alpha + \lambda \left(\sum_\alpha \rho_\alpha - 1 \right) + \gamma \left(\sum_\alpha \rho_\alpha E_{N,\alpha} - U \right), \tag{18.31}$$

where λ and γ are the undetermined multipliers. The probabilities that maximize S_I satisfy the equations

$$\frac{\partial F_{\text{aux}}}{\partial \rho_\alpha} = -k(\ln \rho_\alpha + 1) + \lambda + \gamma E_{N,\alpha} = 0. \tag{18.32}$$

The resulting probabilities are $\rho_\alpha = e^{(\lambda/k-1)}e^{-\beta E_{N,\alpha}}$, where $\beta = -\gamma/k$. By representing $e^{(\lambda/k-1)}$ by $1/Z_N$, the probability that the system is in microstate α becomes

$$\boxed{\rho_\alpha^c = \frac{e^{-\beta E_{N,\alpha}}}{Z_N}.} \tag{18.33}$$

These are the *canonical probabilities*. Z_N is the *partition function*,[7] and its value is the same for all microstates. It is determined by substituting ρ_α^c into the normalization condition $\sum_\alpha \rho_\alpha^c = 1$ and solving for Z_N,

$$\boxed{Z_N = \sum\nolimits_\alpha e^{-\beta E_{N,\alpha}}.} \tag{18.34}$$

Since most systems have no upper bound on their energies $E_{N,\alpha}$ (two-level systems are an exception), the convergence of the sum in (18.34) requires that $\beta > 0$.

The internal energy is obtained by substituting ρ_α^c into (18.30),

$$U = \langle E_N \rangle = \sum_\alpha \rho_\alpha^c E_{N,\alpha} = \sum_\alpha \frac{e^{-\beta E_{N,\alpha}}}{Z_N} E_{N,\alpha}. \tag{18.35}$$

Notice that the derivative of the logarithm of Z_N with respect to β is

$$\frac{\partial \ln Z_N}{\partial \beta} = \frac{1}{Z_N}\sum_\alpha -E_{N,\alpha} e^{-\beta E_{N,\alpha}} = -\sum_\alpha \frac{e^{-\beta E_{N,\alpha}}}{Z_N} E_{N,\alpha}. \tag{18.36}$$

Thus, the internal energy is related to the partition function by

$$U = \langle E_N \rangle = -\frac{\partial \ln Z_N}{\partial \beta}. \tag{18.37}$$

As shown below, the multiplier β is related to temperature by $\beta = 1/k_B T$. Thus, the least biased description of a system in equilibrium at absolute temperature T is given by the *canonical probabilities*

$$\boxed{\rho_\alpha^c = \frac{e^{-E_{N,\alpha}/k_B T}}{Z_N},} \tag{18.38}$$

and the *partition function* is

$$\boxed{Z_N = \sum\nolimits_\alpha e^{-E_{N,\alpha}/k_B T}.} \tag{18.39}$$

Exponentials of the form $e^{-E/k_B T}$ are called *Boltzmann factors*.

---

[7] The "Z" comes from the German word *Zustandssumme*, which means "sum over states." An alternate derivation of the canonical probabilities is given in Section 17.5.

Grand Canonical Ensemble

Although systems in which the number of particle can vary will not be discussed in detail until Chapter 26, the probabilities that describe such open systems are derived here. Since they are obtained by a simple extension of the derivation of the canonical probabilities. We will continue to use index α to identify the microstates of a system containing N particles, which implies that the precise meaning of α depends on the value of N. Since both α and N can vary, the probabilities are represented by $\rho_{N,\alpha}$, and the average energy is

$$U = \langle E_N \rangle = \sum_N \sum_\alpha \rho_{N,\alpha} E_{N,\alpha}, \tag{18.40}$$

where the sum over N goes from zero to infinity and $E_{N,\alpha}$ is the energy of microstate α when the system contains N particles. The macroscopic value of the number of particles in an open system is represented by $\overline{N}$, which is identified with the average number of particles,

$$\overline{N} = \langle N \rangle = \sum_N \sum_\alpha \rho_{N,\alpha} N. \tag{18.41}$$

The probabilities $\rho_{N,\alpha}$ that maximize S_I must satisfy three conditions: the normalization condition plus the requirements that $U = \langle E_N \rangle$ and $\overline{N} = \langle N \rangle$. The auxiliary function needed to find the probabilities with Lagrange's method of undetermined multipliers is

$$\begin{aligned} F_{\text{aux}}(\cdots \rho_\alpha \cdots; \lambda, \gamma) = &- k \sum_N \sum_\alpha \rho_{N,\alpha} \ln \rho_{N,\alpha} \\ &- \lambda \left(\sum_N \sum_\alpha \rho_{N,\alpha} - 1 \right) + \gamma \left(\sum_N \sum_\alpha \rho_{N,\alpha} E_{N,\alpha} - U \right) \\ &+ \eta \left(\sum_N \sum_\alpha \rho_{N,\alpha} N - \overline{N} \right), \end{aligned} \tag{18.42}$$

where λ, γ, and η are the undetermined multipliers. The probabilities that maximize S_I satisfy the equations

$$\frac{\partial F_{\text{aux}}}{\partial \rho_{N,\alpha}} = -k(\ln \rho_{N,\alpha} + 1) + \lambda + \gamma E_{N,\alpha} + \eta N = 0, \tag{18.43}$$

which imply that

$$\rho_{N,\alpha} = e^{(\lambda/k)-1} e^{(\gamma/k)E_{N,\alpha}} e^{(\eta/k)N}. \tag{18.44}$$

The value of $e^{(\lambda/k)-1}$ is independent of N and α and is represented by $1/Q$. By setting $\gamma/k = -\beta$ and $\eta/k = \beta\mu$, the probability of microstate α of a system with N particles becomes

$$\boxed{\rho_{N,\alpha} = \frac{e^{\beta(\mu N - E_{N,\alpha})}}{Q}.} \tag{18.45}$$

These are the probabilities that describe the *grand canonical ensemble*. The *grand partition function* Q is found by substituting $\rho_{N,\alpha}$ into the normalization condition $\sum_N \sum_\alpha \rho_{N,\alpha} = 1$ and solving for Q,

$$Q = \sum_N e^{\beta\mu N} \sum_\alpha e^{-\beta E_{N,\alpha}} = \sum_N e^{\beta\mu N} Z_N. \tag{18.46}$$

The parameters β and μ in $\rho_{N,\alpha}$ are determined by the requirements on U and $\overline{N}$ in (18.40) and (18.41).

18.5 Canonical Thermodynamics

Gibbs Entropy

The canonical ensemble is the most useful ensemble for predicting thermodynamic properties (until we analyze systems of fermion and bosons in Chapter 26). The connection between microscopic and macroscopic properties is obtained by identifying the maximum value of information-entropy S_I with thermodynamic entropy. By using the canonical probabilities ρ^c_α and the Boltzmann constant k_B, the expression for S_I in (18.21) becomes

$$S_G = -k_B \sum_\alpha \rho^c_\alpha \ln \rho^c_\alpha, \tag{18.47}$$

which is called *Gibbs entropy* and sometimes *canonical entropy*. Since S_I is maximized by the canonical probabilities, the least biased description of a system in equilibrium at temperature T is given by Gibbs entropy. As with the microcanonical ensemble, the canonical ensemble indicates that thermodynamic entropy is a measure of our uncertainty about a system's microscopic state.

To determine thermodynamic properties with the canonical ensemble, we need the relationship between entropy and the average energy $\langle E_N \rangle$. It follows from $\rho^c_\alpha = e^{-\beta E_{N,\alpha}}/Z_N$ that $\ln \rho^c_\alpha = (-\beta E_{N,\alpha} - \ln Z_N)$. By substituting this into (18.47) and using the normalization condition $\sum_\alpha \rho^c_\alpha = 1$, we find that

$$S_G = -k_B \sum_\alpha \rho^c_\alpha(-\beta E_{N,\alpha} - \ln Z_N) = k_B \beta \langle E_N \rangle + k_B \ln Z_N, \tag{18.48}$$

where $\langle E_N \rangle = \sum_\alpha \rho^c_\alpha E_{N,\alpha}$.

PVT Systems

By identifying $\langle E_N \rangle$ with the internal energy U of a *PVT* system and identifying S_G with thermodynamic entropy S, equation (18.48) becomes

$$S = k_B(\beta U + \ln Z_N). \tag{18.49}$$

Since volume V enters as a parameter in the eigenvalues $E_{N,\alpha}$, the derivative of S with respect to U with the V fixed is

$$\left(\frac{\partial S}{\partial U}\right)_V = \left(\frac{\partial S}{\partial \beta}\right)_V \left(\frac{\partial \beta}{\partial U}\right)_V = k_B\left(U + \beta\frac{\partial U}{\partial \beta} + \frac{d \ln Z_N}{d\beta}\right)_V \left(\frac{\partial \beta}{\partial U}\right)_V$$
$$= k_B\left[U + \beta\left(\frac{\partial U}{\partial \beta}\right)_V - U\right]\left(\frac{\partial \beta}{\partial U}\right)_V = k_B\beta. \tag{18.50}$$

The thermodynamic relationship between the derivative of entropy and temperature T is $(\partial S/\partial U)_V = 1/T$ (see (12.29)). Combining this with (18.50) implies that

$$\boxed{\beta = 1/k_BT.} \tag{18.51}$$

By using this, the result in (18.49) becomes

$$S = \frac{U}{T} + k_B \ln Z_N, \tag{18.52}$$

which can be rewritten as

$$-k_BT \ln Z_N = U - TS. \tag{18.53}$$

It follows from this and the definition of the Helmholtz free energy in (12.5), $F = U - TS$, that

$$\boxed{F = -k_BT \ln Z_N.} \tag{18.54}$$

This is the *partition function formula for the free energy*.

As discussed in Chapter 12, the Helmholtz free energy $F = F(T, V)$ is a thermodynamic potential and every potential contains a full description of the system's equilibrium properties. Specifically, entropy and pressure are given by

$$S = -\left(\frac{\partial F}{\partial T}\right)_V \quad \text{and} \quad P = -\left(\frac{\partial F}{\partial V}\right)_T. \tag{18.55}$$

Furthermore, the second derivatives of F determine the constant-volume heat capacity C_V, the isothermal bulk modulus B_T, and the volume coefficient of thermal expansion α_V. Since the free energy is determined by the partition function Z_N, the thermodynamic properties of PVT systems can be predicted from their microscopic description by evaluating the partition function.

Paramagnetism[8]

The energy of a paramagnetic system is determined by its interaction with the magnetic field $\mathcal{H}$, which is maintained by an external source of energy. Because of this, the ensemble average of energy $\langle E_N \rangle$ is identified with enthalpy E_{th} (not internal energy U). By setting

[8] Paramagnetic systems are discussed in more detail in Chapter 21.

$\langle E_N \rangle = E_{\text{th}}$ and identifying Gibbs entropy with thermodynamic entropy, equation (18.48) indicats that

$$S = k_B(\beta E_{\text{th}} + \ln Z_N). \tag{18.56}$$

Similarly, equation (18.37) for the average energy is replaced by

$$E_{\text{th}} = \langle E_N \rangle = -\frac{\partial \ln Z_N}{\partial \beta}. \tag{18.57}$$

The derivative of S with respect to E_{th} with $\mathcal{H}$ fixed is

$$\left(\frac{\partial S}{\partial E_{\text{th}}}\right)_{\mathcal{H}} = \left(\frac{\partial S}{\partial \beta}\right)_{\mathcal{H}}\left(\frac{\partial \beta}{\partial E_{\text{th}}}\right)_{\mathcal{H}} = k_B\left(E_{\text{th}} + \beta\frac{\partial E_{\text{th}}}{\partial \beta} + \frac{d \ln Z_N}{d\beta}\right)_{\mathcal{H}}\left(\frac{\partial \beta}{\partial E_{\text{th}}}\right)_{\mathcal{H}} = k_B \beta. \tag{18.58}$$

As found in Table 14.3, the derivative of the enthalpy with respect to entropy is $T = (\partial E_{\text{th}}/\partial S)_{\mathcal{H}}$, which implies that $(\partial S/\partial E_{\text{th}})_{\mathcal{H}} = 1/T$. It then follows from (18.58) that $\beta = 1/k_B T$. As the constant-field heat capacity is $C_{\mathcal{H}} = (đQ/dT)_{\mathcal{H}}$, where $đQ = TdS$, we find that

$$C_{\mathcal{H}} = T\left(\frac{\partial S}{\partial T}\right)_{\mathcal{H}}. \tag{18.59}$$

By using $d\beta/dT = -k_B \beta^2$, which follows from $\beta = 1/k_B T$, the heat capacity $C_{\mathcal{H}}$ can be expressed as

$$C_{\mathcal{H}} = -\beta\left(\frac{\partial S}{\partial \beta}\right)_{\mathcal{H}}. \tag{18.60}$$

Another useful expression for the heat capacity is found by substituting (18.57) into (18.56), which gives

$$S = k_B\left(-\beta\frac{\partial \ln Z_N}{\partial \beta} + \ln Z_N\right). \tag{18.61}$$

Since the derivative of S with respect to β is $\left(\frac{\partial S}{\partial \beta}\right)_{\mathcal{H}} = -k_B\beta\left(\frac{\partial^2 \ln Z_N}{\partial \beta^2}\right)_{\mathcal{H}}$, it follows from this and (18.60) that

$$C_{\mathcal{H}} = k_B \beta^2\left(\frac{\partial^2 \ln Z_N}{\partial \beta^2}\right)_{\mathcal{H}}. \tag{18.62}$$

Two-Level Systems

The importance of the partition function is illustrated by calculating the heat capacity of a collection of independent spin-$\frac{1}{2}$ particles in an external magnetic field, which is an example of two-level systems. The energy eigenvalues of a two-level system of N particles given in (18.10) are

$$E_{N,\alpha} = \sum_{i=1}^{N} n_i \varepsilon_o, \tag{18.63}$$

where α is an abbreviation for the quantum numbers $(n_1, n_2, \cdots, n_N)$ and each n_i has two values, 0 and 1. The difference in the energy of the two states depends on the strength of the magnetic field $\mathcal{H}$. Substituting $E_{N,\alpha}$ into the expression for the partition function in (18.34) gives

$$Z_N = \sum_{\alpha} e^{-\beta E_{N,\alpha}} = \sum_{n_1=0,1} \sum_{n_2=0,1} \cdots \sum_{n_N=0,1} e^{-\beta \sum_i n_i \varepsilon_o}, \tag{18.64}$$

where (18.13) has been used to express the sum over α. It follows from $e^{x_1+x_2+x_3+\cdots} = e^{x_1} e^{x_2} e^{x_3} \cdots$ that

$$\begin{aligned} Z_N &= \sum_{n_1=0,1} \sum_{n_2=0,1} \cdots \sum_{n_N=0,1} e^{-\beta n_1 \varepsilon_o} e^{-\beta n_2 \varepsilon_o} \cdots e^{-\beta n_N \varepsilon_o} \\ &= \sum_{n_1=0,1} e^{-\beta n_1 \varepsilon_o} \sum_{n_2=0,1} e^{-\beta n_2 \varepsilon_o} \cdots \sum_{n_N=0,1} e^{-\beta n_N \varepsilon_o}. \end{aligned} \tag{18.65}$$

Since only the subscripts on the summation indices are different, each sum has the same value. Then, by replacing β with $1/k_B T$, we obtain

$$Z_N = \left(\sum_{n=0,1} e^{-\beta n \varepsilon_o} \right)^N = (1 + e^{-\varepsilon_o / k_B T})^N. \tag{18.66}$$

The logarithm of Z_N and its derivatives are

$$\ln Z_N = N \ln(1 + e^{-\varepsilon_o / k_B T}), \tag{18.67}$$

$$\frac{\partial \ln Z_N}{\partial \beta} = N \frac{e^{-\beta \varepsilon_o}(-\varepsilon_o)}{1 + e^{-\beta \varepsilon_o}} = \frac{-N \varepsilon_o}{e^{+\beta \varepsilon_o} + 1}, \tag{18.68}$$

and

$$\frac{\partial^2 \ln Z_N}{\partial \beta^2} = N \varepsilon_o^2 \frac{e^{\beta \varepsilon_o}}{(e^{\beta \varepsilon_o} + 1)^2} = N \left(\frac{\varepsilon_o}{e^{+\frac{1}{2}\beta \varepsilon_o} + e^{-\frac{1}{2}\beta \varepsilon_o}} \right)^2. \tag{18.69}$$

By substituting $\partial^2 \ln Z_N / \partial \beta^2$ into the expression for $C_{\mathcal{H}}$ and using $\beta = 1/k_B T$, the constant-field heat capacity becomes[9]

$$C_{\mathcal{H}} = N k_B \left(\frac{\varepsilon_o}{k_B T} \right)^2 \left(\frac{1}{e^{\varepsilon_o / 2 k_B T} + e^{-\varepsilon_o / 2 k_B T}} \right)^2. \tag{18.70}$$

18.6 Composite Systems

To further our understanding of the canonical ensemble, we investigate systems consisting of subsystems and consider systems made up of two subsystems, 1 and 2. Results for composite systems with more than two parts are obtained by treating each part as one of the subsystems while treating the other parts as the other subsystem. We focus on *PVT* systems.

[9] A plot of $C_{\mathcal{H}}$ is shown in Figure 21.2.

The microstates of a two-part system are specified by separately indicating the microstate of each subsystem, which are labeled α' and α''. The microstates of the combined (total) system are then identified by $(\alpha'\alpha'')$ (instead of α), and the combined energy is

$$E^{\text{tot}}_{\alpha'\alpha''} = E^{(1)}_{\alpha'} + E^{(2)}_{\alpha''} + I_{\alpha'\alpha''}. \tag{18.71}$$

The exchange of energy as work W changes the parameters that specify the volumes in the subsystem energies $E^{(1)}_{\alpha'}$ and $E^{(2)}_{\alpha''}$. The exchange of energy as heat Q is described by the interaction energies $I_{\alpha'\alpha''}$, which are associated with the random microscopic interactions of the particles. As these interactions take place at the surface where the subsystems are in contact, they are negligibly small compared to the energies $E^{(1)}_{\alpha'}$ and $E^{(2)}_{\alpha''}$. Nevertheless, some of the terms $I_{\alpha'\alpha''}$ are non-zero when the subsystems are in *thermal contact*. When the interaction terms are all zero, the subsystems are *adiabatically isolated.*

Extensive and Intensive Properties[10]

Internal energy U and entropy S are extensive properties, which implies that their values in a composite system are sums of their subsystem values. Temperature T is an important example of an intensive property, which implies it is independent of the system's size. We need to verify that the canonical ensemble correctly predicts the extensive and intensive characteristics of U, S, and T.

The internal energy U_{tot} of the combined system is the ensemble average of $E^{\text{tot}}_{\alpha'\alpha''}$, and the energies U_1 and U_2 of the subsystems are averages of $E^{(1)}_{\alpha'}$ and $E^{(2)}_{\alpha''}$. The extensive (additive) characteristic of internal energy implies that $U_{\text{tot}} = U_1 + U_2$, which in turn requires that the equilibrium average of the interaction terms $I_{\alpha'\alpha''}$ is negligible. Although not exactly true, it is effectively true in macroscopic systems.

When the interaction terms $I_{\alpha'\alpha''}$ are negligibly small (but not exactly zero), the energy $E^{\text{tot}}_{\alpha'\alpha''}$ is effectively $E^{(1)}_{\alpha'} + E^{(2)}_{\alpha''}$. It then follows that $e^{-\beta E^{\text{tot}}_{\alpha'\alpha''}} = e^{-\beta E^{(1)}_{\alpha'}} e^{-\beta E^{(2)}_{\alpha''}}$, so that the canonical probabilities of the combined system become

$$\rho^{\text{tot}}_{\alpha'\alpha''} = Z_N^{-1} e^{-\beta E^{\text{tot}}_{\alpha'\alpha''}} = Z_1^{-1} e^{-\beta E^{(1)}_{\alpha'}} Z_2^{-1} e^{-\beta E^{(2)}_{\alpha''}} = \rho^{(1)}_{\alpha'} \rho^{(2)}_{\alpha''}, \tag{18.72}$$

where the partition function Z_N defined in (18.39) becomes $Z_1 Z_2$. The probabilities for microstates α' and α'' of the subsystems are

$$\rho^{(1)}_{\alpha'} = \frac{e^{-\beta E^{(1)}_{\alpha'}}}{Z_1} \quad \text{and} \quad \rho^{(2)}_{\alpha''} = \frac{e^{-\beta E^{(2)}_{\alpha''}}}{Z_2}. \tag{18.73}$$

The factorization $\rho^{\text{tot}}_{\alpha'\alpha''} = \rho^{(1)}_{\alpha'} \rho^{(2)}_{\alpha''}$ indicates that the subsystems are statistically independent, which implies the subsystems are effectively statistically independent when in thermal equilibrium with each other.

We found in (18.24) that the information-entropy S_I^{tot} of a system with statistically independent parts is the sum of the values for the parts, *i.e.*, $S_I^{\text{tot}} = S_{I,1} + S_{I,2}$. Since S_G is the maximum value of S_I, it follows that

$$S_G^{\text{tot}} = S_{G,1} + S_{G,2}, \tag{18.74}$$

which indicates that *Gibbs entropy is extensive*.

[10] The concepts of intensive and extensive are described in Section 1.5.

Temperature[11]

Since the property obtained by dividing one extensive property by another is intensive and as both entropy and internal energy are extensive, it follows from $\beta = 1/k_BT$ and $(dS/dU)_V = k_B\beta$ that temperature is intensive. The temperatures of two or more thermally interacting subsystems are the same, while adiabatically isolated subsystems can have different temperatures. When adiabatically isolated, the information-entropy of each subsystem is maximized independent of the other. As a result, the subsystem probabilities are

$$\rho_{\alpha'}^{(1)} = \frac{e^{-\beta_1 E_{\alpha'}^{(1)}}}{Z_1} \quad \text{and} \quad \rho_{\alpha''}^{(2)} = \frac{e^{-\beta_2 E_{\alpha''}^{(2)}}}{Z_2}, \tag{18.75}$$

where β_1 and β_2 can have different values. The difference between this result and the earlier result for thermally interacting subsystems is that the value of β in (18.73) is the same in both subsystems. Thus, $\beta = 1/k_BT$ has the same value in thermally interacting subsystems but can be different in adiabatically isolated subsystems, as is required. Combining $\beta = 1/k_BT$ and $(dS/\partial U)_V = k_B\beta$ implies that $T = (dU/dS)_V$, which indicates that T is an absolute temperature.[12]

Entropy

We need to check that Gibbs entropy S_G has the characteristics of thermodynamic entropy, as summarized by the entropy statement of the second law in Section 10.3. Specifically, we need to check (i) that S_G is extensive, (ii) that S_G increases monotonically as internal energy increases when no work is done, and (iii) that S_G never decreases in adiabatically isolated systems. Equation (18.74) indicates that S_G is extensive. From $(\partial S/\partial U)_V = k_B\beta$ and the requirement that $\beta > 0$ mentioned below (18.34) it follows that $(\partial S/\partial U)_V > 0$. As no work is done by PVT systems when V is constant, it follows that S_G increases as internal energy U increases.

That Gibbs entropy cannot decrease in an adiabatically isolated system follows from the identification of S_G with the maximum value of information-entropy S_I. Since entropy is only defined in equilibrium, processes in which the system is in equilibrium in the initial and final states need to be considered. The kind of processes we investigate involves the removal of one or more constraints. The opening the valve between the two containers in the free expansion experiment (described in Section 2.4) is an example of the removal of a constraint. Removing constraints does not eliminate the microstates available to an isolated system but can make additional states available. When the set of states available is changed, the probabilities must be recalculated, which is done by maximizing S_I. Since the probabilities that initially described the system remain a possible choice for the final probabilities, the recalculated maximum value of S_I cannot decrease. Since S_G is identified with the maximum value of S_I, Gibbs entropy S_G never decreases. Notice that, although Gibbs entropy has the characteristics of thermodynamic entropy (at least for the kind of processes investigated), the microscopic interpretation of entropy given by Gibbs entropy (and Boltzmann entropy) tells us nothing about the atomic-level interactions that cause the entropy to change.

---

[11] The defining characteristics of temperature are given in Section 1.4.

[12] That $(dU/dS)_V$ is an absolute temperature is shown in Appendix 10.A.

Problems

18.1 Find the probabilities ρ_1 and ρ_2 that maximize the information entropy S_I consistent with the value of the average $\bar{\varepsilon} = \langle \varepsilon \rangle = \rho_1 \varepsilon_1 + \rho_2 \varepsilon_2$, where ε_1 and ε_2 are the energies of states 1 and 2. Use the resulting probabilities to find $\bar{\varepsilon}$ and compare with the answer to Problem 16.2. Use the method of undetermined multipliers.

18.2 Consider a two-level system with energies $\varepsilon_1 = -\varepsilon_o/2$ and $\varepsilon_2 = +\varepsilon_o/2$. Normalize the Boltzmann factors $e^{-\beta\varepsilon_j}$ to find the probabilities $\rho_j = e^{-\beta\varepsilon_j}/Z$. Set $\beta = 1/k_BT$ and find the internal energy $U = N\langle \varepsilon \rangle$ and heat capacity $C = \partial U/\partial T$ of a system of N two-level particles. Compare with equation (18.70).

Appendix 18.A

Uniqueness Theorem

The uniqueness theorem for information-entropy is proved in this appendix. The proof proceeds by assuming three characteristics of the function $S_I(\rho_1 \dots \rho_j \dots \rho_J)$ and then showing that the function which has these characteristics is $-k\sum_j \rho_j \ln \rho_j$. Since the ρ_j are probabilities, it is required that $\rho_j \geq 0$ and $\sum_j \rho_j = 1$. The function for information-entropy is required to have the following characteristics:

1. $S_I(\rho_1 \dots \rho_j \dots \rho_J)$ is a continuous symmetric function of the probabilities of J possible outcomes. The symmetric property means that $S_I(\rho_1 \dots \rho_j \dots \rho_J)$ is independent of the order of the arguments $\rho_1 \dots \rho_j \dots \rho_J$.
2. $S_I(\rho_1 \dots \rho_j \dots \rho_J)$ is a monotonically increasing function of M when M of the probabilities are equal and the others are zero. In this case it is convenient to set $\rho_j = 1/M$ for $j \leq M$ and $\rho_j = 0$ for $j > M$. Then, by defining

$$I(M) = S_I\left(\frac{1}{M}, \dots, \frac{1}{M}, 0, \dots\right), \tag{18.A.1}$$

the second characteristic becomes

$$I(M_1) > I(M_2) \text{ when } M_1 > M_2. \tag{18.A.2}$$

3. $S_I(\rho_1 \dots \rho_j \dots \rho_J)$ is independent of how the outcomes are grouped. The meaning of this *composition rule* is illustrated by considering the example diagrammed in **Figure 18.A.1**. In the left diagram there are four possible outcomes with probabilities $\rho_1 = 2/10$, $\rho_2 = 3/10$, $\rho_3 = 1/3$, and $\rho_4 = 1/6$. In the right hand diagram we have two outcomes, each of which consists of two of the original outcomes. When normalized with respect to their group, the probabilities ρ_1 and ρ_2 become 2/5 and 3/5 and the probabilities ρ_3 and ρ_4 become 2/3 and 1/3. For the final result to be the same in both situations, we require that

$$S_I\left(\frac{2}{10}, \frac{3}{10}, \frac{1}{3}, \frac{1}{6}\right) = S_I\left(\frac{1}{2}, \frac{1}{2}\right) + \left[\frac{1}{2}S_I\left(\frac{2}{5}, \frac{3}{5}\right) + \frac{1}{2}S_I\left(\frac{2}{3}, \frac{1}{3}\right)\right]. \tag{18.A.3}$$

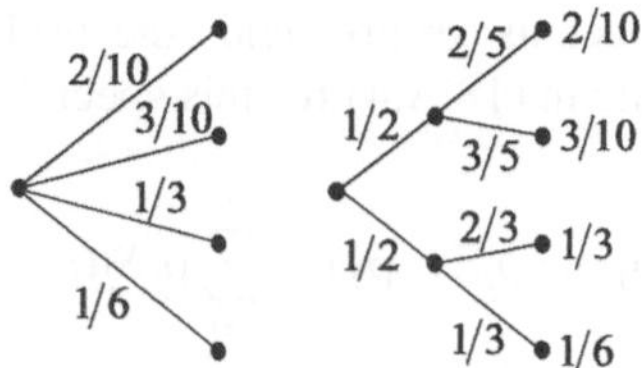

Figure 18.A.1 *Illustration of the composition rule*

When there are J possible outcomes arranged in G groups, the *composition rule* specifies that

$$S_I(\rho_1 \ldots \rho_j \ldots \rho_J) = S_I(P_1 \ldots P_g \ldots P_G) + \sum_{g=1}^{G} P_g S_I(p_{g,1} \ldots p_{g,i} \ldots). \tag{18.A.4}$$

The single subscript j has been replaced by two subscripts, so that ρ_j is replaced by $\rho_{g,i}$, where g identifies the group $(g = 1, 2, \cdots, G)$ and i identifies the outcomes within the group. The total probability P_g of group g is

$$P_g = \rho_{g,1} + \cdots + \rho_{g,i} + \cdots + \rho_{g,n_g}, \tag{18.A.5}$$

where n_g represents the number of outcomes in the group and $\sum_g n_g = J$. By normalizing the probabilities $\rho_{g,i}$ in group g with respect to their group, we obtain the probabilities $p_{g,i} = \rho_{g,i}/P_g$. Then, since $\sum_i p_{g,i} = 1$, the arguments of $S_I(p_{g,1} \cdots p_{g,i} \cdots)$ satisfy the normalization condition $\sum_j \rho_j = 1$.

Proof Since its functional form is the same for all experiments, the form of $S_I(\rho_1 \ldots \rho_j \ldots \rho_J)$ can be found by considering special cases. We start by representing the probabilities ρ_j by rational numbers, so that the probabilities ρ_j become $\rho_j = m_j/D$, where the ρ_j can be represented with arbitrary accuracy by using very large integers m_j and D. It follows from the normalization condition $\sum_j \rho_j = 1$ that the denominator is $D = \sum_j m_j$.

Next, we consider the special case in which there are D equally probable outcomes, which we label by index d, so that $\rho_d = 1/D$ and $d = 1, 2, \cdots, D$. We then group the outcomes so the total probability of each group is equal to one of the probabilities ρ_j in $S_I(\rho_1 \cdots \rho_j \cdots \rho_J)$. When grouped in this way, the groups in the composition rule are identified by j (instead of g), the number of groups is J (instead of G), and the number of outcomes in group j is m_j (instead of n_g). By replacing the single subscript d by two subscripts, the probability ρ_d becomes $\rho_{j,i}$ (instead of $p_{g,i}$), where j identifies the group and i labels the different outcomes within the group. The total probability of group j is ρ_j (instead of P_g), so that (18.A.5) becomes

$$\rho_j = \rho_{j,1} + \cdots + \rho_{j,i} + \cdots + \rho_{j,m_j} = \frac{m_j}{D}. \tag{18.A.6}$$

Normalizing the $\rho_{j,i}$ with respect its group yields the probabilities $p_{j,i} = \rho_{j,i}/\rho_j$, where $\sum_i p_{j,i} = 1$. The composition rule in (18.A.4) for this special case specifies that

$$S_I(p_1 \ \cdots \ p_d \ \cdots \ p_D) = S_I(\rho_1 \ \cdots \ \rho_j \ \cdots \ \rho_J) + \sum_{j=1}^{J} \rho_j S_I(p_{j,1} \ \cdots \ p_{j,i} \ \cdots \ p_{j,m_j}). \tag{18.A.7}$$

Since $\rho_{j,i} = \rho_d = 1/D$ and $\rho_j = m_j/D$, it follows from $p_{j,i} = \rho_{j,i}/\rho_j$ that $p_{j,i} = 1/m_j$. As the probabilities ρ_d are equal, the notation in (18.A.1) can be used to rewrite (18.A.7) as

$$I(D) = S_I(\rho_1 \ \cdots \ \rho_j \ \cdots \ \rho_J) + \sum_{j=1}^{J} \rho_j I(m_j). \tag{18.A.8}$$

We now consider experiments with J outcomes of equal probability. The normalization condition requires that $\rho_j = 1/J$. Since $\rho_j = m_j/D$, this implies that

$$m_1 = \cdots = m_J = m, \tag{18.A.9}$$

where m represents the common value of m_j. It then follows from $D = \sum_j m_j$ that $D = Jm$, so that (18.A.8) becomes

$$I(Jm) = I(J) + \sum_{j=1}^{J} \frac{1}{J} I(m) = I(J) + I(m). \tag{18.A.10}$$

The continuous function with the property $I(Jm) = I(J) + I(m)$ is the logarithm function times a constant. The properties of logarithms imply that $\ln(Jm) = \ln J + \ln m$, which when multiplied by constant k becomes $k \ln(Jm) = k \ln J + k \ln m$. Thus,

$$I(M) = k \ln M. \tag{18.A.11}$$

The second characteristic summarized in (18.A.2) requires that $k > 0$. Using $I(M) = k \ln M$ in (18.A.8) gives

$$k \ln D = S_I(\rho_1 \ \cdots \ \rho_j \ \cdots \ \rho_J) + \sum_{j=1}^{J} \rho_j k \ln m_j. \tag{18.A.12}$$

By solving for $S_I(\rho_1 \ \cdots \ \rho_j \ \cdots \ \rho_J)$ and using $\sum_j \rho_j = 1$, we find that

$$S_I(\rho_1 \ \cdots \ \rho_j \ \cdots \ \rho_J) = \left(\sum_{j=1}^{J} \rho_j \right) k \ln D - \sum_{j=1}^{J} \rho_j k \ln m_j = -k \sum_{j=1}^{J} \rho_j \ln \frac{m_j}{D}. \tag{18.A.13}$$

Since $\rho_j = m_j/D$, it follows that

$$\boxed{S_I(\rho_1 \ \cdots \ \rho_j \ \cdots \ \rho_J) = -k \sum_{j=1}^{J} \rho_j \ln \rho_j.} \tag{18.A.14}$$

19

Partition Function

In Chapter 18 we derived the partition function formula for predicting thermodynamic properties. We now evaluate the partition function, which is determined by the energies of the system's microstates, which in turn are determined by its Hamiltonian. In classical mechanics, the microstate energies are directly related to the Hamiltonian. The quantum mechanical evaluation of the partition function is the subject of Chapter 20.

Systems are described at the atomic level by the same Hamiltonian independent of whether they are in a solid, liquid, or gaseous phase. However, approximations are introduced to facilitate the evaluation of the partition function, and different approximations are appropriate for the different phases. This chapter introduces the Hamiltonian for a typical *PVT* system and discusses the approximations that yield the ideal gas model and the Einstein model for the specific heats of solids.

19.1 Hamiltonians and Phase Space

The Hamiltonian is the function that gives the dependence of a system's energy on the momentum and position variables of its constituents. We are interested in systems of N particles which are identified by subscript i where $i = 1, 2, \ldots, N$. The position of particle i is $\boldsymbol{r}_i = (x_i, y_i, z_i)$, where x_i, y_i, and z_i are Cartesian coordinates, and the components of momentum are $\boldsymbol{p}_i = (p_i^x, p_i^y, p_i^z)$. The Hamiltonian of a system of N particles is a function of the $6N$ momentum and position variables,[1]

$$H_N(p_1^x, p_1^y, p_1^z, \ldots, p_N^z, x_1, \ldots, x_N, y_N, z_N). \tag{19.1}$$

The $6N$ independent variables are conveniently represented by $(\boldsymbol{p}_1 \cdots \boldsymbol{r}_N)$, where

$$(\boldsymbol{p}_1 \cdots \boldsymbol{r}_N) = (p_1^x, p_1^y, p_1^z, \ldots, p_N^z, x_1, \ldots, x_N, y_N, z_N). \tag{19.2}$$

[1] The Cartesian coordinates (x_i, y_i, z_i) of the particles are the natural choice of generalized coordinates when visualizing *PVT* systems as collections of atoms or molecules. Other sets of generalized coordinates are discussed in Chapter 23.

Thermodynamics and Statistical Mechanics: An Integrated Approach, First Edition.
Robert J Hardy and Christian Binek.

We are interested in Hamiltonians of the form

$$H_N(\boldsymbol{p}_1 \cdots \boldsymbol{r}_N) = K_N(\boldsymbol{p}_1 \cdots \boldsymbol{p}_N) + \Phi_N(\boldsymbol{r}_1 \cdots \boldsymbol{r}_N), \tag{19.3}$$

where $K_N(\boldsymbol{p}_1 \cdots \boldsymbol{p}_N)$ is the system's kinetic energy,

$$K_N(\boldsymbol{p}_1 \cdots \boldsymbol{p}_N) = \sum_{i=1}^{N} \frac{1}{2m_i} |\boldsymbol{p}_i|^2, \tag{19.4}$$

and $\Phi_N(\boldsymbol{r}_1 \cdots \boldsymbol{r}_N)$ is its potential energy. Specific *values* of the energy will be distinguished from the *function* $H_N(\boldsymbol{p}_1 \cdots \boldsymbol{r}_N)$ by representing them by E_N.

Microstates and Phase Space

The variables $(\boldsymbol{p}_1 \cdots \boldsymbol{r}_N)$ are the coordinates of an abstract space called *phase space*. Both momentum and position variables are needed because both are required to determine how the system evolves. *In classical mechanics, the microstate of a system is represented by a point in phase space*. This *phase point* traces out a line (or trajectory) in phase space as the system evolves. The evolution is determined by *Hamilton's equations* of motion, which are 6*N*-coupled ordinary differential equations, where the six equations for the momentum and position of particle *i* are summarized as follows:

$$\frac{d\boldsymbol{r}_i}{dt} = \frac{\partial H_N(\boldsymbol{p}_1 \cdots \boldsymbol{r}_N)}{\partial \boldsymbol{p}_i} \quad \text{and} \quad \frac{d\boldsymbol{p}_i}{dt} = -\frac{\partial H_N(\boldsymbol{p}_1 \cdots \boldsymbol{r}_N)}{\partial \boldsymbol{r}_i}. \tag{19.5}$$

From Hamilton's equations and (19.3), we find that $d\boldsymbol{r}_i/dt = \boldsymbol{p}_i/m_i$ and $d\boldsymbol{p}_i/dt = -\partial\Phi_N/\partial\boldsymbol{r}_i$, which imply that

$$\boldsymbol{p}_i = m_i \boldsymbol{v}_i \quad \text{and} \quad m_i \boldsymbol{a}_i = -\frac{\partial \Phi_N}{\partial \boldsymbol{r}_i}, \tag{19.6}$$

where $\boldsymbol{v}_i = d\boldsymbol{r}_i/dt$ is the velocity and $\boldsymbol{a}_i = d\boldsymbol{v}_i/dt$ is the acceleration. The second equation is effectively a statement of Newton's second law $\boldsymbol{F}_i = m_i \boldsymbol{a}_i$, where $\boldsymbol{F}_i = -\partial\Phi/\partial\boldsymbol{r}_i$ is the force on particle *i*.

19.2 Model Hamiltonians

Although the air in a tire, the water in a cup, and a piece of aluminum have very different properties, the same statistical assumptions are used for analyzing them. What distinguishes different types of system is their Hamiltonian. To illustrate the role of the Hamiltonian, we investigate systems of rare gas atoms. These systems, which consist of atoms of neon, argon, krypton, or xenon, are relatively easy to analyze. Since the atoms do not combine to form molecules, the gases are monatomic. Although referred to as *rare gases*, the systems also exist as liquids and solids. Their melting and vaporization temperatures at 1 atm of pressure are given in Table 19.1.[2]

[2] Anderson H. L. (ed.) (1989) *A Physicist's Desk Reference, the Second Edition of Physics Vade Mecum*, ed. H. L. Anderson), American Institute of Physics, New York, p. 341.

Table 19.1 Rare gas systems

		T_m (K)	T_v (K)	ε/k_B (K)	ε ($\times 10^{-21}$J)	r_o (nm)	m (amu)
Neon	Ne	25	27	36	0.50	0.312	20.18
Argon	Ar	84	87	119	1.64	0.381	39.95
Krypton	Kr	116	120	164	2.26	0.408	83.80
Xenon	Xe	161	165	229	3.16	0.444	131.3

T_m is the melting temperature; T_v is the vaporization temperature; 1 amu = 1.6605×10^{-27}kg

Interatomic interactions

The forces between atoms are determined by the behavior of the electrons that surround the nuclei. In many situations, these forces can be idealized as the interactions of small deformable spherical objects. They can then be described by an *interatomic potential* $\phi(r)$, which gives the potential energy of interaction of two atoms when nuclei are a distance r apart. Determining the functional form of $\phi(r)$ is a quantum mechanical problem outside the scope of this book.

It is useful to have an approximate expression for the interatomic potentials which has the correct qualitative features.[3] The *Lennard–Jones potential* is frequently used for this purpose. Its functional form is

$$\phi(r) = 4\varepsilon\left[\left(\frac{\sigma}{r}\right)^{12} - \left(\frac{\sigma}{r}\right)^{6}\right], \tag{19.7}$$

which is shown in **Figure 19.1**. The depth of the potential well is ε, the separation r at the minimum is $r_o = 2^{1/6}\sigma$ where $2^{1/6} = 1.122$, and $\phi(r_o) = -\varepsilon$. The potential passes through zero at $r = \sigma$, so that $\phi(\sigma) = 0$. Either r_o or σ can be used to characterize the diameter of

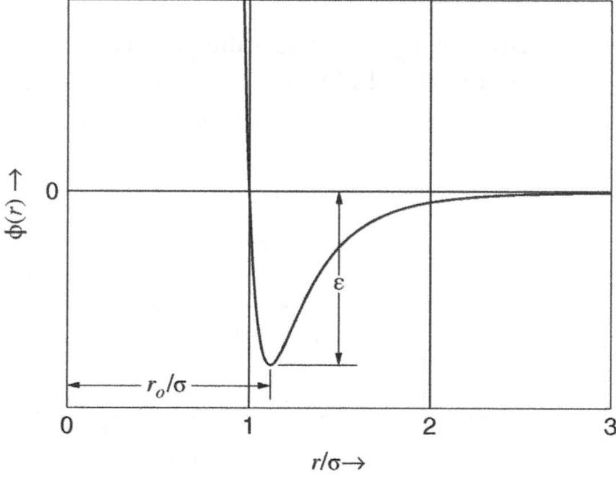

Figure 19.1 *Lennard–Jones Potential*

[3] In a system containing atoms of different types, the interatomic potentials depend on the types of atoms involved.

the atoms. Typical values for ε and r_o for atoms of the same type are given in Table 19.1.[4] Notice that the melting and vaporization temperatures increase as the depth of the potential well increases.

The minimum at $r = r_o$ is associated with a tendency of the atoms to bind to each other. The rapid increase in $\phi(r)$ at $r < r_o$ gives rise to repulsive forces that push atoms apart at small separations. The approach of $\phi(r)$ toward zero at $r > r_o$ indicates that the range of the potential is effectively limited to two or three times the diameters of the atoms. This and the values of r_o in the table indicate that the atoms move independently when separated by more than a few nanometers.

The total contribution of the interatomic potentials to the system's energy is the sum of potentials for each pair of atoms, so that the interatomic interactions energy is

$$\Phi_{\text{int}}(\boldsymbol{r}_1 \cdots \boldsymbol{r}_N) = \sum_{1 \le i < j \le N} \phi(r_{ij}), \tag{19.8}$$

where $r_{ij} = |\boldsymbol{r}_i - \boldsymbol{r}_j|$ is the distance between the nuclei of atoms i and j. Because of the restriction $1 \le i < j \le N$, each pair of atoms is included only once in Φ_{int}. The total number of pairs is $N(N-1)/2$.

Atom–wall interaction

The microscopic origin of pressure is the force exerted on the walls of its container by the atoms in the system. The reactions to these forces are the forces on the atoms exerted by the walls. Although an accurate representation of these forces would be difficult, their effect can be understood by idealizing the walls as smooth flat surfaces. The potential energy associated with the interaction of atom i with the container walls can then be described by an *atom–wall potential* $\phi_{a\text{-}w}(\boldsymbol{r}_i)$, where the components of the force on atom i are

$$F_i^x(r_i) = -\frac{\partial \phi_{a\text{-}w}(r_i)}{\partial x_i}, \quad F_i^y(r_i) = -\frac{\partial \phi_{a\text{-}w}(r_i)}{\partial y_i}, \quad F_i^z(r_i) = -\frac{\partial \phi_{a\text{-}w}(r_i)}{\partial z_i}. \tag{19.9}$$

To illustrate the characteristics of $\phi_{a\text{-}w}$, consider the gas inside a cubic container of volume $V = L^3$ and imagine an atom as it collides with a wall, as indicated in **Figure 19.2a**. Let $w(x_i)$ represent the potential energy associated with the interaction of atom i and the wall

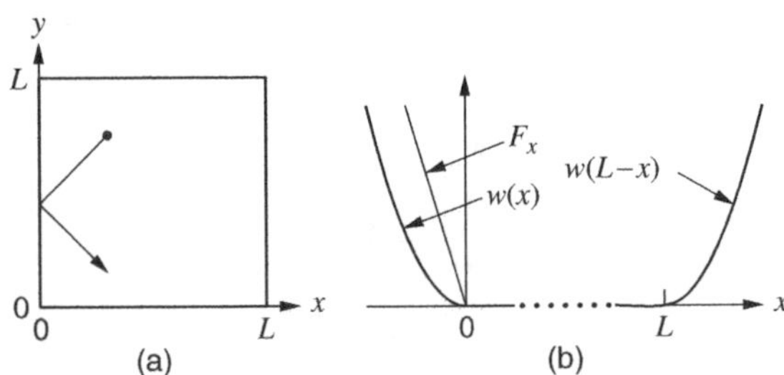

Figure 19.2 *Idealized Atom–Wall Interaction. (a) An Atom–Wall Collision. (b) Potential Energies at* $x = 0$ and L

[4] Horton G. K. (1979), in *Rare Gas Solids*, vol. 1 (eds M. L. Klein and J. A. Venables), Academic, New York, p. 88.

at $x = 0$, so that the x-component of the force on the atom is $F_{x,i} = -dw(x_i)/dx_i$. Since no force is exerted on the atom until it comes in contact with the wall, $w(x_i)$ vanishes when x_i is greater than zero. When the atom hits the wall, it is pushed back into the container within a distance of a nanometer or less. This implies that the potential energy $w(x_i)$ increases rapidly at negative values of x_i. Idealized expressions for $w(x)$ and $F_x = -dw(x)/dx$ are shown in diagram (b). Notice that the range of the x-values in diagram (b) is of the order of a nanometer, while the ranges of the x and y values in diagram (a) are typically of the order of centimeters. For cubic containers, the potential $\phi_{a\text{-}w}(\boldsymbol{r}_i)$ can be expressed as a sum of six terms with each term describing the interaction with a different wall,

$$\phi_{a-w}(r_i) = w(x_i) + w(L - x_i) + w(y_i) + w(L - y_i) + w(z_i) + w(L - z_i). \tag{19.10}$$

The potential energy associated with the force exerted by the wall at $x = L$ is $w(L - x_i)$, which is the composite function formed by using $L - x_i$ as the argument of $w(x)$. Since $w(x)$ increases to the left, $w(L - x)$ increases to the right, as indicated in the figure. The force associated with $w(L - x_i)$ pushes the atom to the left, that is, toward the interior of the container. Since $w(L - x_i)$ is zero for $x_i < L$, the force vanishes in the interior. In a similar manner, the final four terms in (19.10) idealize the effects of the walls at $y = 0$, $y = L$, $z = 0$, and $z = L$.

Although the above discussion is about the atoms in a gas, the general characteristics of the atom–wall interaction are the same for liquids and solids. Specifically, $\phi_{a\text{-}w}(\boldsymbol{r})$ vanishes inside the container and increases very rapidly at its surface. Since atoms only penetrate the walls by a nanometer or less, it is convenient to model atom-wall potential as

$$\phi_{a\text{-}w}(\boldsymbol{r}) = \begin{cases} 0, \text{ if } \boldsymbol{r} \text{ is inside container.} \\ \infty, \text{ if } \boldsymbol{r} \text{ is outside container.} \end{cases} \tag{19.11}$$

The total contribution to the system's energy from the interactions with the container is given by the atom–wall energy

$$\Phi_{a\text{-}w}(\boldsymbol{r}_1 \cdots \boldsymbol{r}_N) = \sum_{i=1}^{N} \phi_{a\text{-}w}(\boldsymbol{r}_i). \tag{19.12}$$

A Model Hamiltonian

Both the interatomic interaction energy Φ_{int} in (19.8) and the atom–wall energy $\Phi_{a\text{-}w}$ in (19.12) contribute to the potential energy $\Phi_N = \Phi_{a\text{-}w} + \Phi_{\text{int}}$. The complete Hamiltonian for systems of rare gas atoms is obtained by combining Φ_N with the kinetic energy K_N from (19.4). The result is

$$\boxed{H_N(\boldsymbol{p}_1 \cdots \boldsymbol{r}_N) = \sum_{i=1}^{N} \frac{1}{2m_i} |\boldsymbol{p}_i|^2 + \sum_{i=1}^{N} \phi_{a\text{-}w}(\boldsymbol{r}_i) + \sum_{i<j} \phi(r_{ij}).} \tag{19.13}$$

The solid, liquid, and gaseous phases are all modeled by this Hamiltonian. It is the system's temperature and volume that determine whether it is a solid, a liquid, or a gas. However, it is difficult to make predictions without making approximations, and different approximations are appropriate for the different phases. The approximation that yields the ideal gas model is given below. The approximations used for modeling solids are discussed in Section 19.6. Since no similar approximation exists for liquids, computer simulations are needed for their analysis.

Monatomic Ideal Gas

The ideal gas model is accurate when the system's specific volume V/N is large. When this is the case, the average distance between atoms is large compared to their diameter. Since the interaction energy $\phi(r_{ij})$ becomes negligible when the distance r_{ij} between atoms i and j is greater than about a nanometer, the interatomic interaction energy Φ_{int} becomes negligibly small and the specific volume becomes large. This indicates that an ideal gas of rare gas atoms can be modeled by neglecting the term Φ_{int} in (19.13). By using (19.4) and (19.12), the resulting ideal gas Hamiltonian $H_N^{ig} = K_N + \Phi_{a\text{-}w}$ becomes

$$\boxed{H_N^{ig}(\boldsymbol{p}_1 \cdots \boldsymbol{r}_N) = \sum_{i=1}^{N} \left[\frac{1}{2m_i} |\boldsymbol{p}_i|^2 + \phi_{a\text{-}w}(\boldsymbol{r}_i) \right].} \tag{19.14}$$

19.3 Classical Canonical Ensemble

The canonical ensemble gives us the tools for predicting thermodynamic behavior from the microscopic description contained in a system's Hamiltonian. As the number of particles in thermodynamic systems is typically of the order of Avogadro's number ($N_A = 6.0 \times 10^{23}$), it would be impossible to solve the equations of motion, even if we wanted to. It is both necessary and convenient to analyze the systems statistically. Because the microstates in classical mechanics are continuous, the statistical description is expressed in terms of probability densities.

Probability densities

Before assigning probabilities to regions in phase space, consider experiments in which the outcomes are specified by a single continuous variable. For example, the probability that the speed v of a particle (molecule) is between v_1 and v_2 is given by the integral

$$\int_{v_1}^{v_2} \rho(v)dv = \text{Probability that the speed is between } v_1 \text{ and } v_2, \tag{19.15}$$

where the *probability density* $\rho(v)$ is the probability per unit of speed. The above relationship is conveniently (but less precisely) summarized by stating that

$$\rho(v)dv = \text{Probability that the speed between } v \text{ and } v + dv. \tag{19.16}$$

Probability densities are required to be *nonnegative* and *normalized*, which implies that

$$\rho(v) \geq 0 \quad \text{and} \quad \int \rho(v)dv = 1. \tag{19.17}$$

The average value of a property X whose dependence on v is given by a function $X(v)$ is

$$\langle X \rangle = \int \rho(v) X(v) dv. \tag{19.18}$$

For example, the average speed of a particles is $\int \rho(v) v dv$.

When the above-mentioned ideas are generalized to experiments (systems) with a multidimensional continuum of outcomes (microstates), equation (19.15) becomes an integration over a volume of phase space, and $\rho(v)dv$ becomes a probability density times a volume element. Specifically, the probability that a system of N particles is in volume element $dp^{3N}dr^{3N}$ is summarized by

$$\rho(\boldsymbol{p}_1 \cdots \boldsymbol{r}_N) dp^{3N} dr^{3N} = \begin{pmatrix} \text{Probability the system is at phase point} \\ (\boldsymbol{p}_1 \cdots \boldsymbol{r}_N) \text{ in volume element } dp^{3N}dr^{3N} \end{pmatrix}. \tag{19.19}$$

The *probability density* $\rho(\boldsymbol{p}_1 \cdots \boldsymbol{r}_N)$ is the probability per unit per unit volume of phase space. It is *nonnegative* and satisfies the *normalization condition*

$$\int dp^{3N} \int dr^{3N} \rho(\boldsymbol{p}_1 \cdots \boldsymbol{r}_N) = 1, \tag{19.20}$$

where the integrations are over all of phase space. In less abbreviated notation, the integrations become

$$\int dp^{3N} \int dr^{3N} \to \int dp_1^3 \int dr_1^3 \cdots \int dp_i^3 \int dr_i^3 \cdots \int dp_N^3 \int dr_N^3, \tag{19.21}$$

where

$$\int dp_i^3 \int dr_i^3 \to \int_{-\infty}^{+\infty} dp_i^x \int_{-\infty}^{+\infty} dp_i^y \int_{-\infty}^{+\infty} dp_i^z \int_{-\infty}^{+\infty} dx_i \int_{-\infty}^{+\infty} dy_i \int_{-\infty}^{+\infty} dz_i. \tag{19.22}$$

Since the normalization condition yields a dimensionless number, the dimensions of $\rho(\boldsymbol{p}_1 \cdots \boldsymbol{r}_N)$ must be the reciprocal of the dimensions of $dp^{3N}dr^{3N}$, which implies that the dimensions of $\rho(\boldsymbol{p}_1 \cdots \boldsymbol{r}_N)$ are $(\text{mass} \times \text{length}^2/\text{time})^{-3N}$.

An ensemble is an imaginary collection of "a great number of independent systems, identical in nature, but differing in phase."[5] In classical mechanics they are described by probability densities. The phase point $(\boldsymbol{p}_1 \cdots \boldsymbol{r}_N)$ that represents a system traces out a trajectory in phase space, and this is true for every member of the ensemble. The motion of the phase points that represent the members of an ensemble can be interpreted as a flow in phase space in analogy to the flow of molecules in an ordinary fluid. Because of this flow, the distribution of points and the associated probability density are in general functions of time. However, since the macroscopic properties of systems in equilibrium do not spontaneously change, the probability densities that describe them should not change. For this to occur, the rate at which phase points flow into any volume element of phase space must equal the rate at which points flow out. Since energy is conserved, functions of the energy are time independent. This suggest that the probability density that describes a system in equilibrium is a function of the Hamiltonian.

[5] Gibbs J. W. (1902) *Elementary Principles of Statistical Mechanics*, Yale University Press, p. 5 (reprinted by Dover Publications, 1960).

Boltzmann Factors

We found in Chapter 18 that the probability that a system in equilibrium is in microstate α is proportional to the Boltzmann factor $e^{-E_{N,\alpha}/k_BT}$, where $E_{N,\alpha}$ is the energy of the state. In both classical and quantum mechanics, the canonical probabilities are proportional to the Boltzmann factors. In classical mechanics, the microstates are represented by points in phase space $(\boldsymbol{p}_1 \cdots \boldsymbol{r}_N)$, instead of energy eigenstates α, and their energies are determined by the Hamiltonian $H_N(p_1 \cdots r_N)$. A system in equilibrium at temperature T is described by the *canonical probability density*,

$$\boxed{\rho^c(\boldsymbol{p}_1 \cdots \boldsymbol{r}_N) = \frac{C_N}{Z_N} e^{-H_N(\boldsymbol{p}_1 \cdots \boldsymbol{r}_N)/k_BT},} \tag{19.23}$$

where C_N is a constant and Z_N is the *classical partition function*. By substituting $\rho^c(\boldsymbol{p}_1 \cdots \boldsymbol{r}_N)$ into the normalization condition in (19.20) and solving for Z_N, we find that

$$\boxed{Z_N(T, V) = C_N \int dp^{3N} \int dr^{3N} e^{-H_N(\boldsymbol{p}_1 \cdots \boldsymbol{r}_N)/k_BT}.} \tag{19.24}$$

The dependence on temperature enters through the T in the Boltzmann factor e^{-H_N/k_BT}, while the dependence on volume V enters through the Hamiltonian. Because of the integration over phase space, Z_N is not a function of the momentum and position variables. Classically, as well as quantum mechanically, the Helmholtz free energy is given by the *partition function formula* in (18.54), which states that

$$\boxed{F(T, V) = -k_BT \ln Z_N(T, V).} \tag{19.25}$$

The constant C_N in the probability density is chosen so that the results predicted classically agree with those obtained quantum mechanically when quantum effects are negligible. This occurs at temperatures that are not too low, where the meaning of "too low" depends on the system considered. In systems of identical particles, C_N is equal to $(h^{3N}N!)^{-1}$, where h is Planck's constant. As discussed in Section 22.3, the factor of $1/N!$ resolves a problem – known as *Gibbs paradox* – that arises in the description of systems of N identical particles. Planck's constant h and $dp\,dr$ both have dimensions of (mass × length^2/time), so that both h^{3N} and $dp^{3N}dr^{3N}$ have the same dimensions. As a result, the classical partition function is dimensionless.

19.4 Thermodynamic Properties and Averages

Mechanical and Statistical Properties

A system's properties can be predicted either by using the partition function formula and standard thermodynamic relationships or by finding canonical averages of its microscopic properties. There are two kinds of thermodynamic properties, *mechanical properties* and *statistical properties*. Mechanical properties have well-defined values independent of the macroscopic condition of the system. In contrast, statistical properties such as temperature, entropy, and the free energies are only defined for systems in thermal equilibrium. They are called *statistical properties* because systems are described statistically when in the macroscopic condition of equilibrium.

Phase Functions

The dependence of a mechanical property on the system's microstate is specified by a *phase function*. The phase function for energy is the Hamiltonian. Pressure and magnetic moment are also mechanical properties. Some mechanical properties are specified by parameters in its Hamiltonian. Volume is an example.

A system's microscopic state is specified classically by the variables $(\boldsymbol{p}_1 \cdots \boldsymbol{r}_N)$, while its thermodynamic state is specified by its thermodynamic coordinates, which are few in number (two for *PVT* systems). Hence, a tremendous reduction in the number of independent variables must occur in the prediction of thermodynamic properties. The reduction is achieved through averages. Classically, the *ensemble average* of mechanical property X is

$$\boxed{\langle X \rangle = \int dp^{3N} \int dr^{3N} \rho(\boldsymbol{p}_1 \cdots \boldsymbol{r}_N) X(\boldsymbol{p}_1 \cdots \boldsymbol{r}_N),} \tag{19.26}$$

where $X(\boldsymbol{p}_1 \cdots \boldsymbol{r}_N)$ is the corresponding phase function. The statistical interpretation of thermodynamics is based on the following principle: *The thermodynamic value of a mechanical property (other than a parameters in the Hamiltonian) is given by the canonical averages of the corresponding phase function.*

Partition Function

The internal energy U and pressure P can be obtained either by averaging the corresponding phase functions or by using the partition function formula for the free energy. We now verify that the identification of thermodynamic properties with averages is consistent with the partition function formula by showing that the two approaches yield the same result.

Some standard thermodynamic relationships from Table 12.1 are

$$F = U - TS, \qquad P = -\left(\frac{\partial F}{\partial V}\right)_T, \qquad S = -\left(\frac{\partial F}{\partial T}\right)_V, \tag{19.27}$$

where F is the Helmholtz free energy and S is the entropy. These relationships indicate that

$$U = F + TS = F - T\left(\frac{\partial F}{\partial T}\right)_V. \tag{19.28}$$

By using the partition function formula $F = -k_B T \ln Z_N$, it follows that

$$U = -k_B T \ln Z_N + T\frac{\partial}{\partial T}(k_B T \ln Z_N) = k_B T^2 \left(\frac{\partial \ln Z_N}{\partial T}\right)_V. \tag{19.29}$$

It also follows that

$$P = -\frac{\partial}{\partial V}(-k_B T \ln Z_N) = k_B T \left(\frac{\partial \ln Z_N}{\partial V}\right)_T. \tag{19.30}$$

Energy

Since the phase function for energy is the Hamiltonian, the average energy is

$$\langle H_N \rangle = \int dp^{3N} \int dr^{3N} \rho^c(\boldsymbol{p}_1 \ldots \boldsymbol{r}_N) H_N(\boldsymbol{p}_1 \ldots \boldsymbol{r}_N), \tag{19.31}$$

and this average is the internal energy U when the probability density describes a system in equilibrium. Thus, $\langle H_N \rangle$ is the internal energy U. By using the canonical probability density in (19.23), we obtain

$$U = \langle H_N \rangle = \frac{C_N}{Z_N} \int dp^{3N} \int dr^{3N} e^{-H_N(\boldsymbol{p}_1 \cdots \boldsymbol{r}_N)/k_B T} H_N(\boldsymbol{p}_1 \cdots \boldsymbol{r}_N). \tag{19.32}$$

It is convenient to express the dependence on temperature through the *inverse temperature* β, where

$$\beta = \frac{1}{k_B T}. \tag{19.33}$$

Then, the probability density and partition function become

$$\rho^c = \frac{C_N}{Z_N} e^{-\beta H_N} \quad \text{and} \quad Z_N = C_N \int dp^{3N} \int dr^{3N} e^{-\beta H_N}. \tag{19.34}$$

The derivative of the logarithm of Z_N with respect to β is

$$\frac{\partial \ln Z_N}{\partial \beta} = \frac{1}{Z_N} \frac{\partial Z_N}{\partial \beta} = \frac{C_N}{Z_N} \int dp^{3N} \int dr^{3N} (-H_N) e^{-\beta H_N}. \tag{19.35}$$

By comparing this with (19.32), the internal energy becomes

$$\boxed{U = \langle H_N \rangle = -\left(\frac{\partial \ln Z_N}{\partial \beta}\right)_V,} \tag{19.36}$$

Which by using $T = 1/k_B \beta$ becomes

$$U = \langle H_N \rangle = k_B T^2 \left(\frac{\partial \ln Z_N}{\partial T}\right)_V, \tag{19.37}$$

which is the same as the expression for U in (19.29).

We have used several different symbols for representing a system's energy. Since each symbol has a different significance, it is useful to review their meanings:

- $H_N(\boldsymbol{p}_1 \cdots \boldsymbol{r}_N)$ is the function that determines the energy of microstate $(\boldsymbol{p}_1 \cdots \boldsymbol{r}_N)$.
- E_N is a specific value of the microscopic energy.

- $\langle H_N \rangle$ is the average value of the energy.
- U is the equilibrium state function for energy.

Pressure

It is not immediately obvious what the phase function for pressure is. We can find it by showing what it must be for the partition function formula $F = -k_B T \ln Z_N$ to be valid. In (19.30), we found that

$$P = k_B T \left(\frac{\partial \ln Z_N}{\partial V} \right)_T . \tag{19.38}$$

Since the volume dependence of the partition function enters through the Hamiltonian, the derivative of its logarithm is

$$\frac{\partial \ln Z_N}{\partial V} = \frac{C_N}{Z_N} \int dp^{3N} \int dr^{3N} e^{-H_N(\boldsymbol{p}_1 \cdots \boldsymbol{r}_N)/k_B T} \left(-\frac{1}{k_B T} \frac{\partial H_N (\boldsymbol{p}_1 \cdots \boldsymbol{r}_N)}{\partial V} \right). \tag{19.39}$$

By using this and the probability density $\rho^c = (C_N/Z_N) e^{-H_N/k_B T}$, the result in (19.38) becomes

$$P = \int dp^{3N} \int dr^{3N} \rho^c(\boldsymbol{p}_1 \cdots \boldsymbol{r}_N) \left(-\frac{\partial H_N (\boldsymbol{p}_1 \cdots \boldsymbol{r}_N)}{\partial V} \right) = \left\langle -\frac{\partial H_N}{\partial V} \right\rangle . \tag{19.40}$$

Since the thermodynamic value of pressure is the average of a phase function, the *phase function for pressure* must be[6]

$$\boxed{P_{\text{ph}}(\boldsymbol{p}_1 \cdots \boldsymbol{r}_N) = -\frac{\partial H_N(\boldsymbol{p}_1 \cdots \boldsymbol{r}_N)}{\partial V}.} \tag{19.41}$$

To show that this phase function is consistent with the significance of pressure as the force per unit area on a surface, we consider a fluid confined in a cylinder by a piston, as diagramed in **Figure 19.3**, and analyze the forces on the piston. The volume of fluid that extends from $x = 0$ to $x = L$ is $V = AL$, where A is the cross-sectional area of the piston.

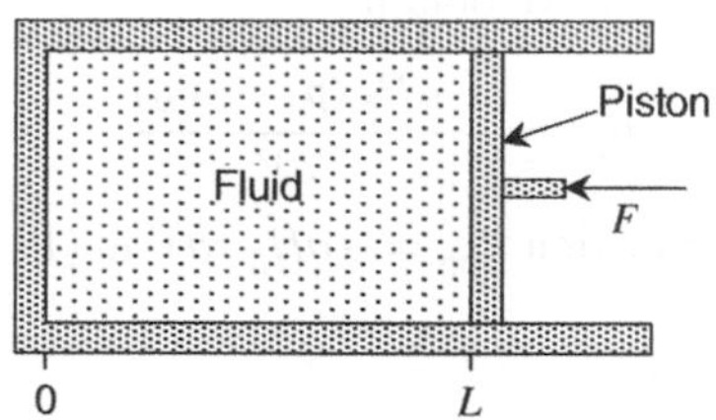

Figure 19.3 *Fluid Confined in a Cylinder*

[6] An equivalent phase functions for pressure, the virial formula, is derived in Chapter 29.

The pressure of the fluid is the normal component (x-component) of the force on the piston divided by area A. This force is balanced by a force that holds the piston in place.

The pressure on the piston is

$$P_{\text{pist}} = \frac{1}{A}\sum_i F^x_{\text{pist},i}, \tag{19.42}$$

where $F^x_{\text{pist},i}$ is the x-component of the force on the piston exerted by atom i. As these forces typically vanish for atoms more than approximately a nanometer from the piston, the majority of the atoms do not contribute to the sum in (19.42). As described in Section 19.2, the x-component of the force on atom i resulting from the atom–wall interaction is $F^x_i = -\partial\phi_{a\text{-}w}/\partial x_i$. The atom-wall potential $\phi_{a\text{-}w}(\boldsymbol{r}_i)$ can be expressed as

$$\phi_{a\text{-}w}(\boldsymbol{r}_i) = w(L - x_i) + \text{other terms}, \tag{19.43}$$

where the "other terms" represent the interactions with parts of the container other than the piston. The potential energy of the interaction between atom i and the piston is represented by $w(L - x_i)$, so that the force on atom i exerted by the piston is

$$F^x_{\text{pist},i} = -\frac{\partial w(L - x_i)}{\partial x_i} = \frac{\partial w(L - x_i)}{\partial L}, \tag{19.44}$$

where we have used the fact that L and x_i appear with opposite signs in the argument of $w(x)$. Since the force $F^x_{\text{pist},i}$ on the piston exerted by atom i opposes the force on the atom exerted by the piston (Newton's third law), the force on the piston is

$$F^x_{\text{pist},i} = -\frac{\partial w(L - x_i)}{\partial L}. \tag{19.45}$$

Since the "other terms" in (19.43) are independent of L, we can combine (19.42) through (19.45) to show that

$$P_{\text{pist}} = -\frac{1}{A}\sum_i \frac{\partial w(L - x_i)}{\partial L} = -\frac{1}{A}\sum_i \frac{\partial \phi_{a\text{-}w}(\boldsymbol{r}_i)}{\partial L} = -\frac{1}{A}\frac{\partial}{\partial L}\Phi_{a\text{-}w}(\boldsymbol{r}_1 \cdots \boldsymbol{r}_N), \tag{19.46}$$

where $\Phi_{a\text{-}w}(\boldsymbol{r}_1 \cdots \boldsymbol{r}_N)$ is the total energy of the atom-wall interaction. Since the Hamiltonian in (19.13) is $H_N = K_N + \Phi_{\text{int}} + \Phi_{a\text{-}w}$ and both the kinetic energy K_N and interatomic interactions energy Φ_{int} are independent of L, the derivative $\partial\Phi_{a\text{-}w}/\partial L$ can be replaced by $\partial H_N/\partial L$. Then, by using $L = V/A$, we obtain

$$P_{\text{pist}} = -\frac{\partial H_N(\boldsymbol{p}_1 \cdots \boldsymbol{r}_N)}{\partial V}. \tag{19.47}$$

This is the same as the phase function $P_{\text{ph}} = -\partial H_N/\partial V$ in (19.41).

19.5 Ideal Gases

Separable Hamiltonians

The ultimate justification for the partition function formula is that it yields predictions in agreement with experiment. In this section we use the ideal gas model to illustrate the procedure for predicting a system's thermodynamic properties from its Hamiltonian. The approximate Hamiltonian for a monatomic ideal gas in (19.14) can be expressed as

$$H_N^{ig}(\boldsymbol{p}_1 \cdots \boldsymbol{r}_N) = \sum_{i=1}^{N} H_i(\boldsymbol{p}_i, \boldsymbol{r}_i), \tag{19.48}$$

where $H_i(\boldsymbol{p}_i, \boldsymbol{r}_i)$ is the *single-particle Hamiltonian*

$$H_i(\boldsymbol{p}_i, \boldsymbol{r}_i) = \frac{1}{2m_i}|\boldsymbol{p}_i|^2 + \phi_{a\text{-}w}(\boldsymbol{r}_i). \tag{19.49}$$

Since the Hamiltonian is a sum of terms in which the momentum and position variables of each particle appear in only one term, it is an example of a *separable Hamiltonian*. As we will find, the evaluation of the partition function is greatly simplified when the Hamiltonian is separable.

The partition function for the system is

$$Z_N = C_N \int dp^{3N} \int dr^{3N} e^{-\beta \sum_i H_i(\boldsymbol{p}_i, \boldsymbol{r}_i)}, \tag{19.50}$$

where $\beta = 1/k_B T$. The key to evaluating these integrals is the relationship $e^{X+Y} = e^X e^Y$. By repeatedly using it, the exponential of a sum can be expressed as a product of exponentials,

$$e^{\sum_i A_i} = e^{(A_1 + A_2 + A_3 + \cdots)} = e^{A_1} e^{A_2} e^{A_3} \ldots. \tag{19.51}$$

Hence,

$$e^{-\beta \sum_i H_i(\boldsymbol{p}_i, \boldsymbol{r}_i)} = e^{-\beta H_1(\boldsymbol{p}_1, \boldsymbol{r}_1)} e^{-\beta H_2(\boldsymbol{p}_2, \boldsymbol{r}_2)} e^{-\beta H_3(\boldsymbol{p}_3, \boldsymbol{r}_3)} \ldots. \tag{19.52}$$

Because the variables for each particle appear in only one term, the integrations for each particle can be associated with the Boltzmann factor for that particle. By using (19.21) to express the integration over phase space, the partition function becomes

$$\begin{aligned} Z_N = & C_N \int dp_1^3 \int dr_1^3 e^{-\beta H_1(\boldsymbol{p}_1, \boldsymbol{r}_1)} \\ & \times \int dp_2^3 \int dr_2^3 e^{-\beta H_2(\boldsymbol{p}_2, \boldsymbol{r}_2)} \\ & \times \cdots \\ & \times \int dp_N^3 \int dr_N^3 e^{-\beta H_N(\boldsymbol{p}_N, \boldsymbol{r}_N)}, \end{aligned} \tag{19.53}$$

where $\times \cdots$ represents the $N - 3$ lines not explicitly shown. Since the integrals for each particle make no reference to the other particles, the contribution of particle i to (19.53) is

$$Z_i = (C_N)^{1/N} \int dp^3 \int dr^3 e^{-\beta H_i(\boldsymbol{p}, \boldsymbol{r})}, \tag{19.54}$$

where the subscripts on the integration variables have been omitted. The resulting partition function for the N particle system now becomes a product of N *single-particle partition functions*,

$$Z_N = Z_1 Z_2 Z_3 \cdots = \prod_{i=1}^{N} Z_i. \tag{19.55}$$

When the particles are all the same, this product simplifies to

$$Z_N = [Z_s]^N, \tag{19.56}$$

where the subscript on Z_s identifies it as a single-particle entity.

The single-particle partition function for a monatomic ideal gas of atoms of the same type is

$$Z_s = (C_N)^{1/N} \int dp^3 \int dr^3 e^{-\beta[(1/2m)|\boldsymbol{p}|^2 + \phi_{a\text{-}w}(\boldsymbol{r})]}. \tag{19.57}$$

Since $|\boldsymbol{p}|^2 = p_x{}^2 + p_y{}^2 + p_z{}^2$, the exponent can be written as the sum of four terms by expressing the exponential of the sum as a product of exponentials,

$$Z_s = (C_N)^{1/N} \int dr^3 e^{-\beta\phi_{a\text{-}w}(\boldsymbol{r})} \times \int_{-\infty}^{+\infty} dp_x e^{-\beta p_x{}^2/2m} \int_{-\infty}^{+\infty} dp_y e^{-\beta p_y{}^2/2m} \int_{-\infty}^{+\infty} dp_z e^{-\beta p_z{}^2/2m}. \tag{19.58}$$

By using the integral of a Gaussian,

$$\int_{-\infty}^{+\infty} du\, e^{-au^2} = \sqrt{\frac{\pi}{a}}, \tag{19.59}$$

the integration over each component of momentum becomes

$$\int_{-\infty}^{+\infty} dp e^{-\beta p^2/2m} = \sqrt{\frac{2\pi m}{\beta}}. \tag{19.60}$$

The atom–wall potential $\phi_{a\text{-}w}(\boldsymbol{r})$ as modeled in (19.11) is zero when $\boldsymbol{r}$ is inside the container and is infinite when $\boldsymbol{r}$ is outside. As $e^0 = 1$ and $e^{-\infty} = 0$, it follows that

$$e^{-\beta\phi_{a\text{-}w}(\boldsymbol{r})} = \begin{cases} 1, \text{if } \boldsymbol{r} \text{ is inside container.} \\ 0, \text{if } \boldsymbol{r} \text{ is outside container.} \end{cases} \tag{19.61}$$

Thus,

$$\int dr^3 e^{-\beta\phi_{a\text{-}w}(\boldsymbol{r})} = V, \tag{19.62}$$

where V is the volume of the container. For example, in a cubic container we have $e^{-\beta\phi_{a\text{-}w}(\boldsymbol{r})} = 1$ when x, y, and z are between zero and L. Otherwise, $e^{-\beta\phi_{a\text{-}w}(\boldsymbol{r})} = 0$, so that the integral becomes

$$\int dr^3 e^{-\beta\phi_{a\text{-}w}(\boldsymbol{r})} = \int_0^L dx \int_0^L dy \int_0^L dz = L^3 = V. \tag{19.63}$$

Substituting the results in (19.60) and (19.62) into (19.58) gives

$$Z_s = (C_N)^{1/N} \left(\frac{2\pi m}{\beta}\right)^{3/2} V. \tag{19.64}$$

By using $Z_N^{ig} = [Z_s]^N$, the partition function for a monatomic ideal gas of N atoms becomes

$$Z_N = C_N \left(\frac{2\pi m}{\beta} \right)^{3N/2} V^N, \tag{19.65}$$

where $\beta = 1/k_B T$. When the particles have different masses, evaluating the expression for Z_i in (19.54) yields a result similar to that of (19.64) but with m replaced by m_i, the mass of the particle. As Z_N equals the product $Z_1 Z_2 Z_3 \cdots$, the m in Z_N would then be replaced by $\overline{m} = (m_1 m_2 \cdots m_N)^{1/N}$.

Example 19.1 Internal energy and heat capacity.

Find the internal energy and constant-volume heat capacity of a monatomic ideal gas.

Solution. Taking the logarithm of the ideal gas expression for the partition function gives

$$\ln Z_N = \ln C_N + \frac{3}{2} N(\ln(2\pi m) - \ln \beta) + N \ln V.$$

As found in (19.36), the internal energy is related to Z_N by $U = -(\partial \ln Z_N / \partial \beta)$. By differentiating $\ln Z_N$ with respect to β, multiplying by -1, and using $\beta = 1/k_B T$, the internal energy becomes

$$U = \frac{3}{2} \frac{N}{\beta} = \frac{3}{2} N k_B T.$$

Since the constant-volume heat capacity is $C_V = (\partial U / \partial T)_V$, it follows that

$$C_V = \frac{3}{2} N k_B. \tag{19.66}$$

By using $N k_B = nR$, this becomes $C_V = \frac{3}{2} nR$, which is the equipartition value for a monatomic gas discussed in Chapter 15.

Helmholtz Free Energy

The above example illustrates a quick way of finding the internal energy and heat capacity when the equation of state is not being sought. The equation of state is obtained from the Helmholtz free energy. Substituting (19.65) into the partition function formula $F = -k_B T \ln Z_N$ and using $\beta = 1/k_B T$ gives

$$F = -k_B T \ln [(C_N)^{1/N} (2\pi m k_B T)^{3/2} V]^N. \tag{19.67}$$

By using the properties of logarithms, the function for the free energy becomes

$$F(T, V) = -N k_B T \left[\frac{3}{2} \ln T + \ln V + c_o \right], \tag{19.68}$$

where

$$c_o = \ln[(C_N)^{1/N} (2\pi m k_B)^{3/2}]. \tag{19.69}$$

From this and $P = -(\partial F/\partial V)_T$, it follows that

$$P = \frac{Nk_BT}{V}, \tag{19.70}$$

which is the equation of state of an ideal gas. Thus, both the equation of state and the heat capacity of a monatomic gas are correctly predicted.

The above expression for the free energy $F(T, V)$ has the same dependence on T and V as the thermodynamic result. By replacing the heat capacity $C_V = nc_V$ with the value for a monatomic gas $1.5nR$ and using $nR = Nk_B$, the thermodynamic result in Chapter 12 becomes (see (12.24))

$$F(T, V) = -Nk_BT\left(-\frac{3}{2} + \frac{3}{2}\ln\frac{T}{T_o} + \ln\frac{V}{V_o} + \frac{s_o}{R}\right), \tag{19.71}$$

where ns_o is the arbitrary value of the entropy at $T = T_o$ and $V = V_o$. Comparing this with (19.68) indicates that

$$c_o = -\frac{3}{2} - \frac{3}{2}\ln T_o - \ln V_o + \frac{s_o}{R}. \tag{19.72}$$

The values of c_o in (19.69) and (19.72) establish a relationship between the parameters s_o, T_o, and V_o in the thermodynamic result and the microscopic parameters k_B, m, and C_N. The constant C_N remains undetermined in the classical canonical ensemble. After analyzing the system quantum mechanically in Chapter 20 and resolving Gibbs paradox in Chapter 22, we find that $C_N = (h^{3N}N!)^{-1}$, where h is Planck's constant (see (22.42)). Notice that the constant c_o affects neither the heat capacity nor the equation of state.

Many Particle versus Single-Particle Concepts

It is important to distinguish the many-particle Hamiltonian and partition function, H_N and Z_N, from the single-particle quantities, H_i and Z_i. The partition function formula $F = -k_BT\ln Z_N$ expresses the fundamental relationship between thermodynamics and the microscopic properties of systems and is valid for all Hamiltonians. In contrast, a single-particle partition function Z_i is based on a single-particle Hamiltonian H_i which only exists in model systems with separable Hamiltonians. Separable Hamiltonians are always approximate. Interparticle interactions are never completely absent. If they were, there would be no mechanism for exchanging energy between particles and thus no mechanism for bringing the system into thermal equilibrium.

19.6 Harmonic Solids

The atoms in solids vibrate about fixed positions in space with amplitudes of vibration that are small compared to the distances between adjacent atoms. The harmonic approximation for solids is obtained by expanding the potential energy in a series of the displacements of the atoms from the positions that minimize the potential energy and neglecting higher order terms.

Let $(\boldsymbol{R}_1 \cdots \boldsymbol{R}_N)$ represent the positions about which the atoms vibrate, where $\boldsymbol{R}_i$ is the atomic position $\boldsymbol{r}_i$ that minimize the potential energy $\Phi_N(\boldsymbol{r}_1 \cdots \boldsymbol{r}_N)$. The minimum energy is $\Phi_{\text{st}} = \Phi_N(\boldsymbol{R}_1 \cdots \boldsymbol{R}_N)$ and is called the *static energy*. Φ_{st} is the energy of the system when the atoms are at rest. The displacement of atom i from its rest position is

$$\boldsymbol{q}_i = \boldsymbol{r}_i - \boldsymbol{R}_i. \tag{19.73}$$

It is convenient to use a single index to identify the components (q_i^x, q_i^y, q_i^z) of $\boldsymbol{q}_i$. When this is done, the components $(q_1^x, q_1^y, q_1^z, q_2^x, \dots, q_N^z)$ become $(q_1, q_2, q_3, q_4, \dots, q_{3N})$. By expanding Φ_N in a Taylor series of the displacements q_j and neglecting terms of higher order than quadratic, we obtain the *harmonic approximation* to the potential energy,

$$\Phi_N(q_1 \cdots q_{3N}) = \Phi_{\text{st}} + \frac{1}{2}\sum_{m=1}^{3N}\sum_{n=1}^{3N} A_{mn} q_m q_n. \tag{19.74}$$

The terms linear in the expansion vanish because the expansion is about a minimum, and the higher order terms are negligible because of the smallness of the displacements. The coefficients A_{mn} are element of the *force constant matrix* and are equal to $\partial^2\Phi_N/(\partial q_m \partial q_n)$. A separable Hamiltonian based on (19.74) is derived in Chapter 24. In this section, we introduce a simplified version of the harmonic approximation which was used by Einstein to demonstrate the effect of quantum mechanics on the specific heats of solids.

Einstein Model

The Hamiltonian for the Einstein model is separable. The model treats a solid as a collection of atoms attached by springs to the positions $(\boldsymbol{R}_1 \cdots \boldsymbol{R}_N)$ that minimize the potential energy. The resulting potential energy is

$$\Phi_N(q_1 \cdots q_{3N}) = \Phi_{\text{st}} + \sum_{i=1}^{N} \frac{1}{2} k_o |\boldsymbol{q}_i|^2 = \Phi_{\text{st}} + \sum_{i=1}^{N} \frac{1}{2} k_o \left[(q_i^x)^2 + (q_i^y)^2 + (q_i^z)^2 \right], \tag{19.75}$$

where Φ_{st} is the static energy and k_o is an effective spring constant. By using a single index to identify them, the components of the displacements $(q_1^x, q_1^y, \dots, q_N^z)$ become $(q_1, q_2, \dots, q_{3N})$ and the components of momentum become $(p_1, p_2, \dots, p_{3N})$. The resulting Hamiltonian is

$$H_N = K_N + \Phi_N = \Phi_{\text{st}} + H_{\text{vib}}, \tag{19.76}$$

where the vibrational Hamiltonian is

$$H_{\text{vib}}(p_1 \cdots q_{3N}) = \sum_{j=1}^{3N} H_j(p_j, q_j) \tag{19.77}$$

and

$$H_j(p_j, q_j) = \frac{1}{2m} p_j^2 + \frac{1}{2} k_o q_j^2. \tag{19.78}$$

This Hamiltonian can also be obtained by neglecting the off-diagonal elements of the force constant matrix of A_{mn}.

Since the variables $(\boldsymbol{r}_1 \cdots \boldsymbol{r}_N)$ in the partition function in (19.24) have been transformed to $(\boldsymbol{q}_1 \cdots \boldsymbol{q}_N)$, which in turn have been represented by $(q_1 \cdots q_{3N})$, the classical partition function for the Einstein model is

$$Z_N = C_N \int dp^{3N} \int dq^{3N} e^{-H_N(p_1 \cdots q_{3N})/k_B T}. \tag{19.79}$$

Since the static energy Φ_{st} is independent of the integration variables, we can use the relationship $e^{-H_N/k_B T} = e^{-\Phi_{\text{st}}/k_B T} e^{-H_{\text{vib}}/k_B T}$ to express Z_N as

$$Z_N = e^{-\Phi_{\text{st}}/k_B T} Z_N^{\text{vib}}, \tag{19.80}$$

where

$$Z_N^{\text{vib}} = C_N \int dp^{3N} \int dq^{3N} e^{-H_{\text{vib}}(p_1 \cdots q_{3N})/k_B T}. \tag{19.81}$$

It is left as a problem to show that the Einstein model when treated classically yields the Dulong and Petit value for the specific heat from Section 15.4. The quantum mechanical treatment of the model is done in Chapter 20.

Problems

19.1 Find the minimum potential energy of clusters of three and four rare gas atoms. In the minimum energy configurations, the three atoms are at the corners of a triangle and 4 atoms are at the corners of a tetrahedron. Include a sketch of the configurations and express the answers in terms of the depth of the potential well.

19.2 Consider a classical monatomic gas of N atoms in a uniform gravitational field confined to an infinitely tall vertical cylinder of cross-sectional area A. This is a model of a planetary atmosphere in which the cylinder represents the effect of the surrounding atmosphere. The Hamiltonian for the model is

$$H_N = \sum_{i=1}^{N} \left(\frac{|\boldsymbol{p}_i|^2}{2m_i} + mgz_i + \phi_A(x_i, y_i) \right),$$

where m is the mass of an atom, g is the acceleration of gravity, and the surface of the planet is at $z = 0$. The potential $\phi_A(x_i, y_i)$ (which is a potential similar to ϕ_{a-w}) confines the atoms within the cylinder. Find the internal energy and constant-area heat capacity $(\partial U/\partial T)_A$ of the gas.

19.3 The vibrational Hamiltonian of an Einstein model with spring constants $k_o(V)$, which depend on volume, is

$$H_{\text{vib}} = \sum_{j=1}^{N} \left[\frac{1}{2m} |\boldsymbol{p}_j|^2 + \frac{1}{2} k_o(V) |\boldsymbol{q}_j|^2 \right].$$

(a) Evaluate the vibrational partition function

$$Z_N^{\text{vib}} = C_N \int dp^{3N} \int dq^{3N} e^{-H_{\text{vib}}(p_1 \cdots q_{3N})/k_B T}.$$

(b) Find the Helmholtz free energy $F(T, V)$. Include the volume dependence of $k_o(V)$ and of the static energy $\Phi_{\text{st}}(V)$.

(c) Find the specific heat c_V and compare with the law of Dulong and Petit.

(d) Find the equation of state.

20

Quantum Systems

The connection between the microscopic and macroscopic properties of a system is made by the partition function, which is obtained by summing the Boltzmann factors e^{-E/k_BT} for the energies of the system's microstates. Classically, microstates correspond to points in phase space. In quantum mechanics, they correspond to energy eigenstates, which are solutions of the time-independent Schrödinger equation. Both classically and quantum mechanically, thermodynamic properties are determined by the Hamiltonian, but the relationship is less direct in quantum mechanics. In this chapter, the properties of the ideal gas and the Einstein models are treated quantum mechanically. We also find the conditions under which the classical approach yields accurate results. We start by reviewing the energy eigenstates of a monatomic gas.

20.1 Energy Eigenstates

The time evolution of systems is determined by the time-dependent Schrödinger equation $\widehat{H}_N\Psi(t) = i\hbar(\partial\Psi(t)/\partial t)$, where $\widehat{H}_N$ is the Hamiltonian operator and $\Psi(t)$ is the system's wave function. The operator $\widehat{H}_N$ for *PVT* systems of *N* particles is obtained from the classical Hamiltonian $H_N(\boldsymbol{p}_1 \cdots \boldsymbol{r}_N)$ by replacing the momentum variable $\boldsymbol{p}_i$ of particle i by the differential operators $(\hbar/i)\nabla_i$. The gradient ∇_i differentiates functions of position variable $\boldsymbol{r}_i$, the i in $(\hbar/i)$ is the square root of -1, and $\hbar$ is Planck's constant divided by 2π ($\hbar = h/2\pi$). By expressing the time dependence of $\Psi(t)$ as $e^{-iE_{N,\alpha}t/\hbar}\Psi_\alpha$ and simplifying, we obtain the *time-independent Schrödinger equation*

$$\widehat{H}_N\Psi_\alpha(\boldsymbol{r}_1 \cdots \boldsymbol{r}_N) = E_{N,\alpha}\Psi_\alpha(\boldsymbol{r}_1 \cdots \boldsymbol{r}_N), \tag{20.1}$$

where index α identifies the different linearly independent solutions. Each independent solution corresponds to an *energy eigenstates* and is associated with an *eigenvalue*

Thermodynamics and Statistical Mechanics: An Integrated Approach, First Edition.
Robert J Hardy and Christian Binek.

$E_{N,\alpha}$ and *eigenfunction* $\Psi_\alpha(\boldsymbol{r}_1 \cdots \boldsymbol{r}_N)$. The energy eigenstates are used as the system's microstates.[1]

Ideal Gases

The eigenstates of a monatomic ideal gas are derived in Appendix 20.A and are summarized below. The confining effect of the container is described in the classical Hamiltonian in (19.14) by the atom-wall potentials, while in quantum mechanics their effect is accounted for by imposing boundary conditions on the eigenfunctions. The Hamiltonian operator for a gas of N particles is obtained by omitting the atom-wall potentials in (19.14) and replacing the variables $\boldsymbol{p}_i$ by operators $(\hbar/i)\nabla_i$. The result is

$$\widehat{H}_N^{ig} = \widehat{H}_1 + \widehat{H}_2 + \cdots + \widehat{H}_N, \tag{20.2}$$

where $\widehat{H}_i$ is the free particle Hamiltonian operator

$$\widehat{H}_i = -\frac{\hbar^2}{2m_i}\nabla_i^2. \tag{20.3}$$

Because of the large number of eigenvalues in the range of energies of significance in gases, we focus on how the eigenvalues are distributed. The distribution is insensitive to the shape of the container, which allows us to choose a shape that simplifies our calculations. A cubic container of volume $V = L^3$ is used.

The eigenvalues $E_{N,\alpha}$ of the gas are sums of *single-particle eigenvalues* $\varepsilon_i(\boldsymbol{n}_i)$, and the eigenfunctions Ψ_α are products of *single-particle eigenfunctions* $\psi_{\boldsymbol{n}_i}(\boldsymbol{r}_i)$. The eigenvalue equation for a single particle (atom) is

$$-\frac{\hbar^2}{2m_i}\nabla_i^2\psi_{\boldsymbol{n}_i}(\boldsymbol{r}_i) = \varepsilon_i(\boldsymbol{n}_i)\psi_{\boldsymbol{n}_i}(\boldsymbol{r}_i), \tag{20.4}$$

which is the Schrödinger equation for Hamiltonian $\widehat{H}_i$. The energies $\varepsilon_i(\boldsymbol{n}_i)$ are determined by the boundary conditions which require that the function $\psi_{\boldsymbol{n}_i}(\boldsymbol{r}_i)$ vanishes at the container walls. The resulting energies are

$$\varepsilon_i(\boldsymbol{n}_i) = \frac{1}{2m_i}\left(\frac{\hbar\pi}{L}\right)^2 |\boldsymbol{n}_i|^2, \tag{20.5}$$

where $|\boldsymbol{n}_i|^2 = n_{x,i}^2 + n_{y,i}^2 + n_{z,i}^2$ and the allowed values of the quantum numbers are

$$n_{x,i} = 1, 2, \ldots, \infty, \qquad n_{y,i} = 1, 2, \ldots, \infty, \qquad n_{z,i} = 1, 2, \ldots, \infty. \tag{20.6}$$

The energy eigenvalues for a monatomic ideal gas of N atoms from (20.A.25) are

$$\boxed{E_{N,\alpha} = \varepsilon_1(\boldsymbol{n}_1) + \varepsilon_2(\boldsymbol{n}_2) + \cdots + \varepsilon_N(\boldsymbol{n}_N),} \tag{20.7}$$

[1] The quantum mechanical description of thermodynamic behavior that uses density matrices instead of energy eigenstates is discussed in Section 23.4.

where index α is an abbreviated notation for the quantum numbers of all N atoms,

$$\alpha = (\boldsymbol{n}_1, \boldsymbol{n}_2, \ldots, \boldsymbol{n}_N) = (n_{x,1}, n_{y,1}, \ldots, n_{z,N}). \tag{20.8}$$

We emphasize that index α labels the system's eigenstates, not its *energy levels*. The number of eigenstates with eigenvalues that equal a specific value of energy is the *degeneracy* of the energy level. The energy levels of gases are highly degenerate.

20.2 Quantum Canonical Ensemble

Different members of an ensemble are in different microstates. When described quantum mechanically, the *microstates* of a system in equilibrium are *energy eigenstates*. As found in Section 18.4, the canonical probability ρ_α^c of microstate α is proportional to the *Boltzmann factor* $e^{-E_{N,\alpha}/k_BT}$, where $E_{N,\alpha}$ is the eigenvalue of the microstate, and the *canonical ensemble* for a system at temperature T is described by the probabilities

$$\boxed{\rho_\alpha^c = \frac{e^{-E_{N,\alpha}/k_BT}}{Z_N}.} \tag{20.9}$$

The probabilities are normalized by Z_N, which is found by substituting ρ_α^c into the normalization condition $\sum_\alpha \rho_\alpha^c = 1$ and solving for Z_N. The resulting expression for the *quantum partition function* is

$$\boxed{Z_N(T, V) = \sum_\alpha e^{-E_{N,\alpha}/k_BT}.} \tag{20.10}$$

The dependence of Z_N on temperature enters through the T in the Boltzmann factor, while the dependence on V enters through the volume dependence of the energies $E_{N,\alpha}$. The thermodynamic properties of PVT systems are related to their microscopic properties by the *partition function formula* for the Helmholtz free energy,

$$\boxed{F(T, V) = -k_BT \ln Z_N(T, V),} \tag{20.11}$$

which is valid both classically and quantum mechanically.

The *ensemble average* of property X, where X_α is the value of the property for a system in microstate α, is[2]

$$\langle X \rangle = \sum_\alpha \rho_\alpha X_\alpha, \tag{20.12}$$

where the sum is over all microstates. The ensemble average of the energy of a system described by the canonical probabilities is

$$\boxed{\langle E_N \rangle = \sum_\alpha \rho_\alpha^c E_{N,\alpha}.} \tag{20.13}$$

[2] This is a special case of the general definition of a quantum mechanical average discussed in Section 23.4.

Since the probabilities ρ^c_α describe a system in equilibrium, the average is identified with the internal energy state function U. It is easily verified that this value for U is the same as that predicted by the partition function formula. By using (20.11) and standard thermodynamic relationships, it follows that

$$U = F - T\left(\frac{\partial F}{\partial T}\right)_V = k_B T^2 \left(\frac{\partial \ln Z_N}{\partial T}\right)_V. \tag{20.14}$$

By introducing $\beta = 1/k_B T$, this becomes

$$\boxed{U = -\left(\frac{\partial \ln Z_N}{\partial \beta}\right)_V,} \tag{20.15}$$

which is the same as (19.36). Differentiating Z_N and using $\rho^c_\alpha = e^{-\beta E_{N,\alpha}}/Z_N$ gives

$$U = \frac{1}{Z_N}\sum_\alpha e^{-\beta E_{N,\alpha}} E_{N,\alpha} = \sum_\alpha \rho^c_\alpha E_{N,\alpha}, \tag{20.16}$$

which is the value of U predicted by the relationship $U = \langle E_N \rangle$.

20.3 Ideal Gases

We start the quantum mechanical analysis of systems by analyzing the ideal gas model. As given in (20.7), the energy eigenvalues monatomic ideal gas of N atoms are $E_{N,\alpha} = \sum_i \varepsilon_i(\boldsymbol{n}_i)$. Since each term in the sum depends on the quantum numbers of a single atom, the energy is *separable*, which simplifies the evaluation of the partition function.[3] (The properties of systems of fermions and bosons are discussed in Chapter 26.) Index α is a abbreviated notation for the set of quantum numbers $(\boldsymbol{n}_1, \boldsymbol{n}_2, \ldots, \boldsymbol{n}_N)$, so that the Boltzmann factors are

$$e^{-E_{N,\alpha}/k_B T} = e^{-[\varepsilon_1(\boldsymbol{n}_1)+\varepsilon_2(\boldsymbol{n}_2)+\cdots+\varepsilon_N(\boldsymbol{n}_N)]/k_B T}. \tag{20.17}$$

The sum over all microstates can be performed by summing over the quantum numbers of each atom, so that

$$\sum_\alpha = \sum_{\boldsymbol{n}_1}\sum_{\boldsymbol{n}_2}\cdots\sum_{\boldsymbol{n}_N}, \tag{20.18}$$

where the sum for each atom consists of three sums,

$$\sum_{\boldsymbol{n}_i} = \sum_{n_{x,i}=1}^{\infty}\ \sum_{n_{y,i}=1}^{\infty}\ \sum_{n_{z,i}=1}^{\infty}. \tag{20.19}$$

The resulting expression for the partition function is

$$Z_N = \sum_\alpha e^{-\beta E_{N,\alpha}} = \sum_{\boldsymbol{n}_1}\sum_{\boldsymbol{n}_2}\cdots\sum_{\boldsymbol{n}_N} e^{-\beta[\varepsilon_1(\boldsymbol{n}_1)+\varepsilon_2(\boldsymbol{n}_2)+\cdots+\varepsilon_N(\boldsymbol{n}_N)]}, \tag{20.20}$$

[3] When the Hamiltonian is separable, the energy eigenvalues are separable until the exchange symmetry of fermions and bosons has been accounted for.

where $\beta = 1/k_B T$. Since the exponential of a sum equals the product of the exponentials, it follows that

$$Z_N = \sum_{n_1}\sum_{n_2}\cdots\sum_{n_N} e^{-\beta\varepsilon_1(n_1)}e^{-\beta\varepsilon_2(n_2)}\ldots e^{-\beta\varepsilon_N(n_N)}, \tag{20.21}$$

which can be rewritten as

$$\boxed{Z_N = \left(\sum\nolimits_{n_1} e^{-\beta\varepsilon_1(n_1)}\right)\left(\sum\nolimits_{n_2} e^{-\beta\varepsilon_2(n_2)}\right)\cdots\left(\sum\nolimits_{n_N} e^{-\beta\varepsilon_N(n_N)}\right).} \tag{20.22}$$

As result of the separability of the energy, the N-particle partition function has separated into a product of N single-particle partition functions.

In a gas of atoms of the same type the dependence of the single-particle eigenvalues is the same for all atoms, so that

$$Z_N = [Z_s]^N, \tag{20.23}$$

where

$$Z_s = \sum_n e^{-\beta\varepsilon(n)}, \tag{20.24}$$

where the subscript s identifies it as a single-particle partition function. The single-particle eigenvalues for atoms of mass m in volume $V = L^3$ are (see (20.5))

$$\varepsilon(\boldsymbol{n}) = \frac{1}{2m}\left(\frac{\hbar\pi}{L}\right)^2 (n_x{}^2 + n_y{}^2 + n_z{}^2). \tag{20.25}$$

This can be expressed as

$$\varepsilon(\boldsymbol{n}) = e(n_x) + e(n_y) + e(n_z), \tag{20.26}$$

where $e(n) = (1/2m)(\hbar\pi n/L)^2$. Substituting this into Z_s gives

$$\begin{aligned} Z_s &= \sum_{\vec{n}} e^{-\beta\varepsilon(n)} = \sum_{n_x=1}^{\infty}\sum_{n_y=1}^{\infty}\sum_{n_z=1}^{\infty} e^{-\beta(e(n_x)+e(n_y)+e(n_z))} \\ &= \sum_{n_x=1}^{\infty} e^{-\beta e(n_x)} \sum_{n_y=1}^{\infty} e^{-\beta e(n_y)} \sum_{n_z=1}^{\infty} e^{-\beta e(n_z)} = \left[\sum_{n=1}^{\infty} e^{-\beta e(n)}\right]^3. \end{aligned} \tag{20.27}$$

Free Energy

Since $Z_N = [Z_s]^N$, finding the partition function Z_N simplifies the problem of evaluating the sum $\sum_n e^{-\beta e(n)}$. This is facilitated by expressing the energies $e(n)$ in (20.26) as

$$e(n) = \frac{1}{2m}\hbar^2 k_n{}^2, \tag{20.28}$$

where

$$k_n = \frac{\pi n}{L}. \tag{20.29}$$

The spacing between the successive values of k_n is

$$\Delta k = k_{n+1} - k_n = \frac{\pi}{L}, \tag{20.30}$$

which implies that $(L/\pi)\Delta k = 1$. By using this, the sum over n becomes

$$\sum_{n=1}^{\infty} e^{-\beta e(n)} = \frac{L}{\pi}\sum_{n=1}^{\infty} e^{-(\beta\hbar^2/2m)k_n^2}\Delta k. \tag{20.31}$$

Estimating the sum by an integral gives

$$\sum_{n=1}^{\infty} e^{-(\beta\hbar^2/2m)k_n^2}\Delta k \approx \int_0^{\infty} e^{-(\beta\hbar^2/2m)k^2}dk. \tag{20.32}$$

The value of this integral is half the value of the integral of a Gaussian from $-\infty$ and $+\infty$ given in (19.59). Since $\int_0^{\infty} e^{-ak^2}dk = \frac{1}{2}\sqrt{\pi/a}$ and $a = \beta\hbar^2/2m$, the value of the sum over n is

$$\frac{1}{2}\sqrt{\frac{2\pi m}{\beta\hbar^2}} = \frac{\pi}{h}\sqrt{2\pi m k_B T}, \tag{20.33}$$

where $\beta = 1/k_BT$ and $h = 2\pi\hbar$. Thus,

$$\sum_{n=1}^{\infty} e^{-\beta e(n)} = \frac{L}{\pi}I_k = \frac{L}{h}\sqrt{2\pi m k_B T}, \tag{20.34}$$

where the length L entered through the boundary conditions on the eigenfunctions. By introducing the volume $V = L^3$, the single-particle partition function becomes

$$Z_s = \left[\sum_n e^{-\beta e(n)}\right]^3 = \left(\frac{2\pi m k_B T}{h^2}\right)^{3/2} V. \tag{20.35}$$

Thus, the N-particle partition function for a monatomic ideal gas is

$$\boxed{Z_N = [Z_s]^N = \left(\frac{2\pi m k_B T}{h^2}\right)^{3N/2} V^N.} \tag{20.36}$$

Substituting this into $F = -k_BT\ln Z_N$ yields the Helmholtz free energy

$$\boxed{F(T,V) = -Nk_BT\left[\frac{3}{2}\ln T + \ln V + c_o\right],} \tag{20.37}$$

where

$$c_o = \ln\frac{(2\pi m k_B)^{3/2}}{h^3}. \tag{20.38}$$

This quantum mechanical expression for $F(T, V)$ is the same as the classical expression in (19.68), although the expressions for c_o are different. Since the value of c_o does not affect

P, U, or C_V, the equation of state ($P = Nk_BT/V$), internal energy ($U = \frac{3}{2}Nk_BT$), and heat capacity predicted quantum mechanically are the same was predicted classically in Section 19.5.

The classical expression for c_o in (19.69) includes a constant C_N that is not determined by the classical theory. Equating the classical and quantum mechanical values of c_o indicates that C_N equals h^{-3N}. However, as discussed in Section 22.3, an additional factor of $1/N!$ is needed to resolve the Gibbs paradox, so that the final value of C_N is $(h^{3N}N!)^{-1}$.

20.4 Einstein Model

In contrast to the results for an ideal gas, the quantum mechanical predictions for the Einstein model differ significantly from the classical results. As discussed in Section 19.6, the Einstein model is a simplified version of the harmonic approximation for the vibrational properties of solids which treats solid as collections of atoms held in place by springs.[4]

By replacing the momentum variables p_j by the operators $\hat{p}_j = (\hbar/i)\partial/\partial q_j$, the model Hamiltonian for the Einstein model in (19.76) through (19.78) becomes

$$\hat{H}_N(\hat{p}_1 \cdots q_{3N}) = \Phi_{\text{st}} + \hat{H}_{\text{vib}}(\hat{p}_1 \cdots q_{3N}), \tag{20.39}$$

where the vibrational Hamiltonian is

$$\hat{H}_{\text{vib}}(\hat{p}_1 \cdots q_{3N}) = \sum_{j=1}^{3N} \hat{H}_j(\hat{p}_j, q_j) \tag{20.40}$$

and

$$\hat{H}_j(\hat{p}_j, q_j) = \frac{1}{2m}\hat{p}_j^{\,2} + \frac{1}{2}k_o q_j^{\,2}. \tag{20.41}$$

The static energy Φ_{st} is the constant term in the expansion of the potential energy, and k_o is an effective spring constant. In (20.39) through (20.41), the components (q_i^x, q_i^y, q_i^z) of the displacement of atom i from its rest position $\boldsymbol{R}_i$ have been identified by a single index j, so that the variables $(q_1^x, q_1^y, q_1^z, q_2^x, \cdots, q_N^z)$ become $(q_1, q_2, q_3, q_4, \cdots, q_{3N})$. The motion associated with different values of j will be referred to as different *modes* of oscillation. The angular frequency of a classical harmonic oscillator of mass m and spring constant k_o is $\omega = \sqrt{k_o/m}$, so that $k_o = m\omega^2$. In terms of ω, the Hamiltonian for mode j is

$$\hat{H}_j(\hat{p}_j, q_j) = \frac{1}{2m}\hat{p}_j^{\,2} + \frac{1}{2}m\omega^2 q_j^{\,2}. \tag{20.42}$$

Since Φ_{st} is a constant, the eigenvalues $E_{N,\alpha}$ of the system are related to the eigenvalues $E_{N,\alpha}^{\text{vib}}$ of the vibrational Hamiltonian by

$$E_{N,\alpha} = \Phi_{\text{st}} + E_{N,\alpha}^{\text{vib}}. \tag{20.43}$$

[4] A more general treatment of the harmonic approximation for solids is outlined in Section 24.3.

Taking $e^{-\beta\Phi_{st}}$ outside of the sum over α in the partition function in (20.10) gives

$$Z_N = \sum_{\alpha} e^{-\beta E_{N,\alpha}} = \sum_{\alpha} e^{-\beta(\Phi_{st}+E_{N,\alpha}^{vib})} = e^{-\beta\Phi_{st}} Z_N^{vib}, \tag{20.44}$$

where $\beta = 1/k_B T$ and

$$Z_N^{vib} = \sum_{\alpha} e^{-\beta E_{N,\alpha}^{vib}}. \tag{20.45}$$

Eigenstates

The Schrödinger equation for mode j is $\widehat{H}_j \psi_{n_j} = \varepsilon(n_j)\psi_{n_j}$, which by introducing the momentum operator $p_j = (\hbar/i)\partial/\partial q_j$ becomes

$$\left(-\frac{\hbar^2}{2m}\frac{\partial^2}{\partial q_j^2} + \frac{m\omega^2}{2} q_j^{\,2} \right) \psi_{n_j}(q_j) = \varepsilon(n_j)\psi_{n_j}(q_j). \tag{20.46}$$

This is the Schrödinger equation for a harmonic oscillator. As shown in most quantum mechanics texts, that the energy eigenvalues are

$$\varepsilon(n_j) = \left(n_j + \frac{1}{2} \right) \hbar\omega, \tag{20.47}$$

where ω is the classical angular frequency and the allowed values of n_j are

$$n_j = 0, 1, 2, \ \ldots \ , \infty. \tag{20.48}$$

The N-particle eigenvalues $E_{N,\alpha}^{vib}$ are sums of the eigenvalues of the different modes,

$$E_{N,\alpha}^{vib} = \varepsilon(n_1) + \varepsilon(n_2) + \cdots + \varepsilon(n_{3N}), \tag{20.49}$$

where α is an abbreviated notation for $3N$ quantum numbers n_j,

$$\alpha = (n_1, n_2, \ \ldots \ , n_{3N}). \tag{20.50}$$

Since each quantum number occurs in only one of the terms in (20.49), the energy is separable.

Partition Function

By substituting $E_{N,\alpha}^{vib}$ into (20.45), the vibrational partition function becomes

$$\begin{aligned} Z_N^{vib} &= \sum_{\alpha} e^{-\beta E_{N,\alpha}^{vib}} = \sum_{n_1=0}^{\infty}\sum_{n_2=0}^{\infty} \cdots \sum_{n_{3N}=0}^{\infty} e^{-\beta[\varepsilon(n_1)+\varepsilon(n_2)+\cdots+\varepsilon(n_{3N})]} \\ &= \sum_{n_1=0}^{\infty} e^{-\beta\varepsilon(n_1)} \sum_{n_2=0}^{\infty} e^{-\beta\varepsilon(n_2)} \ldots \sum_{n_{3N}=0}^{\infty} e^{-\beta\varepsilon(n_{3N})}. \end{aligned} \tag{20.51}$$

Since the frequency ω is the same for all modes, the sums are all the same, so that

$$Z_N^{vib} = [Z_s^{vib}]^{3N}, \tag{20.52}$$

where

$$Z_s^{\text{vib}} = \sum_{n=0}^{\infty} e^{-\beta\varepsilon(n)}. \tag{20.53}$$

It follows from $\varepsilon(n) = \left(n + \frac{1}{2}\right)\hbar\omega$ and the properties of exponentials that

$$e^{-\beta\varepsilon(n)} = e^{-\beta\left(n+\frac{1}{2}\right)\hbar\omega} = e^{-\frac{1}{2}\beta\hbar\omega}(e^{-\beta\hbar\omega})^n, \tag{20.54}$$

so that

$$Z_s^{\text{vib}} = e^{-\frac{1}{2}\beta\hbar\omega}\sum_{n=0}^{\infty}(e^{-\beta\hbar\omega})^n. \tag{20.55}$$

By letting $X = e^{-\beta\hbar\omega}$ and using the expansion

$$\frac{1}{1-X} = 1 + X + X^2 + X^3 + \cdots = \sum_{n=0}^{\infty} X^n, \tag{20.56}$$

we obtain

$$Z_s^{\text{vib}} = \frac{e^{-\frac{1}{2}\beta\hbar\omega}}{1 - e^{-\beta\hbar\omega}}. \tag{20.57}$$

Thus, the vibrational partition function for a solid of N atoms is

$$\boxed{Z_N^{\text{vib}} = [Z_s^{\text{vib}}]^{3N} = \left(\frac{e^{-\frac{1}{2}\beta\hbar\omega}}{1 - e^{-\beta\hbar\omega}}\right)^{3N}.} \tag{20.58}$$

Internal Energy

By combining the expressions for Z_N, ρ_α, and $\langle E_N\rangle$ in (20.10), (20.9), and (20.13), it follows that

$$\begin{aligned} -\frac{\partial \ln Z_N}{\partial\beta} &= -\frac{1}{Z_N}\frac{\partial}{\partial\beta}\sum_\alpha e^{-\beta E_{N,\alpha}} = \sum_\alpha \frac{e^{-\beta E_{N,\alpha}}}{Z_N} E_{N,\alpha} \\ &= \sum_\alpha \rho_\alpha E_{N,\alpha} = \langle E_N\rangle. \end{aligned} \tag{20.59}$$

The equilibrium average $\langle E_N\rangle$ is internal energy, so that

$$U = -\frac{\partial \ln Z_N}{\partial\beta} = \langle E_N\rangle. \tag{20.60}$$

By using $Z_N = e^{-\beta\Phi_{\text{st}}}Z_N^{\text{vib}}$ and $E_{N,\alpha} = \Phi_{\text{st}} + E_{N,\alpha}^{\text{vib}}$, the internal energy of the Einstein model becomes

$$U = -\frac{\partial}{\partial\beta}[-\beta\Phi_{\text{st}} + \ln Z_N^{\text{vib}}] = \Phi_{\text{st}} - \frac{\partial \ln Z_N^{\text{vib}}}{\partial\beta} \tag{20.61}$$

and

$$U = \Phi_{\text{st}} + \langle E_N^{\text{vib}}\rangle. \tag{20.62}$$

Since the logarithm of Z_N^{vib} is

$$\ln Z_N^{\text{vib}} = -3N\left[\ln\left(1 - e^{-\beta\hbar\omega}\right) + \frac{1}{2}\beta\hbar\omega\right], \tag{20.63}$$

the vibrational contribution to the internal energy is

$$U_{\text{vib}} = -\frac{\partial \ln Z_N^{\text{vib}}}{\partial\beta} = 3N\left[\frac{e^{-\beta\hbar\omega}}{1 - e^{-\beta\hbar\omega}} + \frac{1}{2}\right]\hbar\omega, \tag{20.64}$$

which becomes

$$\boxed{U_{\text{vib}} = 3N\left(\frac{1}{e^{\hbar\omega/k_BT} - 1} + \frac{1}{2}\right)\hbar\omega.} \tag{20.65}$$

Since $e^{\hbar\omega/k_BT} \to \infty$ as $T \to 0$, the energy per particle U_{vib}/N at $T = 0$ is $\frac{3}{2}\hbar\omega$. This indicates that when treated quantum mechanically, vibrational motion does not completely disappear at absolute zero. This lowest possible energy is called *zero point energy*.

Specific Heat

The static energy Φ_{st} is not a function of temperature and thus does not contribute to the specific heat. The specific heat expressed as a heat capacity per mode is

$$\frac{C_V}{3N} = \frac{1}{3N}\left(\frac{\partial U}{\partial T}\right)_V = k_B\left(\frac{1}{e^{+\hbar\omega/(2k_BT)} - e^{-\hbar\omega/(2k_BT)}}\right)^2\left(\frac{\hbar\omega}{k_BT}\right)^2. \tag{20.66}$$

This can be simplified to

$$\boxed{\frac{C_V}{3N} = k_B\left[\frac{\theta_E/T}{e^{+\theta_E/2T} - e^{-\theta_E/2T}}\right]^2,} \tag{20.67}$$

where $\theta_E = \hbar\omega/k_B$ is the *Einstein temperature*. The Einstein temperature can be used to characterize the specific heats of solids by adjusting value of θ_E to obtain the best fit of (20.67) to the experimental data. The characteristic Einstein temperature θ_E has roughly the same value as the Debye temperature θ_D discussed in Section 24.4, which has values that range from 100 K to 1000 K. Values of θ_D for a few materials are given in Table 24.1. The characterization of specific heats by θ_D results from the Debye model, which is based on a more detailed description of atomic vibration than the Einstein model.

The Einstein model illustrates the quantum mechanical origin of the drop in the specific heats of solids at low temperatures. The specific heat $C_V/3Nk_B$ versus T/θ_E is plotted in **Figure 20.1**, which indicates that the specific heat goes to zero at absolute zero as required by the third law of thermodynamics. The high temperature limit is found by expanding $e^{\theta_E/T} = e^{\hbar\omega/k_BT}$ in powers $\hbar\omega/k_BT$, where $\hbar\omega$ is the spacing between energy levels. By keeping the first two terms in the expansion $e^{\hbar\omega/k_BT} = 1 + \hbar\omega/k_BT + \cdots$, the vibrational energy U_{vib} in (20.65) becomes

$$U_{\text{vib}} = 3N\left(\frac{1}{(\hbar\omega/k_BT) + \cdots} + \frac{1}{2}\right)\hbar\omega \approx 3N\left(k_BT + \frac{1}{2}\hbar\omega\right) \approx 3Nk_BT. \tag{20.68}$$

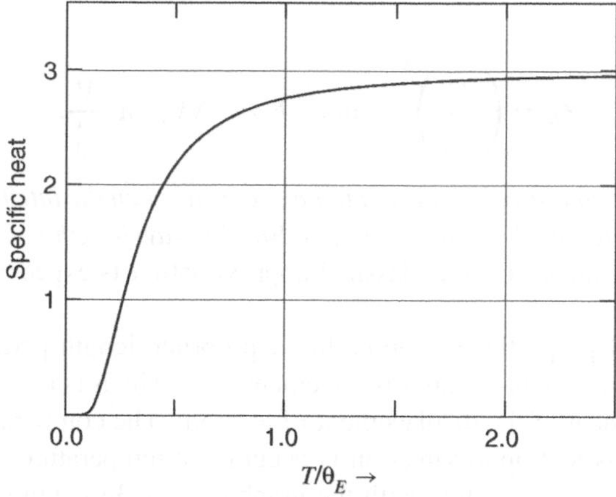

Figure 20.1 *Specific Heat for the Einstein Model*

Thus, at high temperatures C_V tends to the classical value of $3Nk_B$, which is the Dulong and Petit value discussed in Section 15.4.

20.5 Classical Approximation

The accuracy of a classical analysis can be estimated by comparing a length characteristic of quantum effects with the characteristic lengths of the system being studied. The characteristic length for quantum effects is the de Broglie wavelength λ, where $p = h/\lambda$ is momentum and h is Planck's constant. A typical value for the momentum of a system's particles can be obtained from the equipartition hypothesis discussed in Section 15.3, which associates $\frac{1}{2}k_BT$ of energy with each microscopic degree of freedom. The equipartition of energy indicates that the average kinetic energy of a particle moving in three dimensions is

$$\frac{1}{2m}p^2 \approx \frac{3}{2}k_BT. \tag{20.69}$$

Combining this with $\lambda = h/p$ gives

$$\lambda \approx \frac{h}{\sqrt{3mk_BT}}. \tag{20.70}$$

As our interest is in making rough estimates, the significance of λ is unchanged by replacing the 3 by 2π, which gives us the *thermal de Broglie wavelength*

$$\boxed{\lambda_{\text{th}} = \frac{h}{\sqrt{2\pi mk_BT}}.} \tag{20.71}$$

The replacement is convenient because the quantity $2\pi mk_BT$ appears in the classical partition function Z_N and free energy F. The ideal gas values for Z_N and F in (20.36) and (20.37) can be expressed as

$$Z_N = \left(\frac{V}{\lambda_{\text{th}}^3}\right)^N \quad \text{and} \quad F = -Nk_BT \ln \frac{V}{\lambda_{\text{th}}^3}. \tag{20.72}$$

The classical approximation is expected to be accurate when the intrinsic length parameters of a system are greater than the thermal de Broglie wavelength λ_{th}. Since λ_{th} decreases as the temperature T increases, the classical approximation is expected to be accurate at higher temperatures.

For the vibrational properties of solids, the appropriate length parameter for comparing with λ_{th} is the root-mean-square displacement q_{rms}, which is the square root of $\langle {q_j}^2 \rangle$ where q_j is a component of the displacement of an atom. The comparison of λ_{th} with q_{rms} indicates that the classical approximation is accurate at temperatures above the Einstein temperature θ_E, which is consistent with the graph of $C_V/(3Nk_B)$ in Figure 20.1. To verify this statement, we need an estimate for q_{rms}. According to the equipartition theorem (proved in Chapter 21), there is $\frac{1}{2}k_BT$ of energy associated with any variable that appears quadratically in the Hamiltonian. Since the Hamiltonian $\widehat{H}_j$ for mode j in (20.42) contains the term $\frac{1}{2}m\omega^2{q_j}^2$, the equipartition theorem tells us that

$$\frac{1}{2}k_BT = \left\langle \frac{1}{2}m\omega^2{q_j}^2 \right\rangle \quad \text{and} \quad \langle {q_j}^2 \rangle = \frac{k_BT}{m\omega^2}. \tag{20.73}$$

The comparison of $q_{\text{rms}} = \langle {q_j}^2 \rangle^{1/2}$ with the thermal de Broglie length indicates that the classical approximation is expected to be accurate when $q_{\text{rms}} > \lambda_{\text{th}}$, which implies that $q_{\text{rms}}^2 > \lambda_{\text{th}}^2$. Combining this inequality with (20.71) and (20.73) and using $\hbar = h/2\pi$ indicates that

$$\frac{k_BT}{m\omega^2} > \frac{h^2}{2\pi m k_BT} = \frac{2\pi\hbar^2}{mk_BT}. \tag{20.74}$$

Rearranging terms gives

$$T > \sqrt{2\pi}\frac{\hbar\omega}{k_B} = \sqrt{2\pi}\theta_E, \tag{20.75}$$

where $\theta_E = \hbar\omega/k_B$ is the Einstein temperature. Thus, comparing λ_{th} with q_{rms} indicates that the classical approximation becomes accurate at temperatures above θ_E.

The classical analysis of gases is accurate to much lower temperatures than that for solids, because of the large size of the system's length parameters, which are the mean-free-path $\bar{l}$ and the intermolecular separation $\bar{s}$. As found in Example 16.2, the estimated values of $\bar{l}$ and $\bar{s}$ for argon at a temperature and pressure of 273 K and 100 kPa are $\bar{l} = 130\,\text{nm}$ and $\bar{s} = 3.4\,\text{nm}$. Calculating λ_{th} requires a value for the mass m. As the atomic weight of argon is 40, the mass m of an atom (molecule) is $(40)(1.66 \times 10^{-27}\,\text{kg}) = 6.64 \times 10^{-26}\,\text{kg}$. By substituting this value into (20.71), the thermal de Broglie wavelength at $T = 273\,\text{K}$ becomes

$$\lambda_{\text{th}} = \frac{6.62 \times 10^{-34}\,\text{J s}}{\sqrt{2\pi(6.64 \times 10^{-26}\,\text{kg})(1.38 \times 10^{-23}\,\text{J K}^{-1})(273\,\text{K})}} = 0.017\,\text{nm}. \tag{20.76}$$

As this value of λ_{th} is much less than the characteristic length parameters for gases, we expect that quantum effects will only be significant in gases under extreme conditions. However, comparing the thermal de Broglie wavelength to a system's length parameters does not test for the effects of the exchange symmetry of identical fermions and bosons, effects that are significant for the electron gases in metals at ordinary temperatures, as discussed in Chapters 26 and 27.

Since the description of ideal gases obtained quantum mechanically in Section 20.3 is the same as that obtained classically in Section 19.5, an approximation must have been introduced in the quantum mechanical analysis. In fact, the only approximation made was the estimation of the value of a sum by an integral in (20.32). As shown below, this approximation is valid when the thermal de Broglie wavelength is much less than the cube root of volume, $L = V^{1/3}$. Although the mean free path $\bar{l}$ and the intermolecular separation $\bar{s}$ are more appropriate length parameters than L, the only length parameter in the ideal gas model is the L that entered through the single-particle eigenvalues $\varepsilon_i(\boldsymbol{n}_i)$ in (20.5).

The error introduced by approximating the sum by an integral is given by the difference

$$\text{Error} = I_k - \sum_{n=1}^{\infty} e^{-(\beta/2m)\hbar^2 k_n^2} \Delta k, \tag{20.77}$$

where the integral I_k is

$$I_k = \int_0^{\infty} e^{-\beta\hbar^2 k^2/(2m)} dk. \tag{20.78}$$

Expressing the integral as a sum of integrals over intervals of length $\Delta k = \pi/L$ gives

$$I_k = \sum_{n=0}^{\infty} \int_{k_n}^{k_{n+1}} e^{-\beta\hbar^2 k^2/(2m)} dk, \tag{20.79}$$

where $k_n = n\Delta k$. Because the integrand $e^{-\beta\hbar^2 k^2/(2m)}$ increases as k decreases, the integrand of the integral over each interval has its minimum value at the upper limit and its maximum value at the lower limit. By using these minimum and maximum values to estimate the integrals, we obtain the inequality

$$\sum_{n=0}^{\infty} e^{-\beta\hbar^2 k_{n+1}^2/(2m)} \Delta k < I_k < \sum_{n=0}^{\infty} e^{-\beta\hbar^2 k_n^2/(2m)} \Delta k. \tag{20.80}$$

Setting $\bar{n} = n + 1$ in the sum at the left and separating off the $n = 0$ term in the sum at the right gives

$$\sum_{\bar{n}=1}^{\infty} e^{-\beta\hbar^2 k_{\bar{n}}^2/(2m)} \Delta k < I_k < \Delta k + \sum_{n=1}^{\infty} e^{-\beta\hbar^2 k_n^2/(2m)} \Delta k, \tag{20.81}$$

where the sums over $\bar{n}$ and n are the same. By subtracting this sum from each term in (20.81) and using the definition of Error in (20.77), we find that

$$0 < \text{Error} < \Delta k, \tag{20.82}$$

where $\Delta k = \pi/L$. By dividing this by the value of I_k in (20.34), we obtain

$$\frac{\text{Error}}{I_k} < \frac{\Delta k}{I_k} = \frac{(\pi/L)}{(\pi/h)\sqrt{2\pi m k_B T}} = \frac{\lambda_{\text{th}}}{L}, \tag{20.83}$$

where $\lambda_{\text{th}} = h/\sqrt{2\pi m k_B T}$. Thus, the error made in our quantum mechanical analysis is negligible when the characteristic length $L = V^{1/3}$ is much greater than the thermal de Broglie wavelength λ_{th}.

Problems

20.1 Find the degeneracy of each of the four lowest energy levels of three particles confined in a cubic box. (You should find that the total number of eigenstates in these energy levels is 55.)

20.2 Small amounts of gas can be confined at low density and temperature in a potential well. By using k_o to characterize the well, the Hamiltonian of the system becomes

$$\widehat{H}_N(\widehat{\boldsymbol{p}}_1 \,\dots\, \boldsymbol{r}_N) = \sum_{j=1}^{3N}\left(\frac{1}{2m}\widehat{p}_j^2 + \frac{1}{2}k_o x_j^{\,2}\right),$$

where $(\widehat{\boldsymbol{p}}_1, \dots, \widehat{\boldsymbol{p}}_N) = (\widehat{p}_1^x, \widehat{p}_1^y, \dots, p_{3N}^z) = (\widehat{p}_1, \widehat{p}_2, \dots, p_{3N})$ and $(x_1, x_2, \dots, x_{3N})$ are the components of the displacements $(\boldsymbol{r}_1, \dots, \boldsymbol{r}_N)$.

(a) Find the partition function Z_N and internal energy U.
(b) Find the internal energy at $T = 0$.
(c) Show that $U = 3Nk_BT$ at high temperatures.

20.3 Find the thermal de Broglie wave length λ_{th} and the mean-free-path $\bar{l}$ of H_2 gas at $T = 273\,\text{K}$ and $P = 1.0\,\text{atm}$ (see Section 16.5). The diameter of a H_2 molecules is approximately 0.1 nm.

Appendix 20.A

Ideal Gas Eigenstates

Since quantum mechanics texts typically treat systems of only few particles, the derivation of the eigenvalues of an ideal gas is reviewed in this appendix. The microstates used in the description of thermodynamic systems are the eigenvalues of the N-particle time-independent Schrödinger equation,

$$\widehat{H}_N \Psi_\alpha(\boldsymbol{r}_1 \cdots \boldsymbol{r}_N) = E_{N,\alpha}\Psi_\alpha(\boldsymbol{r}_1 \cdots \boldsymbol{r}_N). \tag{20.A.1}$$

The use of the single index α to identify eigenstates enables the elegant description of the canonical ensemble in Section 20.2. However, the apparent simplicity of that description can be deceptive, because in practice index α is an abbreviation for a large number of indices.

The classical Hamiltonian for a monatomic ideal gas from (19.14) is

$$H_N^{ig}(\boldsymbol{p}_1 \cdots \boldsymbol{r}_N) = \sum_{i=1}^{N} \left[\frac{1}{2m_i} |\boldsymbol{p}_i|^2 + \phi_{a\text{-}w}(\boldsymbol{r}_i) \right]. \tag{20.A.2}$$

In quantum mechanics the momentum $\boldsymbol{p}_i$ of atom i is

$$\boldsymbol{p}_i = \frac{\hbar}{i} \nabla_i, \tag{20.A.3}$$

where $\hbar$ is Plank's constant divided by 2π and the i in $(\hbar/i)$ is the imaginary number $\sqrt{-1}$. The kinetic energy operator for atom i is

$$\frac{1}{2m_i} |\boldsymbol{p}_i|^2 = -\frac{\hbar^2}{2m_i} \nabla_i^2 = -\frac{\hbar^2}{2m_i} \left(\frac{\partial^2}{\partial x_i^2} + \frac{\partial^2}{\partial y_i^2} + \frac{\partial^2}{\partial z_i^2} \right), \tag{20.A.4}$$

where $\boldsymbol{r}_i = (x_i, y_i, z_i)$. The only contribution to the potential energy of an ideal gas is from the atom-wall potentials $\phi_{a\text{-}w}(\boldsymbol{r}_i)$, which confine the atoms to a region of space. This confinement is conveniently achieved by imposing boundary conditions on the eigenfunctions. Thus, the Hamiltonian operator $\widehat{H}_N$ for a gas of N atoms is

$$\widehat{H}_N^{ig} = -\sum_{i=1}^{N} \frac{\hbar^2}{2m_i} \nabla_i^2. \tag{20.A.5}$$

The associated *N-particle Schrödinger equation* is

$$\boxed{\sum_{i=1}^{N} \left(-\frac{\hbar^2}{2m_i} \nabla_i^2 \right) \Psi_\alpha(\boldsymbol{r}_1 \ \dots \ \boldsymbol{r}_N) = E_{N,\alpha} \Psi_\alpha(\boldsymbol{r}_1 \ \dots \ \boldsymbol{r}_N).} \tag{20.A.6}$$

The N-particle eigenstates can be obtained by combining the eigenstates of separate single-particle Schrödinger equations. The single-particle equations are obtained by expressing the N-particle eigenfunction Ψ_α as a product of single-particle eigenfunctions,

$$\Psi_\alpha(\boldsymbol{r}_1 \cdots \boldsymbol{r}_N) = \psi_{\boldsymbol{n}_1}(\boldsymbol{r}_1)\psi_{\boldsymbol{n}_2}(\boldsymbol{r}_2) \cdots \psi_{\boldsymbol{n}_N}(\boldsymbol{r}_N) = \prod_{i=1}^{N} \psi_{\boldsymbol{n}_i}(\boldsymbol{r}_i), \tag{20.A.7}$$

and expressing the N-particle eigenvalue $E_{N,\alpha}$ as a sum of single-particle eigenvalues,

$$E_{N,\alpha} = +\varepsilon_2(\boldsymbol{n}_2) + \cdots + \varepsilon_N(\boldsymbol{n}_N) = \sum_{i=1}^{N} \varepsilon_i(\boldsymbol{n}_i). \tag{20.A.8}$$

Substituting (20.A.7) and (20.A.8) into (20.A.6) gives

$$\left(-\sum_{i=1}^{N} \frac{\hbar^2}{2m_i} \nabla_i^2 \right) \psi_{\boldsymbol{n}_1}(\boldsymbol{r}_1) \cdots \psi_{\boldsymbol{n}_N}(\boldsymbol{r}_N) = \left(\sum_{i=1}^{N} \varepsilon_i\left(\boldsymbol{n}_i\right) \right) \psi_{\boldsymbol{n}_1}(\boldsymbol{r}_1) \cdots \psi_{\boldsymbol{n}_N}(\boldsymbol{r}_N). \tag{20.A.9}$$

By rearranging terms, this becomes

$$\sum_{i=1}^{N}\left[\left(-\frac{\hbar^2}{2m_i}\nabla_i^2\psi_{n_i}(\boldsymbol{r}_i)-\varepsilon_i(\boldsymbol{n}_i)\psi_{n_i}(\boldsymbol{r}_i)\right)\cdots\psi_{n_{i-1}}(\boldsymbol{r}_{i-1})\psi_{n_{i+1}}(\boldsymbol{r}_{i+1})\cdots\right]=0, \quad (20.A.10)$$

where the eigenfunction for atom i has been brought inside the parentheses that enclose the ith term. The equation can be satisfied by setting the quantity inside the parenthesis equal to zero for every atom, which yields the requirement that

$$-\frac{\hbar^2}{2m_i}\nabla^2\psi_{n_i}(\boldsymbol{r}_i)-\varepsilon_i(\boldsymbol{n}_i)\psi_{n_i}(\boldsymbol{r}_i)=0. \quad (20.A.11)$$

Omitting subscript i and rearranging terms yield the *single-particle Schrödinger equation* for a free particle,

$$\boxed{-\frac{\hbar^2}{2m}\nabla^2\psi_{\boldsymbol{n}}(\boldsymbol{r})=\varepsilon(\boldsymbol{n})\psi_{\boldsymbol{n}}(\boldsymbol{r}).} \quad (20.A.12)$$

Box Boundary Conditions[5]

Each single-particle eigenvalue can be expressed as a sum of three one-dimensional eigenvalues, and the confinement of the particles (atoms) is achieved by the boundary conditions imposed on the eigenfunctions. For simplicity, we consider a rectangular box of volume $V = L_xL_yL_z$. By placing the origin of coordinates at one corner, the effect of the box is to confine the particles to the region defined by

$$0 \le x \le L_x, \quad 0 \le y \le L_y, \quad 0 \le z \le L_z, \quad (20.A.13)$$

where $\boldsymbol{r} = (x, y, z)$. The boundary conditions then become

$$\psi_{\boldsymbol{n}}(0,y,z)=\psi_{\boldsymbol{n}}(L_x,y,z)=0, \quad \psi_{\boldsymbol{n}}(x,0,z)=\psi_{\boldsymbol{n}}(x,L_y,z)=0, \quad (20.A.14)$$

and similarly for the z-direction. The single-particle Schrödinger equation in (20.A.12) is satisfied by the one-dimensional eigenvalues and eigenfunctions

$$\varepsilon(\boldsymbol{n})=e(n_x)+e(n_y)+e(n_z) \quad (20.A.15)$$

and

$$\psi_{\boldsymbol{n}}(x,y,z)=\psi_{n_x}(x)\psi_{n_y}(y)\psi_{n_z}(z). \quad (20.A.16)$$

By substituting these into (20.A.12) and making arguments similar to those used to obtain (20.A.12), it follows for the x-direction that

$$-\frac{\hbar^2}{2m}\frac{\partial^2}{\partial x^2}\psi_{n_x}(x)=e(n_x)\,\psi_{n_x}(x), \quad (20.A.17)$$

---

[5] The description of the eigenstates based on periodic boundary conditions is given in Section 24.1.

and similarly for the y- and z-directions. It is easily verified that the eigenvalues and eigenfunctions of this are

$$e(n_x) = \frac{1}{2m}\left(\frac{\hbar\pi}{L_x}\right)^2 n_x{}^2 \tag{20.A.18}$$

and

$$\psi_{n_x}(x) = \sqrt{\frac{2}{L_x}}\sin\left(\frac{n_x\pi x}{L_x}\right), \tag{20.A.19}$$

where $\sqrt{2/L_x}$ is introduced to normalize $\psi_{n_x}(x)$. The boundary conditions are satisfied by choosing n_x to be a positive integer,

$$n_x = 1, 2, 3, \cdots, \infty. \tag{20.A.20}$$

The results for the y- and z-directions are similar. The resulting eigenvalues and eigenfunctions of the single-particle Schrödinger equation are

$$\varepsilon(\boldsymbol{n}) = \frac{(\hbar\pi)^2}{2m}\left(\frac{n_x{}^2}{L_x{}^2} + \frac{n_y{}^2}{L_y{}^2} + \frac{n_z{}^2}{L_z{}^2}\right) \tag{20.A.21}$$

and

$$\psi_{\boldsymbol{n}}(\boldsymbol{r}) = \frac{2^{3/2}}{V}\sin\left(\frac{xn_x\pi}{L_x}\right)\sin\left(\frac{yn_y\pi}{L_y}\right)\sin\left(\frac{zn_z\pi}{L_z}\right). \tag{20.A.22}$$

The N-particle eigenfunctions $\Psi_\alpha(\boldsymbol{r}_1\cdots\boldsymbol{r}_N)$ are obtained by introducing subscript i into these eigenfunctions and substituting $\psi_{\boldsymbol{n}_i}(\boldsymbol{r}_i)$ into (20.A.7).

The *single-particle eigenvalues* for particle i in a cubic box of volume $V = L^3$ are

$$\boxed{\varepsilon_i(\boldsymbol{n}_i) = \frac{1}{2m_i}\left(\frac{\hbar\pi}{L}\right)^2 |\boldsymbol{n}_i|^2,} \tag{20.A.23}$$

where $\boldsymbol{n}_i = (n_{x,i}, n_{y,i}, n_{z,i})$ and $|\boldsymbol{n}_i|^2 = n_{x,i}{}^2 + n_{y,i}{}^2 + n_{z,i}{}^2$. The allowed values of the quantum numbers are

$$n_{x,i} = 1, 2, \ldots, \infty, \quad n_{y,i} = 1, 2, \ldots, \infty, \quad n_{z,i} = 1, 2, \ldots, \infty. \tag{20.A.24}$$

As indicated in (20.A.8), the *N-particle eigenvalues* $E_{N,\alpha}$ for N particles in a cubic box, such as a monatomic ideal gas, are

$$\boxed{E_{N,\alpha} = \varepsilon_1(\boldsymbol{n}_1) + \varepsilon_2(\boldsymbol{n}_2) + \cdots + \varepsilon_N(\boldsymbol{n}_N),} \tag{20.A.25}$$

where α is an abbreviated notation for $3N$ quantum numbers,

$$\alpha = (\boldsymbol{n}_1, \boldsymbol{n}_2, \cdots, \boldsymbol{n}_N) = (n_{x,1}, n_{y,1}, n_{z,1}, \ldots, n_{x,N}, n_{y,N}, n_{z,N}). \tag{20.A.26}$$

By combining these results and using $\hbar = h/2\pi$, we obtain

$$E_{N,\alpha} = \sum_{i=1}^{N}(n_{x,i}^2 + n_{y,i}^2 + n_{z,i}^2)\frac{h^2}{8mL^2}, \tag{20.A.27}$$

which are the energies of an ideal gas of N atoms of mass m given in (17.7). Because of the simplicity of the Hamiltonian, the N-particle eigenvalues are sums of single-particle eigenvalues, a simplification that is not possible when interparticle interactions are included.

21

Independent Particles and Paramagnetism

A system's thermodynamic properties can be predicted by using the partition function formula or by evaluating ensemble averages. In both cases, the process is greatly simplified when the system can be idealized as collections of independent parts. A frequent challenge in research is to find model systems that mimic the behavior of a real system while remaining easy to analyze. This challenge is often met by using independent particle models, such as ideal gases and the Einstein model. Paramagnetic systems are another example. After describing the implications of statistical independence, we investigate the atomic-level origin of paramagnetism and deduce the properties of systems of spin-$\frac{1}{2}$ particles.

21.1 Averages

As discussed in Section 19.4, there are two kinds of thermodynamic properties: statistical properties and mechanical properties. Unlike a statistical property, a mechanical property has a microscopic value as well as a macroscopic value. Energy, pressure, and magnetic moment are mechanical properties with microscopic values that are determined by the system's microstate. Their macroscopic values are given by ensemble averages of phase functions. The emphasis in this chapter is on the use of ensemble averages. We begin by reviewing the general properties of averages.

Properties of Averages

1. The average of a constant equals the value of the constant:

$$\boxed{\langle c \rangle = c.} \tag{21.1}$$

Thermodynamics and Statistical Mechanics: An Integrated Approach, First Edition.
Robert J Hardy and Christian Binek.

2. The average of a constant multiplied by a function equals the constant times the average of the function:

$$\boxed{\langle cA \rangle = c\langle A \rangle.} \tag{21.2}$$

3. The average of a sum equals the sum of the averages:

$$\boxed{\langle A + B \rangle = \langle A \rangle + \langle B \rangle.} \tag{21.3}$$

According to (19.26), the classical ensemble average of a property described by phase function $X(\boldsymbol{p}_1 \cdots \boldsymbol{r}_N)$ is

$$\langle X \rangle = \int dp^{3N} \int dr^{3N} \rho(\boldsymbol{p}_1 \cdots \boldsymbol{r}_N) X(\boldsymbol{p}_1 \cdots \boldsymbol{r}_N). \tag{21.4}$$

According to (20.12), the quantum mechanical average is

$$\langle X \rangle = \sum_\alpha \rho_\alpha X_\alpha, \tag{21.5}$$

where X_α is the value of the property in the indicated microstate. The properties in (21.1) through (21.3) are easily verified by expressing the pointed bracket notation in terms of its meaning. As an illustration, consider the average of property X which is the sum of two other properties, A and B. It follows from (21.5) that

$$\langle A + B \rangle = \sum_\alpha \rho_\alpha (A_\alpha + B_\alpha). \tag{21.6}$$

By rearranging terms and again using (21.5), it follows that

$$\sum_\alpha \rho_\alpha (A_\alpha + B_\alpha) = \sum_\alpha (\rho_\alpha A_\alpha + \rho_\alpha B_\alpha) = \sum_\alpha \rho_\alpha A_\alpha + \sum_\alpha \rho_\alpha B_\alpha = \langle A \rangle + \langle B \rangle. \tag{21.7}$$

Hence, $\langle A + B \rangle$ equals $\langle A \rangle + \langle B \rangle$. The same result can also be obtained from the classical expression for averages in (21.4).

Abbreviated Notation

Classical results can often be obtained from quantum mechanical results by interpreting the symbols ρ_α, X_α, etc. as abbreviated notations for the associated classical concepts: The indices α, γ, etc. (which identify microstates) become phase points $(\boldsymbol{p}_1 \cdots \boldsymbol{r}_N)$. The set of values X_α becomes a phase function $X(\boldsymbol{p}_1 \cdots \boldsymbol{r}_N)$. A probability ρ_α becomes a probability density times a volume element, $\rho(\boldsymbol{p}_1 \cdots \boldsymbol{r}_N) dp^{3N} dr^{3N}$. A sum over microstates becomes an integral. These interpretations are summarized as follows:

$$\alpha \rightarrow (\boldsymbol{p}_1 \cdots \boldsymbol{r}_N), \tag{21.8}$$

$$X_\alpha \rightarrow X(\boldsymbol{p}_1 \cdots \boldsymbol{r}_N), \tag{21.9}$$

$$\rho_\alpha \rightarrow \rho(\boldsymbol{p}_1 \cdots \boldsymbol{r}_N) dp^{3N} dr^{3N}, \tag{21.10}$$

$$\sum_{\alpha} \rightarrow \int \cdots \int . \tag{21.11}$$

As an example, consider the definition of an average. By making the indicated replacements, the quantum mechanical average in (21.5) becomes

$$\sum_{\alpha} \rho_{\alpha} X_{\alpha} \rightarrow \int \cdots \int dp^{3N} dr^{3N} \rho(\boldsymbol{p}_1 \cdots \boldsymbol{r}_N) X(\boldsymbol{p}_1 \cdots \boldsymbol{r}_N), \tag{21.12}$$

which is the classical average $\langle X \rangle$ in (21.4). Since general results are often most easily derived in quantum mechanical notation, this interpretation can be quite useful.

21.2 Statistical Independence

Consider an experiment (system) whose outcomes (microstates) are identified by discrete variables α' and α'', where α' identifies the outcomes of part 1 and α'' identifies the outcomes of part 2. The probability of outcome $(\alpha'\alpha'')$ is given by the *joint probability* $\rho_{\alpha'\alpha''}$. When the outcome of the second part is ignored, the probability of the outcome of part 1 is given by the *marginal probability* $\rho^{(1)}_{\alpha'}$, which is obtained by summing over all possible outcomes of the other part,

$$\rho^{(1)}_{\alpha'} = \sum_{\alpha''} \rho_{\alpha'\alpha''}. \tag{21.13}$$

Similarly, the marginal probability for part 2 is $\rho^{(2)}_{\alpha''} = \sum_{\alpha'} \rho_{\alpha'\alpha''}$. When the outcome of each part has no effect on the outcome of the other, the parts are *statistically independent.* In this special case, the probabilities $\rho_{\alpha'\alpha''}$ can be factored into a product of the marginal probabilities,

$$\rho_{\alpha'\alpha''} = \rho^{(1)}_{\alpha'} \rho^{(2)}_{\alpha''}. \tag{21.14}$$

It is convenient to let $\langle \cdots \rangle$ represent averages formed with the joint probabilities $\rho_{\alpha'\alpha''}$ and use $\langle \cdots \rangle_1$ and $\langle \cdots \rangle_2$ represent averages formed with $\rho^{(1)}_{\alpha'}$ and $\rho^{(2)}_{\alpha''}$. The average of a property of a two-part system that depends on outcome α' of part 1 *only* is

$$\langle A \rangle = \sum_{\alpha'\alpha''} \rho_{\alpha'\alpha''} A_{\alpha'} = \sum_{\alpha''} \rho^{(1)}_{\alpha'} \left[\sum_{\alpha''} \rho^{(2)}_{\alpha''} \right] A_{\alpha'} = \sum_{\alpha'} \rho^{(1)}_{\alpha'} A_{\alpha'} = \langle A \rangle_1. \tag{21.15}$$

Only the marginal probability for part 1 is needed when A is a property of that part. Sometimes we are interested in a property whose values are products of properties of the individual parts in the sense that $X_{\alpha'\alpha''} = A_{\alpha'} B_{\alpha''}$. If the parts are statistically independent, the average of such a property becomes

$$\langle AB \rangle = \sum_{\alpha'\alpha''} \rho_{\alpha'\alpha''} X_{\alpha'\alpha''} = \left[\sum_{\alpha'} \rho^{(1)}_{\alpha'} A_{\alpha'} \right] \left[\sum_{\alpha''} \rho^{(2)}_{\alpha''} B_{\alpha''} \right] = \langle A \rangle_1 \langle B \rangle_2. \tag{21.16}$$

Provided the parts are statistically independent, the average of a product of properties of the different parts is the product of the averages.

Quantum Systems

The concept of statistical independence and the extensive characteristic of thermodynamic systems are closely related. Internal energy is the prototypical extensive property. As indicated in (18.71), the combined (total) energy of a two-part system can be expressed quantum mechanically as

$$E^{tot}_{\alpha'\alpha''} = E^{(1)}_{\alpha'} + E^{(2)}_{\alpha''} + I_{\alpha'\alpha''}, \tag{21.17}$$

where $E^{(1)}_{\alpha'}$ and $E^{(2)}_{\alpha''}$ are the energies of subsystems 1 and 2. The random microscopic interactions between the particles in the different subsystems are described by the energies $I_{\alpha'\alpha''}$. Since these interactions occur at the surfaces of the subsystems, the terms $I_{\alpha'\alpha''}$ are negligibly small compared to $E^{(1)}_{\alpha'}$ and $E^{(2)}_{\alpha''}$. Consequently, the equilibrium ensemble averages are

$$U_{tot} = \langle E^{tot}\rangle, \qquad U_1 = \langle E^{(1)}\rangle_1, \qquad U_2 = \langle E^{(2)}\rangle_2, \tag{21.18}$$

where U_{tot}, U_1, and U_2 are the corresponding internal energy functions. By using this and the additive property of averages, the ensemble average of (21.17) becomes $U_{tot} = U_1 + U_2 + \langle I\rangle$. The extensive characteristic of internal energy requires that $U_{tot} = U_1 + U_2$. For this to be true the average $\langle I\rangle$ must be negligible. Although this may not be exactly true for thermally interacting subsystems, it is effectively true for subsystems of macroscopic size. The negligible size of surface effects is essential for the extensive characteristic of thermodynamic properties.

Separable Energies

Energy is separable when the energy eigenvalues can be expressed as a sum of terms that depend on different quantum numbers. When the interaction terms $I_{\alpha'\alpha''}$ are negligible, the eigenvalues of the combined system are

$$E^{tot}_{\alpha'\alpha''} = E^{(1)}_{\alpha'} + E^{(2)}_{\alpha''} \tag{21.19}$$

and the energy is separable. In this case, the probabilities $\rho^{tot}_{\alpha'\alpha''}$ for a system in equilibrium become

$$\rho^{tot}_{\alpha'\alpha''} = \frac{e^{-\beta E^{tot}_{\alpha'\alpha''}}}{Z_{tot}} = \frac{e^{-\beta E^{(1)}_{\alpha'}}}{Z_1}\frac{e^{-\beta E^{(2)}_{\alpha''}}}{Z_2} = \rho^{(1)}_{\alpha'}\rho^{(2)}_{\alpha''}, \tag{21.20}$$

where $\beta = 1/k_BT$. As the probabilities $\rho^{tot}_{\alpha'\alpha''}$ can be factored, the subsystems are statistically independent, and the partition function can be expressed as $Z_{tot} = Z_1Z_2$, where

$$Z_{tot} = \sum_{\alpha'\alpha''} e^{-\beta E^{tot}_{\alpha'\alpha''}}, \qquad Z_1 = \sum_{\alpha'} e^{-\beta E^{(1)}_{\alpha'}}, \qquad Z_2 = \sum_{\alpha''} e^{-\beta E^{(2)}_{\alpha''}}. \tag{21.21}$$

By using $\ln Z_1Z_2 = \ln Z_1 + \ln Z_2$, the Helmholtz free energy $F_{tot} = -k_BT\ln Z_{tot}$ becomes $F_{tot} = F_1 + F_2$, where F_1 and F_2 are the subsystem's free energies. Thus, when in thermal equilibrium, the properties of systems with separable energies simplify as follows:

$$E^{tot}_{\alpha'\alpha''} = E^{(1)}_{\alpha'} + E^{(2)}_{\alpha''}, \quad \rho^{tot}_{\alpha'\alpha''} = \rho^{(1)}_{\alpha'}\rho^{(2)}_{\alpha''}, \quad Z_{tot} = Z_1Z_2, \quad F_{tot} = F_1 + F_2. \tag{21.22}$$

Also, if A and B are properties of the individual subsystems, then $\langle AB\rangle_{tot} = \langle A\rangle_1\langle B\rangle_2$.

These results are easily generalized for systems with more than two parts. When the energy of a combined system consists of subsystems a, b, c, etc. with separable energies, we first treat subsystem a as subsystem 1 and treat the combination b, c, d, etc. as subsystem 2, so that $\rho^{\text{tot}}_{\cdots} = \rho^{(a)}_{\alpha'}\rho^{(b,c,d,\ldots)}_{\cdots}$. We next treat subsystem b as subsystem 1 and treat the combination c, d, e, etc. as subsystem 2, so that $\rho^{(b,c,d,\ldots)}_{\cdots} = \rho^{(b)}_{\alpha''}\rho^{(c,d,e,\ldots)}_{\cdots}$. By continuing in this way and omitting the subscripts that label microstates, we obtain

$$E^{\text{tot}} = E^{(a)} + E^{(b)} + E^{(c)} + \cdots, \tag{21.23}$$

$$\rho^{\text{tot}} = \rho^{(a)}\rho^{(b)}\rho^{(c)}\cdots, \tag{21.24}$$

$$Z_{\text{tot}} = Z_a Z_b Z_c \cdots, \tag{21.25}$$

and

$$F_{\text{tot}} = F_a + F_b + F_c + \cdots. \tag{21.26}$$

Also, when A, B, C, etc. are properties of subsystems a, b, c, etc., we have

$$\langle ABC\cdots\rangle_{\text{tot}} = \langle A\rangle_a \langle B\rangle_b \langle C\rangle_c \cdots. \tag{21.27}$$

These equations summarize the simplicity that results when systems are made up of statistically independent parts. In the analysis of ideal gases and the Einstein model in Chapter 20, the statistically independent parts were individual particles. However, the simplicity that results from using model systems with independent parts is not confined to independent particle models.

21.3 Classical Systems

The effects of statistical independence when systems are analyzed classically are the same as those summarized in (21.23) through (21.27). The principle difference is that the microstates are now associated with probability densities, instead of probabilities. The energy of a two-part classical system is

$$H_N(\boldsymbol{p}_1\cdots\boldsymbol{r}_N) = H_1(\boldsymbol{p}'_1\cdots\boldsymbol{r}'_{N_1}) + H_2(\boldsymbol{p}''_1\cdots\boldsymbol{r}''_{N_2}) + \Phi^{\text{int}}(\cdots\boldsymbol{r}'_{i'}\cdots\boldsymbol{r}''_{i''}\cdots), \tag{21.28}$$

where the momentum and position variables separate into distinct sets, $(\boldsymbol{p}'_1\cdots\boldsymbol{r}'_{N_1})$ and $(\boldsymbol{p}''_1\cdots\boldsymbol{r}''_{N_2})$. The interaction energy Φ^{int} is analogous to the term $I_{\alpha'\alpha''}$ in (21.17). In the model system described in Section 19.2, the interaction energy would be

$$\Phi^{int}(\cdots\boldsymbol{r}'_{i'}\cdots\boldsymbol{r}''_{i''}\cdots) = \sum_{i',i''}\phi(r_{i'i''}), \tag{21.29}$$

where the particles in subsystems 1 and 2 are identified by i' and i'', respectively. The Hamiltonian for subsystem 1 is

$$H_1(\boldsymbol{p}'_1\cdots\boldsymbol{r}'_{N_1}) = \sum_{i'}\frac{1}{2m_{i'}}|\boldsymbol{p}'_{i'}|^2 + \sum_{i'<j'}\phi(r_{i'j'}), \tag{21.30}$$

where only the particle variables of that subsystem appear. (The atom-wall potentials $\phi_{a\text{-}w}$ are omitted.) The expression for H_2 is similar, but with i' and j' replaced by i'' and j''. Since the interatomic potentials $\phi(r_{i'i''})$ vanish when particles i' and i'' are separated by more than a nanometer, only those particles that are located at the surfaces of the subsystems contribute to the sum in (21.29). Consequently, the number of terms that contribute to the sum is much less than the numbers of particles in the subsystems. The interaction energy Φ^{int} can be neglected, so that the Hamiltonian becomes $H_N = H_1 + H_2$, which makes the energy separable.

When the interaction energy is neglected, the canonical probability density for the combined (total) system becomes

$$\rho_{\text{tot}}(\boldsymbol{p}_1 \cdots \boldsymbol{r}_N) = \frac{e^{-\beta H_N(\boldsymbol{p}_1 \cdots \boldsymbol{r}_N)}}{Z_{tot}} = \frac{e^{-\beta H_1(\boldsymbol{p}'_1 \cdots \boldsymbol{r}'_{N_1})}}{Z_1} \frac{e^{-\beta H_2(\boldsymbol{p}''_1 \cdots \boldsymbol{r}''_{N_2})}}{Z_2}$$

$$= \rho_1(\boldsymbol{p}'_1 \cdots \boldsymbol{r}'_{N_1})\rho_2(\boldsymbol{p}''_1 \cdots \boldsymbol{r}''_{N_2}), \tag{21.31}$$

where $\rho_1(\boldsymbol{p}'_1 \cdots \boldsymbol{r}'_{N_1})$ and $\rho_2(\boldsymbol{p}''_1 \cdots \boldsymbol{r}''_{N_2})$ are the probability densities for the subsystems. The partition functions for subsystem 1 is

$$Z_1 = C_{N_1} \int dp^{3N_1} \int dr^{3N_1} e^{-H_1(\boldsymbol{p}'_1 \cdots \boldsymbol{r}'_{N_1})/k_B T} \tag{21.32}$$

and similarly for subsystem 2.

Free Energy

Since the momentum variables $\boldsymbol{p}_i = (p_i^x, p_i^y, p_i^z)$ only appear in the kinetic energy $K_N = \sum_i \frac{1}{2m_i} |\boldsymbol{p}_i|^2$ and the position variables $\boldsymbol{r}_i$ only appear in the potential energy $\Phi(\boldsymbol{r}_1 \cdots \boldsymbol{r}_N)$, the kinetic and potential energies are separable. The Hamiltonian is

$$H_N(\boldsymbol{p}_1 \cdots \boldsymbol{r}_N) = K_N(\boldsymbol{p}_1 \cdots \boldsymbol{p}_N) + \Phi_N(\boldsymbol{r}_1 \cdots \boldsymbol{r}_N), \tag{21.33}$$

so that the classical partition function in (19.24) becomes

$$Z_N = C_N \int dp^{3N} e^{-\beta K_N(\boldsymbol{p}_1 \cdots \boldsymbol{p}_N)} \int dr^{3N} e^{-\beta \Phi_N(\boldsymbol{r}_1 \cdots \boldsymbol{r}_N)}. \tag{21.34}$$

By multiplying the momentum integrations by V^N, dividing the spatial integrations by V^N, and using $\beta = 1/k_B T$, this becomes

$$Z_N = Z_N^{ig} Z_N^{\text{config}}, \tag{21.35}$$

where the *configurational partition function* is

$$\boxed{Z_N^{\text{config}} = V^{-N} \int dr^{3N} e^{-\Phi_N(\boldsymbol{r}_1 \cdots \boldsymbol{r}_N)/k_B T}} \tag{21.36}$$

and the integral is called the *configuration integral.* Z_N^{ig} is the ideal gas partition function found in (19.65),

$$Z_N^{ig} = C_N V^N \int dp^{3N} e^{-K_N(p_1 \cdots p_N)/k_B T} = C_N (2\pi m k_B T)^{3N/2} V^N. \tag{21.37}$$

Since the logarithm of $Z_N^{ig} Z_N^{\text{config}}$ equals $\ln Z_N^{ig} + \ln Z_N^{\text{config}}$, the Helmholtz free energy $F = -k_B T \ln Z_N$ becomes

$$\boxed{F = F^{ig} + F^{\text{config}},} \tag{21.38}$$

where F^{ig} is the free energy of an ideal gas given in (19.68). The *configurational free energy* is

$$\boxed{F^{\text{config}} = -k_B T \ln Z_N^{\text{config}}.} \tag{21.39}$$

This classical result is useful for predicting the properties of a system in all of its phases.

Equipartition Theorem

The equipartition hypothesis was introduced in Chapter 15. Now that we have a theory with which to prove it, the hypothesis becomes a theorem. It is a classical result which asserts that: *The average energy associated with any momentum or position variable that appears quadratically in only one term in the Hamiltonian is* $\frac{1}{2}k_B T$.

To prove the theorem, we consider a separable contribution to the energy of the form

$$\varepsilon(u) = \frac{1}{2} K u^2, \tag{21.40}$$

where u is a momentum or position variable that appears in only one term. When u is a momentum variable, $\varepsilon(u)$ is a contribution to the kinetic energy and $K = 1/m$. When u is a position variable, K is an effective spring constant and $\varepsilon(u)$ is a contribution to the potential energy. The canonical probability density of this contribution to the energy is

$$\rho(u) = \frac{e^{-\beta\varepsilon(u)}}{Z_o}, \tag{21.41}$$

and the average energy is $\langle\varepsilon\rangle = \int \rho_1(u)\varepsilon(u)du$. When the allowed values of u extend from $-\infty$ to $+\infty$, the normalization of $\rho(u)$ requires that

$$Z_o = \int_{-\infty}^{+\infty} e^{-\beta\varepsilon(u)} du = \int_{-\infty}^{+\infty} e^{-\beta\frac{1}{2}Ku^2} du, \tag{21.42}$$

so that the average can be expressed as

$$\langle\varepsilon\rangle = \left\langle \frac{1}{2} K u^2 \right\rangle = \frac{1}{Z_o} \int_{-\infty}^{+\infty} e^{-\beta\frac{1}{2}Ku^2} \left(\frac{1}{2} K u^2 \right) du = -\frac{\partial}{\partial\beta} \ln Z_o. \tag{21.43}$$

By changing the integration variable from u by w, where $w = \beta^{1/2}u$ and $du = \beta^{-1/2}dw$, we obtain $Z_o = \beta^{-1/2}\int e^{-(1/2)Kw^2}dw$. As the integral over w is independent of β, it follows that

$$\left\langle \frac{1}{2}Ku^2 \right\rangle = -\frac{\partial}{\partial\beta}\left[-\frac{1}{2}\ln\beta + \ln\int_{-\infty}^{+\infty} e^{-(1/2)Kw^2}dw\right] = \frac{1}{2\beta}, \tag{21.44}$$

Using $\beta = 1/k_BT$ gives

$$\boxed{\left\langle \frac{1}{2}Ku^2 \right\rangle = \frac{1}{2}k_BT.} \tag{21.45}$$

This summarizes the equipartition theorem, which is a classical result. It does not include quantum effects and thus fails to predict the behavior of systems at low temperatures.

Kinetic Energy

Since there are three momentum variables per particle (p_x, p_y, p_z), the equipartition theorem specifies that the average kinetic energy per particle is

$$\left\langle \frac{1}{2m}|\boldsymbol{p}|^2 \right\rangle = \frac{3}{2}k_BT. \tag{21.46}$$

Although the kinetic energy per particle is a mechanical property, this result does not imply that temperature T is a mechanical property. For example, if the magnitude of the momentum of every particle was p_o, the average kinetic energy per particle would be $\frac{1}{2m}p_o^2$. If this were the case, the system would spontaneously evolve until the Maxwell distributions of velocity and momentum are achieved. Since the system would be spontaneously changing, it would not be in equilibrium and thus would not have temperature. The result in (21.46) is only valid in equilibrium. Even then it is not accurate at low temperature because it does not account for quantum effects. Specifically, it does not account for the fact that the kinetic energy does not go to zero at absolute zero because of zero point energy, which is discussed in Section 20.4.

Maxwell Distribution

The distribution of speeds is determined by the kinetic energies of the particles. The equilibrium probability density $\rho(\boldsymbol{p})$ for the momentum $\boldsymbol{p} = (p_x, p_y, p_z)$ is proportional to the Boltzmann factor,

$$\rho(\boldsymbol{p}) = \frac{e^{-\beta\frac{1}{2m}|\boldsymbol{p}|^2}}{Z_o}. \tag{21.47}$$

The normalization condition requires that $Z_o = (2\pi m/\beta)^{3/2}$. Since $\boldsymbol{p} = m\boldsymbol{v}$, the volume dp^3 of momentum space that corresponds to a volume dv^3 of velocity space is $dp^3 = m^3dv^3$. The probability that a particle is in a particular volume element should be independent of whether its momentum or its velocity is specified. So that the two ways of describing the probability are the same, we require that $\rho_{vel}(\boldsymbol{v})dv^3 = \rho(\boldsymbol{p})dp^3$. By using this, the probability that a particle is in volume element dv^3 becomes

$$\rho_{vel}(\boldsymbol{v})dv^3 = \frac{1}{Z_o} e^{-\beta\frac{1}{2m}|\boldsymbol{p}|^2} dp^3 = \left(\frac{\beta}{2\pi m}\right)^{3/2} e^{-\beta\frac{1}{2}m|\boldsymbol{v}|^2} \left(m^3 dv^3\right). \tag{21.48}$$

The probability density for speed is found by introducing spherical coordinates $\boldsymbol{v} = (v, \theta, \varphi)$, where $v = |\boldsymbol{v}|$ is speed. As the volume between v and $v + dv$ is $dv^3 = 4\pi v^2 dv$ and $\rho_{vel}(\boldsymbol{v})$ is independent of φ and φ, the probability that the particle's speed is between v and $v + dv$ is

$$\rho_{sp}(v)dv = \rho_{vel}(\boldsymbol{v})4\pi v^2 dv = 4\pi\left(\frac{\beta m}{2\pi}\right)^{3/2} v^2 e^{-\beta\frac{1}{2}m|\boldsymbol{v}|^2} dv. \tag{21.49}$$

In systems with particles of the same mass, $\rho_{sp}(v)$ is the velocity distribution of every particle. The probable number of particles with speeds in interval dv in a system of N particles is $N\rho_{sp}(v)dv$. By using $\beta = 1/k_B T$, the number of particles with speeds between v and $v + dv$ becomes

$$n(v)dv = 4\pi N\left(\frac{m}{2\pi k_B T}\right)^{3/2} v^2 e^{-mv^2/(2k_B T)} dv. \tag{21.50}$$

This is the *Maxwell distribution of speeds* plotted earlier in Figure 16.2. As the probability density $\rho(\boldsymbol{p})$ in (21.47) arises from a separable term in the Hamiltonian, the applicability of the result – the Maxwell distribution – is limited only by the requirements that the system be in thermal equilibrium and at a temperature sufficiently high that quantum effects are negligible. It is applicable independent of whether the system is a solid, a liquid, or a gas.

21.4 Paramagnetism

The study of paramagnetism nicely illustrates the use of ensemble averages and the role of statistical independence in predicting thermodynamic properties. Paramagnetism is the response to a magnetic field of particles with permanent magnetic dipole moments which are free to change their orientation. In the absence of a magnetic field, their orientations are random, but they become aligned in the presence of a field.

Since the magnetic moment of a system has both macroscopic and microscopic values, it is a mechanical property, so that its macroscopic value is the ensemble average of the microscopic values. The microscopic value is determined by the orientations of the dipole moments $\boldsymbol{m}_i$ of the constituent magnetic particles,

$$\boldsymbol{M}_N = \sum_{i=1}^{N} \boldsymbol{m}_i. \tag{21.51}$$

The number of magnetic particles N is usually less than the total number of particles. The macroscopic value of the magnetic moment is the ensemble average of $\boldsymbol{M}_N$.

The Hamiltonian for magnetic systems is of the form

$$H_N = H_N^{int} + H_N^{\mathrm{pm}}, \tag{21.52}$$

where H_N^{int} is the energy of the interactions between the magnetic particles and H_N^{pm} is the energy of the interaction with the external magnetic field $\mathcal{H}$. The internal interactions described by H_N^{int} are essential for ferromagnetic behavior but are neglected when studying paramagnetism. The energy of magnetic particle i is $-\mu_0\mathcal{H}\cdot\boldsymbol{m}_i$, where μ_0 is the permeability of free space. The minus sign indicates that the energy is lowest when the dipole moment $\boldsymbol{m}_i$ is aligned parallel to the field. We consider a uniform field in the z-direction, so that the energy of particle i becomes $-\mu_0\mathcal{H}m_i^z$, where $\mathcal{H}$ and m_i^z are the z-components of $\mathcal{H}$ and $\boldsymbol{m}_i$. The resulting paramagnetic Hamiltonian for a system of N particles is

$$H_N^{\mathrm{pm}} = -\sum_{i=1}^{N} \mu_0\mathcal{H}m_i^z. \tag{21.53}$$

The z-component of the microscopic magnetic moment is

$$\boxed{M_N^z = \sum_{i=1}^{N} m_i^z.} \tag{21.54}$$

By using this and taking $\mu_0\mathcal{H}$ outside of the sum in (21.53), the Hamiltonian becomes

$$H_N^{\mathrm{pm}} = -\mu_0\mathcal{H}M_N^z. \tag{21.55}$$

Since each moment m_i^z appears in only one term, the paramagnetic Hamiltonian is separable, so that the equilibrium orientations of the particles are statistically independent. When treated classically, the moment M_N^z and the Hamiltonian H_N^{pm} are functions of the variables $(m_1^z \cdots m_N^z)$. When analyzed quantum mechanically, the energy eigenvalues of the paramagnetic system are

$$\boxed{E_{N,\alpha}(\mathcal{H}) = -\mu_0\mathcal{H}M_{N,\alpha}^z,} \tag{21.56}$$

where $M_{N,\alpha}^z$ is the moment in eigenstate α.

Thermodynamics

The macroscopic value of the magnetic moment is given by the ensemble average of $M_{N,\alpha}^z$,

$$\boxed{M_{\mathrm{th}} = \left\langle M_N^z \right\rangle.} \tag{21.57}$$

The subscript on M_{th} identifies it as the thermodynamic value of the moment, which is related to the *magnetization* $\mathcal{M}$ of a system of volume V by $\mathcal{M} = M_{\mathrm{th}}/V$. We found in Chapter 14 that the energy associated with the interaction with an external source of energy through an electric or magnetic field is the enthalpy E_{th}. Hence, the ensemble average of $E_{N,\alpha}$ is the *enthalpy*,

$$\boxed{E_{\text{th}} = \langle E_N \rangle.} \tag{21.58}$$

According to the thermodynamic relationships for magnetic systems summarized in Table 14.3, the Gibbs free energy G, entropy S, and moment M_{th} are given by

$$G = E_{\text{th}} - TS, \quad S = -\left(\frac{\partial G}{\partial T}\right)_{\mathcal{H}}, \quad M_{\text{th}} = -\frac{1}{\mu_0}\left(\frac{\partial G}{\partial \mathcal{H}}\right)_T. \tag{21.59}$$

The constant-field heat capacity given in (14.39) is

$$C_{\mathcal{H}} = -T\left(\frac{\partial^2 G}{\partial T^2}\right)_{\mathcal{H}} = \left(\frac{\partial E_{\text{th}}}{\partial T}\right)_{\mathcal{H}}. \tag{21.60}$$

The average of the expression for $E_{N,\alpha}$ in (21.56) is

$$E_{\text{th}} = \langle E_N \rangle = -\mu_0 \mathcal{H} \langle M_N^z \rangle = -\mu_0 \mathcal{H} M_{\text{th}}. \tag{21.61}$$

Comparing this with the thermodynamic relationship $E_{\text{th}} = U - \mu_0 \mathcal{H} M_{\text{th}}$ in Table 14.3 indicates that the internal energy U of the system is zero. This is expected, because the energy H_N^{int} of the interactions internal to the system is omitted in the paramagnetic Hamiltonian H_N^{pm}.

Equilibrium properties can be obtained either by finding ensemble averages or by using the partition function formula. The partition function for a paramagnetic system is

$$\boxed{Z(T, \mathcal{H}) = \sum_{\alpha} e^{-E_{N,\alpha}(\mathcal{H})/k_B T}.} \tag{21.62}$$

Since $\mathcal{H}$ is an intensive variable, the free energy that is determined by the partition function formula is the Gibbs free energy, which is the thermodynamic potential with natural coordinates T and $\mathcal{H}$. The partition function formula for magnetic systems is

$$\boxed{G(T, \mathcal{H}) = -k_B T \ln Z(T, \mathcal{H}).} \tag{21.63}$$

The relationships in (21.59) imply that the magnetic moment is related to the partition function by

$$M_{\text{th}} = -\frac{k_B T}{\mu_0} \frac{1}{Z}\left(\frac{\partial Z}{\partial \mathcal{H}}\right)_T. \tag{21.64}$$

The relationships in (21.59) imply that $E_{\text{th}} = G - T(\partial G/\partial T)_{\mathcal{H}}$, which becomes

$$E_{\text{th}} = -k_B \left[T \ln Z - \frac{\partial}{\partial T}(T \ln Z)_{\mathcal{H}}\right] = \frac{k_B T^2}{Z}\left(\frac{\partial Z}{\partial T}\right)_{\mathcal{H}}. \tag{21.65}$$

The validity of the partition function formula can be tested by comparing the predictions for M_{th} and E_{th} with the corresponding ensemble averages. The canonical probability of eigenstate α is

$$\rho_\alpha = \frac{e^{-E_{N,\alpha}(\mathcal{H})/k_B T}}{Z_{\text{pm}}}, \tag{21.66}$$

and the partition function is

$$Z_{\text{pm}} = \sum_{\alpha} e^{-E_{N,\alpha}(\mathcal{H})/k_B T}. \tag{21.67}$$

Since $E_{N,\alpha}(\mathcal{H}) = -\mu_0 \mathcal{H} M^z_{N,\alpha}$, the moment predicted with this is

$$M_{\text{th}} = -\frac{k_B T}{\mu_0}\frac{1}{Z_{\text{pm}}}\left(\frac{\partial Z_{\text{pm}}}{\partial \mathcal{H}}\right)_T = \sum_{\alpha} \frac{e^{\mu_0 \mathcal{H} M^z_{N,\alpha}/k_B T}}{Z_{\text{pm}}} M^z_{N,\alpha} = \sum_{\alpha} \rho_\alpha M^z_{N,\alpha}, \tag{21.68}$$

so that $M_{\text{th}} = \langle M^z_N \rangle$. The predicted enthalpy is

$$E_{\text{th}} = \frac{k_B T^2}{Z_{\text{pm}}}\left(\frac{\partial Z_{\text{pm}}}{\partial T}\right)_{\mathcal{H}} = k_B T^2 \sum_{\alpha} \frac{e^{-E^z_{N,\alpha}(\mathcal{H})/k_B T}}{Z_{\text{pm}}}\left(\frac{E^z_{N,\alpha}(\mathcal{H})}{k_B T^2}\right) = \langle E_N \rangle. \tag{21.69}$$

Thus, the partition function formula yields the correct ensemble averages.

21.5 Spin Systems

The magnetic dipole moment of an atom is proportional to its angular momentum, which has contributions from both the orbital motion of the electrons and their intrinsic spin. The magnitude of the moment is $g\mu_B\sqrt{j(j+1)}$, where j is the total angular momentum quantum number and μ_B is the *Bohr magneton* $\mu_B = e\hbar/2m$. The g factor is a dimensionless number of the order of unity whose value depends on the material considered. For electrons its value is $g = 2$. When the z-axis is in the direction of the magnetic field, the z-component of the dipole moment is $g\mu_B m_j$, where m_j is the *magnetic quantum number*, which has values

$$m_j = -j, -j+1, \cdots, +j-1, +j. \tag{21.70}$$

Classical behavior is obtained in the limit as j goes to infinity.

In this section, we investigate spin-$\frac{1}{2}$ systems,[1] which are examples of the two-level systems discussed in Section 18.2. In this system we have $j = \frac{1}{2}$, so that the magnetic quantum number m_j has two allowed values, $-\frac{1}{2}$ and $+\frac{1}{2}$. The z-component of a magnetic moment is $m^z_i = m_o\sigma_i$, where $\sigma_i = \pm 1$ and m_o is the magnitude of the moment. The magnetic moment of N spin-$\frac{1}{2}$ particles is

$$M^z_{N,\alpha} = \sum_{i=1}^{N} m_o\sigma_i, \tag{21.71}$$

where $\sigma_i = +1$ indicates spin "up" and $\sigma_i = -1$ indicates spin "down."

[1] The particles that contribute to paramagnetic behavior are localized on atoms. In contrast, electrons in metals are free to move throughout the system and obey the Fermi–Dirac distribution this system discussed in Chapters 26 and 27.

Since the energy of a magnetic particle is $-\mu_0\mathcal{H}m_i^z$, the energy of each spin is $-\mu_0\mathcal{H}m_o\sigma_i$ and the energy of N spins is

$$E_{N,\alpha}(\mathcal{H}) = \sum_{i=1}^{N}(-\mu_0\mathcal{H}m_o\sigma_i) = -\varepsilon_{\mathcal{H}}\sum_{i=1}^{N}\sigma_i, \tag{21.72}$$

where $\varepsilon_{\mathcal{H}} = \mu_0\mathcal{H}m_o$. By using σ_i as the quantum number for spin i (instead of its magnetic quantum number), index α becomes an abbreviation for the N quantum numbers,

$$\alpha = (\sigma_1, \sigma_2, \ldots, \sigma_N). \tag{21.73}$$

The canonical probability is

$$\rho_\alpha = \frac{e^{-E_{N,\alpha}(\mathcal{H})/k_BT}}{Z_{\text{pm}}} = \frac{e^{(\varepsilon_{\mathcal{H}}/k_BT)\sum_i\sigma_i}}{Z_{\text{pm}}}. \tag{21.74}$$

By using the properties of exponentials, this becomes

$$\rho_\alpha = \frac{e^{(\varepsilon_{\mathcal{H}}/k_BT)\sigma_1}\ldots e^{(\varepsilon_{\mathcal{H}}/k_BT)\sigma_N}}{Z_{\text{pm}}} = \rho_{\sigma_1}\cdots\rho_{\sigma_N}, \tag{21.75}$$

where ρ_{σ_i} is the probability that spin i is in state σ_i.

Magnetization

The equilibrium properties of systems with statistically independent parts can be found directly by using the properties of averages and the distribution of probability described by Boltzmann factors. The magnetic moment of a spin-$\frac{1}{2}$ systems is

$$M_{\text{th}} = \langle M_N^z\rangle = \left\langle \sum_i m_o\sigma_i \right\rangle = m_o\sum_i\langle\sigma_i\rangle. \tag{21.76}$$

Since the individual spins are statistically independent, the average $\langle\sigma_i\rangle$ is determined by the probability ρ_{σ_i} of state σ_i, which is proportional to the Boltzmann factor associated for energy $-\varepsilon_{\mathcal{H}}\sigma_i$

$$\rho_{\sigma_i} = Ce^{\beta\varepsilon_{\mathcal{H}}\sigma_i}, \tag{21.77}$$

where $\beta = 1/k_BT$ and C is a proportionality constant. Since σ_i has only two values, the normalization of ρ_{σ_i} requires that

$$\sum_{\sigma_i}\rho_{\sigma_i} = \rho_{+1} + \rho_{-1} = Ce^{+\beta\varepsilon_{\mathcal{H}}} + Ce^{-\beta\varepsilon_{\mathcal{H}}} = 1, \tag{21.78}$$

which implies that $C = (e^{+\beta\varepsilon_{\mathcal{H}}} + e^{-\beta\varepsilon_{\mathcal{H}}})^{-1}$. Thus,

$$\rho_{\sigma_i} = \frac{e^{\beta\varepsilon_{\mathcal{H}}\sigma_i}}{e^{+\beta\varepsilon_{\mathcal{H}}} + e^{-\beta\varepsilon_{\mathcal{H}}}}. \tag{21.79}$$

The average value of σ_i is

$$\langle \sigma_i \rangle = \sum_{\sigma_i} \rho_{\sigma_i} \sigma_i = \frac{e^{+\beta \varepsilon_{\mathcal{H}}}(+1) + e^{-\beta \varepsilon_{\mathcal{H}}}(-1)}{e^{+\beta \varepsilon_{\mathcal{H}}} + e^{-\beta \varepsilon_{\mathcal{H}}}}. \tag{21.80}$$

By using the hyperbolic tangent function $\tanh x = (e^{+x} - e^{-x})/(e^{+x} + e^{-x})$, this becomes

$$\langle \sigma_i \rangle = \tanh(\beta \varepsilon_{\mathcal{H}}). \tag{21.81}$$

Since the averages are the same for all N spins, the equilibrium value of the magnetic moment is

$$\boxed{M_{\text{th}} = N m_o \tanh x,} \tag{21.82}$$

where $x = \beta \varepsilon_{\mathcal{H}} = \mu_0 \mathcal{H} m_o / k_B T$.

The magnetization $\mathcal{M} = M_{\text{th}}/V$ of a spin-$\frac{1}{2}$ system is shown in **Figure 21.1**, where $M_{\text{th}}/N m_o$ is plotted versus $x = \mu_0 \mathcal{H} m_o / k_B T$. Since $\varepsilon_{\mathcal{H}}$ is proportional to $\mathcal{H}$, the figure illustrates the tendency of the individual moments to align along the direction of the magnetic field and for the system's magnetization to tend to a maximum value in strong fields.

The derivative of magnetization with respect to magnetic field $(\partial \mathcal{M}/\partial \mathcal{H})_T$ is the isothermal susceptibility χ_T. By using $\tanh x = x + \cdots$, the magnetic moment in (21.82) when x is small is $N m_o x = N \mu_0 \mathcal{H} m_o^2 / k_B T$. In weak fields the susceptibility is

$$\chi_T = \left(\frac{\partial \mathcal{M}}{\partial \mathcal{H}} \right)_T = \frac{N {m_o}^2 \mu_0}{V k_B T}. \tag{21.83}$$

This is an expression of *Curie's law*, which states that magnetic susceptibility is inversely proportional to temperature.

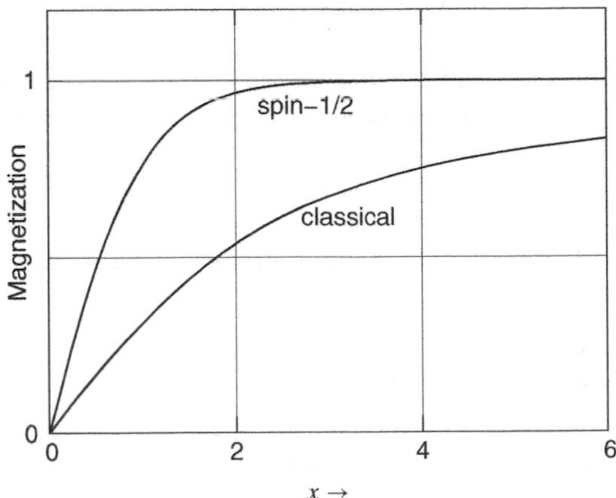

Figure 21.1 *Dependence of Magnetization on Field Strength*

Enthalpy and Heat Capacity

The constant-field heat capacity from (21.60) is $C_{\mathcal{H}} = (\partial E_{\text{th}}/\partial T)_{\mathcal{H}}$. The enthalpy from (21.61) is $E_{\text{th}} = -\mu_0 \mathcal{H} M_{\text{th}}$, so that

$$E_{\text{th}} = -N\mu_0 \mathcal{H} m_o \tanh x, \tag{21.84}$$

where $x = \mu_0 \mathcal{H} m_o / k_B T$ and $\tanh x = (e^{+x} - e^{-x})/(e^{+x} + e^{-x})$. Differentiating this with respect to T gives

$$C_{\mathcal{H}} = \left(\frac{\partial E_{\text{th}}}{\partial T}\right)_{\mathcal{H}} = -N\mu_0 \mathcal{H} m_o \frac{4}{(e^{+x} + e^{-x})^2} \frac{\partial x}{\partial T}. \tag{21.85}$$

By using $\partial x/\partial T = -x/T$, it follows that

$$\boxed{C_{\mathcal{H}} = Nk_B \left(\frac{2x}{e^{+x} + e^{-x}}\right)^2 = Nk_B \left[\frac{x}{\cosh x}\right]^2,} \tag{21.86}$$

where $\cosh x = \frac{1}{2}(e^{+x} + e^{-x})$ is the hyperbolic cosine function.

The dependence of the heat capacity per particle (specific heat) on temperature is shown in **Figure 21.2**, where $C_{\mathcal{H}}/Nk_B$ is plotted versus temperature. The numbers on the horizontal axis are values of $k_B T/(\mu_0 \mathcal{H} m_o)$. As required by the third law of thermodynamics, the specific heat goes to zero as absolute zero is approached. The specific heat is peaked at intermediate temperatures and goes to zero at high temperatures. This type of behavior is typical of systems in which an ordering process is occurring and is qualitatively different from *PVT* systems in which heat capacities tend to constant values at high temperatures. The ordering process in paramagnetic systems is the aligning of the individual magnetic moments along the direction of the magnetic field.

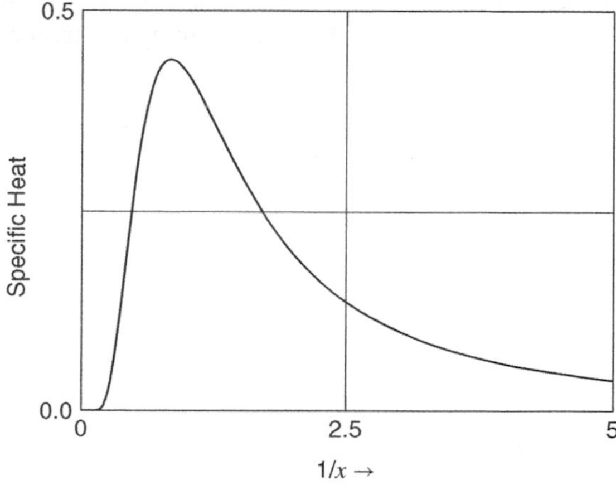

Figure 21.2 *Dependence of Specific Heat on Temperature*

Partition Function and Entropy

Since the eigenvalues of a spin-$\frac{1}{2}$ systems is $E_{N,\alpha}(\mathcal{H}) = -\varepsilon_{\mathcal{H}} \sum_i \sigma_i$, the partition function is

$$Z_{pm} = \sum_{\alpha} e^{-E_{N,\alpha}(\mathcal{H})/k_B T} = \sum_{\sigma_1=\pm 1} \cdots \sum_{\sigma_N=\pm 1} \left(e^{(\varepsilon_{\mathcal{H}}/k_B T)\sigma_1} \ldots e^{(\varepsilon_{\mathcal{H}}/k_B T)\sigma_N} \right)$$

$$= \left(e^{+\varepsilon_{\mathcal{H}}/k_B T} + e^{-\varepsilon_{\mathcal{H}}/k_B T} \right)^N = (2\cosh x)^N, \tag{21.87}$$

where $x = \varepsilon_{\mathcal{H}}/k_B T$ and $\varepsilon_{\mathcal{H}} = \mu_0 \mathcal{H} m_o$. The system's entropy is obtained by combining $S = -(\partial G/\partial T)_{\mathcal{H}}$ with $G = -k_B T \ln Z$ and using $\partial x/\partial T = -x/T$. Thus,

$$S = -\left(\frac{\partial G}{\partial T}\right)_{\mathcal{H}} = Nk_B \frac{\partial}{\partial T}[T \ln(e^{+x} + e^{-x})]$$

$$= Nk_B[\ln(e^{+x} + e^{-x}) - x \tanh x], \tag{21.88}$$

where $x = \mu_0 \mathcal{H} m_o/k_B T$. Entropy is a monotonically increasing function of x, so that it increases as the ratio $T/\mathcal{H}$ increases. The dependence of entropy on temperature is shown in **Figure 21.3**, where S/Nk_B is plotted versus temperature.

At low temperatures ($k_B T < \mu_0 \mathcal{H} m_o$), the individual magnetic moments become aligned along the magnetic field, and the large amount of ordering is reflected in low entropy. At high temperatures (or weak fields), the two orientations of the spins become equally likely, and the associated disorder is associated with high entropy. Since $\ln(e^{+x} + e^{-x})$ becomes $\ln 2$ at $x = 0$, it follows from the proportionality of x to $\mathcal{H}$ indicates that the entropy tends to $Nk_B \ln 2$ as the magnetic field goes to zero ($S \to Nk_B \ln 2$ as $\mathcal{H} \to 0$). The first term in (21.88) tends to $+Nk_B x$ as $\mathcal{H}$ tends to infinity and the second term tends to $-Nk_B x$, so that the entropy tends to zero in strong magnetic fields ($S \to 0$ as $\mathcal{H} \to \infty$).

Magnetic Cooling

Paramagnetic systems are useful for obtaining low temperatures because of the dependence of entropy on the ratio $T/\mathcal{H}$. A procedure for obtaining low temperatures utilizes two

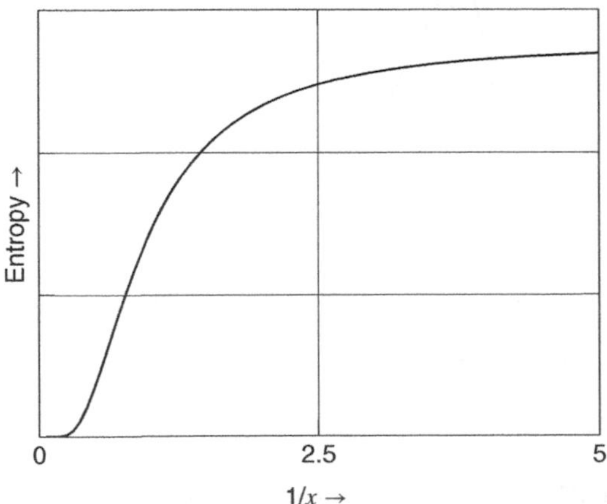

Figure 21.3 Dependence of Entropy on Temperature

systems and a heat reservoir. System 1 is the paramagnetic system and system 2 is a system whose temperature is to be lowered. For example, the goal might be to reach temperatures lower than 4.2 K, the temperature of liquid helium at ordinary pressures. The systems are placed in contact with the reservoir while system 1 is in a strong magnetic field. Without changing the field, the systems are isolated from the reservoir. The magnetic field is then reversibly (slowly) taken to zero with the two systems in thermal contact. Since the process is adiabatic and reversible, the total entropy is constant. Because the entropy of the paramagnetic system increases when $T/\mathcal{H}$ increases, the decrease in $\mathcal{H}$ will be compensated by the desired decrease in the temperature T.

21.6 Classical Dipoles

Classical Limit

When treated classically, the paramagnetic Hamiltonian is a function of the angles θ_i between the magnetic dipole moments $\boldsymbol{m}_i$ and the field $\mathcal{H}$. Thus,

$$H_{\text{pm}}(\theta_1 \cdots \theta_N) = -\sum_{i=1}^{N} \varepsilon_{\mathcal{H}} \cos\theta_i, \tag{21.89}$$

where $\varepsilon_{\mathcal{H}} = \mu_0 \mathcal{H} m_o$ and the field is in the z-direction. When the magnitudes m_o of the moments are the same, the z-component of the system's microscopic magnetic moment is

$$M_N^z(\theta_1 \cdots \theta_N) = \sum_{i=1}^{N} m_o \cos\theta_i, \tag{21.90}$$

and the macroscopic magnetic moment is

$$M_{\text{th}} = \langle M_N^z \rangle = \sum_{i=1}^{N} m_o \langle \cos\theta_i \rangle = N m_o \langle \cos\theta \rangle. \tag{21.91}$$

The average $\langle \cos\theta \rangle$ is the same for all of the magnetic particles,

$$\langle \cos\theta \rangle = \int_0^{2\pi} d\phi \int_0^{\pi} d\theta \sin\theta \rho(\theta) \cos\theta. \tag{21.92}$$

The probability density $\rho(\theta)$ is proportional to the Boltzmann factor for energy $-\varepsilon_{\mathcal{H}} \cos\theta$,

$$\rho(\theta) = \frac{e^{-\beta(-\varepsilon_{\mathcal{H}} \cos\theta)}}{Z_o} = \frac{e^{+\beta\varepsilon_{\mathcal{H}} \cos\theta}}{Z_o}. \tag{21.93}$$

The constant Z_o is determined by the normalization condition

$$\int_0^{2\pi} d\phi \int_0^{\pi} d\theta \sin\theta \rho(\theta) = \frac{2\pi}{Z_o} \int_0^{\pi} d\theta \sin\theta e^{\beta\varepsilon_{\mathcal{H}} \cos\theta} = 1, \tag{21.94}$$

where the 2π comes from the integral over ϕ gives. Solving for Z_o gives

$$Z_o = 2\pi \int_0^{\pi} d\theta \sin\theta \, e^{\beta\varepsilon_{\mathcal{H}}\cos\theta}. \tag{21.95}$$

The average can be expressed as

$$\langle\cos\theta\rangle = \frac{\int_0^{\pi} d\theta \sin\theta e^{\beta\varepsilon_{\mathcal{H}}\cos\theta}\cos\theta}{\int_0^{\pi} d\theta \sin\theta e^{\beta\varepsilon_{\mathcal{H}}\cos\theta}} = \frac{d}{d(\beta\varepsilon_m)}\ln\left(\int_0^{\pi} d\theta \sin\theta e^{\beta\varepsilon_{\mathcal{H}}\cos\theta}\right). \tag{21.96}$$

Changing integration variable from θ to $u = \cos\theta$ gives

$$\int_0^{\pi} d\theta \sin\theta e^{\beta\varepsilon_{\mathcal{H}}\cos\theta} = -\int_{+1}^{-1} du e^{\beta\varepsilon_{\mathcal{H}}u} = -\frac{e^{-\beta\varepsilon_{\mathcal{H}}}}{\beta\varepsilon_{\mathcal{H}}} + \frac{e^{+\beta\varepsilon_{\mathcal{H}}}}{\beta\varepsilon_{\mathcal{H}}} = \frac{2\sinh\beta\varepsilon_{\mathcal{H}}}{\beta\varepsilon_{\mathcal{H}}}, \tag{21.97}$$

where $\sinh u$ is the hyperbolic sine function. Thus,

$$\langle\cos\theta\rangle = \frac{d}{d(\beta\varepsilon_{\mathcal{H}})}[\ln(2\sinh\beta\varepsilon_{\mathcal{H}}) - \ln\beta\varepsilon_{\mathcal{H}}] = \frac{\cosh\beta\varepsilon_{\mathcal{H}}}{\sinh\beta\varepsilon_{\mathcal{H}}} - \frac{1}{\beta\varepsilon_{\mathcal{H}}}. \tag{21.98}$$

The resulting classical expression for the equilibrium magnetic moment is

$$\boxed{M_{\text{th}} = Nm_o L(x),} \tag{21.99}$$

where

$$x = \beta\varepsilon_{\mathcal{H}} = \frac{\mu_0\mathcal{H}m_o}{k_B T} \quad \text{and} \quad L(x) = \frac{1}{\tanh x} - \frac{1}{x}. \tag{21.100}$$

$L(x)$ is called the *Langevin function*. The equilibrium moment in (21.99) is given by the curve labeled as classical in Figure 21.1, where M_{th}/Nm_o is plotted versus x.

As the first few terms in the expansion of $(\tanh x)^{-1}$ are $x^{-1} + \frac{1}{3}x + \cdots$, the lowest order term in the expansion of $L(x)$ is $\frac{1}{3}x$. Hence, the weak-field magnetic moment is $M_{\text{th}} = \frac{1}{3}Nm_o\beta\varepsilon_m$. By using $\mathcal{M} = M_{\text{th}}/V$ and $x = \mu_0\mathcal{H}m_o/k_BT$, the weak-field magnetic susceptibility in the classical limit becomes

$$\chi_T = \left(\frac{\partial\mathcal{M}}{\partial\mathcal{H}}\right)_T = \frac{Nm_o{}^2\mu_0}{3Vk_BT}. \tag{21.101}$$

This is one-third the value for spin-$\frac{1}{2}$ systems.

Dielectrics

Many of the properties of dielectric systems can be obtained from the properties of paramagnetic systems by replacing $\mu_0\mathcal{H}$ with the electric field $\mathcal{E}$ and replacing the magnetic dipole moment $\boldsymbol{m}_i$ with the electric dipole moment $\boldsymbol{p}_i^{\text{el}}$. For molecules containing two or more atoms, the orientations of the dipole moments are determined by orientations of the molecules. The energy associated with the interaction of a dipole with the electric field is $-\boldsymbol{p}_i^{\text{el}} \cdot \mathcal{E}$, and the *electric dipole moment* of a system is $\sum_i \boldsymbol{p}_i^{\text{el}}$.

Problems

21.1 Use the equipartition theorem to estimate the root-mean-square velocity of Rb atoms at the temperature of 10^{-8} K for Bose-Einstein condensation mentioned in Table 1.1. The atomic weight of the rubidium is 87.

21.2 The vibrational potential energy of an atom in the Einstein model is $\frac{1}{2}k_o(q_x^2 + q_y^2 + q_z^2)$, where $\boldsymbol{q} = (q_x, q_y, q_z)$ is displacement. Find the classical probability density $\rho(q_x, q_y, q_z)$ for the displacement.

21.3 A Hamiltonian for modeling planetary atmospheres (introduced in Problem 19.2) is

$$H_N = \sum_{i=1}^{N} \left(\frac{|\boldsymbol{p}_i|^2}{2m_i} + m_i g z_i + \phi_A(x_i, y_i) \right).$$

The surface of the planet is at $z = 0$ and g is the acceleration of gravity. Assume there are atoms of two different masses, m_A and m_B.

(a) What are the probability densities $\rho_A(z)$ and $\rho_B(z)$ that the atoms of mass m_A and m_B are at height z.

(b) Use statistical averages $\langle \cdots \rangle$ to find the average heights of the atoms of both types.

21.4 Consider a system of N independent spin-1 particles with quantum numbers m_i, where $m_i = -1, 0, \text{or} + 1$. The magnetic energy of particle i is

$$\varepsilon(m_i) = \begin{cases} -\mu\mathcal{H}, & \text{if } m_i = +1. \\ 0, & \text{if } m_i = 0. \\ +\mu\mathcal{H}, & \text{if } m_i = -1. \end{cases}$$

(a) Find the partition function Z_N and Gibbs free energy $G = -k_B T \ln Z_N$.

(b) Find the magnetic moment $M_{\text{th}} = -(1/\mu_0)(\partial G/\partial' \mathcal{H})_T$

Appendix 21.A

Negative Temperature

The concept of negative temperature is briefly discussed in this appendix. The motivation for the concept comes from the thermodynamic relationship between entropy and temperature in magnetic systems and the association of entropy with the number of accessible states $\Omega(E)$,

$$\frac{\partial S}{\partial E} = \frac{1}{T} \quad \text{and} \quad S = k_B \ln \Omega(E). \tag{21.A.1}$$

When the thermodynamic coordinates are temperature and magnetic field, energy E is the enthalpy, and the relationship $\partial S/\partial E = 1/T$ follows from the result $T = (\partial E_{\text{th}}/\partial S)_{\mathcal{H}}$ in Table 14.3. (The subscript on E_{th} will be omitted here.) Negative temperatures are an extension of the equilibrium concept of temperature to nonequilibrium situations in which some aspects of a system is almost in thermal equilibrium. Sometimes the interaction between a subsystem and its environment is sufficiently weak that the subsystem approaches a condition of internal equilibrium much faster than it comes to equilibrium with the remainder of the system. Although the final equilibrium state of the complete system is described by a single temperature, it may be practical to characterize the weakly interacting parts by different temperatures. The use of negative temperatures does not alter the significance of absolute zero as the coldest possible temperature because a system described by a negative temperature is *hotter* than systems with positive temperatures.

When in a strong external magnetic field, the spin degrees of freedom of the atomic nuclei of ordinary matter are an example of an (almost) isolated subsystem. The higher energy states of the spins can be populated by radio frequency techniques, which are nonthermal in nature. When energy is exchanged between the spins more rapidly than with the remainder of the system, the spin subsystem can be described as having negative temperature. The population inversion in lasers is another situation described by negative temperatures.

Most systems have no upper bound on their energies $E_{N,\alpha}$, but isolated spin systems are an exception. The spin-$\frac{1}{2}$ systems discussed in Section 21.5 will be used as an illustration. Each spin has only two energy states, and the energy of an N-particle system has lower and upper bounds of $-N\varepsilon_{\mathcal{H}}$ and $+N\varepsilon_{\mathcal{H}}$, where $\varepsilon_{\mathcal{H}} = \mu_0 \mathcal{H} m_o$. Since there are no microstates with energies greater than $N\varepsilon_{\mathcal{H}}$, the number of accessible states $\Omega(E)$ vanishes at high energies. Furthermore, $\Omega(E)$ decreases as E approaches the upper bound, which implies that the derivative $\partial S/\partial E$ is negative, which in turn implies that T is negative.

Each spin-$\frac{1}{2}$ particle is a two-level system of the type discussed in Section 18.2. As given in (18.17), the logarithm of the number of accessible states in a system of N particles is

$$\ln \Omega(E) = N[-f \ln f - (1-f)\ln(1-f)], \tag{21.A.2}$$

where f is the fraction of the particles in the higher energy state. Since the minimum energy is $-N\varepsilon_{\mathcal{H}}$ and since each spin in the higher energy state adds $2\varepsilon_{\mathcal{H}}$ to the energy, the energy E of the N spins is related to the fraction f by

$$E = (fN)(2\varepsilon_{\mathcal{H}}) - N\varepsilon_{\mathcal{H}} \quad \text{and} \quad f = \frac{1}{2}\left(\frac{E}{N\varepsilon_{\mathcal{H}}} + 1\right). \tag{21.A.3}$$

The dependence of the entropy on energy is shown in **Figure 21.A.1**, where $S/k_B = \ln \Omega(E)$ is plotted versus $E/N\varepsilon_{\mathcal{H}} = 2f - 1$. The figure can also be obtained by using equations (21.84) and (21.88), which express the system's enthalpy and entropy as functions of $\beta\varepsilon_{\mathcal{H}}$, and assigning both positive and negative values to β.

It follows from (21.A.1) that the derivative of the entropy is proportional to $\partial \ln \Omega/\partial E = \beta$, where $\beta = 1/k_B T$. The dependence of inverse temperature β and absolute temperature T on energy is shown in **Figure 21.A.2**. Specifically, $\beta = \partial \ln \Omega/\partial E$ and $T = 1/k_B\beta$ are plotted versus $E/N\varepsilon_{\mathcal{H}}$.

The interactions of the nuclear spins with their environment, although weak, will transfer energy to other parts of the system and ultimately bring the complete system to equilibrium with a single temperature. When systems with different initial temperatures are brought into thermal contact, the exchange of energy is characterized by stating that the hotter system loses energy and the colder system gains energy. As indicated by the figure, if a system

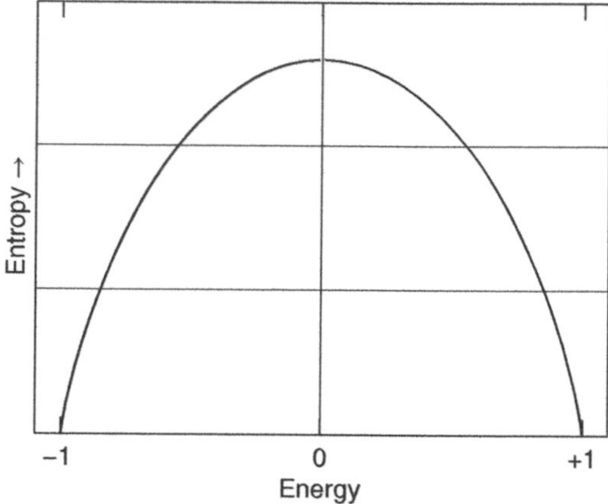

Figure 21.A.1 *Entropy versus Energy*

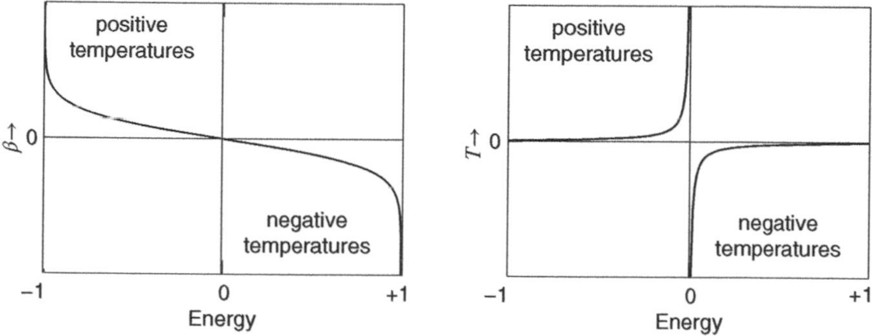

Figure 21.A.2 *β and T versus Energy*

with a negative temperature looses energy, its final temperature will either become more negative or become positive. Since the heat capacities of the kind of systems that can be described by a negative temperature are typically small compared to the heat capacity of the ordinary matter, the final temperature of the combined system will be positive. As we go from colder to hotter, the succession of temperatures T goes from zero to plus infinity, jumps discontinuously to minus infinity, and then approaches zero from negative values. Note that there is no discontinuity in the succession of inverse temperatures β, which go continuously from plus infinity to minus infinity as we go from colder to hotter.

Figure 21.A.1 Entropy versus Energy

22
Fluctuations and Energy Distributions

Now that the essential tools for predicting the equilibrium properties have been outlined, we focus on the effect of the large number of particles on the properties of systems and find out why accurate results can be obtained by averaging over many microstates when at any instant the system is in just one of those states. Also, we resolve a paradox that arises in the microscopic interpretation of entropy and investigate why seemingly different ensembles yield identical predictions.

22.1 Standard Deviation

The thermodynamic value of mechanical property X can be expressed as $\langle X \rangle \pm \sigma_X$, where $\langle X \rangle$ is its average and σ_X is the *standard deviation*. The existence of σ_X reflects the fact that equilibrium properties are not exactly constant but possess small random variations called *fluctuations*. The average $\langle X \rangle$ is the thermodynamic value of a property, and the fluctuations in its value are described by the standard deviation. Statistical mechanics predicts both thermodynamic properties and their fluctuations. We find that the size of fluctuations is reduced from the average value by a factor of $1/\sqrt{N}$, where N is the number of particles or, more precisely, it is the number of particles that are essential to the property being investigated.

Standard deviation σ_X is the square root of the variance, where the *variance* of property X is the average of the square of the deviation from the average,

$$\mathrm{Var}[X] = \langle [X - \langle X \rangle]^2 \rangle. \tag{22.1}$$

Thermodynamics and Statistical Mechanics: An Integrated Approach, First Edition.
Robert J Hardy and Christian Binek.

Variance measures the spread or width of the distribution. By using the properties of averages and the normalization condition $\sum_\alpha \rho_\alpha = 1$ it follows that

$$\begin{aligned}\left\langle [X - \langle X\rangle]^2 \right\rangle &= \sum_\alpha \rho_\alpha [X_\alpha - \langle X\rangle]^2 = \sum_\alpha \rho_\alpha \left[X_\alpha{}^2 - 2X_\alpha \langle X\rangle + \langle X\rangle^2\right] \\ &= \left[\sum_\alpha \rho_\alpha X_\alpha{}^2\right] - 2\left[\sum_\alpha \rho_\alpha X_\alpha\right]\langle X\rangle + \left[\sum_\alpha \rho_\alpha\right]\langle X\rangle^2 \\ &= \langle X^2\rangle - 2\langle X\rangle\langle X\rangle + \langle X\rangle^2 = \langle X^2\rangle - \langle X\rangle^2. \end{aligned} \tag{22.2}$$

Thus,

$$\mathrm{Var}[X] = \langle X^2\rangle - \langle X\rangle^2. \tag{22.3}$$

This is also true for systems described by probability densities. The resulting expression for the *standard deviation* is

$$\boxed{\sigma_X = \sqrt{\mathrm{Var}[X]} = \sqrt{\langle X^2\rangle - \langle X\rangle^2}.} \tag{22.4}$$

Statistically Independent Subsystems

Consider a system of N statistically independent subsystems, such as the particles in an ideal gas, an Einstein solid, or a paramagnet, and assume the particles are identical. The properties of such systems are given by sums of the properties of the individual subsystems. Consider a property X_{tot} of the complete system that is the sum of the properties X_i of identical subsystems,

$$X_{\mathrm{tot}} = \sum_{i=1}^{N} X_i. \tag{22.5}$$

Since the average of a sum equals the sum of the averages, the average of X_{tot} is

$$\langle X_{\mathrm{tot}}\rangle = \sum_{i=1}^{N} \langle X_i\rangle. \tag{22.6}$$

When the subsystems (particles) are identical, the averages $\langle X_i\rangle$ are the same for all N particles i, so that

$$\langle X_{\mathrm{tot}}\rangle = N\langle X_i\rangle, \tag{22.7}$$

where i is the index of any one of them.

The square of X_{tot} is a product of two sums. By using different indices to label terms from different sums, it follows that

$$X_{\mathrm{tot}}^2 = \left(\sum_{i=1}^{N} X_i\right)^2 = \left(\sum_{i=1}^{N} X_i\right)\left(\sum_{j=1}^{N} X_j\right) = \sum_{i=1}^{N}\sum_{j=1}^{N} X_i X_j. \tag{22.8}$$

The average of the square is

$$\langle X_{\mathrm{tot}}^2\rangle = \sum_{i=1}^{N}\sum_{j=1}^{N} \langle X_i X_j\rangle, \tag{22.9}$$

and the square of the average is

$$\langle X_{\text{tot}}\rangle^2 = \langle X_{\text{tot}}\rangle\langle X_{\text{tot}}\rangle = \left\langle \sum_{i=1}^{N} X_i \right\rangle \left\langle \sum_{j=1}^{N} X_j \right\rangle = \sum_{i=1}^{N}\sum_{j=1}^{N} \langle X_i\rangle\langle X_j\rangle. \tag{22.10}$$

Thus, the variance of X_{tot} is

$$\text{Var}[X_{\text{tot}}] = \langle X_{\text{tot}}^2\rangle - \langle X_{\text{tot}}\rangle^2 = \sum_{i=1}^{N}\sum_{j=1}^{N} [\langle X_iX_j\rangle - \langle X_i\rangle\langle X_j\rangle]. \tag{22.11}$$

As indicated in (21.16), when the individual particles are statistically independent, the average $\langle X_iX_j\rangle$ is equals to $\langle X_i\rangle\langle X_j\rangle$ when $i \neq j$. Consequently,

$$\langle X_iX_j\rangle = \begin{cases} \langle X_i^2\rangle, \text{if } i = j. \\ \langle X_i\rangle\langle X_j\rangle, \text{if } i \neq j. \end{cases} \tag{22.12}$$

As a result, the terms with $i \neq j$ in (22.11) vanish, so that the variance of X_{tot} becomes

$$\text{Var}\left[X_{\text{tot}}\right] = \sum_{i=1}^{N} \left[\langle X_i^2\rangle - \langle X_i\rangle^2\right]. \tag{22.13}$$

For identical particles, the individual averages are the same for all i, so that

$$\text{Var}[X_{\text{tot}}] = N\text{Var}[X_i], \tag{22.14}$$

where

$$\text{Var}[X_i] = \langle X_i^2\rangle - \langle X_i\rangle^2 \tag{22.15}$$

is the variance for an individual particle. Since the standard deviation is the square root of the variance, it follows that

$$\boxed{\sigma_{X_{\text{tot}}} = \sqrt{N}\sigma_{X_i},} \tag{22.16}$$

where

$$\sigma_{X_i} = \sqrt{\langle X_i^2\rangle - \langle X_i\rangle^2}. \tag{22.17}$$

Since $\langle X_{\text{tot}}\rangle = N\langle X_i\rangle$, the *relative uncertainty* in property X_{tot} that results from thermal fluctuations is

$$\frac{\sigma_{X_{\text{tot}}}}{\langle X_{\text{tot}}\rangle} = \frac{1}{\sqrt{N}}\frac{\sigma_{X_i}}{\langle X_i\rangle}. \tag{22.18}$$

Thus, in systems of independent particles, the size of the fluctuations is less than thermodynamic value of the property by a factor of $1/\sqrt{N}$.

The above discussion indicates that the values of the mechanical properties of systems of independent particles are essentially the same in the majority of a system's microstates. Although this conclusion was obtained for systems of independent particles, it is applicable to any system that can be modeled as a collection of statistically independent subsystems.

Example 22.1 Magnetic moment

Find the standard deviation of the magnetic moment in a system of N independent spin-$\frac{1}{2}$ particles at temperature T.

Solution. As discussed in Section 21.5, the magnetic moment of N independent spins is the ensemble average of the z-component of the moment $M^z_{N,\alpha}$, where

$$M^z_{N,\alpha} = \sum_{i=1}^{N}(-m_o\sigma_i).$$

$\sigma_i = +1$ indicates spin "up" and $\sigma_i = -1$ indicates spin "down." The average value of σ_i from (21.81) is

$$\langle\sigma_i\rangle = -\tanh(\beta\varepsilon_{\mathcal{H}}),$$

where $\varepsilon_{\mathcal{H}} = \mu_0\mathcal{H}m_o$ and $\beta = 1/k_BT$. According to (21.82), the equilibrium value of the magnetic moment is

$$M_{\text{th}} = \langle M^z_N\rangle = Nm_o\tanh(\beta\varepsilon_{\mathcal{H}}).$$

Since all of the particles have the same moment, the variance of the system's magnetic moment is $\text{Var}[M^z_N] = N\text{Var}[-m_o\sigma_i]$, where the variance of the individual moments is

$$\text{Var}\left[-m_o\sigma_i\right] = \left\langle(-m_o\sigma_i)^2\right\rangle - \langle -m_o\sigma_i\rangle^2 = m_o^{\ 2}\left[\langle\sigma_i^{\ 2}\rangle - \langle\sigma_i\rangle^2\right].$$

Since $\sigma_i^2 = 1$ for both $\sigma_i = +1$ and $\sigma_i = -1$, we find by using the normalization condition that

$$\left\langle\sigma_i^{\ 2}\right\rangle = \sum_{\sigma_i}\rho_{\sigma_i}\sigma_i^{\ 2} = \sum_{\sigma_i}\rho_{\sigma_i} = 1.$$

By using the above values of $\langle\sigma_i\rangle$ and $\langle\sigma_i^{\ 2}\rangle$, the single-particle variance becomes

$$\text{Var}\left[-m_o\sigma_i\right] = m_o^{\ 2}[1 - \tanh^2(\beta\varepsilon_{\mathcal{H}})].$$

The system's variance is $\text{Var}[M^z_N]$ which is N times this. Thus, the standard deviation of the magnetic moment of N spin-$\frac{1}{2}$ particles is

$$\sigma_M = k_B\sqrt{N}m_o\sqrt{1-\tanh^2 x}, \tag{22.19}$$

where $x = \beta\varepsilon_{\mathcal{H}} = \mu_0\mathcal{H}m_o/k_BT$.

As expected, the equilibrium magnetic moment is proportional to N and the fluctuations are proportional to $\sqrt{N}$. The relative uncertainty is especially interesting because of the way in which x depends on temperature and magnetic field. Since $\tanh x \to 1$ as $x \to \infty$ and $\tanh 0 = 1$, the function $\tanh x$ tends to $+1$ at low temperatures or strong magnetic fields ($k_BT \ll \mu_0\mathcal{H}m_o$) and tends to be zero at high temperatures or weak magnetic fields ($k_BT \gg \mu_0\mathcal{H}m_o$). Hence, at low temperatures or strong magnetic fields, the magnetic moment tends to a maximum value with all spins aligned in the same direction with insignificant fluctuations ($M_{\text{th}} = Nm_o \pm 0$). At high temperatures or weak magnetic fields, the moment tends to be zero with significant fluctuations ($M_{\text{th}} = 0 \pm \sqrt{N}m_o$).

22.2 Energy Fluctuations

That fluctuations in the energy are of the order of $1/\sqrt{N}$ can be shown without any assumption about statistically independence. To show this, we start by expressing the constant-volume heat capacity C_V in terms of β, where $\beta = 1/k_BT$,

$$C_V = \left(\frac{\partial U}{\partial T}\right)_V = \left(\frac{\partial U}{\partial \beta}\right)_V \frac{d\beta}{dT} = \left(\frac{\partial U}{\partial \beta}\right)_V \left(\frac{-1}{k_BT^2}\right). \tag{22.20}$$

By using the relationship $U = -\partial \ln Z_N/\partial \beta$ between the internal energy U and the partition function Z_N from (20.15), the constant-volume heat capacity becomes

$$\begin{aligned} C_V &= \frac{1}{k_BT^2}\frac{\partial^2 \ln Z_N}{\partial \beta^2} = \frac{1}{k_BT^2}\frac{\partial}{\partial \beta}\left[\frac{1}{Z_N}\frac{\partial Z_N}{\partial \beta}\right] \\ &= \frac{1}{k_BT^2}\left[\frac{1}{Z_N}\frac{\partial^2 Z_N}{\partial \beta^2} - \left(\frac{1}{Z_N}\frac{\partial Z_N}{\partial \beta}\right)^2\right]. \end{aligned} \tag{22.21}$$

It follows from this and $Z_N = \sum_\alpha e^{-\beta E_{N,\alpha}}$ that

$$\begin{aligned} C_V &= \frac{1}{k_BT^2}\left[\sum_\alpha \frac{e^{-\beta E_{N,\alpha}}}{Z_N}(-E_{N,\alpha})^2 - \left(\sum_\alpha \frac{e^{-\beta E_{N,\alpha}}}{Z_N}(-E_{N,\alpha})\right)^2\right] \\ &= \frac{1}{k_BT^2}\left[\sum_\alpha \rho_\alpha E_{N,\alpha}{}^2 - \left(\sum_\alpha \rho_\alpha E_{N,\alpha}\right)^2\right]. \end{aligned} \tag{22.22}$$

In the notation for averages, this becomes

$$C_V = \frac{1}{k_BT^2}\left[\langle E_N{}^2\rangle - \langle E_N\rangle^2\right]. \tag{22.23}$$

Thus, the variance of the energy is

$$\mathrm{Var}[E_N] = k_BT^2 C_V, \tag{22.24}$$

and the standard deviation is

$$\boxed{\sigma_{E_N} = T\sqrt{k_BC_V}.} \tag{22.25}$$

As an example, we find the standard deviation of energy in an ideal gas. According to the equipartition theorem in Chapter 21, the average energy of a gas of N molecules is $\langle E_N\rangle = \frac{1}{2}Nfk_BT$, where f is the number of separable quadratic terms in the Hamiltonian. Since the heat capacity is $C_V = \frac{1}{2}Nfk_B$, the internal energy is

$$U = \langle E_N\rangle \pm \sigma_E = \frac{1}{2}fNk_BT \pm k_BT\sqrt{\frac{1}{2}fN} = \frac{1}{2}fNk_BT\left(1 \pm \sqrt{\frac{2}{fN}}\right). \tag{22.26}$$

The relative uncertainty $\sqrt{2/(fN)}$ for one mole ($N = 6 \times 10^{23}$) of a diatomic gas ($f = 5$) is 8×10^{-13}, which clearly indicates the negligible size of fluctuations.

Thermodynamic Limit

The logic of mathematics yields exact relationships, so that uncertainties in the values of thermodynamic properties can lead to difficulties when proving things. Since the size of the uncertainties decreases as N increases, exact results can often be obtained by taking the limit $N \to \infty$, where the limit is taken after the extensive variables have been divided by N to prevent them from diverging. This limit, known as the *thermodynamic limit*, yields valid results for real systems because N, although finite, is usually very large. Thermal fluctuations in macroscopic systems are usually insignificant compared to experimental uncertainties except in the vicinity of a critical point.

22.3 Gibbs Paradox

A complication arises in the microscopic interpretation of entropy for systems of identical particles. When different gases such as helium and oxygen are allowed to mix, the mixing is irreversible, but when the molecules in the gases are of the same type, no observable change occurs. This difference should be reflected in a change in entropy.

The Helmholtz free energy of a monatomic ideal gas, given in (19.68) and (20.37), is

$$F(T, V) = -Nk_B T \left[\ln V + \frac{3}{2} \ln T + c_o \right], \tag{22.27}$$

where

$$c_o = \ln \frac{(2\pi m k_B)^{3/2}}{h^3}. \tag{22.28}$$

The equation of state and heat capacity are not affected by the value of the constant c_o, but the entropy is. As determined by the thermodynamic relationship $S = -(\partial F/\partial T)_V$, the entropy of the gas is

$$S(T, V) = Nk_B \left[\ln V + \frac{3}{2} \ln T + c_o + \frac{3}{2} \right]. \tag{22.29}$$

The paradox is illustrated by the adiabatic mixing of the gases in the two containers diagrammed in **Figure 22.1**. The temperatures and pressures of gas 1 and gas 2 are the same and are initially separated by an impermeable wall. An opening is made in the wall and the gases are allowed to mix and come to equilibrium. Since the volume occupied by the molecules of gas 1 is initially V_1 but becomes $V_1 + V_2$ after the mixing, the change in entropy according to (22.29) is

$$\Delta S_1 = N_1 k_B \ln(V_1 + V_2) - N_1 k_B \ln V_1 = N_1 k_B \ln \left(\frac{V_1 + V_2}{V_1} \right), \tag{22.30}$$

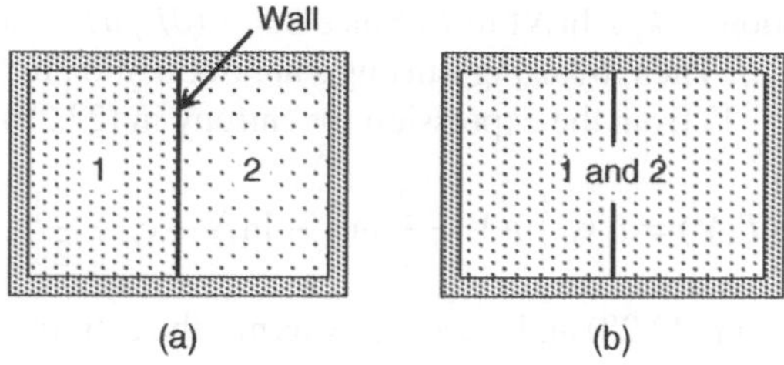

Figure 22.1 *Mixing of Two Gases. (a) Initial State, (b) Final State*

where N_1 is the number of particles (atoms) in gas 1. A similar entropy change occurs in gas 2, so that the total change in the entropy of the combined systems is

$$\Delta S = N_1 k_B \ln\left(\frac{V_1 + V_2}{V_1}\right) + N_2 k_B \ln\left(\frac{V_1 + V_2}{V_2}\right). \tag{22.31}$$

This is the *entropy of mixing* ΔS_{mix} discussed in Section 11.6. *Gibbs paradox* arises because this expression does not account for differences in the types of particles involved. When the particles in the two gases are different, the process is irreversible and the entropy of the combined system increases. However, if the particles in gas 1 are indistinguishable from those in gas 2, the system can be returned to its initial configuration by simply closing the opening. This implies that the process is reversible, which means that the entropy should remain unchanged.

Another difficulty is that the expression for entropy in (22.29) is not extensive, as it should be, because of the contribution $Nk_B \ln V$. As discussed in Section 1.5, when the size of a homogeneous system is scaled by a factor of λ, the values of its extensive properties are proportional to λ, whereas its intensive properties are unchanged. Since both N and V are extensive variables, scaling the size of a gas by a factor of λ, changes N to λN and V to λV, which implies that the scaled value of $Nk_B \ln V$ is $\lambda N k_B \ln \lambda V$. Because of the λ in the argument of the logarithm, its scaled value is not proportional to λ.

Resolution

Gibbs resolved both of these difficulties by accounting for the indistinguishable rearrangements of particles. When N identical particles are identified by indices $i = 1, 2, \cdots, N$, there are many different ways in which we can assign the indices to the particles without changing the collection of momentum and position variables that describe the system. As given in (16.5), the number of ways that N distinct objects can be arranged is $N!$. Hence, there are $N!$ different ways in which we can assign the indices to the particles without changing the description of the system.

In the calculations of the free energy in Chapters 19 and 20, the N particles in the system were considered to be distinguishable. To correct this, we divide the partition function Z_N by $N!$. Since the free energy F is given by the formula $F = -k_B T \ln Z_N$, the division of Z_N

by $N!$ causes the addition of $k_BT \ln N!$ to F. Since $S = -(\partial F/\partial T)_V$, adding this term to F subtracts $k_B \ln N!$ from S. Then, by using Stirling's approximation $\ln N! = N \ln N - N$, we can subtract $Nk_B(\ln N - 1)$ from the expression for entropy in (22.29). The result is

$$S(T, V) = Nk_B \left[\ln V + \frac{3}{2}\ln T - \ln N + c_o + \frac{5}{2}\right]. \tag{22.32}$$

By using the value of c_o in (22.28) and rearranging terms, the entropy of a monatomic ideal gas of N identical particles becomes

$$\boxed{S(T, V) = Nk_B \left[\ln \frac{V}{N} + \frac{3}{2}\ln\left(k_BT\right) + \frac{3}{2}\ln\left(\frac{2\pi m}{h^2}\right) + \frac{5}{2}\right].} \tag{22.33}$$

This result is known as the *Sackur–Tetrode equation*. It is extensive and expresses the entropy without any undetermined constants. It was developed independently by Otto Sackur and Hugo Tetrode. It can be expressed as

$$S(T, V) = \ln \frac{V}{N\lambda_{\text{th}}^3} + \frac{5}{2}Nk_B, \tag{22.34}$$

where $\lambda_{\text{th}} = h/\sqrt{2\pi m k_BT}$ is the thermal de Broglie wave length introduced in (20.71).

Mixtures

We now consider systems that consist of mixtures of N_A particles of type A, N_B particles of type B, and so forth. The total number of particles is $N_{\text{tot}} = N_A + N_B + \cdots$. There are $N_A!$ different arrangements of the particles of type A, $N_B!$ different arrangements of particles of type B, and so forth, so that the total number of indistinguishable arrangements is $N_A!N_B!\cdots$. Dividing the partition function by this adds $k_BT\ln(N_A!N_B!\cdots)$ to the free energy F and subtracts $k_B(\ln N_A! + \ln N_B! + \cdots)$ from the entropy $S = -(\partial F/\partial T)_V$. The corrected expression for the entropy of a mixture, which is found by modifying (22.33), is

$$\boxed{S(T, V) = \sum_k N_k k_B \left[\ln \frac{V}{N_k} + \frac{3}{2}\ln m_k\right] - N_{\text{tot}} f_S(T),} \tag{22.35}$$

where

$$f_S(T) = k_B \left[\frac{3}{2}\ln \frac{2\pi k_BT}{h^2} + \frac{5}{2}\right]. \tag{22.36}$$

Index k identifies different types of particles, m_k is the mass of a particle of type k, and the function $f_S(T)$ is independent of k.

The experiment diagrammed in **Figure 22.1** can be used to test that (22.35) resolves Gibbs paradox. Since the gases are initially separated and consist of particles of a single type, the initial entropy of the combined system is

$$S_0 = N_1k_B \left[\ln \frac{V_1}{N_1} + \frac{3}{2}\ln m_1\right] + N_2k_B \left[\ln \frac{V_2}{N_2} + \frac{3}{2}\ln m_2\right] - N_{\text{tot}} f_S(T), \tag{22.37}$$

where $N_{tot} = N_1 + N_2$. In the final state, the volume occupied by both gas 1 and gas 2 is $V_1 + V_2$, so that the final entropy with two types of particles present is

$$S_f^{two} = N_1 k_B \left[\ln \frac{V_1 + V_2}{N_1} + \frac{3}{2} \ln m_1\right] + N_2 k_B \left[\ln \frac{V_1 + V_2}{N_2} + \frac{3}{2} \ln m_2\right] - N_{tot} f_S(T). \tag{22.38}$$

The change $S_f^{two} - S_0$ is

$$\Delta S^{two} = N_1 k_B \ln \frac{V_1 + V_2}{V_1} + N_2 k_B \ln \frac{V_1 + V_2}{V_2}, \tag{22.39}$$

which is the change in entropy found in (22.31). When the particles in gas 1 and gas 2 are identical, the final state has $N_1 + N_2$ particles of just one type in volume $V_1 + V_2$. Hence, the final entropy is

$$S_f^{one} = (N_1 + N_2) k_B \left[\ln \frac{V_1 + V_2}{N_1 + N_2} + \frac{3}{2} \ln m\right] - N_{tot} f_S(T), \tag{22.40}$$

where m_1 and m_2 have been replaced by m. Since the temperatures and pressures of the gases are the same, the specific volumes V_1/N_1, V_2/N_2, and $(V_1 + V_2)/(N_1 + N_2)$ are the same and can be represented by V/N. Hence, the change in entropy with one type of particle present is

$$\Delta S^{one} = (N_1 + N_2) k_B \ln \frac{V}{N} - \left[N_1 k_B \ln \frac{V}{N} + N_2 k_B \ln \frac{V}{N}\right] = 0. \tag{22.41}$$

As ΔS^{one} is zero while ΔS^{two} is positive, the expression for entropy in (22.35) does indeed account for the expected dependence of entropy on whether or not the particles are identical.

Conclusion

Although the expression for entropy in (22.33) was obtained by dividing the partition function by $N!$, we will continue to express the free energy as $F = -k_B T \ln Z_N$ and include the division by $N!$ in corrected expressions for the partition function. As a result, the constant C_N in the classical partition function in (19.24) becomes

$$C_N = \frac{1}{h^{3N} N!}. \tag{22.42}$$

As discussed in Section 20.3, the factor of $1/h^{3N}$ is needed to make the classical partition function consistent with the quantum mechanical result. Thus, the correct classical partition function is

$$\boxed{Z_N = \frac{1}{h^{3N} N!} \int dp^{3N} \int dr^{3N} e^{-H_N(p_1 \cdots r_N)/k_B T}.} \tag{22.43}$$

The quantum partition function becomes[1]

$$\boxed{Z_N = \frac{1}{N!} \sum_{\alpha} e^{-E_{N,\alpha}/k_B T}.} \tag{22.44}$$

[1] The factor of $1/N!$ in (22.44) is not required after the exchange symmetry of identical fermions and bosons is accounted for in Chapter 26.

The corrected expression for the Boltzmann entropy is obtained by dividing the number of accessible states $\Omega(E)$ by the number of indistinguishable rearrangements,

$$\boxed{S_B = k_B \ln \frac{\Omega(E)}{N!}.} \tag{22.45}$$

Equations (22.42) through (22.45) are for systems of N identical particles. The corresponding expressions for systems with several types of particles are obtained by replacing $N!$ with $N_1!N_2!N_3!\cdots$.

22.4 Microcanonical Ensemble

We have been primarily concerned with the canonical ensemble. Unlike the canonical ensemble, which assigns a probability proportional to e^{-E/k_BT} to each microstate, the microcanonical ensemble includes only microstates with a single value of energy. Although the two ensembles predict the same macroscopic properties, they appear quite different at the microscopic level. Seeing how this comes about contributes to our understanding of the atomic-level origins of thermodynamic behavior. Quantum mechanical concepts were used to describe the microcanonical ensemble in Chapter 17. To take advantage of the direct relationship between energy and the Hamiltonian in classical mechanics, we now treat it classically.

The connection between a system's macroscopic properties and its microscopic description is made through the number of accessible microstates $\Omega(E)$ and the Boltzmann formula for entropy in (22.45). Classically, a accessible microstate is specified by a point $(\boldsymbol{p}_1 \cdots \boldsymbol{r}_N)$ in phase space and its energy is given by the Hamiltonian $H_N(\boldsymbol{p}_1 \cdots \boldsymbol{r}_N)$. Since there is a continuum of microstates, the microcanonical ensemble is described by the probability density

$$\boxed{\rho_E^{\text{mc}}(\boldsymbol{p}_1 \cdots \boldsymbol{r}_N) = \frac{\delta\left(E - H_N\left(\boldsymbol{p}_1 \cdots \boldsymbol{r}_N\right)\right)}{\omega_{\text{ph}}(E)},} \tag{22.46}$$

where the delta function $\delta(E - H_N)$ selects out the microstates with energy E and $\omega_{\text{ph}}(E)$ normalizes the probability density.

A *delta function*, also called a *Dirac delta function*, has the following properties:

$$\delta(x) = \begin{cases} \infty, \text{if } x = 0, \\ 0, \text{if } x \neq 0, \end{cases} \tag{22.47}$$

$$\int \delta(x) dx = 1, \tag{22.48}$$

$$\int \delta(x - a) f(x) dx = f(a). \tag{22.49}$$

The second and third properties indicate that $\delta(x)$ has the same units as $1/x$. From a strict mathematical point of view, $\delta(x)$ is not a function but represents a sequence of functions

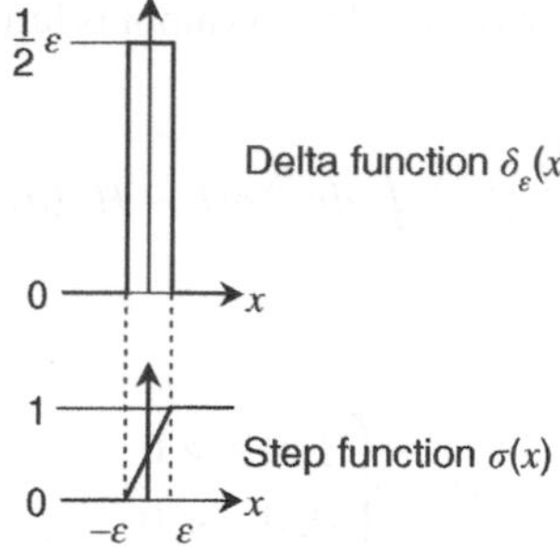

Figure 22.2 *Sequences of Functions*

which are peaked at $x = 0$ and approach the above-mentioned properties as one proceeds along the sequence. One possible choice of functions is diagrammed in **Figure 22.2**, where the function $\delta_\varepsilon(x)$ is

$$\delta_\varepsilon(x) = \frac{1}{2\varepsilon}\begin{cases} 1, \text{if } -\varepsilon \leq x \leq +\varepsilon, \\ 0, \text{otherwise,} \end{cases} \tag{22.50}$$

and ε goes to zero as one proceeds along the sequence.

The probability assigned to volume element $dr^{3N}dp^{3N}$ in phase space is given by the product $\rho_E^{\text{mc}}dr^{3N}dp^{3N}$. The microcanonical ensemble assigns equal probabilities to equal volumes of phase space, which can be seen by representing the delta function in ρ_E^{mc} by $\delta_\varepsilon(x)$ and setting $2\varepsilon = \delta E$. We then have $\rho_E^{\text{mc}} = 1/(\delta E\omega_{\text{ph}})$, which is the same for all points $(\boldsymbol{p}_1 \cdots \boldsymbol{r}_N)$ with energy $H_N(\boldsymbol{p}_1 \cdots \boldsymbol{r}_N)$ between $E - \frac{1}{2}\delta E$ and $E + \frac{1}{2}\delta E$. Note that just as $x^2 + y^2 + z^2 = r^2$ defines the surface of a sphere, the equation $H_N(\boldsymbol{p}_1 \cdots \boldsymbol{r}_N) = E$ defines a constant energy surface in phase space.

The function $\omega_{\text{ph}}(E)$ is found by substituting (22.46) into the normalization condition $\int dr^{3N} \int dp^{3N} \rho_E^{\text{mc}} = 1$ and solving for $\omega_{\text{ph}}(E)$,

$$\omega_{\text{ph}}(E) = \int dr^{3N} \int dp^{3N} \delta\left(E - H_N\left(\boldsymbol{p}_1 \cdots \boldsymbol{r}_N\right)\right). \tag{22.51}$$

When the delta function is represented by the functions in (22.50) with $2\varepsilon = \delta E$, the integrations in (22.51) give the volume of phase space with energy between $E - \frac{1}{2}\delta E$ and $E + \frac{1}{2}\delta E$ and the division by 2ε divides the volume by energy δE, which indicates that $\omega_{\text{ph}}(E)$ is the *volume of phase space per unit of energy*.

Another useful characteristic of a system is the volume of phase space inside the surface with energy E. The *volume of phase space with energy less than E* is

$$\Sigma_{\text{ph}}(E) = \underbrace{\int dr^{3N} \int dp^{3N}}_{H_N(\boldsymbol{p}_1 \cdots \boldsymbol{r}_N) \leq E}(+1). \tag{22.52}$$

(The "or equal to" in the "less than or equal to" symbol is left implicit.) The volume $\Sigma_{ph}(E)$ can also be written as

$$\Sigma_{ph}(E) = \int dr^{3N} \int dp^{3N} \sigma(E - H_N(\boldsymbol{p}_1 \cdots \boldsymbol{r}_N)), \tag{22.53}$$

where $\sigma(x)$ is the *unit step function*

$$\sigma(x) = \begin{cases} 1, \text{if } x \geq 0. \\ 0, \text{if } x < 0. \end{cases} \tag{22.54}$$

The step function can be interpreted as a sequence of continuous functions that rise steeply at $x = 0$ and approach a unit step as one proceeds along the sequence. A possible sequence is shown in Figure 22.2. As illustrated by the figure, the derivative of a step function is essentially a delta function,

$$\frac{d\sigma(x)}{dx} = \delta(x). \tag{22.55}$$

By using this to find the derivative $\partial\Sigma_{ph}(E)/\partial E$ and comparing the result with (22.51), it follows that

$$\frac{\partial\Sigma_{ph}(E)}{\partial E} = \int dr^{3N} \int dp^{3N} \delta(E - H_N(\boldsymbol{p}_1 \cdots \boldsymbol{r}_N)) = \omega_{ph}(E). \tag{22.56}$$

The classical analogs of the *number of microstates with energy less than E*, and the *number of accessible states* defined in Chapter 17 is obtained by associating a phase space volume of h^{3N} with every eigenstate, where h is Planck's constant. The resulting classical interpretations $\Sigma(E)$ and $\Omega(E)$ are

$$\Sigma(E) = \frac{1}{h^{3N}} \Sigma_{ph}(E) \quad \text{and} \quad \Omega(E) = \omega_{ph}(E) \frac{\delta E}{h^{3N}}. \tag{22.57}$$

The predictions of the classical ensemble tend to those of the quantum ensemble at higher temperatures, where the meaning of higher depends on the system involved.

Ideal Gas

The nature of the dependence of $\Sigma_{ph}(E)$ and $\omega_{ph}(E)$ on energy E has consequences that are illustrated by the ideal gas model. By identifying the components of momentum by a single subscript j, the momentum variables $(p_1^x, p_1^y, \ldots, p_N^z)$ of an N-particle gas becomes $(p_1, p_2, \ldots, p_{3N})$ so that the Hamiltonian of a monatomic system is

$$H_N(\boldsymbol{p}_1 \cdots \boldsymbol{r}_N) = \sum_j \frac{1}{2m} p_j^2, \tag{22.58}$$

where atom-wall interaction has been left implicit. The volume of phase space $\Sigma_{ph}(E)$ is found by integrating over the regions of phase space in which $H_N(\boldsymbol{p}_1 \ldots \boldsymbol{r}_N) \leq E$. For an ideal gas the definition in (22.52) becomes

$$\Sigma_{\text{ph}}(E) = \int dr^{3N} \int dp^{3N} \begin{Bmatrix} 1, \text{if } \sum_j \frac{1}{2m} p_j^2 \le E \\ 0, \text{otherwise} \end{Bmatrix}. \tag{22.59}$$

There is no dependence on the coordinates $(\boldsymbol{r}_1, \boldsymbol{r}_2, \ldots, \boldsymbol{r}_N)$ in the integrand because the potential energy is zero when the particles are inside the container. Since the potential energy is effectively infinite when any particle is outside the container, the integrations over coordinates only contribute to $\Sigma_{\text{ph}}(E)$ when every particle is in the container. As a result, the integrations over $(\boldsymbol{r}_1, \boldsymbol{r}_2, \ldots, \boldsymbol{r}_N)$ yield a factor of V^N, where V is the volume of the container. The integrations over $(p_1, p_2, \ldots, p_{3N})$ can be simplified by changing p_j to u_j, where $p_j = u_j(2mE)^{1/2}$. The result is

$$\Sigma_{\text{ph}}(E) = V^N (2mE)^{3N/2} \int_{-\infty}^{+\infty} du_1 \cdots \int_{-\infty}^{+\infty} du_{3N} \begin{pmatrix} 1, \text{if} \sum_j u_j^2 \le 1 \\ 0, \text{otherwise} \end{pmatrix}. \tag{22.60}$$

By using the value of the integrals over $(u_1, \ldots, u_{3N})$ found in Appendix 17.A, we find that

$$\boxed{\Sigma_{\text{ph}}(E) = V^N \frac{E^{aN}}{(3N/2)!} (2\pi m)^{aN},} \tag{22.61}$$

where $a = 3/2$. Also, the volume of phase space per unit of energy is

$$\omega_{\text{ph}}(E) = \frac{\partial \Sigma_{\text{ph}}(E)}{\partial E} = V^N \frac{E^{aN-1}}{(aN-1)!} (2\pi m)^{aN}. \tag{22.62}$$

As indicated in (22.57), the number of microstates with energy less than E is obtained by dividing $\Sigma_{\text{ph}}(E)$ by h^{3N}, which yields the result obtained quantum mechanically in (17.15).

The dependence of $\Sigma_{\text{ph}}(E)$ on energy E for a monatomic gas is

$$\Sigma_{\text{ph}}(E) = K E^{aN}, \tag{22.63}$$

where $a = 3/2$ and K is a proportionality constant independent of E. This type of energy dependence is characteristic of systems with constant heat capacity and occurs in other systems with various values of a. The significance of this type of dependence on E can be seen by plotting X^n, which is done for a few values of n in **Figure 22.3**. Notice that there are no curves with $N = 1000$ and larger because the resulting curves would be indistinguishable from a vertical line at $X = 1$.

The implications of the figure for equilibrium behavior is seen by setting $X = E/E_{\text{sys}}$, where E is the energy of the microstates with energy less than E_{sys} and E_{sys} is identified with the system's internal energy. When $n = aN$ and N is the number of constituent particles, the function X^n becomes

$$X^n = \left(\frac{E}{E_{\text{sys}}}\right)^{aN} = \frac{\Sigma_{\text{ph}}(E)}{\Sigma_{\text{ph}}(E_{\text{sys}})}, \tag{22.64}$$

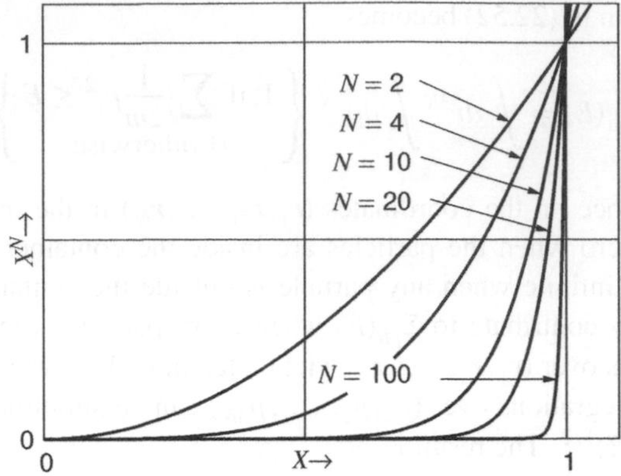

Figure 22.3 *Powers of X*

where $\Sigma_{ph}(E)/\Sigma_{ph}(E_{sys})$ is the ratio of the volume of phase space (number of microstates) with energy less than E over the volume less than E_{sys}. The extremely rapid increase in the value of X^n as X approaches $X = 1$ indicates that the number of microstates increases extremely rapidly as energy E approaches the internal energy. Hence, almost all of the microstates represented by $\Sigma_{ph}(E_{sys})$ have energies E that are very close to the internal energy.

Entropy

In the microcanonical ensemble the connection between thermodynamics and the microscopic properties of a system is made through the thermodynamic potential $S(U, V)$ and the Boltzmann formula for entropy,

$$\boxed{S_B(E) = k_B \ln \frac{\Omega(E)}{N!}.} \tag{22.65}$$

$\Omega(E) = \omega_{ph}(E)\delta E h^{-3N}$, where E is identified with the internal energy. Because of the extremely rapid increase of $\Sigma_{ph}(E)$ with increasing energy E, the entropy is also given by

$$\boxed{S_\Sigma(E) = k_B \ln \frac{\Sigma(E)}{N!},} \tag{22.66}$$

where $\Sigma(E) = \Sigma_{ph}(E)h^{-3N}$. Since $\omega_{ph}(E)$ is the derivative of $\Sigma_{ph}(E)$, it is somewhat surprising that $S_B(E)$ and $S_\Sigma(E)$ are effectively equivalent expressions for entropy. We can test this assertion by using $\Sigma_{ph}(E) = KE^{aN}$ to compare the two expressions. It follows from $\omega_{ph} = \partial\Sigma_{ph}/\partial E$ that

$$\frac{\Omega(E)}{N!} = \omega_{ph}(E)\frac{\delta E}{h^{3N}N!} = aNKE^{aN-1}\frac{\delta E}{h^{3N}N!} = \frac{\Sigma_{ph}(E)}{h^{3N}N!}\frac{a\delta E}{E/N}. \tag{22.67}$$

Hence, $S_B(E) - S_\Sigma(E) = k_B[\ln a\delta E - \ln(E/N)]$. It is left as a problem to show that when the expression for $\sum_{\text{ph}}(E)$ in (22.61) and the internal energy for a monatomic gas are used, the entropy $S_\Sigma(E)$ is given by the Sackur–Tetrode equation in (22.33). As $S_\Sigma(E)$ is proportional to N while the difference $S_B(E) - S_\Sigma(E)$ is independent of N, the difference $S_B(E) - S_\Sigma(E)$ is negligible when N has values typical of macroscopic systems.

Energy and Pressure

We found in Section 19.4 that internal energy U and pressure P are given by canonical averages of phase functions. We need to check that U and P are also given by the corresponding microcanonical averages. As the Hamiltonian $H_N(\boldsymbol{p}_1 \cdots \boldsymbol{r}_N)$ is the phase function for energy and the microcanonical probability density in (22.46) only includes points $(\boldsymbol{p}_1 \cdots \boldsymbol{r}_N)$ that represent microstates of energy E, the microcanonical average of H_N equals E. Since E is identified with internal energy U, it follows that $U = \langle H_N \rangle$.

The phase function for pressure from (19.38) is $P_{\text{ph}} = -\partial H_N/\partial V$. To show that pressure is given by the average of P_{ph}, we use the thermodynamic relationships

$$\frac{1}{T} = \left(\frac{\partial S}{\partial U}\right)_V \quad \text{and} \quad \frac{P}{T} = \left(\frac{\partial S}{\partial V}\right)_U \tag{22.68}$$

from (12.29). By combining these equations, we find that

$$P = \frac{(\partial S/\partial V)_U}{(\partial S/\partial U)_V}, \tag{22.69}$$

where P is the function of U and V. This ratio of derivatives is conveniently evaluated by using the entropy $S_\Sigma(E)$ from (22.66). By using $\Sigma(E) = \Sigma_{\text{ph}}(E)h^{-3N}$ and $E = U$, the entropy as a function of U and V becomes

$$S(U,V) = k_B \ln\left(\frac{\Sigma_{\text{ph}}(U,V)}{h^{3N}N!}\right), \tag{22.70}$$

where

$$\Sigma_{\text{ph}}(U,V) = \int dr^{3N} \int dp^{3N} \sigma(U - H_N). \tag{22.71}$$

Substituting $S(U,V)$ into (22.69) gives

$$P = \frac{(\partial \Sigma_{\text{ph}}/\partial V)_U}{(\partial \Sigma_{\text{ph}}/\partial U)_V}. \tag{22.72}$$

As the dependence on V enters through the Hamiltonian, it follows from $d\sigma/dx = \delta(x)$ that $\partial\sigma(U - H_N)/\partial U = \delta(U - H_N)$ and that $\partial\sigma(U - H_N)/\partial V = \delta(U - H_N)(-\partial H_N/\partial V)$, so that

$$P = \frac{\int dr^{3N} \int dp^{3N} \delta(U - H_N)(-\partial H_N/\partial V)}{\int dr^{3N} \int dp^{3N} \delta(U - H_N)}. \tag{22.73}$$

By using the definitions of ρ_E^{mc} and ω_{ph} in (22.46) and (22.51) with $E = U$, the microcanonical average for pressure becomes

$$P = \int dr^{3N} \int dp^{3N} \rho_E^{mc}(\boldsymbol{p}_1 \cdots \boldsymbol{r}_N)\left(-\frac{\partial H_N(\boldsymbol{p}_1 \cdots \boldsymbol{r}_N)}{\partial V}\right) = \left\langle -\frac{\partial H_N}{\partial V} \right\rangle, \quad (22.74)$$

so that $P = \langle P_{ph} \rangle$.

Notice that different ensembles have been used to obtain the averages $\langle H_N \rangle$ and $\langle P_{ph} \rangle$. One ensemble is a collection of systems with a single value of energy, while the other is a collection of systems with energies less than or equal to some maximum energy. The first ensemble gives us the Boltzmann entropy S_B and $\langle H_N \rangle$, while the other gives us S_Σ and $\langle P_{ph} \rangle$. We are allowed to use both the ensembles because of the extremely rapid rate at which the volume of phase space increases with increasing energy.

22.5 Comparison of Ensembles

The microcanonical ensemble describes systems with a definite energy, whereas the canonical ensemble describes systems with a definite temperature. Because temperature is a statistical property that describes how systems interact when placed in thermal contact, we expect the canonical ensemble to include a range of energies. As we will see, the range is so small that it is not surprising that the predictions of the two ensembles are the same.

The distribution of energy associated with a probability density $\rho(\boldsymbol{p}_1 \cdots \boldsymbol{r}_N)$ in phase space is given by

$$P(E) = \int dp^{3N} \int dr^{3N} \rho(\boldsymbol{p}_1 \cdots \boldsymbol{r}_N)\delta(E - H_N(\boldsymbol{p}_1 \cdots \boldsymbol{r}_N)), \quad (22.75)$$

where $P(E)$ is the probability per unit of energy and $P(E)\,dE$ is the probability the energy E is in interval dE. The delta function $\delta(E - H_N)$ selects the region of phase space in which $H_N(\boldsymbol{p}_1 \cdots \boldsymbol{r}_N)$ is equal to E.

Canonical Ensemble

The canonical ensemble modifies the idea that equal volumes of phase space are equally probable, so that it describes a system with a definite temperature. As given in (19.23), the canonical probability density is

$$\rho^c(\boldsymbol{p}_1 \cdots \boldsymbol{r}_N) = \frac{C_N}{Z_N} e^{-\beta H_N(\boldsymbol{p}_1 \cdots \boldsymbol{r}_N)}, \quad (22.76)$$

where $\beta = 1/k_B T$. The canonical distributions of energy is obtained by substituting $\rho^c(\boldsymbol{p}_1 \cdots \boldsymbol{r}_N)$ into (22.75),

$$P^c(E) = \frac{C_N}{Z_N} \int dp^{3N} \int dr^{3N} e^{-\beta H_N(\boldsymbol{p}_1 \cdots \boldsymbol{r}_N)} \delta(E - H_N(\boldsymbol{p}_1 \cdots \boldsymbol{r}_N)). \quad (22.77)$$

Since the delta function selects points in phase space for which $H_N(\boldsymbol{p}_1 \cdots \boldsymbol{r}_N) = E$, the Hamiltonian in the integrand can be replaced by E, so that the Boltzmann factor $e^{-\beta H_N}$ becomes $e^{-\beta E}$. Taking $e^{-\beta E}$ outside the integrations gives

$$P^c(E) = \frac{C_N}{Z_N} e^{-\beta E} \int dp^{3N} \int dr^{3N} \delta(E - H_N(\boldsymbol{p}_1 \cdots \boldsymbol{r}_N)), \tag{22.78}$$

where the integrations over the delta function is the quantity $\omega_{\text{ph}}(E)$ defined in (22.51). Thus, the distributions of energy in the canonical ensemble is

$$P^c(E) = \frac{C_N}{Z_N} e^{-\beta E} \omega_{\text{ph}}(E). \tag{22.79}$$

Ideal Gas

The characteristics of the canonical distribution of energy can be seen from the ideal gas model. The partition function Z_N for the model is given by (19.65), which by rearranging terms becomes

$$\frac{C_N}{Z_N} = V^{-N}(2\pi m)^{-3N/2}\beta^{3N/2}. \tag{22.80}$$

Combining this with the ideal gas expression for $\omega_{\text{ph}}(E)$ in (22.62) gives

$$\boxed{P^c(E) = \beta e^{-\beta E} \frac{(\beta E)^{aN-1}}{(aN-1)!},} \tag{22.81}$$

where $a = 3/2$. As mentioned earlier, this type of energy distribution is typical of systems with a constant specific heat. The most probable energy $\widehat{E}$ is the value of E that satisfies $\partial P^c(E)/\partial E = 0$. The result is $\widehat{E} = (aN-1)/\beta$, where $\beta = 1/k_B T$. Since minus one is negligible compared to aN, the most probable energy for a monatomic gas is

$$\widehat{E} = \frac{3}{2} N k_B T, \tag{22.82}$$

which is the internal energy of the gas predicted by both ensembles. $P^c(E)$ is peaked at $\widehat{E}$ because the rapid increase of E^{aN} at lower energies is terminated by a rapid decrease of $e^{-\beta E}$ at higher energies.

To emphasize the concentration of energy in the canonical ensemble, we express the probability per unit of energy as

$$P^c(E) = \frac{\beta}{n}\delta_n(z-1), \tag{22.83}$$

where

$$\delta_n(z-1) = n e^{-nz}\frac{(nz)^n}{n!}. \tag{22.84}$$

Setting $z = \beta E/n$ and $n = aN - 1$ makes this equivalent to (22.81). The function $\delta_n(z-1)$ is plotted versus z in **Figure 22.4** for three values of n. As indicated by the notation, the

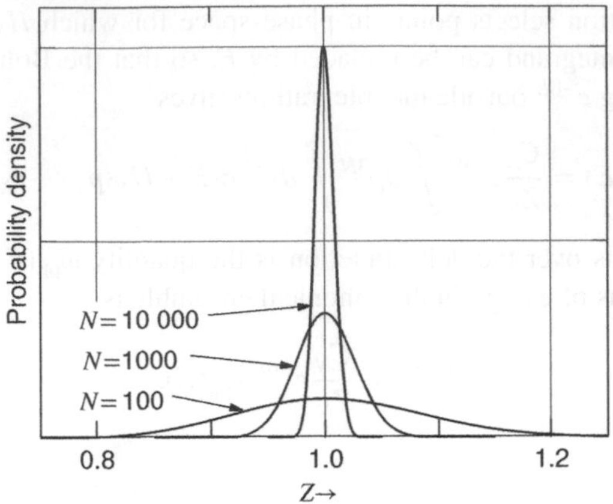

Figure 22.4 *Dependence of the Canonical Probability Density on Energy*

sequence of functions $\delta_n(x)$ with increasing values of n is another representation of a delta function. A delta function is peaked at $x = 0$, which implies that $\delta(z-1)$ is peaked at $z = 1$, as shown in the figure. Substituting $z = \beta E/n$ into $z = 1$ and using $n = a - 1$ indicate that $P^c(E)$ is peaked where $\beta E/(aN-1) = 1$. By omitting the -1 and using $\beta = 1/k_B T$ and $a = 3/2$, this means that $P^c(E)$ is peaked at $E = \frac{3}{2}Nk_BT$. Since the number of particles N in macroscopic systems is very much greater than the values plotted, the figure clearly shows the concentration of probability at a single energy. Thus, the distribution of energy in the canonical ensemble is essentially the same as the distribution in the microcanonical ensemble.

As a final note on the sequence of functions $\delta_n(x)$, we show that the resulting delta function has the property $\int \delta(x)dx = 1$ specified in (22.48). By using the definition of $\delta_n(x)$ in (22.84) and changing integration variables from x to z to t, where $t = nz$ and $dz = dt/n$, we find that

$$\int \delta_n(x)dx = \int \delta_n(z-1)dz = \int \frac{ne^{-nz}(nz)^n}{n!}dz = \frac{1}{n!}\int_0^\infty e^{-t}t^n dt. \qquad (22.85)$$

The integral $\int_0^\infty e^{-t}t^n dt$ is equal to $n!$, so that the property in (22.48) is possessed by every function in the sequence. (The integral is closely related to the standard gamma function.)

Thermodynamic Properties

The connection between the microscopic and thermodynamics behavior of systems in the canonical ensemble is made through the partition function formula for the free energy $F(T, V)$, while in the microcanonical ensemble the connection is made through the Boltzmann formula for the entropy $S(U, V)$. Since both the free energy $F(T, V)$ and the entropy $S(U, V)$ are thermodynamic potentials, both ensembles give full descriptions of a system's properties. The connections established by the two ensembles are

$$F = -k_B T \ln Z_N \quad \text{and} \quad S = k_B \ln \frac{\Omega(U)}{N!}. \tag{22.86}$$

Both the partition function Z_N and the number of accessible states Ω transform descriptions in terms of microscopic variables $(\boldsymbol{p}_1 \cdots \boldsymbol{r}_N)$ to descriptions in terms of thermodynamic variables, and both are determined by the Hamiltonian. Since the equilibrium properties obtained from $F(T, V)$ and $S(U, V)$ are the same, the thermodynamic relationship between the two potentials can be used to verify that the ensembles yield the same predictions. The relationship between the free energy and the internal energy is $F = U - TS$. Substituting the microscopic formulas for the potentials into this gives

$$-k_B T \ln Z_N = U - Tk_B \ln \frac{\Omega(U)}{N!}, \tag{22.87}$$

where $1/T = (\partial S/\partial U)_V$.

To test the accuracy of (22.87), we rewrite the expression for partition function Z_N in (22.43). By using the properties of delta functions, the Boltzmann factor $e^{-\beta H_N}$ can be expressed as

$$e^{-\beta H_N(\boldsymbol{p}_1 \cdots \boldsymbol{r}_N)} = \int dE e^{-\beta E} \delta(E - H_N(\boldsymbol{p}_1 \cdots \boldsymbol{r}_N)). \tag{22.88}$$

For simplicity, let the microscopic variables $(\boldsymbol{p}_1 \cdots \boldsymbol{r}_N)$ be implicit. After changing the order of integration, the partition function becomes

$$Z_N = \frac{1}{h^{3N} N!} \int dE e^{-\beta E} \int dp^{3N} \int dr^{3N} \delta(E - H_N), \tag{22.89}$$

where $\beta = 1/k_B T$. The phase space integration of the delta function equals $\omega_{\text{ph}}(E)$, as given in (22.51). By using $\Omega(E) = \omega_{\text{ph}}(E)\delta E h^{-3N}$ from (22.57), we obtain

$$Z_N = \frac{1}{N!\delta E} \int dE e^{-\beta E} \Omega(E). \tag{22.90}$$

By expressing Ω as $e^{\ln \Omega}$, this becomes

$$Z_N = \frac{1}{N!\delta E} \int dE e^{-(\beta E - \ln \Omega(E))}. \tag{22.91}$$

So far, no approximations have been made. We now expand the function $(\beta E - \ln \Omega(E))$ in a Taylor series about the maximum energy $\hat{E}$. Since the linear term vanishes, it follows that

$$(\beta E - \ln \Omega(E)) = \left[\beta \hat{E} - \ln \Omega\left(\hat{E}\right)\right] + \frac{1}{2} X_2\left(\hat{E}\right)\left(E - \hat{E}\right)^2 + \cdots, \tag{22.92}$$

where

$$X_2(\hat{E}) = -\frac{\partial^2 \ln \Omega(\hat{E})}{\partial E^2}. \tag{22.93}$$

Substituting (22.92) into (22.91) and neglecting higher order terms give

$$Z_N = \frac{e^{-\beta \hat{E} + \ln \Omega(\hat{E})}}{N!\delta E} \int dE e^{-\frac{1}{2} X_2(\hat{E})(E - \hat{E})^2}, \tag{22.94}$$

where the integral over E is the integral of a Gaussian. When the limits of integration are extended to $\pm\infty$, its value is $(2\pi/X_2)^{1/2}$, so that the logarithm of Z_N becomes

$$\ln Z_N = -\beta\hat{E} + \ln(\Omega(\hat{E})/N!) + D_N, \tag{22.95}$$

where

$$D_N = \frac{1}{2}\ln\frac{2\pi}{\delta E^2 X_2(\hat{E})}. \tag{22.96}$$

By identifying $\hat{E}$ with internal energy U and multiplying by $-k_BT$, we obtain

$$-k_BT\ln Z_N = U - k_BT\ln\frac{\Omega(U)}{N!} - k_BT D_N, \tag{22.97}$$

which becomes $F = U - TS - k_BT D_N$. This result indicates that the predictions of the canonical and microcanonical ensembles are the same to the extent that the term $k_BT D_N$ is negligible compared to F, U, and TS.

The value of D_N can be estimated by using the thermodynamic relationships $1/T = (\partial S/\partial U)_V$ and $S = k_B\ln(\Omega(U)/N!)$ to evaluate the function $X_2(\hat{E})$. When $\hat{E}$ is replaced by U, we find that

$$\begin{aligned} X_2(U) &= -\frac{\partial^2 \ln\Omega(U)}{\partial U^2} = -\frac{1}{k_B}\left(\frac{\partial^2 S}{\partial U^2}\right)_V = -\frac{1}{k_B}\left(\frac{\partial}{\partial U}\frac{1}{T}\right)_V \\ &= \frac{1}{k_BT^2}\left(\frac{\partial T}{\partial U}\right)_V = \frac{1}{k_BT^2C_V}, \end{aligned} \tag{22.98}$$

where $C_V = (\partial U/\partial T)_V$ is the constant-volume heat capacity. It is interesting to note that the square root of $1/X_2$ equals the variance of the energy given in (22.24). Substituting X_2 into D_N gives

$$D_N = \frac{1}{2}\ln\frac{2\pi k_BT^2C_V}{\delta E^2}. \tag{22.99}$$

The term $k_BT D_N$ is negligible because of the dependence of the other quantities involved on the number of particles N. The intensive property T is independent of N while F, U, S, and C_V are extensive, which implies that they are proportional to N. The size of $k_BT D_N$ is controlled by the logarithm of C_V and behaves like $\ln N$, while F, U, and TS are proportional to N. Because of this, $k_BT D_N$ is negligible compared to F, U, and TS. If we use Avogadro's number ($N_A = 6.0 \times 10^{23}$) as an estimate for the value of N, the logarithm of N is $\ln N = 55$, which is certainly negligible compared to 10^{23}. Thus, even when F and S are obtained from different ensembles, the relationship $F = U - TS$ is accurate to the extent that 100 is negligible compared to 10^{23}.

There is nothing to prevent us from using ensembles that assigns nonzero probability to microstates that are counterintuitive, provided the probability assigned to them is negligibly small. Furthermore, including such states may simplify calculations without affecting the predictions obtained.

Problems

22.1 Consider a classical system in equilibrium at temperature T. Find the average value $\langle p_x \rangle$ and the standard deviation σ_{p_x} of the x-component of the momentum of an atom.

22.2 Assume that the value of σ_E/E for a small cluster of atoms is 10^{-3}. Treat the cluster as a classical Einstein solid.

- **(a)** How many atoms are there in the cluster.
- **(b)** Estimate the volume of the cluster on the assumption that the typical distance between atoms is approximately 0.4 nm.
- **(c)** Estimate the diameter of the cluster by treating it as a sphere.

22.3 **(a)** Use the probability density $P^c(E)$ for a monatomic ideal gas to find the averages $\langle E \rangle$ and $\langle E^2 \rangle$, where E is internal energy.

- **(b)** Find σ_E and $\sigma_E/\langle E \rangle$.
(The integral $\int_0^\infty x^n e^{-x} dx = n!$ may be useful.)

22.4 Derive the Sackur–Tetrode equation for a monatomic ideal gas by starting with the microcanonical entropy $S_\Sigma(E) = k_B \ln \Sigma(E)/N!$, where $\Sigma(E) = \Sigma_{\text{ph}}(E)/h^{3N}$.

23

Generalizations and Diatomic Gases

When treated classically, the microstates of systems have been specified by the momentum and position variables of the constituent particles. We now introduce the use of the generalized coordinates of analytical mechanics to specify microstates and illustrate their value by predicting the properties of diatomic gases. When treated quantum mechanically, the microstates have been energy eigenstates. We now generalize the description of the canonical ensemble by employing the density matrix formalism of quantum mechanics. We begin by showing that the assignment of probability implicit in the ensembles is independent of the choice of coordinates.

23.1 Generalized Coordinates

The probability densities used in classical statistical mechanics give a probability per unit volume of phase space. When the coordinates (variables) that describe the microstates are transformed, we require that the probability assigned to any region in phase space is unaffected by a change of variables. The rule for transforming probability densities follows from a general requirement that applies whenever outcomes are described by continuous variables.

For example, consider an experiment in which the outcomes are specified by a single variable. Specifically, consider outcomes that can be specified either by x or y, where $y = y(x)$ and $dy/dx \geq 0$. The probability that the outcome is between x and $x + dx$ is $\rho_x(x)dx$. In terms of y, the probability is $\rho_y(y)dy$, where the interval dy is between $y(x)$ and $y(x + dx)$. When dx is sufficiently small that dy equals $(dy/dx)dx$, the requirement that the probability be unaffected by the change from x to y implies that

$$\rho_x(x) = \rho_y(y)\frac{dy}{dx}. \tag{23.1}$$

Thermodynamics and Statistical Mechanics: An Integrated Approach, First Edition.
Robert J Hardy and Christian Binek.

By using the rule for changing integration variables, it follows that

$$\int \rho_y(y)dy = \int \rho_y(y(x))\frac{dy}{dx}dx = \int \rho_x(x)dx. \tag{23.2}$$

This result indicates that the normalization condition for $\rho_y(y)$ follows from the normalization condition for $\rho_x(x)$ and tells us that *probability densities transform like the integrands of integrals.*

The rule for changing variables in a multidimensional integral is

$$\int dxdy \ldots F(x, y, \ldots) = \int dudv \cdots F(x(u, v, \cdots), y(u, v, \ldots), \ldots)\frac{\partial(x, y, \ldots)}{\partial(u, v, \ldots)}, \tag{23.3}$$

where $\partial(x, y, \ldots)/\partial(u, v, \ldots)$ is the Jacobian of the transformation from variables $(x, y, \ldots)$ to variables $(u, v, \ldots)$. This result indicates that the rule for transforming probability densities defined on multidimensional spaces is

$$\rho_{uv}(u, v, \ldots) = \rho_{xy}(x(u, v, \ldots), y(u, v, \ldots), \ldots)\frac{\partial(x, y, \ldots)}{\partial(u, v, \ldots)}. \tag{23.4}$$

An important example of this rule is the transformation from Cartesian coordinates (x, y, z) to spherical coordinates (r, θ, φ). The equations that relate the two coordinate systems are

$$x = r\sin\theta\cos\varphi, \quad y = r\sin\theta\sin\varphi, \quad z = r\cos\theta. \tag{23.5}$$

The Jacobian J of a transformation is the determinant of the matrix of partial derivatives,

$$J = \frac{\partial(x, y, z)}{\partial(r, \theta, \varphi)} = \begin{vmatrix} \partial x/\partial r & \partial x/\partial\theta & \partial x/\partial\varphi \\ \partial y/\partial r & \partial y/\partial\theta & \partial y/\partial\varphi \\ \partial z/\partial r & \partial z/\partial\theta & \partial\varphi/\partial\varphi \end{vmatrix}. \tag{23.6}$$

By using the identity $\sin^2 u + \cos^2 u = 1$, the Jacobian becomes

$$J = \begin{vmatrix} \sin\theta\cos\varphi & r\cos\theta\cos\varphi & -r\sin\theta\sin\varphi \\ \sin\theta\sin\varphi & r\cos\theta\sin\varphi & r\sin\theta\cos\varphi \\ \cos\theta & -r\sin\theta & 0 \end{vmatrix} = r^2\sin\theta. \tag{23.7}$$

Thus,

$$\rho_{r\theta\varphi}(r, \theta, \varphi) = \rho_{xyz}(x, y, z)r^2\sin\theta. \tag{23.8}$$

Generalized Coordinates and Momenta

Since probability densities are multiplied by a Jacobian when the coordinates are transformed, we need to know what kind of coordinates can be used for describing thermodynamic systems. In fact, any set of *generalized coordinates and momenta* can be used. This freedom of choice is possible because the Jacobian of the transformation between different sets of coordinates and momenta is one ($J = 1$).

Once the coordinates have been selected, the generalized momenta are obtained by differentiating the Lagrangian. A system's Lagrangian is its kinetic energy minus the potential energy. When Cartesian coordinates (x, y, z) are used, the Lagrangian for a particle of mass m in a spherically symmetric potential $\phi(r)$ is

$$L(\dot{x}, \dot{y}, \dot{z}, x, y, z) = \frac{1}{2}m(\dot{x}^2 + \dot{y}^2 + \dot{z}^2) - \phi(r), \tag{23.9}$$

where $r = r(x, y, z)$. The velocities are $(\dot{x}, \dot{y}, \dot{z})$, and $\phi(r)$ is the potential energy of the particle at a distance $r(x, y, z)$ from point $(0, 0, 0)$. The generalized momenta are

$$p_x = \frac{\partial L}{\partial \dot{x}} = m\dot{x}, \quad p_y = \frac{\partial L}{\partial \dot{y}} = m\dot{y}, \quad p_z = \frac{\partial L}{\partial \dot{z}} = m\dot{z}. \tag{23.10}$$

The Hamiltonian function is $H = \dot{p}_x\dot{x} + \dot{p}_y\dot{y} + \dot{p}_z\dot{z} - L$, where $(\dot{x}, \dot{y}, \dot{z})$ are expressed as functions of (p_x, p_y, p_z). By using (23.10), we find that

$$H(p_x, p_y, p_z, x, y, z) = \frac{1}{2m}{p_x}^2 + \frac{1}{2m}{p_y}^2 + \frac{1}{2m}{p_z}^2 + \phi(r(x, y, z)). \tag{23.11}$$

This is the Hamiltonian for a single particle in Cartesian coordinates. We now find the corresponding Hamiltonian when spherical coordinates (r, θ, φ) are used.

The equations that relate the Cartesian coordinate to spherical coordinates are $x = r\sin\theta\cos\varphi$, $y = r\sin\theta\sin\varphi$, and $z = r\cos\theta$, and their time derivatives are

$$\begin{aligned}
\dot{x} &= \dot{r}\sin\theta\cos\varphi + \dot{\theta}r\cos\theta\cos\varphi - \dot{\varphi}r\sin\theta\sin\varphi, \\
\dot{y} &= \dot{r}\sin\theta\sin\phi + \dot{\theta}r\cos\theta\sin\varphi + \dot{\varphi}r\sin\theta\cos\varphi, \\
\dot{z} &= \dot{r}\cos\theta - \dot{\theta}r\sin\theta.
\end{aligned} \tag{23.12}$$

The kinetic energy in spherical coordinates is obtained by substituting these expressions for $(\dot{x}, \dot{y}, \dot{z})$ into the kinetic energy $\frac{1}{2}m(\dot{x}^2 + \dot{y}^2 + \dot{z}^2)$. By using the identity $\sin^2 u + \cos^2 u = 1$, we obtain

$$K(\dot{r}, \dot{\theta}, \dot{\varphi}, r, \theta, \varphi) = \frac{1}{2}m(\dot{r}^2 + \dot{\theta}^2 r^2 + \dot{\varphi}^2 r^2 \sin^2\theta). \tag{23.13}$$

Thus, in spherical coordinates the Lagrangian is

$$L(\dot{r}, \dot{\theta}, \dot{\varphi}, r, \theta, \varphi) = \frac{1}{2}m(\dot{r}^2 + \dot{\theta}^2 r^2 + \dot{\varphi}^2 r^2 \sin^2\theta) - \phi(r), \tag{23.14}$$

and the generalized momenta are

$$p_r = \frac{\partial L}{\partial \dot{r}} = m\dot{r}, \quad p_\theta = \frac{\partial L}{\partial \dot{\theta}} = mr^2\dot{\theta}, \quad p_\varphi = \frac{\partial L}{\partial \dot{\varphi}} = mr^2\sin^2\theta\dot{\varphi}. \tag{23.15}$$

The Hamiltonian function is $H = p_r\dot{r} + p_\theta\dot{\theta} + p_\varphi\dot{\varphi} - L$, where $(\dot{r}, \dot{\theta}, \dot{\varphi})$ are expressed as functions of $(p_r, p_\theta, p_\varphi, r, \theta, \varphi)$. It follows from (23.15) that

$$\dot{r} = \frac{p_r}{m}, \quad \dot{\theta} = \frac{p_\theta}{mr^2}, \quad \dot{\varphi} = \frac{p_\varphi}{mr^2\sin^2\theta}. \tag{23.16}$$

The resulting Hamiltonian for a particle in a spherical symmetric potential is

$$H(p_r, p_\theta, p_\varphi, r, \theta, \varphi) = \frac{1}{2m}p_r^2 + \frac{1}{2mr^2}p_\theta^2 + \frac{1}{2mr^2\sin^2\theta}p_\varphi^2 + \phi(r). \tag{23.17}$$

Notice that the generalized momenta $(p_r, p_\theta, p_\varphi)$ are not equal to mass times velocity and that the kinetic energy is a function of both the momentum variables and the position variables.

The Jacobian of the transformation from (p_x, p_y, p_z, x, y, z) to $(p_r, p_\theta, p_\varphi, r, \theta, \varphi)$ is

$$J = \begin{vmatrix} \partial p_x/\partial p_r & \partial p_x/\partial p_\theta & \partial p_x/\partial p_\varphi & \partial p_x/\partial r & \partial p_x/\partial \theta & \partial p_x/\partial \varphi \\ \partial p_y/\partial p_r & \partial p_y/\partial p_\theta & \partial p_y/\partial p_\varphi & \partial p_y/\partial r & \partial p_y/\partial \theta & \partial p_y/\partial \varphi \\ \partial p_z/\partial p_r & \partial p_z/\partial p_\theta & \partial p_z/\partial p_\varphi & \partial p_z/\partial r & \partial p_z/\partial \theta & \partial p_z/\partial \varphi \\ 0 & 0 & 0 & \partial x/\partial r & \partial x/\partial \theta & \partial x/\partial \varphi \\ 0 & 0 & 0 & \partial y/\partial r & \partial y/\partial \theta & \partial y/\partial \varphi \\ 0 & 0 & 0 & \partial z/\partial r & \partial z/\partial \theta & \partial z/\partial \varphi \end{vmatrix}. \tag{23.18}$$

The nine zeros at the lower left of this reflects that (x, y, z) are not functions of $(p_r, p_\theta, p_\varphi)$. Because of these zeros, the 6×6 determinant can be factored into a product of two 3×3 determinants,

$$J = \begin{vmatrix} \partial p_x/\partial p_r & \partial p_x/\partial p_\theta & \partial p_x/\partial p_\varphi \\ \partial p_y/\partial p_r & \partial p_y/\partial p_\theta & \partial p_y/\partial p_\varphi \\ \partial p_z/\partial p_r & \partial p_z/\partial p_\theta & \partial p_z/\partial p_\varphi \end{vmatrix} \times \begin{vmatrix} \partial x/\partial r & \partial x/\partial \theta & \partial x/\partial \varphi \\ \partial y/\partial r & \partial y/\partial \theta & \partial y/\partial \varphi \\ \partial z/\partial r & \partial z/\partial \theta & \partial \varphi/\partial \varphi \end{vmatrix} = J_m J_c, \tag{23.19}$$

where J_m and J_c are the determinants for the momentum and positions variables, respectively. The determinant for the position variables from (23.7) is $r^2 \sin\theta$. To find the determinant for the momentum variables, we express (p_x, p_y, p_z) as functions of $(p_r, p_\theta, p_\varphi, r, \theta, \varphi)$ by multiplying the equations for $(\dot{x}, \dot{y}, \dot{z})$ in (23.12) by m, which gives us equations for (p_x, p_y, p_z). Equation (23.16) can then be used to replace the dependence on $(\dot{r}, \dot{\theta}, \dot{\varphi})$ by a dependence on $(p_r, p_\theta, p_\varphi, r, \theta, \varphi)$. After some algebraic manipulations, we find that

$$J_m = \frac{1}{r^2 \sin\theta}. \tag{23.20}$$

Thus, the Jacobian $J = J_m J_c$ for the transformation from the generalized coordinates and momenta (p_x, p_y, p_z, x, y, z) to the set $(p_r, p_\theta, p_\varphi, r, \theta, \varphi)$ is one ($J = 1$), as indicated above.

Generalized coordinates are commonly represented in analytical mechanics by q's and the generalized momenta by p_j. Since any set of generalized coordinates and momenta can be used to describe a thermodynamic system, the general expression for the canonical probability density for systems of N unconstrained particles in three-space is

$$\rho(\cdots p_j \cdots q_j \cdots) = e^{-H_N(\cdots p_j \cdots q_j \cdots)/k_B T} \frac{C_N}{Z_N}, \tag{23.21}$$

and the partition function is

$$Z_N = C_N \int dp^{3N} \int dq^{3N} e^{-H_N(\cdots p_i \cdots q_i \cdots)/k_B T}. \tag{23.22}$$

Constraints such as those for the rigid dumbbell model of diatomic gases reduce the number of generalized coordinates and momenta.

23.2 Diatomic Gases

The ideal gas model has been used to predict the properties of monatomic gases. The study of diatomic gases further illustrates how the principles of statistical mechanics are applied and demonstrates the value of generalized coordinates. We can still consider each molecule to be statistically independent, but now we must account for interactions between the two atoms within a molecule. The Hamiltonian for a gas of N diatomic molecules is

$$H_N(\boldsymbol{p}_{i,1},\boldsymbol{p}_{i,2},\ \ldots\ ,\boldsymbol{r}_{N,1},\boldsymbol{r}_{N,2}) = \sum_{i=1}^{N} H_i(\boldsymbol{p}_{i,1},\boldsymbol{p}_{i,2},\boldsymbol{r}_{i,1},\boldsymbol{r}_{i,2}), \tag{23.23}$$

where the first subscript labels the molecules involved and the second subscript identifies the two atoms. In Cartesian coordinates, the Hamiltonian for molecule i is

$$H_i(\boldsymbol{p}_{i,1},\boldsymbol{p}_{i,2},\boldsymbol{r}_{i,1},\boldsymbol{r}_{i,2}) = \frac{1}{2m}|\boldsymbol{p}_{i,1}|^2 + \frac{1}{2m}|\boldsymbol{p}_{i,2}|^2 + \phi(|\boldsymbol{r}_{i,2} - \boldsymbol{r}_{i,1}|) + \phi_{a\text{-}w}. \tag{23.24}$$

The potential energy that binds the two atoms together is $\phi(|\boldsymbol{r}_{i,2} - \boldsymbol{r}_{i,1}|)$, and the interaction of the molecule with the walls of the container is represented by $\phi_{a\text{-}w}$.

Since there are $6N$ coordinates and $6N$ momenta, the classical partition function in (22.43) becomes

$$Z_N = \frac{1}{h^{6N}N!}\int dp^{6N}\int dr^{6N}e^{-H_N/k_BT}. \tag{23.25}$$

When the molecules are of the same type, the partition function can be factorized as $Z_N = [Z_s]^N$, where

$$Z_s = \frac{1}{h^6}\left(\frac{1}{N!}\right)^{1/N}\int dp_1^3\int dp_2^3\int dr_1^3\int dr_2^3 e^{-H_s(\boldsymbol{p}_1,\boldsymbol{p}_2,\boldsymbol{r}_1,\boldsymbol{r}_2)/k_BT}, \tag{23.26}$$

where subscript s identifies Z_s and H_s as *single molecule* properties. The Helmholtz free energy of the gas is $F = -Nk_BT\ln Z_s$.

The Lagrangian of a molecule in Cartesian coordinates is

$$L_s(\dot{\boldsymbol{r}}_1,\dot{\boldsymbol{r}}_2,\boldsymbol{r}_1,\boldsymbol{r}_2) = \frac{1}{2}m_1|\dot{\boldsymbol{r}}_1|^2 + \frac{1}{2}m_2|\dot{\boldsymbol{r}}_2|^2 - \phi(|\boldsymbol{r}_2 - \boldsymbol{r}_1|) - \phi_{a\text{-}w}. \tag{23.27}$$

The generalized momenta are $\boldsymbol{p}_1 = m_1\dot{\boldsymbol{r}}_1$ and $\boldsymbol{p}_2 = m_2\dot{\boldsymbol{r}}_2$, and the Hamiltonian is

$$H_s(\boldsymbol{p}_1,\boldsymbol{p}_2,\boldsymbol{r}_1,\boldsymbol{r}_2) = \frac{1}{2m_1}|\boldsymbol{p}_1|^2 + \frac{1}{2m_2}|\boldsymbol{p}_2|^2 + \phi(|\boldsymbol{r}_2 - \boldsymbol{r}_1|) + \phi_{a\text{-}w}. \tag{23.28}$$

Center of Mass Motion

The evaluation of the single-molecule partition function is simplified by using the *center of mass* $\boldsymbol{R}$ and *relative displacement* $\boldsymbol{r}$, where $\boldsymbol{R} = (X, Y, Z)$ and $\boldsymbol{r} = (x, y, z)$, which are related to the positions $\boldsymbol{r}_1 = (x_1, y_1, z_1)$ and $\boldsymbol{r}_2 = (x_2, y_2, z_2)$ of the individual atoms by

$$\boldsymbol{R} = \frac{m_1\boldsymbol{r}_1 + m_2\boldsymbol{r}_2}{M} \quad \text{and} \quad \boldsymbol{r} = \boldsymbol{r}_2 - \boldsymbol{r}_1, \tag{23.29}$$

where $M = m_1 + m_2$ is the mass of the molecule. Solving for $\boldsymbol{r}_1$ and $\boldsymbol{r}_2$ gives

$$\boldsymbol{r}_1 = \boldsymbol{R} + \frac{m_2}{M}\boldsymbol{r} \quad \text{and} \quad \boldsymbol{r}_2 = \boldsymbol{R} - \frac{m_1}{M}\boldsymbol{r}, \tag{23.30}$$

and their time derivatives are

$$\dot{\boldsymbol{r}}_1 = \dot{\boldsymbol{R}} + \frac{m_2}{M}\dot{\boldsymbol{r}} \quad \text{and} \quad \dot{\boldsymbol{r}}_2 = \dot{\boldsymbol{R}} - \frac{m_1}{M}\dot{\boldsymbol{r}}. \tag{23.31}$$

Substituting these into the above Lagrangian and rearranging of terms gives

$$L_s(\dot{\boldsymbol{R}}, \dot{\boldsymbol{r}}, \boldsymbol{R}, \boldsymbol{r}) = \frac{1}{2}M|\dot{\boldsymbol{R}}|^2 + \frac{1}{2}\mu|\dot{\boldsymbol{r}}|^2 - \phi(|\boldsymbol{r}|) - \phi_{a\text{-}w}, \tag{23.32}$$

where $\mu = m_1 m_2/(m_1 + m_2)$ is the *reduced mass*. The generalized momenta are

$$\boldsymbol{P} = M\dot{\boldsymbol{R}} \quad \text{and} \quad \boldsymbol{p} = \mu\dot{\boldsymbol{r}}. \tag{23.33}$$

The *total momentum* is $\boldsymbol{P} = (P_x, P_y, P_z)$ and *relative momentum* is $\boldsymbol{p} = (p_x, p_y, p_z)$. By expressing $(\dot{\boldsymbol{R}}, \dot{\boldsymbol{r}})$ as functions of $(\boldsymbol{P}, \boldsymbol{p})$, the Hamiltonian $H_s = \boldsymbol{P} \cdot \dot{\boldsymbol{R}} + \boldsymbol{p} \cdot \dot{\boldsymbol{r}} - L_s$ becomes

$$H_s(\boldsymbol{P}, \boldsymbol{p}, \boldsymbol{R}, \boldsymbol{r}) = \frac{1}{2M}|\boldsymbol{P}|^2 + \frac{1}{2\mu}|\boldsymbol{p}|^2 + \phi(|\boldsymbol{r}|) + \phi_{a\text{-}w}, \tag{23.34}$$

where $r = |\boldsymbol{r}|$ is the distance between atoms.

As the distance between the atoms in a molecule is much less than the dimensions of the container, the interaction between a molecule and the container walls can be treated as a function of the center of mass position $\boldsymbol{R}$. The Hamiltonian now separates into relative motion and center of mass Hamiltonians,

$$H_s(\boldsymbol{P}, \boldsymbol{p}, \boldsymbol{R}, \boldsymbol{r}) = H_{\text{cm}}(\boldsymbol{P}, \boldsymbol{R}) + H_{\text{rel}}(\boldsymbol{p}, \boldsymbol{r}), \tag{23.35}$$

where

$$H_{\text{rel}}(\boldsymbol{p}, \boldsymbol{r}) = \frac{1}{2\mu}|\boldsymbol{p}|^2 + \phi(|\boldsymbol{r}|) \tag{23.36}$$

and

$$H_{\text{cm}}(\boldsymbol{P}, \boldsymbol{R}) = \frac{1}{2M}|\boldsymbol{P}|^2 + \phi_{a\text{-}w}(\boldsymbol{R}). \tag{23.37}$$

The single-molecule partition function becomes

$$Z_s(T, V) = Z_{\text{cm}}(T, V) Z_{\text{rel}}(T), \tag{23.38}$$

where

$$Z_{\rm rel}(T) = \frac{1}{h^3}\int dp^3 \int dr^3 e^{-H_{\rm rel}(\boldsymbol{p},\boldsymbol{r})/k_BT} \tag{23.39}$$

and

$$Z_{\rm cm}(T,V) = (C_N)^{1/N}\int dP^3 \int dR^3 e^{-H_{\rm cm}(\boldsymbol{P},\boldsymbol{R})/k_BT}, \tag{23.40}$$

where $C_N = 1/(h^{3N}N!)$. The volume dependence of $Z_{\rm cm}(T,V)$ enters through $\phi_{a\text{-}w}(\boldsymbol{R})$, and $Z_{\rm rel}(T)$ is independent of volume. The free energy of the gas is

$$F = -Nk_BT\ln Z_s = -Nk_BT(\ln Z_{\rm rel} + \ln Z_{\rm cm}). \tag{23.41}$$

Equation of State

The center of mass partition function $Z_{\rm cm}$ is the same as the partition function Z_s for a monatomic ideal gas discussed in Section 19.5 except that the integration variables are $\boldsymbol{P}$ and $\boldsymbol{R}$ instead of $\boldsymbol{p}$ and $\boldsymbol{r}$, and the mass is M instead of m. By replacing m with M in (19.64) and using $\beta = 1/k_BT$, the center of mass partition function becomes

$$Z_{\rm cm} = (C_N)^{1/N}(2\pi Mk_BT)^{3/2}V. \tag{23.42}$$

Since $Z_{\rm rel}$ is independent of V, it follows from $F = -Nk_BT\ln Z_s$ that the pressure of the gas is

$$P = -\left(\frac{\partial F}{\partial V}\right)_T = Nk_BT\left(\frac{\partial \ln Z_{\rm cm}}{\partial V}\right)_T = \frac{Nk_BT}{V}, \tag{23.43}$$

so that $PV = Nk_BT$. Thus, the system obeys the idea gas equation of state, a result that is valid for molecules with any number of atoms, provided the interactions between different molecules can be neglected.

Specific Heat

The relative motion of the atoms within the molecules does not contribute to the equation of state but does contribute to the specific heat. The effect can be found by using the relationships

$$C_V = \left(\frac{\partial U}{\partial T}\right)_V \quad \text{and} \quad U = -\left(\frac{\partial \ln Z_N}{\partial \beta}\right)_V. \tag{23.44}$$

Since $Z_N = [Z_s]^N$ and $Z_s = Z_{\rm cm}Z_{\rm rel}$, the partition function of the gas is $Z_N = [Z_{\rm cm}Z_{\rm rel}]^N$ and the internal energy is $U = N(u_{\rm cm} + u_{\rm rel})$, where

$$u_{\rm cm} = -\frac{\partial \ln Z_{\rm cm}}{\partial \beta} \quad \text{and} \quad u_{\rm rel} = -\frac{\partial \ln Z_{\rm rel}}{\partial \beta}. \tag{23.45}$$

The constant-volume heat capacity of the system is $C_V = N(c_V^{\rm cm} + c_V^{\rm rel})$, where

$$c_V^{\rm cm} = \left(\frac{\partial u_{\rm cm}}{\partial T}\right)_V \quad \text{and} \quad c_V^{\rm rel} = \left(\frac{\partial u_{\rm rel}}{\partial T}\right)_V. \tag{23.46}$$

Since $Z_{\rm cm}$ is proportional to $(k_B T)^{\frac{3}{2}} = \beta^{-\frac{3}{2}}$ and $\beta = 1/k_B T$, the contributions of the center of mass motion to the internal energy per molecule and heat capacity per molecule (specific heat) are

$$u_{\rm cm} = \frac{3}{2\beta} = \frac{3}{2}k_B T \quad \text{and} \quad c_V^{\rm cm} = \frac{3}{2}k_B. \tag{23.47}$$

Rotations and Vibrations

The relative motion Hamiltonian $H_{\rm rel}$ in (23.36) is the same as the Hamiltonian for a particle in a spherical symmetric potential in (23.17) except that m is replaced by μ,

$$H_{\rm rel}(p_r, p_\theta, p_\varphi, r, \theta, \varphi) = \frac{1}{2\mu}p_r^2 + \frac{1}{2\mu r^2}p_\theta^2 + \frac{1}{2\mu r^2 \sin^2\theta}p_\varphi^2 + \phi(r). \tag{23.48}$$

As shown in Figure 19.1, a typical interatomic potential $\phi(r)$ has a minimum at $r = r_o$, becomes large at small distances ($r < r_o$), and goes to zero at large distances ($r > r_o$). Because of the rapid increase in the value of $\phi(r)$ away from its minimum, the distance between the atoms r is approximately r_o. This suggests that we approximate the r in the kinetic energy terms by r_o. When this is done, the Hamiltonian separates into independent rotational and vibrational parts, so that

$$H_{\rm rel}(p_r, p_\theta, p_\varphi, r, \theta) = H_{\rm rot}(p_\theta, p_\varphi, \theta, \varphi) + H_r(p_r, r), \tag{23.49}$$

where

$$H_{\rm rot}(p_\theta, p_\varphi, \theta, \varphi) = \frac{p_\theta^2}{2\mu r_o{}^2} + \frac{p_\varphi^2}{2\mu r_o{}^2 \sin^2\theta} \tag{23.50}$$

and

$$H_{\rm vib}(p_r, r) = \frac{1}{2\mu}p_r^2 + \phi(r). \tag{23.51}$$

When expressed in terms of $\beta = 1/k_B T$, the relative motion partition function becomes

$$Z_{\rm rel} = Z_{\rm rot} Z_{\rm vib}, \tag{23.52}$$

where

$$Z_{\rm vib} = \frac{1}{h}\int_0^{+\infty} dr e^{-\beta\phi(r)} \int_{-\infty}^{+\infty} dp_r e^{-\beta\frac{1}{2\mu}p_r^2} \tag{23.53}$$

and

$$Z_{\rm rot} = \frac{1}{h^2}\int_0^{\pi} d\theta \int_0^{2\pi} d\varphi \int_{-\infty}^{+\infty} dp_\theta e^{-\beta p_\theta^2/(2\mu r_o{}^2)} \int_{-\infty}^{+\infty} dp_\varphi e^{-\beta p_\varphi^2/(2\mu r_o^2 \sin^2\theta)}. \tag{23.54}$$

Evaluating the integral over p_θ in the rotational partition function $Z_{\rm rot}$ gives $(2\pi\mu/\beta)^{1/2} r_o$. Then, the integral over p_φ gives $(2\pi\mu/\beta)^{1/2} r_o \sin\theta$, so that

$$Z_{\rm rot} = \left(\frac{2\pi\mu}{\beta}\right)\left(\frac{r_o}{h}\right)^2 \int_0^{\pi} d\theta \int_0^{2\pi} d\varphi \sin\theta. \tag{23.55}$$

The integrations over θ or φ yield a factor of 4π,

$$Z_{\text{rot}} = 8\pi^2\mu\left(\frac{r_o}{h}\right)^2\frac{1}{\beta}. \tag{23.56}$$

Thus, the contributions of rotations to the internal energy and the specific heat are

$$u_{\text{rot}} = -\frac{\partial \ln Z_{\text{rot}}}{\partial\beta} = \beta^{-1} = k_B T \quad \text{and} \quad c_V^{\text{rot}} = \frac{\partial u_{\text{rot}}}{\partial T} = k_B. \tag{23.57}$$

Rigid Dumbbell Model

In the rigid dumbbell model, the distance between the atoms is fixed at r_o. As a result, p_r and r are no longer integration variables in Z_{vib}. By evaluating $\phi(r)$ at r_o and setting $p_r = 0$, the vibrational partition function in (23.53) becomes $Z_{\text{vib}} = e^{\beta\varepsilon_o}$, where $\varepsilon_o = -\phi(r_o)$. The contribution to the internal energy is

$$u_{\text{vib}} = -\varepsilon_o \quad \text{and} \quad c_V^{\text{vib}} = 0. \tag{23.58}$$

Combining this with the values for c_V^{cm} and $c_V^{\text{rel}} = c_V^{\text{rot}} + c_V^{\text{vib}}$ gives

$$C_V = N(c_V^{\text{cm}} + c_V^{\text{rot}} + c_V^{\text{vib}}) = \frac{5}{2}Nk_B T. \tag{23.59}$$

This is the equipartition value for the rigid dumbbell model, which assigns three degrees of freedom to the translational motion of a molecule and two degrees of freedom to the rotational motion (see Section 15.3).

Vibrating Molecules

Because of the rapid increase in the interatomic potential $\phi(r)$ away from its minimum, it can be approximated by expanding in a Taylor series and neglecting higher order terms. Keeping terms through $(r - r_o)^2$ gives the *harmonic approximation*

$$\phi(r) \approx -\varepsilon_o + \frac{1}{2}k_o(r - r_o)^2. \tag{23.60}$$

The depth of the potential well is $\varepsilon_o = -\phi(r_o)$, the first derivative vanishes because the expansion is about the minimum, and the second derivative $k_o = d^2\phi/dr^2$ is an effective force constant (spring constant). By using the harmonic approximation, the vibrational partition function in (23.53) becomes

$$Z_{\text{vib}} = \frac{e^{\beta\varepsilon_o}}{h}\int_0^{\infty} dr e^{-\frac{1}{2}\beta k_o (r-a)^2}\int_{-\infty}^{+\infty} dp_r e^{-\beta\frac{1}{2\mu}p_r^2}. \tag{23.61}$$

The lower integration limit can be extended to $-\infty$, so that both integrations are integrals of Gaussians from $-\infty$ to $+\infty$. Thus,

$$Z_{\text{vib}} = \frac{e^{\beta\varepsilon_o}}{h}\sqrt{\frac{2\pi}{\beta k_o}}\sqrt{\frac{2\pi\mu}{\beta}} = \frac{2\pi}{h\omega_o}\frac{e^{\beta\varepsilon_o}}{\beta}, \tag{23.62}$$

where $\omega_o = \sqrt{k_o/\mu}$ is the angular frequency of a classical oscillator. The contributions of vibrations to the internal energy and the specific heat are

$$u_{\text{vib}} = -\frac{\partial \ln Z_{\text{vib}}}{\partial \beta} = -\varepsilon_o + \frac{1}{\beta} \quad \text{and} \quad c_V^{\text{vib}} = \frac{\partial u_{\text{vib}}}{\partial T} = k_B. \tag{23.63}$$

Combining c_V^{vib} with c_V^{cm} and c_V^{rot} gives

$$C_V = N(c_V^{\text{cm}} + c_V^{\text{rot}} + c_V^{\text{vib}}) = \frac{7}{2}Nk_B. \tag{23.64}$$

This is the equipartition value of C_V for a gas of vibrating molecules, which assigns two degrees of freedom to the vibrational motion of each molecule.

23.3 Quantum Effects

Classically, there is no theoretical reason for the rigid dumbbell model. Nevertheless, it is needed to obtain agreement with the experimental data in Table 15.2. Its justification comes from quantum mechanics. To demonstrate this, we represent the amount by which the atoms are separated from r_o by $q = r - r_o$, so that the harmonic approximation in (23.60) becomes $\phi(q) = -\varepsilon_o + \frac{1}{2}m\omega_o^2 q^2$, where $\omega_o = \sqrt{k_o/\mu}$. Then, the vibrational Hamiltonian in (23.51) becomes

$$\widehat{H}_{\text{vib}} = -\varepsilon_o + \frac{1}{2\mu}p^2 + \frac{1}{2}\mu\omega_o^2 q^2. \tag{23.65}$$

This is the Hamiltonian for a harmonic oscillator of mass μ and classical frequency ω_o with its energy shifted by $-\varepsilon_o$ As in (20.47), the energy eigenvalues of a harmonic oscillator are $\left(n + \frac{1}{2}\right)\hbar\omega_o$, where the allowed values of quantum number n are $n = 0, 1, 2, \ldots$ Thus, the energies of a system described by $\widehat{H}_{\text{vib}}$ are

$$\varepsilon_n = -\varepsilon_o + \left(n + \frac{1}{2}\right)\hbar\omega_o, \tag{23.66}$$

where $\hbar = h/2\pi$ and the spacing between eigenvalues is $\hbar\omega_o$. The quantum partition function is

$$Z_{\text{vib}} = \sum_{n=0}^{\infty} e^{-\beta\varepsilon_n} = \sum_{n=0}^{\infty} e^{-\beta\left[-\varepsilon_o + \left(n+\frac{1}{2}\right)\hbar\omega_o\right]} = e^{\beta\left(\varepsilon_o - \frac{1}{2}\hbar\omega_o\right)} \sum_{n=0}^{\infty} (e^{-\beta\hbar\omega_o})^n. \tag{23.67}$$

The sum over n can be evaluated by using the expansion $(1 - X)^{-1} = \sum_n X^n$, so that

$$Z_{\text{vib}} = e^{\beta\left(\varepsilon_o - \frac{1}{2}\hbar\omega_o\right)}(1 - e^{-\beta\hbar\omega_o})^{-1}. \tag{23.68}$$

By using $\beta = 1/k_B T$, the vibrational contribution to the internal energy becomes

$$u_{\text{vib}} = -\frac{\partial \ln Z_{\text{vib}}}{\partial \beta} = -\varepsilon_o + \frac{1}{2}\hbar\omega_o + \frac{e^{-\beta\hbar\omega_o}\hbar\omega_o}{1 - e^{-\beta\hbar\omega_o}} = -\varepsilon_o + \frac{1}{2}\hbar\omega_o + \frac{\hbar\omega_o}{e^{\hbar\omega_o/k_B T} - 1}, \tag{23.69}$$

and the specific heat is

$$c_V^{\text{vib}} = \frac{\partial u_{\text{vib}}}{\partial T} = \left(\frac{\hbar\omega_o}{e^{\hbar\omega_o/k_BT} - 1}\right)^2 \frac{e^{\hbar\omega_o/k_BT}}{k_BT^2} = 4k_B\left(\frac{x_o}{e^{+x_o} - e^{-x_o}}\right)^2, \tag{23.70}$$

where $x_o = \hbar\omega_o/(2k_BT)$.

At high temperatures ($k_BT \gg \hbar\omega_o$), the exponentials $e^{\pm x_o}$ can be approximated by $e^{\pm x_o} = 1 \pm x_o + \cdots \approx 1$. Thus, at high temperatures, this contribution to the specific heat becomes $c_V^{\text{vib}} \approx k_B$, which is the classical contribution of vibrating molecules given in (23.63). At low temperatures ($k_BT \ll \hbar\omega_o$), the exponential e^{+x_o} becomes large, so that the contribution to the specific heat vanishes ($c_V^{\text{vib}} \approx 0$), which is the classical value of c_V^{vib} in the rigid dumbbell model. Thus, when the spacing between eigenvalues $\hbar\omega_o$ is significantly greater than the thermal energy k_BT, the vibrational motion of the molecules is frozen out.

Hydrogen Gas

Because of the small mass of the atoms, the transition to classical behavior occurs at higher temperatures in hydrogen than in other diatomic gases. Quantum effects can be seen in the contributions to the specific heat from both the vibrational and rotational motion of the molecules. The rotational contribution is found by using the energy eigenvalues of a rigid rotator, which are

$$\varepsilon_{j\text{m}} = \frac{\hbar^2}{2I} j(j+1), \tag{23.71}$$

where I is the moment of inertia of the molecule. The quantum numbers are j and m and their allowed values are $j = 0, 1, 2, \ldots$ and $m = -j, \ldots, 0, \ldots, +j$. The energy is determined by j and is independent of m. When treated quantum mechanically, the rotational partition function is

$$Z_{\text{rot}} = \sum_{jm} e^{-\beta\varepsilon_{jm}}. \tag{23.72}$$

Since m has $2j+1$ different values for each j, the degeneracy of energy level j is $2j+1$, so that summing over m gives

$$Z_{\text{rot}} = \sum_{j=0}^{\infty} e^{-\frac{1}{2}\beta\varepsilon_o j(j+1)}(2j+1), \tag{23.73}$$

where $\varepsilon_o = \hbar^2/I$ is the energy difference between the two lowest energy levels. The value of ε_o/k_B for hydrogen is 175 K. For two atoms of the same mass, the moment of inertia for rotation about the center of mass is $I = \mu r_o^2$, where the reduced mass μ is half the mass of an atom. Notice that Z_{rot} is a sum of a product of the Boltzmann factors of the energy levels times their degeneracy, which is a common feature of the quantum partition function. By using $\beta = 1/k_BT$, the rotational specific heat $\partial u_{\text{rot}}/\partial T$ becomes $-k_B\beta^2(\partial u_{\text{rot}}/\partial\beta)$. Then, from $u_{\text{rot}} = -\partial \ln Z_{\text{rot}}/\partial\beta$, we obtain

$$c_V^{\text{rot}} = k_B\beta^2\left[\frac{1}{Z_{\text{rot}}}\frac{\partial^2 Z_{\text{rot}}}{\partial\beta^2} - \left(\frac{1}{Z_{\text{rot}}}\frac{\partial Z_{\text{rot}}}{\partial\beta}\right)^2\right]. \tag{23.74}$$

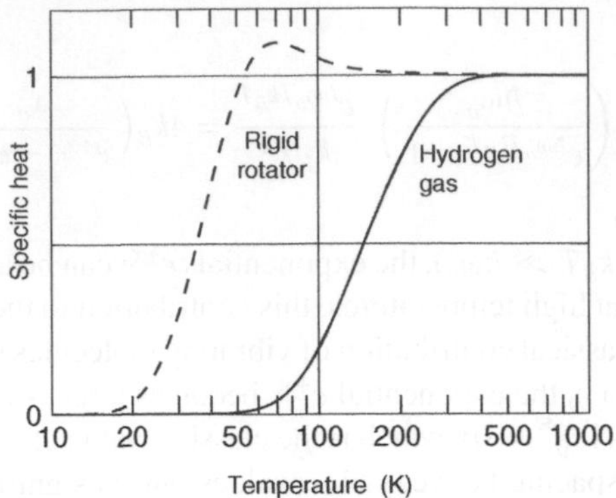

Figure 23.1 *Rotational Contribution c_V^{rot}/k_B to the Specific Heat*

The temperature dependence of c_V^{rot}/k_B predicted by (23.73) and (23.74) is represented by the dashed line in **Figure 23.1**.

It was found in the early twentieth century that the specific heat indicated by the dashed line does not agree with the experimental data. The inconsistency results from a property of quantum mechanical systems that was not understood at the time. As the molecule consists of two indistinguishable particles, its wave function is required to be either symmetric or antisymmetric, as discussed in Chapter 26. Agreement with the data is obtained when the energy levels and their degeneracies were correctly accounted for. The corrected values are indicated by the solid curve. Both curves indicate that the rotational motion of the molecules is frozen out at low temperatures.

The temperature dependence of the specific heat of hydrogen is shown **Figure 23.2** with the effects of exchange symmetry accounted for. The value of $(c_V^{\text{cm}} + c_V^{\text{rot}} + c_V^{\text{vib}})/k_B$ is

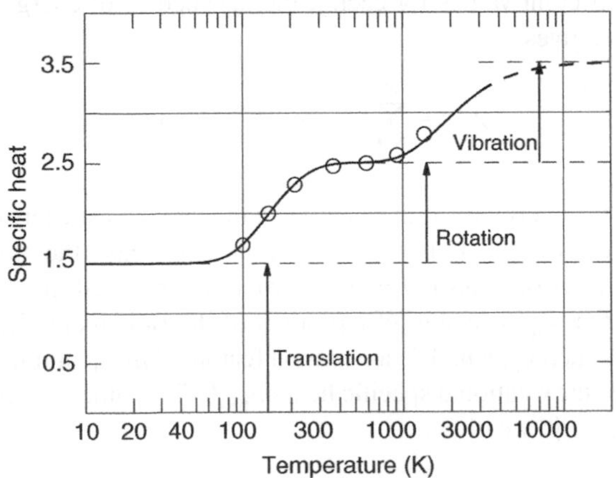

Figure 23.2 *Specific Heat of Hydrogen Gas*

plotted versus the logarithm of temperature and the circles are experimental data. The figure indicates that hydrogen gas behaves like a monatomic gas below 70 K and that the molecules behaves like rigid dumbbells between 300 and 600 K. Vibrational motion begins to be apparent at about 1000 K, but the molecules start to dissociate before their behavior becomes fully classical in character.

23.4 Density Matrices

The use of energy eigenstates to evaluate the partition function does not imply that the systems are actually in one of those states. Determining the energy of a thermodynamic system is not a measurement in the sense of the quantum mechanical measurement theory. Energy eigenstates are used because they are a convenient representation of the more general relationship between thermodynamic properties and the Hamiltonian expressed in the density matrix formalism.

The concept of a density matrix is conveniently introduced by considering an ensemble of systems described by wave functions, instead of eigenstates. For example, the members of an ensemble could be characterized by wave functions $\Psi_j(\boldsymbol{r}_1 \cdots \boldsymbol{r}_N)$, where j labels the members of the ensemble ($j = 1, 2, \ldots, J$). Index j is used (instead of α, μ, and so on) to indicate that the wave functions may not be the members of a complete set and may not be orthogonal to each other. Each $\Psi_j(\boldsymbol{r}_1 \cdots \boldsymbol{r}_N)$ is assumed to be normalized. In Dirac *bra* and *ket* notation, the expectation value (average value) of a quantum mechanical observable of a system with wave function Ψ_j is $\langle \Psi_j | \hat{X} | \Psi_j \rangle$, where $\hat{X}$ is the Hermitian operator associated with the property. The *ensemble average* $\langle X \rangle$ of property X is obtained by averaging the expectation values of the J members of the ensemble,

$$\langle X \rangle = \frac{1}{J} \sum_{j=1}^{J} \left\langle \Psi_j \middle| \hat{X} \middle| \Psi_j \right\rangle. \tag{23.75}$$

To introduce density matrices, we expand each wave function Ψ_j in a complete series of orthonormal functions Ψ_μ defined on the space used to characterize the system, which for N-particle systems are functions of the position variable $(\boldsymbol{r}_1 \cdots \boldsymbol{r}_N)$. The *completeness* property indicates that any function defined on this space can be expanded in a series of Ψ_μ, so that

$$\Psi_j(\boldsymbol{r}_1 \cdots \boldsymbol{r}_N) = \sum_\mu c_{j\mu} \Psi_\mu(\boldsymbol{r}_1 \cdots \boldsymbol{r}_N), \tag{20.76}$$

where the coefficients $c_{j\mu}$ are in general complex numbers. In Dirac notation, the expansion becomes

$$|\Psi_j\rangle = \sum_\mu c_{j\mu} |\Psi_\mu\rangle \quad \text{and} \quad \langle \Psi_j | = \sum_\mu c_{j\mu}^* \langle \Psi_\mu |, \tag{23.77}$$

where $c_{j\mu}^*$ is the complex conjugate of $c_{j\mu}$. Substituting these expansions into (23.76) gives

$$\langle X \rangle = \frac{1}{J} \sum_j \sum_{\gamma\mu} c_{j\gamma}^* c_{j\mu} \langle \Psi_\gamma | \hat{X} | \Psi_\mu \rangle, \tag{23.78}$$

which can be expressed as

$$\langle X\rangle = \sum_{\mu\gamma} \rho_{\mu\gamma}\langle\Psi_\gamma|\widehat{X}|\Psi_\mu\rangle, \tag{23.79}$$

where

$$\rho_{\mu\gamma} = \frac{1}{J}\sum_j c^*_{j\gamma}c_{j\mu}. \tag{23.80}$$

The quantity $\rho_{\gamma\mu}$ is a *density matrix*.

The *matrix elements* $\langle\Psi_\gamma|\widehat{X}|\Psi_\mu\rangle$ are often written as $\langle\gamma|\widehat{X}|\mu\rangle$ or $X_{\gamma\mu}$, and the numbers $\rho_{\mu\gamma}$ can be treated as the matrix element of an operator $\widehat{\rho}$, the *density operator*. The matrix product of $\rho_{\mu\gamma}$ and $X_{\gamma\sigma}$ is $(\widehat{\rho}\widehat{X})_{\mu\sigma} = \sum_\gamma \rho_{\mu\gamma}X_{\gamma\sigma}$, so that the average $\langle X\rangle$ is the sum of the diagonal elements of $(\widehat{\rho}\widehat{X})_{\mu\sigma}$. Since the sum of the diagonal elements of a matrix is its *trace*, the ensemble average of a property X is[1]

$$\langle X\rangle = Tr\widehat{\rho}\widehat{X}. \tag{23.81}$$

The description of systems by density matrices is the most general form of the quantum mechanical description of thermodynamic systems. The same density matrix and thus the same averages $\langle X\rangle$ can be obtained from ensembles whose members are described by different collections of wave functions Ψ_j.

Properties

A density matrix has special properties that can be understood from (23.80). It is immediately apparent that $\rho_{\mu\gamma} = \rho^*_{\gamma\mu}$, which indicates that $\rho_{\mu\gamma}$ is a Hermitian matrix. *Density matrices are Hermitian.* The diagonal elements of $\rho_{\mu\gamma}$ are

$$\rho_{\mu\mu} = \frac{1}{J}\sum_j c_{j\mu}c^*_{j\mu} = \frac{1}{J}\sum_j |c_{j\mu}|^2, \tag{23.82}$$

where $|c_{j\mu}|$ is the magnitude of $c_{j\mu}$. *The diagonal elements of the density matrix are non-negative.* The orthonormal (*i.e.*, orthogonal and normal) property of Ψ_γ and Ψ_μ implies that

$$\langle\Psi_\mu|\Psi_\nu\rangle = \delta_{\mu\nu} = \begin{cases} 1, & \text{if } \mu=\nu, \\ 0, & \text{otherwise,}\end{cases} \tag{20.83}$$

where $\delta_{\mu\nu}$ is a Kronecker delta. Since the normalization of the Ψ_j implies that $\langle\Psi_j|\Psi_j\rangle = 1$, expanding Ψ_j in a series of functions Ψ_μ implies that

$$\langle\Psi_j|\Psi_j\rangle = \sum_{\gamma\mu} c^*_{j\gamma}c_{j\mu}\langle\Psi_\gamma|\Psi_\mu\rangle = \sum_{\gamma\mu} c^*_{j\gamma}c_{j\mu}\delta_{\gamma\mu} = \sum_\mu |c_{j\mu}|^2 = 1. \tag{23.84}$$

[1] It follows from this definition of an ensemble average that $\langle c\rangle = c$, $\langle c\widehat{A}\rangle = c\langle\widehat{A}\rangle$, and $\langle\widehat{A}+\widehat{B}\rangle = \langle\widehat{A}\rangle + \langle\widehat{B}\rangle$, where $\langle c\rangle$ is the average of a scalar times the identity operator and $\widehat{A}$ and $\widehat{B}$ are the Hermitian operators.

By using this, the trace of the density matrix becomes

$$\sum_{\mu} \rho_{\mu\mu} = \sum_{\mu} \frac{1}{J} \sum_{j} |c_{j\mu}|^2 = \frac{1}{J} \sum_{j} \sum_{\mu} |c_{j\mu}|^2 = \frac{1}{J} \sum_{j} 1 = 1. \tag{23.85}$$

The trace of the density matrix equals one ($\mathrm{Tr}\hat{\rho} = 1$), which is the normalization condition for density matrices.

Different complete sets of orthonormal functions are related by unitary transformations. In Dirac notation, the transformation that relates the set Ψ_μ to set $\overline{\Psi}_{\overline{\mu}}$ is expressed by

$$\langle \overline{\Psi}_{\overline{\mu}}| = \sum_{\alpha} \langle \overline{\Psi}_{\overline{\mu}} | \Psi_\alpha \rangle \langle \Psi_\alpha | \quad \text{and} \quad |\overline{\Psi}_{\overline{\mu}}\rangle = \sum_{\alpha} |\Psi_\alpha\rangle \langle \Psi_\alpha | \overline{\Psi}_{\overline{\mu}} \rangle, \tag{23.86}$$

where $\sum_{\overline{\mu}} \langle \Psi_\gamma | \overline{\Psi}_{\overline{\mu}} \rangle \langle \overline{\Psi}_{\overline{\mu}} | \Psi_\alpha \rangle = \delta_{\gamma\alpha}$. The matrix elements of any operator $\hat{O}$ in the two sets of functions are related by

$$O_{\overline{\alpha}\overline{\nu}} = \sum_{\gamma\mu} \langle \overline{\Psi}_{\overline{\alpha}} | \Psi_\gamma \rangle O_{\gamma\mu} \langle \Psi_\mu | \overline{\Psi}_{\overline{\nu}} \rangle. \tag{23.87}$$

The different choices of complete sets are referred to as different *representations*. By transforming the trace from $\overline{\Psi}_{\overline{\mu}}$ representation to Ψ_μ representation, we find that

$$\begin{aligned} \mathrm{Tr}\,\hat{O} &= \sum_{\overline{\alpha}} O_{\overline{\alpha}\overline{\alpha}} = \sum_{\gamma\mu} \langle \overline{\Psi}_{\overline{\alpha}} | \Psi_\gamma \rangle O_{\gamma\mu} \langle \Psi_\mu | \overline{\Psi}_{\overline{\alpha}} \rangle \\ &= \sum_{\gamma\mu} O_{\gamma\mu} \left(\sum_{\overline{\alpha}} \left\langle \Psi_\mu | \overline{\Psi}_{\overline{\alpha}} \right\rangle \left\langle \overline{\Psi}_{\overline{\alpha}} | \Psi_\gamma \right\rangle \right) = \sum_{\gamma\mu} O_{\gamma\mu} \delta_{\mu\gamma} = \sum_{\gamma} O_{\gamma\gamma}. \end{aligned} \tag{23.88}$$

Thus, the trace of an operator is representation independent. By setting $\hat{O} = \hat{\rho}\hat{X}$, it follows that ensemble averages $\langle X \rangle$ are *representation independent*.

Probabilities

The density matrix can be diagonalized by a unitary transformation. In the representation that diagonalizes it, the density matrix is

$$\rho_{\overline{\alpha}\overline{\gamma}} = \rho_{\overline{\alpha}\overline{\alpha}} \delta_{\overline{\alpha}\overline{\gamma}}, \tag{23.89}$$

where $\delta_{\overline{\alpha}\overline{\gamma}}$ is a Kronecker delta. Since the diagonal elements of the density matrix are nonnegative and its trace equals one, the diagonal elements $\rho_{\overline{\alpha}\overline{\alpha}}$ satisfy the nonnegativity condition ($\rho_{\overline{\alpha}\overline{\alpha}} \geq 0$) and normalization condition ($\sum_{\overline{\alpha}} \rho_{\overline{\alpha}\overline{\alpha}} = 1$) for probabilities. *In the representation that diagonalizes the density matrix, the diagonal element $\rho_{\overline{\alpha}\overline{\alpha}}$ is the probability that the system is in the state described by wave function $\Psi_{\overline{\alpha}}$.*

When all of the elements of a density matrix are zero except diagonal element $\rho_{\alpha\alpha}$, the normalization condition $\mathrm{Tr}\hat{\rho} = 1$ requires that $\rho_{\alpha\alpha} = 1$, which implies that the system is in the state described by wave function Ψ_α. When is described by a single wave function, the system is said to be in a *pure state*. When not in a pure state, it is in a *mixed state*. When $\rho_{\alpha\alpha} = 1$ is the only non-zero element in a pure-state density matrix, it follows that

$(\widehat{\rho}\widehat{\rho})_{\mu\sigma} = \delta_{\mu\alpha}\delta_{\gamma\alpha}$. Thus, *the density matrix of a system in a pure state is equal to its square, i.e.*, $\widehat{\rho} = \widehat{\rho}^2$, and the property is preserved by unitary transformations.

23.5 Canonical Ensemble

In density matrix formalism, the canonical ensemble is described by

$$\boxed{\widehat{\rho}^c = \frac{e^{-\widehat{H}/k_BT}}{Z_{qm}},} \tag{23.90}$$

where $\widehat{H}$ is the Hamiltonian operator. It follows from the normalization condition $\text{Tr}\widehat{\rho} = 1$ that the general expression for the quantum partition function is

$$\boxed{Z_{qm} = \text{Tr}e^{-\widehat{H}/k_BT}.} \tag{23.91}$$

The thermodynamic properties of the systems are obtained from the partition function formula $F = -k_BT \ln Z_{qm}$.

An operator like $e^{-\widehat{H}/k_BT}$, which is a function of another operator, can be defined either as a power series or in terms of eigenfunctions and eigenvalues. When the operator $f(\widehat{O})$ is constructed from a function $f(x)$ that can be represented by the series $f(x) = \sum_n a_n x^n$, the operator becomes

$$f(\widehat{O}) = \sum_{n=0}^{\infty} a_n \widehat{O}^n, \tag{23.92}$$

where $\widehat{O}^n$ is obtained by multiplying $\widehat{O}$ by itself n times. The operator $f(\widehat{O})$ can also be defined in terms of the solutions of eigenvalue equation $\widehat{O}\Psi_\gamma = O_\gamma \Psi_\gamma$. Provided the eigenfunctions Ψ_γ form a complete orthonormal set, the operator $f(\widehat{O})$ is

$$f(\widehat{O}) = \sum_{\gamma} |\Psi_\gamma\rangle f(O_\gamma)\langle\Psi_\gamma|, \tag{23.93}$$

where O_γ is an eigenvalue associated with eigenfunction Φ_γ.

Energy Representation

The use of energy eigenstates to evaluate the partition function is a convenient special case of the relationship between thermodynamic properties and the Hamiltonian implicit the canonical density operator $\widehat{\rho}^c$. To show this, we use energy representation, which uses the eigenfunctions of the time-independent Schrödinger equation. As given in Chapter 20, the Schrödinger equation for a system of N particles is $\widehat{H}_N\Psi_\alpha = E_{N,\alpha}\Psi_\alpha$, where $\widehat{H}_N$ is the Hamiltonian operator, $E_{N,\alpha}$ is an energy eigenvalue, and Ψ_α is the associated eigenfunction.

The matrix elements of the Hamiltonian operator in energy representation are

$$\begin{aligned}\langle\Psi_\gamma|\widehat{H}_N|\Psi_\alpha\rangle &= \int dr^{3N}\Psi_\gamma^*\widehat{H}_N\Psi_\alpha = E_{N,\alpha}\int dr^{3N}\Psi_\gamma^*\Psi_\alpha \\ &= E_{N,\alpha}\langle\Psi_\gamma|\Psi_\alpha\rangle = E_{N,\alpha}\delta_{\gamma\alpha},\end{aligned} \tag{20.94}$$

where the Kronecker delta $\delta_{\gamma\alpha}$ indicates that the matrix is diagonal. By using the definition of the function of an operator in (23.93), the matrix elements of the operator $e^{-\widehat{H}_N/k_BT}$ become

$$\begin{aligned}\langle\Psi_\alpha|e^{-\widehat{H}_N/k_BT}|\Psi_\gamma\rangle &= \sum_\gamma\langle\Psi_\alpha|\Psi_\gamma\rangle e^{-E_{N,\gamma}/k_BT}\langle\Psi_\gamma|\Psi_\alpha\rangle \\ &= \sum_\gamma \delta_{\alpha\gamma}e^{-E_{N,\gamma}/k_BT}\delta_{\gamma\alpha} = e^{-E_{N,\alpha}/k_BT}\delta_{\alpha\gamma}.\end{aligned} \tag{23.95}$$

This implies that the canonical density matrix $\rho^c_{\alpha\gamma} = \langle\Psi_\alpha|\widehat{\rho}^c|\Psi_\gamma\rangle$ is diagonal in energy representation. The partition function in (23.91) becomes

$$Z_{qm} = \mathrm{Tr}e^{-\widehat{H}_N/k_BT} = \sum_\gamma\left\langle\Psi_\gamma\middle|e^{-\widehat{H}_N/k_BT}\middle|\Psi_\gamma\right\rangle = \sum_\gamma e^{-E_{N,\alpha}/k_BT}\delta_{\alpha\gamma}, \tag{23.96}$$

so that

$$\boxed{Z_{qm} = \sum_\alpha e^{-E_{N,\alpha}/k_BT}.} \tag{23.97}$$

Furthermore, the canonical ensemble average of property X becomes

$$\begin{aligned}\langle X\rangle &= \mathrm{Tr}\widehat{\rho}^c\widehat{X} = \sum_\alpha\langle\Psi_\alpha|\widehat{\rho}^c\widehat{X}|\Psi_\alpha\rangle \\ &= \sum_{\alpha\gamma}\frac{1}{Z_{qm}}\langle\Psi_\alpha|\widehat{\rho}^c|\Psi_\gamma\rangle\langle\Psi_\gamma|\widehat{X}|\Psi_\alpha\rangle = \sum_{\alpha\gamma}\rho^c_{\alpha\gamma}X_{\gamma\alpha}.\end{aligned} \tag{23.98}$$

When the canonical density matrix is diagonal, the ensemble average of X is

$$\boxed{\langle X\rangle = \sum_\alpha \rho^c_{\alpha\alpha}X_{\alpha\alpha}.} \tag{23.99}$$

The above formula for the partition function Z_{qm} is the same as the expression for the quantum partition function in Chapter 20. The above expression for the ensemble average of X is the same as (20.12), except that there the diagonal elements $\rho^c_{\alpha\alpha}$ and $X_{\alpha\alpha}$ are expressed as ρ^c_α and X_α. Although an assignment of probability to a system's microstates is implicit in the density-matrix approach, the meaning of the assignment is certainly more transparent when presented in energy representation. Nevertheless, the possibility of using other representation can be useful. This is indicated in the derivation of the classical approximation in Appendix 23.A.

Problems

23.1 The classical partition function for an ideal gas confined in a sphere of radius R is $Z_N = C_N I(\beta, R)^N$, where the integral $I(\beta, R)$ is

$$I(\beta,R) = \int_0^\infty dr \int_0^\pi d\theta \int_0^{2\pi} d\varphi \int_{-\infty}^\infty dp_r \int_{-\infty}^\infty dp_\theta \int_{-\infty}^\infty dp_\varphi e^{-\beta H(p_r,p_\theta,p_\varphi,r,\theta,\varphi)}.$$

(Notice that the factor $r^2 \sin\theta$ that enters when integration variables are changed for (x, y, z) to (r, θ, φ) does not appear in $I(\beta, R)$ but is supplied by the integration over p_r, p_θ, and p_φ.) The single particle Hamiltonian is

$$H(p_r, p_\theta, p_\varphi, r, \theta, \varphi) = \frac{1}{2m} p_r^2 + \frac{1}{2mr^2} p_\theta^2 + \frac{1}{2mr^2 \sin^2\theta} p_\varphi^2 + \phi_{a-w}(r),$$

where $\phi_{a-w}(r) = 0$ for $r < R$ and $\phi_{a-w}(r) = \infty$ for $r > R$.

(a) Evaluate $I(\beta, R)$ and find the partition function Z_N.

(b) By using $\beta = 1/k_B T$ and $V = \frac{4}{3}\pi R^3$, find the Helmholtz free energy $F(T, V)$. Compare with the free energy found in Section 19.5.

Appendix 23.A

Classical Approximation

We found in Chapter 20 that classical statistical mechanics gives accurate results when the temperature is not too low. This appendix presents a general derivation of this result by deducing the classical partition function from the quantum partition function by expanding in powers of $\hbar$ and keeping the lowest order contribution.

The classical Hamiltonian for an N-particle system from (19.3) is the sum of kinetic and potential terms,

$$H_{\text{cl}}(\boldsymbol{p}_1 \cdots \boldsymbol{r}_N) = K_{\text{cl}}(\boldsymbol{p}_1 \cdots \boldsymbol{p}_N) + \Phi_N(\boldsymbol{r}_1 \cdots \boldsymbol{r}_N). \tag{23.A.1}$$

The potential energy describes interparticle interactions and is a function of the position variables $(\boldsymbol{r}_1 \cdots \boldsymbol{r}_N)$. Classically, kinetic energy is a function of the momentum *variables* $(\boldsymbol{p}_1 \cdots \boldsymbol{p}_N)$,

$$K_{\text{cl}} = \sum_j \frac{1}{2m_j} |\boldsymbol{p}_j|^2. \tag{23.A.2}$$

Quantum mechanically, it is given by the *operator*

$$\widehat{K}_{qm} = \sum_j \frac{-1}{2m_j} \hbar^2 \nabla_j^2. \tag{23.A.3}$$

The momentum operator for particle j is $(\hbar/i)\nabla_j$, where ∇_j differentiates functions of its position variable $\boldsymbol{r}_j$. The Hamiltonian operator is

$$\widehat{H} = \widehat{K}_{qm} + \Phi_N. \tag{23.A.4}$$

Momentum Representation

An expression for expanding in powers of $\hbar$ is obtained by evaluating the quantum partition function $Z_{qm} = \mathrm{Tr}e^{-\hat{H}/k_BT}$ from (23.91) in momentum representation. The eigenvalue equation for the momentum of particle j is

$$(\hbar/i)\nabla_j e^{i\boldsymbol{k}_j\cdot\boldsymbol{r}_j} = \hbar\boldsymbol{k}_j e^{i\boldsymbol{k}_j\cdot\boldsymbol{r}_j}, \qquad (23.A.5)$$

where the eigenfunction $e^{i\boldsymbol{k}_j\cdot\boldsymbol{r}_j}$ describe a plane wave of wavelength of $\lambda = 2\pi/k_j$ with wave fronts perpendicular to wave vector $\boldsymbol{k}_j$ and the eigenvalue is $\hbar\boldsymbol{k}_j$. The normalized eigenfunctions of the momentum operator for the system, $\sum_j(\hbar/i)\nabla_j$, are

$$\Theta_{\boldsymbol{k}_1\cdots\boldsymbol{k}_N} = \frac{e^{i\boldsymbol{k}_1\cdot\boldsymbol{r}_1}\ldots e^{i\boldsymbol{k}_N\cdot\boldsymbol{r}_N}}{V^{N/2}}, \qquad (23.A.6)$$

where V is the system's volume.[2] After using the series $e^x = \sum_n x^n/n!$ to expand the operator $e^{-\hat{H}/k_BT}$ in powers of the Hamiltonian operator, the quantum partition function becomes

$$\begin{aligned} Z_{qm} &= \mathrm{Tr}\,e^{-\beta\hat{H}} = \sum_{\boldsymbol{k}_1}\cdots\sum_{\boldsymbol{k}_N}\left\langle\Theta_{\boldsymbol{k}_1\ldots\boldsymbol{k}_N}\right| e^{-\beta\hat{H}}\left|\Theta_{\boldsymbol{k}_1\ldots\boldsymbol{k}_N}\right\rangle \\ &= \sum_{n=0}^{\infty}\frac{(-\beta)^n}{n!}\left[\sum_{\boldsymbol{k}_1}\cdots\sum_{\boldsymbol{k}_N}\left\langle\Theta_{\boldsymbol{k}_1\cdots\boldsymbol{k}_N}\right|\hat{H}^n\left|\Theta_{\boldsymbol{k}_1\cdots\boldsymbol{k}_N}\right\rangle\right], \end{aligned} \qquad (23.A.7)$$

where $\beta = 1/k_BT$.

The first step in evaluating $\hat{H}^n\Theta_{\boldsymbol{k}_1\cdots\boldsymbol{k}_N}$ is finding the value of $\hat{K}_{qm}F_{\boldsymbol{k}_1\cdots\boldsymbol{k}_N}\Theta_{\boldsymbol{k}_1\cdots\boldsymbol{k}_N}$, where $F_{\boldsymbol{k}_1\cdots\boldsymbol{k}_N}$ is any function of $(\boldsymbol{r}_1\cdots\boldsymbol{r}_N)$ that depends on the values of $(\boldsymbol{k}_1\cdots\boldsymbol{k}_N)$. The functional dependence of $\Theta_{\boldsymbol{k}_1\cdots\boldsymbol{k}_N}$ and $F_{\boldsymbol{k}_1\cdots\boldsymbol{k}_N}$ on $(\boldsymbol{r}_1\cdots\boldsymbol{r}_N)$ is understood but not explicitly indicated. By using the relationship $\nabla_j e^{i\boldsymbol{k}\cdot\boldsymbol{r}} = i\boldsymbol{k}\,e^{i\boldsymbol{k}\cdot\boldsymbol{r}}$ and evaluating the derivatives indicated by ∇_j^2, it follows that

$$\begin{aligned} \hat{K}_{qm}F_{\boldsymbol{k}_1\cdots\boldsymbol{k}_N}\Theta_{\boldsymbol{k}_1\cdots\boldsymbol{k}_N} &= -\sum_j\frac{1}{2m_j}\hbar^2\nabla_j^2F_{\boldsymbol{k}_1\cdots\boldsymbol{k}_N}\Theta_{\boldsymbol{k}_1\cdots\boldsymbol{k}_N} \\ &= -\sum_j\frac{1}{2m_j}\begin{pmatrix} F_{\boldsymbol{k}_1\cdots\boldsymbol{k}_N}\left|i\hbar\boldsymbol{k}_j\right|^2\Theta_{\boldsymbol{k}_1\cdots\boldsymbol{k}_N} \\ +\hbar\nabla_jF_{\boldsymbol{k}_1\cdots\boldsymbol{k}_N}\cdot(i\hbar\boldsymbol{k}_j)\Theta_{\boldsymbol{k}_1\cdots\boldsymbol{k}_N} + \hbar^2(\nabla_j^2F_{\boldsymbol{k}_1\cdots\boldsymbol{k}_N})\Theta_{\boldsymbol{k}_1\cdots\boldsymbol{k}_N} \end{pmatrix}. \end{aligned} \qquad (23.A.8)$$

The transition from operators to variables is obtained by representing the eigenvalues of the momentum operator of particle j as $\boldsymbol{p}_j = \hbar\boldsymbol{k}_j$, so that $\sum_j\frac{1}{2m_j}\left|i\hbar\boldsymbol{k}_j\right|^2$ becomes $\sum_j\frac{1}{2m_j}\left|\boldsymbol{p}_j\right|^2 = K_{cl}$. The above result then becomes

$$\hat{K}_{qm}F_{\boldsymbol{k}_1\cdots\boldsymbol{k}_N}\Theta_{\boldsymbol{k}_1\cdots\boldsymbol{k}_N} = K_{cl}F_{\boldsymbol{k}_1\cdots\boldsymbol{k}_N}\Theta_{\boldsymbol{k}_1\cdots\boldsymbol{k}_N} + O(\hbar), \qquad (23.A.9)$$

[2] Plane wave eigenfunction and eigenvector are discussed in more detail in Section 24.1.

where $O(\hbar)$ represents the terms proportional to $\hbar$ and $\hbar^2$. K_{cl} is the classical kinetic energy of a system with momentum variables $(\hbar \boldsymbol{k}_1 \cdots \hbar \boldsymbol{k}_N)$. By adding the product $\Phi_N F_{k_1 \cdots k_N} \Theta_{k_1 \cdots k_N}$ to $\widehat{K}_{qm} F_{k_1 \cdots k_N} \Theta_{k_1 \cdots k_N}$, we obtain

$$\widehat{H} F_{k_1 \cdots k_N} \Theta_{k_1 \cdots k_N} = (\widehat{K}_{qm} + \Phi_N) F_{k_1 \cdots k_N} \Theta = H_{cl} F_{k_1 \cdots k_N} \Theta_{k_1 \cdots k_N} + O(\hbar), \tag{23.A.10}$$

where H_{cl} is the classical Hamiltonian $H_{cl} = K_{cl} + \Phi_N$. In the special case when $F_{k_1 \cdots k_N} = 1$, the above result becomes

$$\widehat{H} \Theta_{k_1 \cdots k_N} = H_{cl} \Theta_{k_1 \cdots k_N} + O(\hbar). \tag{23.A.11}$$

The effect of operating on $\Theta_{k_1 \cdots k_N}$ by $\widehat{H}^n$ is obtained by successively using (23.A.11) and letting $O(\hbar)$ represent all terms proportional to $\hbar$ or higher power of $\hbar$. The successive replacement of $\widehat{H}$ by H_{cl} gives

$$\begin{aligned} \widehat{H}^n \Theta_{k_1 \cdots k_N} &= \widehat{H}^{n-1} H_{cl} \Theta_{k_1 \cdots k_N} + O(\hbar) \\ \cdots &= \widehat{H}^m H_{cl}^{n-m} \Theta_{k_1 \cdots k_N} + O(\hbar) = \cdots = H_{cl}^n \Theta_{k_1 \cdots k_N} + O(\hbar). \end{aligned} \tag{23.A.12}$$

By using this, the matrix elements of $\widehat{H}^n$ in momentum representations become

$$\begin{aligned} \left\langle \Theta_{k_1 \cdots k_N} \middle| \widehat{H}^n \middle| \Theta_{k_1 \cdots k_N} \right\rangle &= \int dr^{3N} \Theta^*_{k_1 \cdots k_N} \widehat{H}^n \Theta_{k_1 \cdots k_N} \\ &= \int dr^{3N} \Theta^*_{k_1 \cdots k_N} \left[H_{cl} \left(\vec{p}_1 \cdots \vec{r}_N \right) \right]^n \Theta_{k_1 \cdots k_N} + O(\hbar), \end{aligned} \tag{23.A.13}$$

where $\Theta^*_{k_1 \cdots k_N}$ is the complex conjugate of $\Theta_{k_1 \cdots k_N}$ and $O(\hbar)$ now represents the terms proportional to $\hbar$ or higher power of $\hbar$ in equation (23.A.12). By using $e^{-i\boldsymbol{k}_j \cdot \boldsymbol{r}_j} e^{i\boldsymbol{k}_j \cdot \boldsymbol{r}_j} = 1$, it follows from (23.A.6) that $\Theta^*_{k_1 \cdots k_N} \Theta_{k_1 \cdots k_N} = 1/V^N$, so that

$$\langle \Theta_{k_1 \cdots k_N} | \widehat{H}^n | \Theta_{k_1 \cdots k_N} \rangle = \frac{1}{V^N} \int dr^{3N} [H_{cl}(\boldsymbol{p}_1 \cdots \boldsymbol{r}_N)]^n + O(\hbar). \tag{23.A.14}$$

To evaluate the sums $\sum_{\mathbf{k}_1} \cdots \sum_{\mathbf{k}_N}$ in (23.A.7), we use the sum over the allowed values of $\boldsymbol{k}_j$ for each particle. Each sum can be found by using the number of allowed values per unit volume of k-space, which, as found in Chapter 24, is $V/(2\pi)^3$. We then convert the sums over k-vectors into integrals over k-space, which by using $\boldsymbol{p} = \hbar \boldsymbol{k}$ become integrals over momentum space, as summarized by

$$\sum_{\vec{k}} \to \frac{V}{(2\pi)^3} \int dk^3 \to \frac{V}{(2\pi\hbar)^3} \int dp^3 \to \frac{V}{h^3} \int dp^3. \tag{23.A.15}$$

By treating each particle in this way, the sum of the matrix elements of $\widehat{H}^n$ in (23.A.14) becomes

$$\sum_{k_1} \cdots \sum_{k_N} \langle \Theta_{k_1 \cdots k_N} | \widehat{H}^n | \Theta_{k_1 \cdots k_N} \rangle = h^{-3N} \int dp^{3N} \int dr^{3N} \left[H_{cl} \left(\boldsymbol{p}_1 \cdots \boldsymbol{r}_N \right) \right]^n + O(\hbar). \tag{23.A.16}$$

Substituting this into the expression for the quantum partition function in (23.A.7) and neglecting the terms of order $\hbar$ and higher gives

$$Z_{\rm cl} = h^{-3N}\sum_{n=0}^{\infty}\frac{(-\beta)^n}{n!}\int dp^{3N}\int dr^{3N}[H_{\rm cl}(\boldsymbol{p}_1\cdots\boldsymbol{r}_N)]^n. \tag{23.A.17}$$

By using the expansion $e^x = \sum_n x^n/n!$, this becomes the classical partition function

$$Z_{\rm cl} = h^{-3N}\int dp^{3N}\int dr^{3N}e^{-\beta H_{\rm cl}(\boldsymbol{p}_1\cdots\boldsymbol{r}_N)}. \tag{23.A.18}$$

This is the classical expression in (19.24) with $C_N = h^{-3N}$ and $\beta = 1/k_BT$. When the system consists of N indistinguishable particles, an additional factor of $1/N!$ is needed as discussed in Section 22.3.

An expansion of the Helmholtz free energy F in powers of $\hbar$ can be obtained by including the contributions to Z_{qm} of the neglected terms represented by $O(\hbar)$. By keeping terms proportional to $\hbar$ and $\hbar^2$, the resulting expression for the free energy becomes[3]

$$F = -k_BT\ln Z_{\rm cl} + \frac{1}{24}\left(\frac{\hbar}{k_BT}\right)^2\sum_j\frac{1}{m_j}\langle(\nabla_j\Phi_N)^2\rangle + \cdots, \tag{23.A.19}$$

where the contribution proportional to $\hbar$ is zero. The pointed brackets $\langle\cdots\rangle$ indicate the classical canonical average of the square of the gradient of the potential energy Φ_N. The quantum result limits to the classical approximation at high temperature limit, and the $\hbar^2$ term can be used to find quantum corrections.

[3] This expansion of the free energy is called the *Wigner–Kirkwood expansion*. It is derived in Landau and Lifshitz, *Statistical Mechanics*, pp. 93–97. The original references are Wigner E. P. (1932), *Phys. Rev.*, **40**, 749 and Kirkwood J. G. (1933), *Phys. Rev.*, **44**, 31.

Part V
Statistical Mechanics II

The statistical methods used in Part IV are modified here for the analysis of electromagnetic radiation, atomic vibrations in solids, and collections of identical fermions and bosons. We investigate the properties of blackbody radiation, derive the Fermi-Dirac and Bose-Einstein distributions, and predict the low temperature phenomena known as Bose-Einstein condensation. Finally, some of the effects determined by the interactions neglected in independent particle models are considered, and the role of computer simulations in statistical mechanics is discussed.

Thermodynamics and Statistical Mechanics: An Integrated Approach, First Edition.
Robert J Hardy and Christian Binek.

24

Photons and Phonons

Our quantum mechanical analysis has described a system's microstates as collections of eigenstates of the individual particles. An alternate approach is to specify the microstates by giving the number of particles (or excitations) in the different single-particle states. The alternate approach is especially useful in studies of electromagnetic radiation, atomic vibrations in solids, and systems of identical particles. In many situations, the single-particle states are plane waves with a definite wavelength and frequency. The focus of this chapter is on the excitations of the electromagnetic fields – *photons* – and vibrational motion in solids – *phonons*. We begin by reviewing the plane wave description of systems of free particles.

24.1 Plane Wave Eigenstates

The one-dimensional waves in violin strings, two-dimensional waves on a drumheads, and three-dimensional electromagnetic waves have many characteristics in common. Waves in a three-dimensional continuum are described by the *wave equation*

$$\nabla^2\Theta(\boldsymbol{r},t) - \frac{1}{c^2}\frac{\partial^2\Theta(\boldsymbol{r},t)}{\partial t^2} = 0, \tag{24.1}$$

where $\boldsymbol{r} = (x, y, z)$. For electromagnetic radiation, the *wave velocity* c is the speed of light, and the electric and magnetic fields are proportional to $\Theta(\boldsymbol{r}, t)$. The oscillations of the electromagnetic field within a cavity can be either traveling waves or standing waves. The standing wave description is obtained by assuming a time dependence of the form

$$\Theta(\boldsymbol{r},t) = \psi(\boldsymbol{r})e^{-i\omega t}, \tag{24.2}$$

Thermodynamics and Statistical Mechanics: An Integrated Approach, First Edition.
Robert J Hardy and Christian Binek.

where ω is angular frequency (radians per second). Substituting this into the wave equation and canceling factors of $e^{-i\omega t}$ yield the eigenvalue equation

$$-\nabla^2\psi(\boldsymbol{r}) = k^2\psi(\boldsymbol{r}), \tag{24.3}$$

where $k = \omega/c$. The eigenvalues k^2 are determined by the boundary conditions on the eigenfunctions $\psi(\boldsymbol{r})$.

The time-independent Schrödinger equation for a free particle is obtained by multiplying (24.3) by $\hbar^2/2m$, which gives

$$-\frac{\hbar^2}{2m}\nabla^2\psi(\boldsymbol{r}) = \varepsilon\psi(\boldsymbol{r}), \tag{24.4}$$

where $\varepsilon = \hbar^2k^2/2m$. Although the particles are free in that they do not interact with each other, they are spatially confined by interactions with a container of volume V. The eigenvalues of (24.4) are calculated in Appendix 20.A by requiring that $\psi(\boldsymbol{r})$ vanishes at the walls of a rectangular box. These are *box boundary conditions*. The eigenfunctions for a box with walls at $x = 0$, $y = 0$, $z = 0$, $x = L_x$, $y = L_y$, and $z = L_z$ are

$$\psi(x, y, z) = A\sin(k_x x)\sin(k_y y)\sin(k_z z), \tag{24.5}$$

where the allowed values of k_x, k_y, and k_z are

$$k_x = \frac{\pi n_x}{L_x}, \quad k_y = \frac{\pi n_y}{L_y}, \quad k_z = \frac{\pi n_z}{L_z}. \tag{24.6}$$

The linearly independent solutions of the eigenvalue equations are given by *positive integer values* of n_x, n_y, and n_z. Functions such as the instantaneous values of the electric and magnetic fields can be expressed as sums of these linearly independent solutions, and each solution describes an independent mode of oscillation.

Periodic Boundary Conditions

Periodic boundaries are often more convenient for describing blackbody radiation than box boundary conditions. *Periodic boundary conditions* require that eigenfunctions have the same values at opposite sides of the container. For a rectangular box of volume $V = L_xL_yL_z$, this implies that

$$\psi(0, y, z) = \psi(L_x, y, z), \quad \psi(x, 0, z) = \psi(x, L_y, z), \quad \psi(x, y, 0) = \psi(x, y, L_z). \tag{24.7}$$

These requirements are satisfied by

$$\psi_{\boldsymbol{k}}(\boldsymbol{r}) = Ae^{i(k_x x + k_y y + k_z z)} = Ae^{i\boldsymbol{k}\cdot\boldsymbol{r}}, \tag{24.8}$$

where $\boldsymbol{k} = (k_x, k_y, k_z)$ is a wave vector and A is a normalization constant. The allowed values of k_x, k_y, and k_z are

$$k_x = \frac{2\pi n_x}{L_x}, \quad k_y = \frac{2\pi n_y}{L_y}, \quad k_z = \frac{2\pi n_z}{L_z}. \tag{24.9}$$

These values for k_x, k_y, and k_z are different from those in (24.6), and the integers n_x, n_y, and n_z are not confined to positive values but can be either positive or negative. Each eigenfunction $\psi_k(\boldsymbol{r})$ describes a *plane wave* of wavelength of $\lambda = 2\pi/k$ with wave fronts perpendicular to *wave vector* $\boldsymbol{k}$. As indicated in (24.2), a solution to the wave equation is obtained by multiplying $e^{i\boldsymbol{k}\cdot\boldsymbol{r}}$ by $e^{-i\omega t}$. The result $\Theta(\boldsymbol{r}, t) = Ae^{i(\boldsymbol{k}\cdot\boldsymbol{r}-\omega t)}$ describes a traveling plane wave with wave velocity $c = \omega/k$.

The distribution of eigenvalues is often more important than their specific values. When the waves that dominate a phenomenon have wavelengths that are small compared to the dimensions of the container, as they are in blackbody radiation and in the quantum mechanical analysis of ideal gases, the distribution of eigenvalues is conveniently described by a density of states. Since the energy of a free particles is $\varepsilon = \hbar^2k^2/2m$, the distribution of the allowed values of $\boldsymbol{k}$ is needed. By using $k_x = 2\pi n_x/L_x$, the change Δk_x between adjacent values of k_x is found by changing n_x to $n_x + 1$, which indicates that $\Delta k_x = 2\pi/L_x$. Similarly, the changes in k_y and k_z are $\Delta k_y = 2\pi/L_y$ and $\Delta k_z = 2\pi/L_z$, so that the k-space volume associated with each wave vector is

$$\Delta k_x \Delta k_y \Delta k_z = \frac{8\pi^3}{L_x L_y L_z}. \tag{24.10}$$

The reciprocal of this is the *density of wave vectors* D_0, which is the number of wave vectors per unit volume of k-space. Since $V = L_x L_y L_z$ is the real-space volume of the system, it follows that

$$D_0 = \frac{V}{8\pi^3}. \tag{24.11}$$

The number of wave vectors with magnitudes between k and $k + \Delta k$ equals the product of the density of wave vectors times the volume $4\pi k^2\Delta k$ of a spherical shell of thickness Δk and volume $4\pi k^2\Delta k$. Thus,

$$D_k\Delta k = D_0 4\pi k^2\Delta k = \frac{V}{2\pi^2}k^2\Delta k, \tag{24.12}$$

where $D_k = (V/2\pi^2)k^2$ is the *number of wave vectors per unit of magnitude*. Although this result was obtained by using the spacing between $\boldsymbol{k}$ vectors when periodic boundary conditions are used, it is the same for box boundary conditions. This can be seen by comparing the values of (k_x, k_y, k_z) in (24.6) with those in (24.9). Although the density of wave vectors for box boundary conditions is eight times greater than that for periodic boundaries, the increase is compensated by the one-eighth reduction in volume between k and $k + \Delta k$ when only positive values of k_x, k_y and k_z are allowed.

Free Particles

The number of eigenvalues with energy between ε and $\varepsilon + \Delta\varepsilon$ is represented by $D(\varepsilon)\Delta\varepsilon$ and is related to the number wave vectors with magnitude between k and $k + \Delta k$ by

$$D(\varepsilon)\Delta\varepsilon = D_k\Delta k = \frac{V}{2\pi^2}k^2\frac{dk}{d\varepsilon}\Delta\varepsilon, \tag{24.13}$$

where $\Delta k = (dk/d\varepsilon)\Delta\varepsilon$. Since the energy of a free particle is $\varepsilon = \hbar^2k^2/2m$, it follows that $k = \sqrt{2m\varepsilon/\hbar^2}\varepsilon^{1/2}$. This implies that $dk/d\varepsilon = \frac{1}{2}\sqrt{2m\varepsilon/\hbar^2}\,\varepsilon^{-1/2}$. Combining these results gives

$$D(\varepsilon) = \frac{V}{4\pi^2}\left(\frac{2m}{\hbar^2}\right)^{3/2}\varepsilon^{1/2}, \tag{24.14}$$

where $D(\varepsilon)$ is the free particle density of states, which is the number of plane waves states per unit of energy.

When periodic boundary conditions are used, the eigenfunctions and eigenvalues of the free particle Schrödinger equation in (24.4) are

$$\psi_k(\boldsymbol{r}) = \frac{e^{i\boldsymbol{k}\cdot\boldsymbol{r}}}{\sqrt{V}} \quad \text{and} \quad \varepsilon_k = \frac{\hbar^2 k^2}{2m}, \tag{24.15}$$

where $\psi_k(\boldsymbol{r})$ has been normalized by setting A in (24.8) equal to $1/\sqrt{V}$. The eigenfunctions $\psi_k(\boldsymbol{r})$ are also eigenfunctions of the momentum operator $(\hbar/i)\nabla$. Thus,

$$\frac{\hbar}{i}\nabla\psi_k(\boldsymbol{r}) = \hbar\boldsymbol{k}\psi_k(\boldsymbol{r}), \tag{24.16}$$

where the eigenvalues of momentum are $\hbar\boldsymbol{k}$.

Ideal Gas[1]

The monatomic ideal gas described by the Hamiltonian in (19.14) is a collection of N independent free particles. Since the Hamiltonian is separable, the N-particle eigenvalues are

$$E_{N,\alpha} = \varepsilon_{k_1} + \varepsilon_{k_2} + \cdots + \varepsilon_{k_N}, \tag{24.17}$$

where index α is an abbreviation for N wave vectors,

$$\alpha = (\boldsymbol{k}_1, \boldsymbol{k}_2, \ldots, \boldsymbol{k}_N). \tag{24.18}$$

As found in Section 20.3, the N-particle partition function Z_N for atoms of the same mass is $Z_N = [Z_s]^N$. When the eigenstates are identified by $\boldsymbol{k}$, instead of $\boldsymbol{n}$ as in Chapter 20, the single-particle partition function is

$$Z_s = \sum_{\boldsymbol{k}} e^{-\beta\varepsilon_k}. \tag{24.19}$$

To evaluate the sum over $\boldsymbol{k}$, we sort the energies ε_k into intervals of width $\Delta\varepsilon$. Since the number of eigenvalues with energy in the interval between $\varepsilon_n = n\,\Delta\varepsilon$ and $\varepsilon_n + \Delta\varepsilon$ is $D(\varepsilon_n)\Delta\varepsilon$, we can rewrite (24.19) as

$$Z_s = \sum_{n=0}^{\infty} e^{-\beta\varepsilon_n} D(\varepsilon_n)\Delta\varepsilon. \tag{24.20}$$

By taking the limit $\Delta\varepsilon \to 0$ and using the density of states in (24.14), we obtain

$$Z_s = \int_0^\infty e^{-\beta\varepsilon} D(\varepsilon)d\varepsilon = \frac{V}{4\pi^2}\left(\frac{2m}{\hbar^2}\right)^{3/2}\int_0^\infty e^{-\beta\varepsilon}\varepsilon^{1/2}d\varepsilon. \tag{24.21}$$

The value of the integral is $\frac{1}{2}\pi^{1/2}\beta^{-3/2}$. By using $\beta = 1/k_B T$, the single-particle partition function becomes

$$Z_s = \frac{(2\pi m k_B T)^{3/2}}{h^3} V. \tag{24.22}$$

[1] The properties of ideal gases are affected by the principle of indistinguishability, whose effects are significant at low temperatures and for electron gases. The effects on systems of identical fermions and bosons are described in Chapters 26 and 27.

Thus, the resulting partition function for a gas of N atoms is

$$Z_N = [Z_s]^N = \frac{(2\pi m k_B T)^{3N/2}}{h^3} V^N. \tag{24.23}$$

This result is identical to the partition function for a monatomic ideal gas found in Chapter 20 by using box boundary conditions (see (20.36)). Since equilibrium properties are determined by the partition function, the properties of the gas are unaffected by the choice of boundary conditions.

24.2 Photons

The electromagnetic radiation in an evacuated chamber, conventionally called a *cavity*, is the system being investigated. Experimentally, the system is studied by analyzing the radiation – called *blackbody radiation* – emitted through a small hole in the cavity wall. The electromagnetic fields can be expressed as sums of independent modes of oscillation similar to the plane wave eigenfunctions for free particles. Periodic boundary conditions can be used with each mode identified by a wave vector $\boldsymbol{k}$. Since electromagnetic waves have two independent polarizations, a polarization index s with two values is needed $(s = 1, 2)$.

Classically, the energy of the radiation in mode $\boldsymbol{k}s$ is proportional to the square of the amplitude of the fields. Einstein, in his quantum mechanical treatment of the photoelectric effect, postulated that light (electromagnetic radiation) is made up of particle-like entities called *photons* and that the energy of a photon of angular frequency ω is $\hbar\omega$. This implies that the energy of the radiation in microstate α is a sum of the energies in the different modes, so that

$$\boxed{E_\alpha = \sum_{\boldsymbol{k}s} N_{\boldsymbol{k}s} \hbar\omega_{\boldsymbol{k}s},} \tag{24.24}$$

where $N_{\boldsymbol{k}s}$ *is number of photons in mode* $\boldsymbol{k}s$. The energy of a photon in mode $\boldsymbol{k}s$ is $\hbar\omega_{\boldsymbol{k}s}$ and its frequency is $\omega_{\boldsymbol{k}s} = ck$, where c is the speed of light and k is the magnitude of $\boldsymbol{k}$. The system's microstates are specified by the numbers of photons $N_{\boldsymbol{k}s}$ in every mode,

$$\alpha = (\ldots, N_{\boldsymbol{k}s}, \ldots). \tag{24.25}$$

Any number of photons can be in each mode $(N_{\boldsymbol{k}s} = 1, 2, \ldots, \infty)$, and there are an infinite number of modes. The equilibrium properties of the radiation are determined by the canonical ensemble.

Average Number of Photons

The energy E_α is separable with each mode acting as an independent subsystem. The notation can be simplified by identifying the modes by a single subscript j,

$$j = (\boldsymbol{k}, s), \tag{24.26}$$

so that the system's energy becomes

$$E_\alpha = \sum_j N_j \hbar\omega_j, \tag{24.27}$$

where $\alpha = (\ldots, N_j, \ldots)$. The equilibrium probability $\rho_j(N_j)$ for mode j is proportional to the Boltzmann factor formed with the energy of the mode,

$$\rho_j(N_j) = \frac{1}{Z_j} e^{-\beta N_j \hbar\omega_j}. \tag{24.28}$$

The constant Z_j is determined by the normalization condition $\sum_{N_j} \rho_j(N_j) = 1$, which implies that

$$Z_j = \sum_{N_j=0}^{\infty} e^{-\beta N_j \hbar\omega_j}. \tag{24.29}$$

By using $1 + X + X^2 + X^3 + \cdots = (1 - X)^{-1}$, we find that

$$Z_j = \sum_{N_j=0}^{\infty} (e^{-\beta\hbar\omega_j})^{N_j} = (1 - e^{-\beta\hbar\omega_j})^{-1}. \tag{24.30}$$

The average number of photons can be expressed as

$$\langle N_j \rangle = \sum_{N_j=0}^{\infty} \rho_j(N_j) N_j = \frac{1}{Z_j} \sum_{N_j=0}^{\infty} N_j e^{-\beta N_j \hbar\omega_j}. \tag{24.31}$$

By treating $\beta\hbar\omega_j$ as a single variable, we find that

$$\langle N_j \rangle = -\frac{d}{d(\beta\hbar\omega_j)} \ln Z_j = \frac{d}{d(\beta\hbar\omega_j)} \ln(1 - e^{-\beta\hbar\omega_j}) = \frac{e^{-\beta\hbar\omega_j}}{1 - e^{-\beta\hbar\omega_j}}. \tag{24.32}$$

Since $\beta = 1/k_B T$, the *average number of photons* in mode $j = (\boldsymbol{k}, s)$ is

$$\boxed{\langle N_{\boldsymbol{k}s} \rangle = \frac{1}{e^{\hbar\omega_{\boldsymbol{k}s}/k_B T} - 1}.} \tag{24.33}$$

Partition Function and Internal Energy

By using the above results for Z_j, the partition function for total system becomes

$$\begin{aligned} Z^{\text{tot}} &= \sum_{\alpha} e^{-\beta E_\alpha} = \sum_{N_1} \sum_{N_2} \cdots e^{-\beta(N_1\hbar\omega_1 + N_2\hbar\omega_2 + \cdots)} \\ &= \sum_{N_1} e^{-\beta N_1 \hbar\omega_1} \sum_{N_2} e^{-\beta N_2 \hbar\omega_2} \cdots = Z_1 Z_2 \cdots = \left(\frac{1}{1 - e^{-\beta\hbar\omega_1}}\right)\left(\frac{1}{1 - e^{-\beta\hbar\omega_2}}\right) \cdots. \end{aligned} \tag{24.34}$$

The internal energy is

$$\begin{aligned} U &= -\frac{\partial \ln Z^{\text{tot}}}{\partial\beta} = -\frac{\partial}{\partial\beta} \sum_j \ln Z_j = \sum_j \frac{\partial}{\partial\beta} \ln(1 - e^{-\beta\hbar\omega_j}) \\ &= \sum_j \frac{e^{-\beta\hbar\omega_j}}{1 - e^{-\beta\hbar\omega_j}} \hbar\omega_j = \sum_j \frac{\hbar\omega_j}{e^{\beta\hbar\omega_j} - 1}, \end{aligned} \tag{24.35}$$

which is also equal to the ensemble average

$$\langle E\rangle = \left\langle \sum_j N_j \hbar\omega_j \right\rangle = \sum_j \langle N_j\rangle \hbar\omega_j. \tag{24.36}$$

Energy Density

Since $\omega_{ks} = ck$ is independent of the polarization index s and the direction of wave vector $\boldsymbol{k}$, it is convenient to express ω_{ks} as ω_k, where k is the magnitude of $\boldsymbol{k}$. After replacing j by ks, the internal energy becomes

$$U = \langle E\rangle = \sum_{ks} \frac{\hbar\omega_k}{e^{\beta\hbar\omega_k} - 1} = 2\sum_k \frac{\hbar\omega_k}{e^{\beta\hbar\omega_k} - 1}, \tag{24.37}$$

where the factor of 2 comes from the sum over s. By using the density of states D_0 to replace the sum by an integral, we obtain

$$U = 2\int dk^3 D_0 \frac{\hbar\omega_k}{e^{\beta\hbar\omega_k} - 1} = \frac{V}{4\pi^3}\int_0^\infty 4\pi k^2 dk \frac{\hbar\omega_k}{e^{\beta\hbar\omega_k} - 1}, \tag{24.38}$$

where $\omega_k = ck$. We now change the integration variable from k to x, where $x = \beta\hbar\omega$ and $k = x/(\beta\hbar c)$, so that

$$\frac{U}{V} = \frac{1}{\pi^2\hbar^3 c^3 \beta^4}\int_0^\infty \frac{x^3}{e^x - 1} dx. \tag{24.39}$$

Since $\beta = 1/k_B T$ and the value of the integral is $\pi^4/15$, the *energy density* of the radiation is

$$u = \frac{U}{V} = \left(\frac{\pi^2}{15}\frac{k_B{}^4}{\hbar^3 c^3}\right) T^4. \tag{24.40}$$

This is the energy density of the system which is often referred to as a *photon gas*.

Planck's Radiation Law

Since the radiation law is conventionally expressed in terms of $\nu = \omega/2\pi$, where ν is the frequency in cycles per unit time, we re-express (24.39) by changing the integration variable from x to ν, where $x = h\nu/k_B T$. Then, the energy density is

$$u = \frac{U}{V} = \frac{8\pi h}{c^3}\int_0^\infty \frac{\nu^3}{e^{h\nu/k_B T} - 1} d\nu, \tag{24.41}$$

where the energy of the radiation is distributed over a range of frequencies. This can be expressed as

$$u = \int_0^\infty u(\nu) d\nu, \tag{24.42}$$

where $u(\nu)$ is the *energy density per unit of frequency*. Comparing (24.42) with (24.41) indicates that

$$\boxed{u(\nu) = \frac{8\pi h}{c^3}\frac{\nu^3}{e^{h\nu/k_B T} - 1}.} \tag{24.43}$$

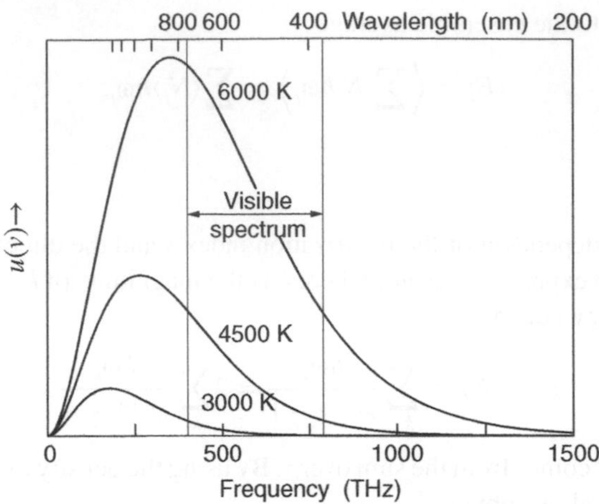

Figure 24.1 *Planck's Radiation Law*

This is the *Planck's radiation law* for blackbody radiation, a result of crucial importance in the development of quantum mechanics. The energy density per unit of frequency is plotted for three temperatures in **Figure 24.1** with the frequency given in units of THz (1 THz $= 10^{12}$Hz). For comparison, the approximate temperature of the surface of the sun is 5800 K.

Blackbody Radiation

A blackbody is an idealized object that absorbs 100% of the radiation incident on it and radiates energy with a characteristic spectrum that depends on its temperature. The practical realization of radiation from a blackbody is the radiation emitted from a small hole in the side of a cavity with walls at a uniform temperature. The *intensity* I is the energy radiated per unit area per unit time is

$$I = \frac{\text{Energy}}{\text{Area} \cdot \text{Time}}. \tag{24.44}$$

A hole of area A in the wall of a cavity is diagrammed in **Figure 24.2**. The energy traveling in direction θ that passes through the hole in time Δt is indicated by shading. The volume of the shaded region is $A\,c\cos\theta\,\Delta t$, where c is the speed of light in vacuum and $c\cos\theta$ is the z-component of the velocity of waves moving in direction θ. The fraction of the radiation in the cavity that is traveling in direction (θ, φ) within the range of solid angle $\sin\theta d\theta d\varphi$ is $(1/4\pi)\sin\theta d\theta d\varphi$, so that the energy that passes through the hole in time Δt in direction θ is

$$\frac{U}{V}(A\,c\cos\theta\,\Delta t)\frac{\sin\theta\, d\theta d\varphi}{4\pi} = \frac{u\,c\,A\Delta t}{4\pi}\cos\theta\sin\theta\, d\theta d\varphi, \tag{24.45}$$

where $u = U/V$ is the energy density within the cavity. As only radiation traveling toward the wall passes through the hole $(0 \le \theta \le \pi/2)$, the intensity is

$$I = \frac{1}{A\Delta t}\frac{ucA\Delta t}{4\pi}\int_0^{\pi/2}\cos\theta\sin\theta\, d\theta\int_0^{2\pi} d\varphi = \frac{c}{4}u. \tag{24.46}$$

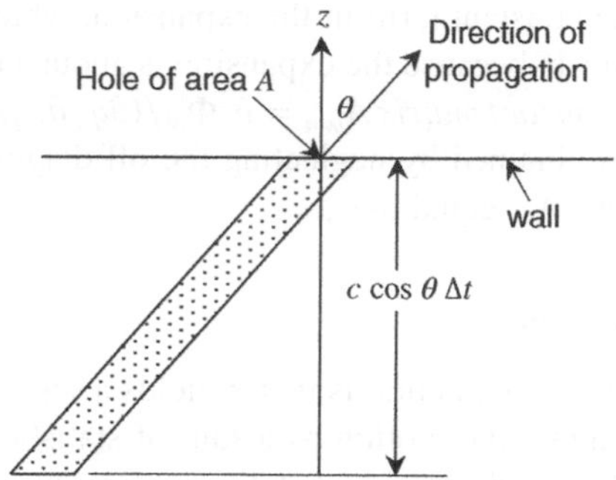

Figure 24.2 *Radiation Passing Through a Hole*

Combining this with the expression for the energy density u in (24.40) gives

$$\boxed{I = \sigma T^4,} \tag{24.47}$$

where

$$\sigma = \frac{cK_u}{4} = \frac{\pi^2}{60}\frac{k_B{}^4}{\hbar^3 c^2}. \tag{24.48}$$

This is the *Stefan–Boltzmann law*, which states that the total energy radiated per unit time per unit surface area by a blackbody is proportional to the fourth power of temperature. The value of the *Stefan–Boltzmann constant* is $\sigma = 5.670 \times 10^{-8}\,\mathrm{J\,m^{-2}\,s^{-1}\,K^{-4}}$.

24.3 Harmonic Approximation

Although the oscillations of the electromagnetic field and the vibrations of atoms in crystalline solids are very different phenomena, their quantum mechanical descriptions have many similarities. To understand these similarities, we need to express the motion of the atoms in terms of plane waves similar to those of the electromagnetic fields. The harmonic approximation does that for us. In this section, we develop the approximation outlined in Section 19.6 without making the simplifications implicit in the Einstein model. The approximation is used in Section 24.4 to introduce the excitations called *phonons*.

The atoms in solids vibrate about *sites*, which are the positions of the atoms that minimize the potential energy. We consider systems of N atoms and represent the locations of the of the sites by $(\boldsymbol{R}_1, \dots, \boldsymbol{R}_N)$. Then, the position of atom i is $\boldsymbol{q}_i + \boldsymbol{R}_i$, where $\boldsymbol{q}_i = (q_i^x, q_i^y, q_i^z)$ is the displacement of atom i from site $\boldsymbol{R}_i$. The harmonic approximation is obtained by expanding the potential energy Φ_N in a Taylor series of the components (q_i^x, q_i^y, q_i^z) of the displacements of the atoms. The components of the displacements are conveniently identified by a single index, so that the variables $(q_1^x, q_1^y, q_1^z, q_2^x, \dots, q_N^z)$ become $(q_1, q_2, q_3, q_4, \dots, q_{3N})$. By neglecting terms of higher order than quadratic, the potential energy in the *harmonic approximation* becomes

$$\Phi_N(q_1, \dots, q_{3N}) = \Phi_{\text{st}} + \frac{1}{2}\sum_m \sum_n A_{mn} q_m q_n. \tag{24.49}$$

The *static energy* Φ_{st} is the constant term in the expansion, which is the minimum value of Φ_N. The linear terms vanish, because the expansion is about a minimum. The quadratic terms depend on the *force constant matrix* $A_{mn} = \partial^2\Phi_N/(\partial q_m \partial q_n)$. The approximation that gives the Einstein model is obtained by neglecting the off diagonal elements of A_{mn} and setting the diagonal elements A_{nn} equal to k_o.

Normal Mode Transformation

The virtue of the harmonic approximation is that it yields a separable Hamiltonian, which allows us to analyze the vibrational motion as a sum of statistically independent modes. This result is achieved by the transformation that diagonalizes the force constant matrix. We begin by multiplying the components q_n by the square root of the mass, which facilitates the study of solids with different kinds of atoms. This introduces new coordinates $q'_n = m_n^{1/2} q_n$, so that the quadratic terms in (24.49) become

$$\sum_m \sum_n A_{mn} q_m q_n = \sum_m \sum_n A'_{mn} q'_m q'_n, \tag{24.50}$$

where $A'_{mn} = A_{mn}/(m_m m_n)^{1/2}$. The matrix A'_{mn} is real and symmetric and can be diagonalized by an orthogonal transformation. The transformation changes the variables $(q'_1, q'_2, \ldots, q'_{3N})$ to a new set of variables $(x_1, x_2, \ldots, x_{3N})$, where the two sets of variables are related by

$$q'_n = \sum_j T_{nj} x_j \quad \text{and} \quad x_j = \sum_n q'_n T_{nj}. \tag{24.51}$$

The transformation matrix T_{nj} is such that

$$\sum_m \sum_n A'_{mn} T_{mj} T_{nl} = \omega_j^2 \delta_{jl}, \tag{24.52}$$

where ω_j^2 is the nth diagonal element of the transformed matrix and δ_{jl} is the Kronecker delta symbol ($\delta_{jj} = 1$ and $\delta_{jl} = 0$ if $j \neq l$). The transformation is orthogonal, which means that the inverse of T_{nj} is equal to its transpose,

$$\sum_n T_{jn} T_{nl}^{-1} = \sum_n T_{jn} T_{\text{ln}} = \delta_{jl}. \tag{24.53}$$

The quadratic terms now become

$$\begin{aligned}\sum_m \sum_n A'_{mn} q'_m q'_n &= \sum_m \sum_n A'_{mn} \sum_j T_{mj} x_j \sum_l T_{nl} x_l \\ &= \sum_j \sum_l \omega_j^2 \delta_{jl} x_j x_l = \sum_j \omega_j^2 x_j^2.\end{aligned} \tag{24.54}$$

The $3N$ variables $(x_1, x_2, \ldots, x_{3N})$ are called *normal coordinates*, and each coordinate x_j describes a collective displacement of many atoms. The *normal mode transformation* from the $(q_1, q_2, \ldots, q_{3N})$ to normal coordinates $(x_1, x_2, \ldots, x_{3N})$ transforms the potential energy in (24.49) to

$$\Phi_N = \Phi_{\text{st}} + \frac{1}{2}\sum_j \omega_j^2 x_j^2, \tag{24.55}$$

where $j = 1, 2, \ldots, 3N$. Index j identifies the different *normal modes* of the vibrational motion. As the units of x_j are $\text{kg}^{1/2}$ m and the units of energy are kg m^2 s^{-2}, the units of ω_j are the units of frequency (s^{-1}).

Harmonic Hamiltonian

By indicating time derivatives by dots over the variables and transforming to normal coordinates, the kinetic energy of the system becomes

$$\begin{aligned} K_N &= \sum_n \left(\frac{1}{2} m_n \dot{q}_n^{\,2}\right) = \frac{1}{2}\sum_n \dot{q}_n'^{\,2} = \frac{1}{2}\sum_n \sum_j T_{nj}\dot{x}_j \sum_l T_{nl}\dot{x}_l \\ &= \frac{1}{2}\sum_j \sum_l \delta_{jl}\dot{x}_j\dot{x}_l = \frac{1}{2}\sum_j \dot{x}_j^{\,2}. \end{aligned} \tag{24.56}$$

The Lagrangian of the system is $L = K_N - \Phi_N$, which in the harmonic approximation is

$$L(\dot{x}_1, \ldots, x_{3N}) = \sum_j \left(\frac{1}{2}\dot{x}_j^2 - \frac{1}{2}\omega_j^2 x_j^2\right) - \Phi_{\text{st}}. \tag{24.57}$$

The jth generalized momentum $p_j = \partial L/\partial \dot{x}_j$ is $p_j = \dot{x}_j$, so that the harmonic Hamiltonian $H_N = \sum_j p_j \dot{x}_j - L$ is

$$H_N(p_1, \ldots, x_{3N}) = \Phi_{\text{st}} + H_{\text{vib}}(p_1, \ldots, x_{3N}), \tag{24.58}$$

where

$$H_{\text{vib}}(p_1, \ldots, x_{3N}) = \sum_{j=1}^{3N} H_j(p_j, x_j) \tag{24.59}$$

and

$$H_j(p_j, x_j) = \frac{1}{2}p_j^2 + \frac{1}{2}\omega_j^2 x_j^2. \tag{24.60}$$

After replacing the variables p_j by the operators $\hat{p}_j = (\hbar/i)\partial/\partial x_j$, the harmonic Hamiltonian is very similar to the Hamiltonian for the Einstein model analyzed in Section 20.4, but in the harmonic approximation the frequencies ω_j of different modes are different. Also, the displacement of each atom is given by a sum of the normal mode displacements. Specifically, the components $(q_1, q_2, \ldots, q_{3N})$ of particle displacements are

$$q_n = m_n^{-1/2} q'_n = m_n^{-1/2} \sum_j T_{nj} x_j. \tag{24.61}$$

In analogy to (20.47), the energy eigenvalues of Hamiltonian $\hat{H}_j$ are

$$\varepsilon(N_j) = \left(N_j + \frac{1}{2}\right)\hbar\omega_j, \tag{24.62}$$

where the allowed values of N_j are $N_j = 0, 1, 2, \dots, \infty$, so that the energy of a solid of N atoms is

$$E_\alpha^{\rm vib} = \sum_j \varepsilon(N_j) = \sum_j \left(N_j + \frac{1}{2}\right)\hbar\omega_j, \tag{24.63}$$

where α is an abbreviation for the quantum numbers $(N_1, N_2, \dots, N_{3N})$.

Equilibrium Properties

The average value of N_j is determined by a probability proportional to a Boltzmann factor,

$$\rho(N_j) = \frac{1}{C_j} e^{-\beta\left(N_j+\frac{1}{2}\right)\hbar\omega_j} = \frac{e^{-\frac{1}{2}\beta\hbar\omega_j}}{C_j} e^{-\beta N_j \hbar\omega_j}. \tag{24.64}$$

The constant C_j is determined by the normalization condition, which implies that

$$C_j = e^{-\frac{1}{2}\beta\hbar\omega_j} \sum_{N_j=0}^{\infty} e^{-\beta N_j \hbar\omega_j} = e^{-\frac{1}{2}\beta\hbar\omega_j} Z_j, \tag{24.65}$$

where $Z_j = \sum_{N_j} e^{-\beta N_j \hbar\omega_j}$. As given in (24.30), its value is $Z_j = (1 - e^{-\beta\hbar\omega_j})^{-1}$. The factor of $e^{-\frac{1}{2}\beta\hbar\omega_j}$ in C_j cancels the factor in $\rho(N_j)$, so that $\rho(N_j) = e^{-\beta N_j \hbar\omega_j}/Z_j$. Thus, the average value of N_j is

$$\langle N_j \rangle = \sum_{N_j=0}^{\infty} \rho_j(N_j) N_j = \frac{1}{Z_j} \sum_{N_j=0}^{\infty} N_j e^{-\beta N_j \hbar\omega_j}, \tag{24.66}$$

which is the same as the expression for photons in (24.31). It then follows from (24.33) that

$$\boxed{\langle N_j \rangle = \frac{1}{e^{\hbar\omega_j/k_B T} - 1}.} \tag{24.67}$$

The transformation to normal coordinates allows us to treat the strongly interacting particles in solids as a collection of statistically independent subsystems in which each normal mode is one of the subsystems. Results for the harmonic approximation can be obtained from the results for the Einstein model in Section 20.4 by replacing the single frequency ω used there by frequencies ω_j, which depend on index j. As in (20.62), the internal energy can be expressed as $U = \Phi_{\rm st} + \langle E_N^{\rm vib} \rangle$, where $\langle E_N^{\rm vib} \rangle$ is the equilibrium average of $E_\alpha^{\rm vib}$. It then follows from (24.63) that the vibrational contribution to the internal energy is

$$\boxed{U_{\rm vib} = \langle E_N^{\rm vib} \rangle = \sum_{j=1}^{3N} \left(\langle N_j \rangle + \frac{1}{2}\right)\hbar\omega_j.} \tag{24.68}$$

This simplifies to the Einstein model result when all frequencies are the same. At $T = 0$, the energy becomes $U_{\text{vib}} = \sum_j \frac{1}{2}\hbar\omega_j$. This is the system's *zero point energy*. Its existence indicates that the motion of the atoms does not completely disappear at absolute zero.

The harmonic approximation is useful for studying both amorphous solids and crystalline solids. Amorphous solids, also called as *glasses*, are similar to liquids in that there is no long range order in the locations of the atoms, but the atoms vibrate about sites fixed in space. In crystals, the normal modes are plane waves and the vibrational excitations are called *phonons*.

24.4 Phonons

The atoms in crystalline solids vibrate about sites that form a periodic lattice of points in space. Because of the periodicity, the normal coordinates that describe the atomic displacements $\boldsymbol{q}_i = (q_i^x, q_i^y, q_i^z)$ have the appearance of plane waves which can be characterized by wave vectors $\boldsymbol{k} = (k_x, k_y, k_z)$. The atomic displacements associated with a typical normal mode in a monatomic crystal are diagrammed in **Figure 24.3**: The atoms in an xy-plane are represented by circles, which for clarity are smaller than the atoms would be. The lattice sites are at the intersections of the straight lines which indicate the underlying symmetry of the structure. The atoms are displaced up and down from their sites in the wave-like pattern indicated by the curved lines. The wave vector for the diagrammed mode is in the x-direction ($k_y = k_z = 0$), and the wave has transverse polarization with atoms displaced in the y-direction ($q_i^x = q_i^z = 0$).

When describing crystalline solids, the normal mode index j is replaced by $\boldsymbol{k}s$, where $\boldsymbol{k}$ is the wave vector and index s that identifies the polarization of wave. As there are two transverse polarizations and one longitudinal polarization, the index has three possible values ($s = 1, 2, 3$). Complex numbers are commonly used to describe the normal coordinates, in which cases the displacement of atom i is

$$\boldsymbol{q}_i = \sum_{\boldsymbol{k}s} \text{Re}(\boldsymbol{A}_{\boldsymbol{k}s} e^{i\boldsymbol{k}\cdot\boldsymbol{R}_i}), \tag{24.69}$$

where $\boldsymbol{A}_{\boldsymbol{k}s}$ is the amplitude of the wave in mode $\boldsymbol{k}s$ and Re takes real part of $\boldsymbol{A}_{\boldsymbol{k}s} e^{i\boldsymbol{k}\cdot\boldsymbol{R}_i}$. The $3N$ normal modes of a monatomic crystal are identified by N values of $\boldsymbol{k}$ and the three values of s.

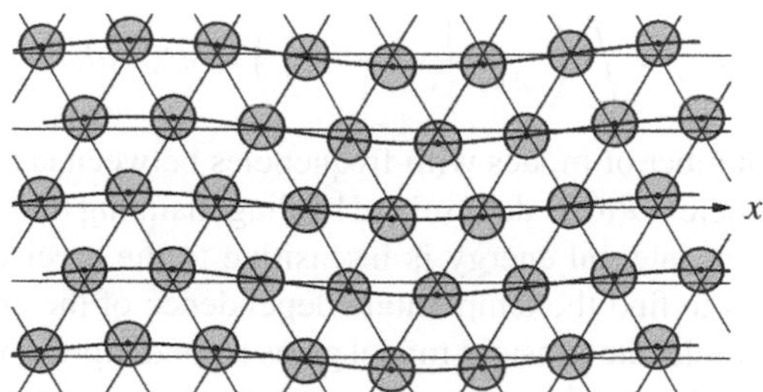

Figure 24.3 *Normal Mode Displacements*

When index j is replaced by $\boldsymbol{ks}$, the energy eigenvalues in (24.63) become

$$E_\alpha^{\text{vib}} = \sum_{\boldsymbol{ks}} \left(N_{\boldsymbol{ks}} + \frac{1}{2}\right) \hbar\omega_{\boldsymbol{ks}}, \tag{24.70}$$

and the allowed values of the quantum numbers are $N_{\boldsymbol{ks}} = 1, 2, \ldots, \infty$. Except for the contributions $\frac{1}{2}\hbar\omega_{\boldsymbol{ks}}$, the eigenvalues E_α^{vib} are the same as those of electromagnetic field in (24.24). Just as $N_{\boldsymbol{ks}}$ is the number of photons in mode $\boldsymbol{k}s$ of the electromagnetic field, *$N_{\boldsymbol{ks}}$ is the number of phonons in vibrational mode $\boldsymbol{k}s$, and the energy of each phonon is $\hbar\omega_{\boldsymbol{ks}}$.* Although the quantum mechanical descriptions of atomic vibrations and electromagnetic fields are similar, there are also significant differences: The exponentials $e^{i\boldsymbol{k}\cdot\boldsymbol{R}_i}$ describe the displacements of individual atoms, whereas the exponentials $e^{i\boldsymbol{k}\cdot\boldsymbol{r}}$ that describe the electromagnetic fields have values at every point $\boldsymbol{r}$ in space. Also, the number of vibrational modes is finite ($3N$ in a monatomic crystals), whereas the number of modes of electromagnetic radiation is infinite.

Internal Energy

The plane wave nature of the normal modes have interesting implications for the temperature dependence of the internal energy $U = \Phi_{\text{st}} + U_{\text{vib}}$ and heat capacity $C_V = (\partial U/\partial T)_V$. When expressed in terms of phonons, the vibrational energy in (24.68) becomes

$$U_{\text{vib}} = \sum_{ks} \left(\langle N_{ks}\rangle + \frac{1}{2}\right) \hbar\omega_{ks}. \tag{24.71}$$

The average number of phonons in mode $\boldsymbol{ks}$ is

$$\langle N_{ks}\rangle = \frac{1}{e^{\hbar\omega_{ks}/k_B T} - 1}, \tag{24.72}$$

and the *zero point energy* is

$$U_0 = \sum_{ks} \frac{1}{2}\hbar\omega_{ks}. \tag{24.73}$$

The sum over normal modes can be expressed as an integral by introducing the number of normal modes per unit frequency $D(\omega)$. Then, by combining the above equations, the vibrational energy becomes

$$U_{\text{vib}} = \int \left(\frac{1}{e^{\hbar\omega/k_B T} - 1} + \frac{1}{2}\right) \hbar\omega D(\omega)\, d\omega, \tag{24.74}$$

where $D(\omega)\Delta\omega$ is the number of modes with frequencies between ω and $\omega + \Delta\omega$.

The density of frequencies $D(\omega)$ is determined by diagonalizing the force constant matrix A_{mn} in (24.49). However, internal energy is insensitive to the precise functional form of $D(\omega)$, and that allows us to find the temperature dependence of the energy with relatively simple models. For example, the Einstein model gives a good representation of the specific heats of solids at intermediate and high temperatures by using a single frequency. Results that are also accurate at low temperatures can be obtained with the Debye model.

Debye Model

Since the exponential in the denominator of (24.72) is large when k_BT is less than $\hbar\omega_{ks}$, the averages $\langle N_{ks}\rangle$ for the higher frequency modes become unimportant at low temperatures. As a result, the low frequencies modes dominate the vibrational energy. The Debye model takes advantage of this. The low frequencies modes are long wavelength excitations, which correspond to sound waves, and the velocity of sound v_s is related to wavelength λ and frequency ν by $v_s = \lambda\nu$. By using $k = 2\pi/\lambda$ and $\nu = 2\pi\omega$, this becomes $\omega = v_s k$, where ω is the angular frequency and k is the magnitude of $\boldsymbol{k}$. Since the velocity of sound in solids depends on polarization, the velocities v_s depend on index s. The relationship $\omega = v_s k$, which is accurate for small values of k, is used in the Debye model for all modes. When the same values of v_s are used for all modes, the density of frequencies $D(\omega)$ is determined by the density of wave vectors D_0, and the range of frequencies is determined by the range of possible wave vectors.

All wave vectors within a spherical region of k-space are included in the Debye model, and the density of wave vectors in k-space is the same as the density for waves in a three-dimensional continuum, which according to (24.11) is $D_0 = V/8\pi^3$. As the normal modes are identified by N different wave vectors, the volume $\frac{4}{3}\pi k_{\max}^3$ of the spherical region must be

$$D_0\frac{4}{3}\pi k_{\max}^3 = \frac{V}{6\pi^2}k_{\max}^3 = N, \tag{24.75}$$

where V is the volume of the solid. Solving for the radius $k_{\max}$ of the region of allowed wave vectors gives $k_{\max} = (6\pi^2 N/V)^{1/3}$.

When the dependence of v_s on polarization is ignored, the relationship between the number of normal modes with frequencies between ω and $\omega + \Delta\omega$, and the number of wave vectors with magnitudes between k and $k + \Delta k$, becomes

$$D(\omega)\Delta\omega = 3D_k\Delta k = \frac{3V}{2\pi^2}k^2\Delta k, \tag{24.76}$$

where the factor of 3 accounts for the three different polarizations and $\Delta k = (dk/d\omega)\Delta\omega$. The relationship between ω and k becomes $\omega = \overline{v}k$, or equivalently $k = \omega/\overline{v}$, where $\overline{v}$ is an average velocity. By using $dk/d\omega = 1/\overline{v}$, the above result becomes

$$D(\omega)\Delta\omega = \frac{3V\omega^2}{2\pi^2\overline{v}^3}\Delta\omega. \tag{24.77}$$

The maximum frequency included in the model is

$$\omega_{\max} = \overline{v}k_{\max} = \overline{v}\left(\frac{6\pi^2 N}{V}\right)^{1/3}. \tag{24.78}$$

By substituting (24.77) into (24.74), the vibrational contribution to the internal energy becomes

$$U_{\text{vib}} = \frac{3V\hbar}{2\pi^2\overline{v}^3}\left[\int_0^{\omega_{\max}} \frac{\omega^3}{e^{\hbar\omega/k_BT} - 1}d\omega + \frac{\omega_{\max}^4}{8}\right], \tag{24.79}$$

where $\frac{1}{8}\omega_{max}^4$ is the integral of $\int \frac{1}{2}\omega^3 d\omega$. This result can be simplified by introducing the *Debye temperature* Θ_D through the relationship

$$\hbar\omega_{max} = k_B\Theta_D. \tag{24.80}$$

It then follows that

$$\omega_{max} = k_B\Theta_D/\hbar \quad \text{and} \quad \Theta_D = \frac{\hbar\omega_{max}}{k_B} = \frac{\hbar\bar{v}}{k_B}\left(\frac{6\pi^2 N}{V}\right)^{1/3}. \tag{24.81}$$

When expressed in terms of θ_D, the factor $3V\hbar/(2\pi^2\bar{v}^3)$ in (24.79) becomes $9N\hbar^4/(k_B\theta_D)^3$. Also, the integral can be simplified by changing variables from ω to x, where $x = \hbar\omega/k_BT$. With these changes, the Debye approximation for the vibrational energy becomes

$$U_{vib} = \frac{9N\hbar^4}{k_B^3\Theta_D^3}\left[\left(\frac{k_BT}{\hbar}\right)^4 \int_0^{x_{max}} \frac{x^3}{e^x - 1}dx + \frac{1}{8}\left(\frac{k_B\Theta_D}{\hbar}\right)^4\right], \tag{24.82}$$

where $x_{max} = \theta_D/T$. Rearranging terms gives

$$\boxed{U_{vib} = 9Nk_B\Theta_D\left[\left(\frac{T}{\Theta_D}\right)^4 \int_0^{\Theta_D/T} \frac{x^3}{e^x - 1}dx + \frac{1}{8}\right].} \tag{24.83}$$

At high temperature ($T \gg \theta_D$) the upper limit $x_{max} = \theta_D/T$ is small, so that we can estimate the integral by using the expansion $e^x = 1 + x + \cdots$. By keeping the lowest order term, the integrand becomes x^2, so that the integral becomes $\frac{1}{3}(\theta_D/T)^3$. By combining this with (24.83), the high temperature limit becomes

$$U_{vib} = 3Nk_BT. \tag{24.84}$$

The Debye approximation for the internal energy U_{vib} is plotted in **Figure 24.4** along with the classical prediction of the equipartition theorem.[2] Notice that the energy approaches a minimum value at absolute zero but does not disappear because of the zero point energy in (24.73). Because the vibrational energy in the harmonic approximation is equally divided between kinetic and potential energies, the kinetic energy of the atoms is only proportional to absolute temperature when the temperature is not too low.

The heat capacity is obtained by differentiating the expression for U_{vib} in (24.79) and simplifying, which gives

$$C_V = \left(\frac{\partial U}{\partial T}\right)_V = 9Nk_B\left(\frac{T}{\theta_D}\right)^3 \int_0^{\Theta_D/T} \frac{x^4 e^x}{(e^x - 1)^2}dx. \tag{24.85}$$

At high temperatures the heat capacity approaches the value predicted by the equipartition theorem,

$$C_V = 3Nk_B. \tag{24.86}$$

[2] The equipartition theorem and its implications for solids are discussed in Sections 21.3 and 15.4.

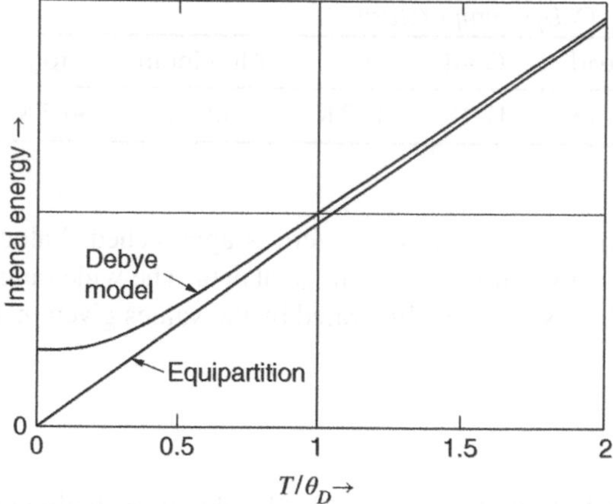

Figure 24.4 *Temperature Dependence of the Internal Energy of Solids*

The heat capacity at very low temperatures ($T \ll \theta_D$) is obtained by taking the upper integration limit in (24.85) to infinity, in which case the integral is equal to $4\pi^4/15$. The resulting low temperature approximation is

$$C_V = \left(\frac{\partial U}{\partial T}\right)_V = \frac{12\pi^4 N k_B}{5}\left(\frac{T}{\theta_D}\right)^3. \tag{24.87}$$

This T^3 approximation is accurate because the long wavelength normal modes that dominate at low temperatures are accurately treated in the Debye model.

The Debye specific heat $C_V/(3Nk_B)$ is plotted in **Figure 24.5** along with the T^3 approximation and the prediction of the Einstein model from (20.67). (The Einstein temperature θ_E used in the figure is 77% of the Debye temperature θ_D.) The Einstein model correctly predicts the drop in the specific heats of solids at intermediate temperatures but predicts

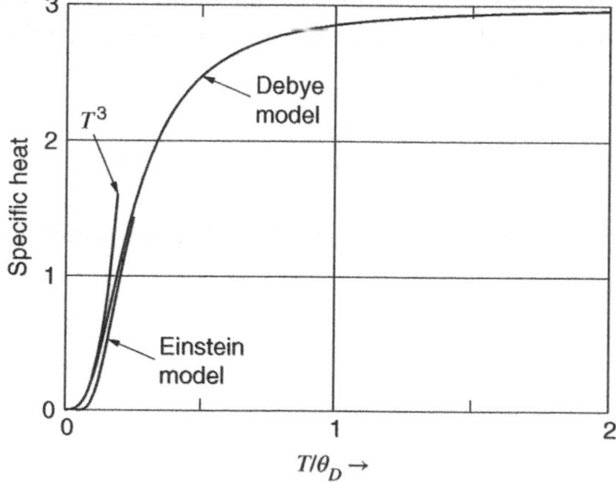

Figure 24.5 *Temperature Dependence of Specific Heat*

Table 24.1 Debye temperatures

Neon	Lead	Gold	Ice	Aluminum	Iron	Silicon
75 K	105 K	170 K	192 K	428 K	467 K	640 K

too rapid an approach to zero as absolute zero is approached. Values of θ_D for specific substances are obtained by fitting to experimental data. The wide range of the Debye temperatures for different substances is illustrated by the values given in Table 24.1.[3]

Problems

24.1 **(a)** Find the Helmholtz free energy F of blackbody radiation in a cavity. Express F as a sum over mode indices j, where $j = (\boldsymbol{k}, s)$.

(b) Show that $d\omega_j/dV = -\omega_j/(3V)$.

(c) Show that pressure P is proportional to the energy density U/V and find the proportionality constant.

(d) For comparison, express the pressure of a monatomic ideal gas as a constant times U/V.

24.2 The vibrational energy in the harmonic approximation is $E_\alpha^{\text{vib}} = \sum_j \left(N_j + \frac{1}{2}\right) \hbar\omega_j$, where α is an abbreviation for $(N_1, N_2, \ldots, N_{3N})$ and $N_j = 0, 1, 2, \ldots, \infty$.

(a) The vibrational contribution to the free energy is $F_{\text{vib}} = -k_B T \ln \sum_\alpha e^{-\beta E_\alpha^{\text{vib}}}$. Show that $F_{\text{vib}} = -k_B T \sum_j \ln\left(\sinh \frac{1}{2}\beta\hbar\omega_j\right)$.

(b) The average value of N_j is $\langle N_j \rangle = (e^{\beta\hbar\omega_j} - 1)^{-1}$. Show that

$$\langle N_j \rangle + \frac{1}{2} = \left[2 \tanh \frac{1}{2}\beta\hbar\omega_j\right]^{-1}.$$

(c) In the *quasi-harmonic approximation* the normal mode frequencies ω_j depend on the volume, which depends of the pressure. Show that the vibrational contribution to the pressure is

$$P = \sum_j \left(\langle N_j \rangle + \frac{1}{2}\right) \hbar\omega_j \gamma_j,$$

where $\gamma_j = -(\partial \ln \omega_j / \partial \ln V)$ is called a *Grüneisen parameter.*

[3] Gray D. E. (1972) *American Institute of Physics Handbook*, 3rd edn, McGraw-Hill, New York, pp. 4–115.

25

Grand Canonical Ensemble

The canonical ensemble describes closed systems, *i.e.*, systems with fixed numbers of particles. Open systems are described by the grand canonical ensemble, and the number of particles is controlled by the chemical potential. Although open systems can exchange particles, the fluctuations in the number of particles are sufficiently small that most systems can be treated as either open or closed. This allows us to choose the ensemble that simplifies the analysis of the system being studied. As discussed in Chapter 26, systems of identical fermions and bosons are best analyzed with the grand canonical ensemble.

25.1 Thermodynamics of Open Systems

A familiar example of an open system is the water in a drinking glass. It is open because of the continual exchange molecules of water, nitrogen, and oxygen between the liquid and the air in the room. We begin our study of open systems by considering single component systems and reviewing the relevant thermodynamic concepts introduced in Chapter 13. Since the instantaneous microscopic value of the number of particles can fluctuate, its thermodynamic value needs to be distinguished from the microscopic value denoted by N. The thermodynamic value is represented by $\overline{N}$, which is the ensemble average of N. At the microscopic level, it is convenient to use $\overline{N}$ to represent the amount of matter in a system, instead of the number of moles n, so that

$$\overline{N} = \langle N \rangle = nN_A, \tag{25.1}$$

where N_A is Avogadro's number.

The thermodynamic potential whose natural coordinates are temperature, pressure, and number of particles is the Gibbs free energy which is given by a function $G(T, P, \overline{N})$. Its differential is

$$dG = \left(\frac{\partial G}{\partial T}\right)_{P\overline{N}} dT + \left(\frac{\partial G}{\partial P}\right)_{T\overline{N}} dP + \left(\frac{\partial G}{\partial \overline{N}}\right)_{TP} d\overline{N}, \tag{25.2}$$

Thermodynamics and Statistical Mechanics: An Integrated Approach, First Edition.
Robert J Hardy and Christian Binek.

where the N in (13.14) is now represented by $\overline{N}$. The relationships $S = -(\partial G/\partial T)_P$ and $V = (\partial G/\partial P)_T$ obtained in Chapter 12 were restricted to systems with fixed numbers of particles. When that restriction is made explicit by the subscripts on the derivatives, entropy and volume are given by

$$S = -\left(\frac{\partial G}{\partial T}\right)_{P\overline{N}} \quad \text{and} \quad V = \left(\frac{\partial G}{\partial P}\right)_{T\overline{N}}. \tag{25.3}$$

The Gibbs free energy per particle is called the *chemical potential*, which is given by the derivative[1]

$$\mu = \left(\frac{\partial G}{\partial \overline{N}}\right)_{TP}. \tag{25.4}$$

The chemical potential μ describes the energy required to introduce one molecule into a system held at constant temperature and pressure. It is an intensive property related to Gibbs free energy by

$$G(T,P,\overline{N}) = \overline{N}\mu(T,P). \tag{25.5}$$

As indicated, μ is independent of $\overline{N}$. As shown in Section 13.2, this is a consequence of G and $\overline{N}$ being extensive, whereas T and P are intensive. By combining (25.2) through (25.4), the differential of $G(T,P,\overline{N})$ becomes

$$dG = -SdT + VdP + \mu d\overline{N}. \tag{25.6}$$

The equilibrium conditions for open systems from (13.28) are

$$\mu_1 = \mu_2, \quad T_1 = T_2, \quad P_1 = P_2. \tag{25.7}$$

Grand Potential

The grand canonical ensemble describes systems in terms of T, V, and μ, which suggests that the thermodynamic potential whose natural variables are T, V, and μ will be useful. Such a potential is obtained by a Legendre transformation of $G(T,P,\overline{N})$ from variables $(T,P,\overline{N})$ to variables (T,P,μ) (see Section (12.2)). This introduces the *grand potential*

$$\Phi_{\text{th}} = G - PV - \mu\overline{N}. \tag{25.8}$$

The differential of the grand potential is

$$d\Phi_{\text{th}} = dG - PdV - VdP - \mu d\overline{N} - \overline{N}d\mu. \tag{25.9}$$

Combining this with $dG = -SdT + VdP + \mu d\overline{N}$ gives

$$\boxed{d\Phi_{\text{th}} = -SdT - PdV - \overline{N}d\mu.} \tag{25.10}$$

Since $d\Phi_{\text{th}}$ is the differential of a function, it follows that

$$\boxed{S = -\left(\frac{\partial \Phi_{\text{th}}}{\partial T}\right)_{V\mu}, \quad P = -\left(\frac{\partial \Phi_{\text{th}}}{\partial V}\right)_{T\mu}, \quad \overline{N} = -\left(\frac{\partial \Phi_{\text{th}}}{\partial \mu}\right)_{TV}.} \tag{25.11}$$

[1] The chemical potential is also defined as the Gibbs energy per mole, which is larger than $\partial G/\partial\overline{N}$ by a factor of N_A.

Combining the definition of Φ_{th} with the result $G = \mu\overline{N}$ in (25.5) implies that

$$\boxed{\Phi_{\text{th}}(T, V, \mu) = -P(T, \mu)V.} \tag{25.12}$$

This result can also be obtained by making the kind of arguments used in Section 13.1 to show that $G = \mu\overline{N}$. The grand potential has an essential role in establishing the relationship between the macroscopic and microscopic descriptions of systems in the grand canonical ensemble.

25.2 Grand Canonical Ensemble

The grand canonical ensemble was derived in Section 18.4 by finding the probability distribution that gives the least biased description of a system when the number of particles can vary. A different way of thinking about systems is appropriate when the number of particles is not fixed. Instead of identifying the system with the matter within a container, we should consider the system to be the matter within a fixed region of space that is part of a larger sample of matter. Since particles can enter and leave the fixed region, we need to specify the number of particles within the region when specifying the system's microstate.

Classical Systems

Since the number of particles is a discrete variable, the classical probability distribution is described by a combined probability and probability density,

$$\begin{aligned} \rho(N; \boldsymbol{p}_1 \cdots \boldsymbol{r}_N) = {} & \text{probability a system has } N \text{ particles with} \\ & \text{momenta and positions in volume } dp^{3N} dr^{3N}. \end{aligned} \tag{25.13}$$

The probability distribution from (18.45) expressed in more detail is

$$\rho(N; \boldsymbol{p}_1 \cdots \boldsymbol{r}_N) = \frac{1}{Q} e^{\mu N / k_B T} \frac{1}{h^{3N} N!} e^{-H_N(\boldsymbol{p}_1 \cdots \boldsymbol{r}_N)/k_B T}. \tag{25.14}$$

The role of the chemical potential μ in the exchange of particles is similar to the role of temperature in the exchange of heat. The value of Q is determined by the normalization condition, which requires that

$$\sum_{N=0}^{\infty} \int dp^{3N} \int dr^{3N} \rho(N; \boldsymbol{p}_1 \cdots \boldsymbol{r}_N) = 1, \tag{25.15}$$

where the position variables $\boldsymbol{r}_i$ are integrated over a region of space of volume V. As the number of momentum and position variables depends on N, the integrations are carried out before the sum. Substituting (25.14) into (25.15) and rearranging terms give

$$\boxed{Q(T, V, \mu) = \sum_{N=0}^{\infty} \frac{e^{\mu N / k_B T}}{h^{3N} N!} \int dp^{3N} \int dr^{3N} e^{-H_N(\boldsymbol{p}_1 \cdots \boldsymbol{r}_N)/k_B T}.} \tag{25.16}$$

This is the classical expression for the *grand partition function*. It can be simplified by using the partition function Z_N for a system of N particles, which as given in (22.43) is

$$Z_N(T, V) = \frac{1}{h^{3N}N!}\int dp^{3N}\int dr^{3N}\, e^{-H_N(p_1\cdots r_N)/k_BT}. \tag{25.17}$$

Combining this with (25.16) gives

$$\boxed{Q(T, V, \mu) = \sum_{N=0}^{\infty} e^{\mu N/k_BT} Z_N(T, V).} \tag{25.18}$$

The factors of $1/N!$ are included in ρ, Q, and Z_N to resolve Gibbs paradox, discussed in Section 22.3, and the factor of h^{-3N} establishes agreement with the quantum ensemble when quantum effects are negligible.

Quantum Systems

The probabilities for open systems when described quantum mechanically are

$$\rho_{N,\alpha} = \frac{1}{Q} e^{\mu N/k_BT} e^{-E_{N,\alpha}/k_BT}, \tag{25.19}$$

where the subscripts on $E_{N,\alpha}$ emphasize that it is an energy eigenvalue of a system of N particles. Normalization requires that

$$\sum_{N=0}^{\infty}\sum_{\alpha} \rho_{N,\alpha} = 1. \tag{25.20}$$

Substituting $\rho_{N,\alpha}$ into this and rearranging terms gives

$$\boxed{Q = \sum_{N=0}^{\infty} e^{\mu N/k_BT} \sum_{\alpha} e^{-E_{N,\alpha}/k_BT},} \tag{25.21}$$

which is the quantum mechanical expression for the *grand partition function*. By using the N-particle quantum partition function from (20.5), we obtain

$$\boxed{Q = \sum_{N=0}^{\infty} e^{\mu N/k_BT} Z_N,} \tag{25.22}$$

where $Z_N = \sum_{\alpha} e^{-E_{N,\alpha}/k_BT}$. The factor of $1/N!$ introduced to resolve Gibbs paradox for classical systems is not included for quantum systems because it is supplied by the concept that identical particles are indistinguishable, as discussed in Chapter 26.

25.3 Properties and Fluctuations

Thermodynamic properties can be obtained either from the grand potential or by evaluating ensemble averages. The grand potential Φ_{th} is related to the grand partition function by $\Phi_{\text{th}} = -k_B T \ln Q$. To justify this assertion, we show that the number of particle $\overline{N}$ and internal energy U deduced from Φ_{th} are the same as the values predicted by the ensemble averages of N and $E_{N,\alpha}$. Combining $\Phi_{\text{th}} = -k_B T \ln Q$ with the relationship $\Phi_{\text{th}} = -PV$ from (25.12) indicates that

$$\boxed{P(T, V, \mu)V = -\Phi_{\text{th}}(T, V, \mu) = k_B T \ln Q(T, V, \mu).} \tag{25.23}$$

By using the probabilities $\rho_{N,\alpha}$ in (25.19), the ensemble average for $\overline{N}$ becomes

$$\overline{N} = \langle N \rangle = \sum_{N=0}^{\infty} \sum_{\alpha} \rho_{N,\alpha} N = \frac{1}{Q} \sum_{N=0}^{\infty} \sum_{\alpha} e^{\mu N/k_B T} e^{-E_{N,\alpha}/k_B T} N, \tag{25.24}$$

where the sum over α is performed before the sum over N. To find the value of $\overline{N}$ from the grand potential, we combine $\Phi_{\text{th}} = -k_B T \ln Q$ with the thermodynamic relationship $\overline{N} = -(\partial \Phi_{\text{th}}/\partial \mu)_{TV}$ to show that

$$\overline{N} = k_B T \frac{\partial \ln Q(T, V, \mu)}{\partial \mu} = \frac{k_B T}{Q} \frac{\partial Q(T, V, \mu)}{\partial \mu}. \tag{25.25}$$

From this and the expression for Q in (25.21), it follows that

$$\overline{N} = \frac{k_B T}{Q} \sum_{N=0}^{\infty} \frac{\partial e^{\mu N/k_B T}}{\partial \mu} \sum_{\alpha} e^{-E_{N,\alpha}/k_B T}. \tag{25.26}$$

By using $(\partial e^{\mu N/k_B T}/\partial \mu) = e^{\mu N/k_B T}(N/k_B T)$, we find that

$$\overline{N} = \frac{1}{Q} \sum_{N=0}^{\infty} \sum_{\alpha} e^{\mu N/k_B T} e^{-E_{N,\alpha}/k_B T} N, \tag{25.27}$$

which is the same as the ensemble average in (25.24).

To find the internal energy from the grand potential, we combine the result $G = \overline{N}\mu$ from (25.5) with the relationship $G = U - TS + PV$ from Table 12.1 and solve for U, which gives $U = \mu\overline{N} + TS - PV$. By using $\Phi_{\text{th}} = -PV$ and the thermodynamic relationships for $\overline{N}$ and S in (25.11), we obtain

$$U = -\mu \left(\frac{\partial \Phi_{\text{th}}}{\partial \mu} \right)_{TV} - T \left(\frac{\partial \Phi_{\text{th}}}{\partial T} \right)_{V\mu} + \Phi_{\text{th}}. \tag{25.28}$$

It follows from this and $\Phi_{\text{th}} = -k_B T \ln Q$ that

$$U = \frac{k_B T}{Q} \left[\mu \left(\frac{\partial Q}{\partial \mu} \right)_{TV} + T \left(\frac{\partial Q}{\partial T} \right)_{V\mu} \right]. \tag{25.29}$$

By substituting the expression for Q in (25.21) into this and using

$$\mu \frac{\partial e^{\mu N/k_B T}}{\partial \mu} = -T \frac{\partial e^{\mu N/k_B T}}{\partial T} \quad \text{and} \quad \frac{\partial e^{-E_{N,\alpha}/k_B T}}{\partial T} = \frac{E_{N,\alpha}}{k_B T^2} e^{-E_{N,\alpha}/k_B T}, \tag{25.30}$$

we find that

$$U = \frac{1}{Q}\sum_{N=0}^{\infty}\sum_{\alpha} e^{\mu N/k_BT} e^{-E_{N,\alpha}/k_BT} E_{N,\alpha} = \sum_{N=0}^{\infty}\sum_{\alpha} \rho_{N,\alpha} E_{N,\alpha} = \langle E\rangle. \tag{25.31}$$

Thus, the internal energy U deduced from $\Phi_{\text{th}} = -k_BT \ln Q$ is the same as the ensemble averages of $E_{N,\alpha}$ calculated with the probabilities $\rho_{N,\alpha}$.

Fluctuations

The probability $\rho(N)$ that a system contains N particles is obtained by summing over the microstates consistent with the value of N,

$$\rho(N) = \sum_{\alpha} \rho_{N,\alpha}. \tag{25.32}$$

The corresponding classical expression is

$$\rho(N) = \int dp^{3N} \int dr^{3N} \rho(N; \boldsymbol{p}_1 \cdots \boldsymbol{r}_N). \tag{25.33}$$

By using the expressions for $\rho(N; \boldsymbol{p}_1 \cdots \boldsymbol{r}_N)$ and $\rho_{N,\alpha}$ in (25.14) and (25.19), we find both classically and quantum mechanically that

$$\rho(N) = \frac{1}{Q} e^{\mu N/k_BT} Z_N. \tag{25.34}$$

The variance of N is found by taking the second derivative of $\ln Q$,

$$\left(\frac{\partial^2 \ln Q}{\partial \mu^2}\right)_{TV} = \frac{\partial}{\partial \mu}\left(\frac{1}{Q}\frac{\partial Q}{\partial \mu}\right)_{TV} = \frac{1}{Q}\left(\frac{\partial^2 Q}{\partial \mu^2}\right)_{TV} - \frac{1}{Q^2}\left(\frac{\partial Q}{\partial \mu}\right)_{TV}^2. \tag{25.35}$$

It follows from $Q = \sum_N e^{\mu N/k_BT} Z_N$ that the first two derivatives of Q are

$$\left(\frac{\partial Q}{\partial \mu}\right)_{TV} = \sum_{N=0}^{\infty} \frac{N}{k_BT} e^{\mu N/k_BT} Z_N = \frac{Q}{k_BT} \sum_{N=0}^{\infty} \rho(N) N \tag{25.36}$$

and

$$\left(\frac{\partial^2 Q}{\partial \mu^2}\right)_{TV} = \sum_{N=0}^{\infty} \left(\frac{N}{k_BT}\right)^2 e^{\mu N/k_BT} Z_N = \frac{Q}{(k_BT)^2} \sum_{N=0}^{\infty} \rho(N) N^2, \tag{25.37}$$

where (25.34) has been used. Thus,

$$\frac{1}{Q}\left(\frac{\partial Q}{\partial \mu}\right)_{TV} = \frac{1}{k_BT}\langle N\rangle \quad \text{and} \quad \frac{1}{Q}\left(\frac{\partial^2 Q}{\partial \mu^2}\right)_{TV} = \frac{1}{(k_BT)^2}\langle N^2\rangle. \tag{25.38}$$

Combining these with (25.35) gives

$$\left(\frac{\partial^2 \ln Q}{\partial \mu^2}\right)_{TV} = \frac{1}{(k_BT)^2}\left[\langle N^2\rangle - \langle N\rangle^2\right] = \frac{\text{Var}[N]}{(k_BT)^2}, \tag{25.39}$$

where $\text{Var}[N] = \langle N^2 \rangle - \langle N \rangle^2$ is the variance in N. Thus,

$$\text{Var}[N] = (k_B T)^2 \left(\frac{\partial^2 \ln Q}{\partial \mu^2} \right)_{TV}. \tag{25.40}$$

The size of fluctuations is measured by the standard deviation, which is the square root of the variance,

$$\boxed{\sigma_N = k_B T \left(\frac{\partial^2 \ln Q}{\partial \mu^2} \right)^{1/2}.} \tag{25.41}$$

25.4 Ideal Gases

The ultimate justification of the grand canonical ensemble is the accuracy of its predictions. To test its predictive powers, we check its ability to predict the properties of monatomic gases, which obey the ideal gas equation of state $PV = \overline{N} k_B T$ and have heat capacity $C_V = \frac{3}{2}\overline{N} k_B$. To find the properties of the gas, we use the general expressions for the grand partition function from (25.18) and (25.22),

$$Q = \sum_N e^{\mu N / k_B T} Z_N. \tag{25.42}$$

When energy separates into a sum of independent contributions, the N-particle partition function Z_N becomes a product of single-particle partition functions. For particles of the same type, we have

$$Z_N = [Z_s]^N. \tag{25.43}$$

The factor of $1/N!$ needed to resolve Gibbs paradox is not included in this value. Including the factor of $1/N!$ and substituting (25.43) into the expression for Q give

$$Q = \sum_N e^{\mu N / k_B T} \frac{1}{N!} [Z_s]^N. \tag{25.44}$$

By using $Z_s = e^{\ln Z_s}$ and $\exp x = \sum_n (1/n!) x^n$, we obtain

$$Q = \sum_N \frac{1}{N!} \left(e^{(\mu/k_B T) + \ln Z_s} \right)^N = \exp\left(e^{(\mu/k_B T) + \ln Z_s} \right). \tag{25.45}$$

The logarithm of this is

$$\ln Q = e^{(\mu/k_B T) + \ln Z_s} = e^{\mu/k_B T} Z_s. \tag{25.46}$$

By combining this with $\Phi_{\text{th}} = -k_B T \ln Q$ from (26.23), the grand potential becomes

$$\Phi_{\text{th}} = -k_B T e^{(\mu/k_B T) + \ln Z_s} = -k_B T e^{\mu/k_B T} Z_s. \tag{25.47}$$

From this and the thermodynamic relationship for the average number of particles, we obtain

$$\overline{N} = -\left(\frac{\partial \Phi_{\text{th}}}{\partial \mu} \right)_{TV} = e^{\mu/k_B T} Z_s = -\frac{\Phi_{\text{th}}}{k_B T} = \ln Q. \tag{25.48}$$

To check that the internal energy and heat capacity of a monatomic ideal gas are correctly predicted, the system's single-particle partition function is needed. Its value from (20.36) is

$$Z_s = \left(\frac{2\pi m k_B T}{h^2}\right)^{3/2} V. \tag{25.49}$$

By combining this with (25.47), the grand potential becomes

$$\Phi_{\text{th}}(T, V, \mu) = -K_o e^{\mu/k_B T} (k_B T)^{5/2} V, \tag{25.50}$$

where $K_o = (2\pi m/h^2)^{3/2}$. We found in (25.28) that the internal energy is given by

$$U = \Phi_{\text{th}} - \mu \left(\frac{\partial \Phi_{\text{th}}}{\partial \mu}\right)_{TV} - T\left(\frac{\partial \Phi_{\text{th}}}{\partial T}\right)_{V\mu}. \tag{25.51}$$

The derivatives of Φ_{th} are

$$\frac{\partial \Phi_{\text{th}}}{\partial \mu} = \frac{1}{k_B T}\Phi_{\text{th}} \quad \text{and} \quad \frac{\partial \Phi_{\text{th}}}{\partial T} = \left(\frac{5}{2}\frac{1}{T} - \frac{\mu}{k_B T^2}\right)\Phi_{\text{th}}, \tag{25.52}$$

so that

$$U = \left[1 - \frac{\mu}{k_B T} - T\left(\frac{5}{2}\frac{1}{T} - \frac{\mu}{k_B T^2}\right)\right]\Phi_{\text{th}} = -\frac{3}{2}\Phi_{\text{th}}. \tag{25.53}$$

Combining this with $\overline{N} = -\Phi_{\text{th}}/k_B T$ from (25.48) gives

$$U = \frac{3}{2}\overline{N} k_B T, \tag{25.54}$$

and the constant-volume heat capacity is

$$C_V = \left(\frac{\partial U}{\partial T}\right)_V = \frac{3}{2}\overline{N} k_B. \tag{25.55}$$

This is the correct value of the heat capacity of a monatomic ideal gas. The equation of state is found by combining the relationship $\Phi_{\text{th}} = -PV$ from (25.12) with $\overline{N} = -\Phi_{\text{th}}/k_B T$ to show that

$$PV = \overline{N} k_B T, \tag{25.56}$$

which is the familiar ideal gas equation of state.

We found in the previous section that the size of the fluctuation in N is determined by the second derivative of the logarithm of the grand partition function. If follows from the result $\ln Q = e^{\mu/k_B T} Z_s$ in (25.46) that

$$\frac{\partial^2 \ln Q}{\partial \mu^2} = \frac{\ln Q}{(k_B T)^2}. \tag{25.57}$$

By substituting this into the equation for the variance in (25.40) and using $\overline{N} = \ln Q$, we find that

$$\text{Var}[N] = (k_B T)^2 \left(\frac{\partial^2 \ln Q}{\partial \mu^2}\right)_{TV} = \ln Q = \overline{N}. \tag{25.58}$$

Thus, the standard deviation for the number of particles is $\sigma_N = \sqrt{\overline{N}}$, so that the number of particles (atoms) in the gas can be expresses as

$$\overline{N} \pm \sigma_N = \overline{N} \pm \sqrt{\overline{N}} = \overline{N}\left(1 \pm 1/\sqrt{\overline{N}}\right). \tag{25.59}$$

For example, consider a gas of $\overline{N} = 10^{20}$ atoms at $T = 300\,\text{K}$ and $P = 100\,\text{kPa}$, which has a volume of about $4\,\text{cm}^3$. The relative uncertainty in the number of particles in this gas is $\pm 10^{-10}$, which is small compared to the uncertainty in most experimental measurements.

The grand canonical ensemble correctly predicts the properties of monatomic gases. Nevertheless, the same predictions can be obtained more straightforwardly by using the canonical ensemble, as was done in Chapter 20. This is not the case when we investigate systems of identical fermion and bosons in Chapter 26.

Problems

25.1 **(a)** Show that $U = \Phi_{\text{th}} + TS + \mu\overline{N}$. (The definition of G in Table 12.4 is useful.)

(b) Find the entropy $S(T, V, \mu)$, the number atoms $\overline{N}(T, V, \mu)$, and the internal energy $U(T, V, \mu)$ of a monatomic ideal gas. Use the potential Φ_{th} of the gas given in Section 25.4. (The results are simpler when expressed in terms of Φ_{th}.)

(c) Re-express the internal energy $U(T, V, \mu)$ as $U(T, \overline{N})$.

26

Fermions and Bosons

The indistinguishability of identical particles has different implications in quantum mechanics than in classical mechanics. When a particle is treated classically and has an index assigned to it, it can be distinguished at later times by following its trajectory back to where it was earlier. This cannot be done in quantum mechanics because the concept of a trajectory – a line traced out in space – is no longer applicable. Furthermore, the concepts of indistinguishability and exchange symmetry alter the way in which collections of identical particles are studied. The application of these concepts are best analyzed in the grand canonical ensemble with the microstates described by the numbers of particles in different single-particle states. The two kinds of particles, *fermions* and *bosons*, behave differently because of the restrictions imposed on these numbers.

26.1 Identical Particles

Particles that are identical in every way are *indistinguishable*, otherwise they are *distinguishable*. For example, the different isotopes of an element are distinguishable. A helium atom with a nucleus containing two protons and two neutrons (^{4}He) is distinguishable from an ^{3}He atom, whose nucleus has only one neutron. However, all ^{4}He atoms, all ^{3}He atoms, and all electrons are indistinguishable. Electrons and ^{3}He atoms are fermions, while ^{4}He atoms are bosons.

The Hamiltonian of a system of noninteracting identical particles is a sum of terms,

$$H_N(\boldsymbol{p}_1 \cdots \boldsymbol{r}_N) = H_s(\boldsymbol{p}_1, \boldsymbol{r}_1) + \cdots + H_s(\boldsymbol{p}_i, \boldsymbol{r}_i) + \cdots + H_s(\boldsymbol{p}_N, \boldsymbol{r}_N), \tag{26.1}$$

where the only differences among the single-particle Hamiltonians $H_s(\boldsymbol{p}_i, \boldsymbol{r}_i)$ is in the subscripts on the momentum and position variables. When analyzed quantum mechanically the eigenfunctions of the N-particle Schrödinger equation $\widehat{H}_N \Psi_\alpha = E_{N,\alpha} \Psi_\alpha$ were expressed in Chapter 20 as products of the eigenfunctions of the single-particle Schrödinger

Thermodynamics and Statistical Mechanics: An Integrated Approach, First Edition.
Robert J Hardy and Christian Binek.

equation $\widehat{H}_s \psi_k = \varepsilon_k \psi_k$, which is appropriate when the particles are distinguishable. When they are indistinguishable, the N-particle eigenfunctions are sums of products of the single-particle eigenfunctions, as will be described. For free particles in systems with periodic boundary conditions, the single-particle Hamiltonian and Schrödinger equation are

$$\widehat{H}_s(\widehat{\boldsymbol{p}}_i, \boldsymbol{r}_i) = -\frac{\hbar^2}{2m}\nabla_i^2 \quad \text{and} \quad -\frac{\hbar^2}{2m}\nabla^2 \psi_k(\boldsymbol{r}) = \varepsilon_k \psi_k(\boldsymbol{r}), \tag{26.2}$$

where the eigenfunctions and eigenvalues from (24.15) are

$$\psi_k(\boldsymbol{r}) = \frac{1}{V} e^{i\boldsymbol{k}\cdot\boldsymbol{r}} \quad \text{and} \quad \varepsilon_k = \frac{\hbar^2}{2m} k^2. \tag{26.3}$$

The intrinsic angular momentum, or *spin*, is not included in this. Because of spin, each single-particle state is specified by both a wave vector $\boldsymbol{k}$ and a *spin quantum number* s, whose allowed values are

$$s = -S, -S+1, \cdots, +S-1, +S. \tag{26.4}$$

S is the *magnitude of spin*, which is an intrinsic property of a particle. For bosons, it is an integer ($S = 1, 2, \ldots$). For fermions, it is a half odd integer ($S = \frac{1}{2}, \frac{3}{2}, \ldots$). The discussion here is limited to spin-0 bosons ($S = 0$) and spin-$\frac{1}{2}$ fermions ($S = \frac{1}{2}$). As s has only one allowed value for spin-0 particles, the spin quantum number can be ignored. For spin-$\frac{1}{2}$ particles, the spin quantum number has two values, $s = -\frac{1}{2}$ and $s = +\frac{1}{2}$.

The property of spin introduces a fourth coordinate into the description of a particle. For example, the wave function for an electron has four independent variables x, y, z, and σ, where σ is the *spin coordinate*, which is not to be confused with the spin quantum number. The three position coordinates (x, y, z) are continuous, while the spin coordinate is discrete with only two values, $\sigma = +1$ and $\sigma = -1$. These values can be used to identify the elements of a column vector, so that an electron's wave function becomes

$$\psi(\boldsymbol{r}, \sigma) = \begin{pmatrix} \psi(\boldsymbol{r}, +1) \\ \psi(\boldsymbol{r}, -1) \end{pmatrix}. \tag{26.5}$$

The energy of an electron in a magnetic field is $\mu_0 \mathcal{H} \mu_B \sigma$ (see Section 21.5), where μ_0 is the permeability of free space, $\mathcal{H}$ is the magnitude of the field, and μ_B is the Bohr magneton. In the absence of a magnetic field, the energy is independent of σ, the single-particle Schrödinger equation is unchanged from (26.2), and the eigenfunctions are

$$\psi_{ks}(\boldsymbol{r}, \sigma) = \psi_k(\boldsymbol{r}) \chi_s(\sigma), \tag{26.6}$$

where $\psi_k(\boldsymbol{r})$ is given by (26.3) and the spin functions are

$$\chi_{+\frac{1}{2}}(\sigma) = \begin{pmatrix} 1 \\ 0 \end{pmatrix} \quad \text{and} \quad \chi_{-\frac{1}{2}}(\sigma) = \begin{pmatrix} 0 \\ 1 \end{pmatrix}. \tag{26.7}$$

The energy of state $\boldsymbol{k}s$ is independent of the direction of $\boldsymbol{k}$ and the value of s, so that its energy eigenvalue is

$$\varepsilon_k = \frac{\hbar^2}{2m} k^2. \tag{26.8}$$

Abbreviated Notation

The eigenfunctions of systems of distinguishable particles were expressed in Chapter 20 as sums of single-particle eigenfunctions. When spin is included they are

$$\Psi^0_\alpha(\mathbf{r}_1\sigma_1,\mathbf{r}_2\sigma_2,\ \ldots\ ,\mathbf{r}_N\sigma_N) = \psi_{\mathbf{k}_1s_1}(\mathbf{r}_1\sigma_1)\psi_{\mathbf{k}_2s_2}(\mathbf{r}_2\sigma_2)\cdots\psi_{\mathbf{k}_Ns_N}(\mathbf{r}_N\sigma_N), \tag{26.9}$$

where the associated eigenvalues are

$$E_{N,\alpha} = \varepsilon_{k_1} + \varepsilon_{k_2} + \cdots + \varepsilon_{k_N}, \tag{26.10}$$

and α is an abbreviation for $(\mathbf{k}_1s_1,\mathbf{k}_2s_2,\cdots,\mathbf{k}_Ns_N)$. To prevent important concepts from being obscured by a cloud symbols, it is convenient to use a single symbol j to identify single-particle states and a single index i to identify the coordinates of individual particles,

$$j = (\mathbf{k}, s) \quad \text{and} \quad i = (\mathbf{r}, \sigma). \tag{26.11}$$

In this abbreviated notation, the eigenfunctions and eigenvalues of a system of N distinguishable particles are

$$\Psi^0_\alpha(1, 2,\ \ldots\ , N) = \psi_{j_1}(1)\psi_{j_2}(2)\cdots\psi_{j_N}(N) \tag{26.12}$$

and

$$E_{N,\alpha} = \varepsilon_{j_1} + \varepsilon_{j_2} + \cdots + \varepsilon_{j_N}. \tag{26.13}$$

Since the single-particle states are discrete, they can be put into one-to-one correspondence with the integers ($j = 0, 1, 2,\ \ldots\ , \infty$). It is intuitive to associate the integer with the states so that $j < j'$ implies $\varepsilon_j \leq \varepsilon_{j'}$, which means that $\varepsilon_0 \leq \varepsilon_1 \leq \varepsilon_2 \leq \cdots$.

26.2 Exchange Symmetry

We should be able to exchange the indices (labels) assigned to indistinguishable particles without changing a system's observable properties. This characteristic is achieved by requiring that exchanging the indices of identical particles satisfies the requirement of *exchange symmetry*. When expressed in the abbreviated notation, exchange symmetry requires that

$$\Psi(1,\ \ldots\ , m,\ \ldots\ , n,\ \ldots\ , N) = \pm\Psi(1,\ \ldots\ , n,\ \ldots\ , m,\ \ldots\ , N), \tag{26.14}$$

where the plus sign is for *bosons* and the minus sign for *fermions*.

The significance of exchange symmetry can be seen by considering a system of two particles ($N = 2$). Consider the wave function in (26.12), which becomes

$$\Psi^0_\alpha(1, 2) = \psi_{j'}(1)\psi_{j''}(2). \tag{26.15}$$

Exchanging particle coordinates gives $\Psi^0_\alpha(2, 1) = \psi_{j'}(2)\psi_{j''}(1)$. Since $\Psi^0_\alpha(2, 1)$ is in general *not* equal to either $+\Psi^0_\alpha(1, 2)$ or $-\Psi^0_\alpha(1, 2)$, the function $\Psi^0_\alpha(1, 2)$ does *not* possess exchange symmetry. However, the following linear combination of $\Psi^0_\alpha(1,2)$ and $\Psi^0_\alpha(2,1)$ does have the required symmetry:

$$\Psi^{\pm}_{j'j''}(1,2) = \frac{1}{\sqrt{2}}\left[\psi_{j'}(1)\,\psi_{j''}(2) \pm \psi_{j'}(2)\psi_{j''}(1)\right]. \tag{26.16}$$

The factor of $1/\sqrt{2}$ normalizes the combination. The Hamiltonian for a system of two free particles is $\hat{H}_s(1) + \hat{H}_s(2)$, where $\hat{H}_s(i)$ operates on particle coordinates i and

$$\hat{H}_s(i)\psi_j(i) = \varepsilon_j\psi_j(i) \tag{26.17}$$

Since both $\Psi^0_\alpha(1,2) = \psi_{j'}(1)\psi_{j''}(2)$ and $\Psi^0_\alpha(2,1) = \psi_{j'}(2)\psi_{j''}(1)$ are associated with the eigenvalue $(\varepsilon_{j'} + \varepsilon_{j''})$, the eigenvalue associated with $\Psi^{\pm}_{j'j''}(1,2)$ is also $(\varepsilon_{j'} + \varepsilon_{j''})$. Choosing the minus sign in (26.16) yields $\Psi^-_{j'j''}(1,2)$, which is a fermion eigenfunction because $\Psi^-_{j'j''}(1,2) = -\Psi^-_{j'j''}(2,1)$. Choosing the plus sign yields $\Psi^+_{j'j''}(1,2)$, which is a boson eigenfunction because $\Psi^+_{j'j''}(1,2) = +\Psi^+_{j'j''}(2,1)$.

Imposing exchange symmetry removes the association of the indices j' and j'' with specific particles. It also reduces the number of eigenstates. In the absence of exchange symmetry, the functions $\Psi_\alpha(1,2)$ and $\Psi_\alpha(2,1)$ are associated with *different* eigenstates. In contrast, when exchange symmetry is required, $\Psi^-_{j'j''}$ is the only eigenfunction for two fermions that can be formed from $\psi_{j'}$ and $\psi_{j''}$. Similarly, $\Psi^+_{j'j''}$ is the only eigenfunction for a system of two bosons.

The *Pauli exclusion principle* is another consequence of exchange symmetry. The eigenfunction for two identical fermions is

$$\Psi^-_{j'j''}(1,2) = \frac{1}{\sqrt{2}}\left[\psi_{j'}(1)\,\psi_{j''}(2) - \psi_{j'}(2)\,\psi_{j''}(1)\right]. \tag{26.18}$$

Because $\Psi^-_{j'j''}$ vanishes when $j' = j''$, there is no eigenfunction for two fermions with both particles in the same single-particle state. Thus, *two identical fermions cannot occupy the same single-particle state.*

Systems of N Fermions

The eigenfunctions for systems of N identical fermions can be made consistent with exchange symmetry by representing them as *Slater determinants*

$$\Psi^-_{j_1 j_2,\ldots j_N}(1,2,\ldots,N) = \frac{1}{\sqrt{N!}}\begin{vmatrix} \psi_{j_1}(1) & \psi_{j_1}(2) & \cdots & \psi_{j_1}(N) \\ \psi_{j_2}(1) & \psi_{j_2}(2) & \cdots & \psi_{j_2}(N) \\ \vdots & \vdots & \ddots & \vdots \\ \psi_{j_N}(1) & \psi_{j_N}(2) & \cdots & \psi_{j_N}(N) \end{vmatrix}, \tag{26.19}$$

where the factor of $1/\sqrt{N!}$ normalizes the eigenfunction. The required antisymmetry is a consequence of the fact that exchanging two columns (or two rows) of a determinant multiplies its value by −1. Since exchanging two columns of a Slater determinant exchanges the coordinates of two particles, the exchange of columns m and n yields the result

$$\Psi^-_{j_1 j_2,\dots j_N}(1,\ \dots\ ,m,\ \dots\ ,n,\ \dots\ ,N) = -\Psi^-_{j_1 j_2,\dots j_N}(1,\ \dots\ ,n,\ \dots\ ,m,\ \dots\ ,N), \tag{26.20}$$

which is consistent with (26.14). The Pauli exclusion principle follows from the fact that a determinant is zero when two or more of its rows (or columns) are the same. As each row of a Slater determinant contains the same single-particle function ψ_j, it follows that two (or more) identical fermions cannot be in the same single-particle state.

Exchange symmetry changes neither the system's Hamiltonian nor the energy of its eigenstates, but it significantly reduces the degeneracy of the energy levels. This can be seen by expanding the determinant. The expansion of the determinant for $N = 2$ is

$$\begin{vmatrix} \psi_{j_1}(1) & \psi_{j_1}(2) \\ \psi_{j_2}(1) & \psi_{j_2}(2) \end{vmatrix} = \psi_{j_1}(1)\psi_{j_2}(2) - \psi_{j_1}(2)\psi_{j_2}(1). \tag{26.21}$$

The expansion of the determinant for N particles yields a sum of $N!$ products of N single-particle functions. Each of the $N!$ functions $\psi_{j_1} \cdots \psi_{j_N}$ in (26.19) is a linearly independent eigenfunction associated with energy $(\varepsilon_{j_1} + \varepsilon_{j_2} + \cdots + \varepsilon_{j_N})$. When exchange symmetry is *not* required, this would contribute a degeneracy of $N!$ to the energy level with this energy. In contrast, when exchange symmetry is required, the only eigenfunction that can be formed from these $N!$ single-particle functions is the one in (25.19). Thus, satisfying the requirement of exchange symmetry reduces the degeneracy of the energy levels by a factor of $1/N!$.

As changing the order of the subscripts j_1 through j_N on the Slater determinant $\Psi^-_{j_1 j_2,\dots j_N}$ either leaves it unchanged or multiplies it by −1, the identity of $\Psi^-_{j_1 j_2,\dots j_N}$ as an eigenstate of the system is independent of the ordering of subscripts. This suggests that we identify the eigenstates by specifying the number of times that each index appears in the list $j_1, j_2,\ \dots\ ,j_N$ without indicating its position in the list, so that $\Psi^-_{j_1 j_2,\dots j_N}$ becomes

$$\Psi^-_{N_0,N_1,N_2,\dots}(1,2,\ \dots\ ,N) = \Psi^-_{j_1 j_2,\dots j_N}(1,2,\ \dots\ ,N), \tag{26.22}$$

where N_j is the number of times ψ_j is used to form a row of the Slater determinant and N_j is the *occupation numbers* of single-particle state j.

For fermions, the occupation numbers have only two possible values,

$$\boxed{N_j = 0, 1.} \tag{26.23}$$

When a single index α is used to identify eigenstates of N identical fermions, it is an abbreviation for the occupation numbers of the single-particle states,

$$\alpha = (N_0, N_1, N_2, \dots). \tag{26.24}$$

Since N_j is the number of times that the energy of single-particle state j appears in the sum $(\varepsilon_{j_1} + \varepsilon_{j_2} + \cdots + \varepsilon_{j_N})$, the energy eigenvalue of state α is

$$\boxed{E_{N,\alpha} = \sum_{j=0}^{\infty} N_j \varepsilon_j,} \tag{26.25}$$

where it is require that

$$\boxed{\sum_{j=0}^{\infty} N_j = N.} \tag{26.26}$$

For an ideal gas of fermions, j is an abbreviation for $\boldsymbol{k}s$, and the single-particle energies are $\varepsilon_k = \hbar^2 k^2/2m$.

Systems of *N* Bosons

The eigenfunctions for systems of identical bosons can be made consistent with exchange symmetry in a manner similar to that used for systems of fermions. The difference is that when forming the linear combination of the functions obtained by rearranging the arguments $(1, \dots, N)$ of the functions $\psi_{j_1}(1) \cdots \psi_{j_N}(N)$, the terms are all multiplied by $+1$. The resulting N-particle boson eigenfunctions can be thought of as Slater determinants in which all of the -1's have been replaced by $+1$. Since no minus signs are involved, there is no exclusion principle. Specifically, any number of bosons can be in the same single-particle state. When a single index α is used to identify the eigenstates of systems of bosons, it is again an abbreviation for the occupation numbers of the single-particle states, as in (26.24). The principal difference between bosons and fermions is the values that the occupation numbers can have. For bosons, the occupation numbers can have any value from 0 through N,

$$\boxed{N_j = 0, 1, \dots, N.} \tag{26.27}$$

The energy eigenvalue of state α is

$$\boxed{E_{N,\alpha} = \sum_{j=0}^{\infty} N_j \varepsilon_j,} \tag{26.28}$$

where it is required that

$$\boxed{\sum_{j=0}^{\infty} N_j = N.} \tag{26.29}$$

As a spin quantum number is not needed with spin-0 bosons, index j is an abbreviation for $\boldsymbol{k}$.

Comparisons

The quantum mechanical descriptions of five systems are summarized in Table 26.1. As described in Section 24.1, the microstates of a monatomic ideal gas are specified by the wave vectors $\boldsymbol{k}_i$ of the individual particles ($i = 1, 2, \ldots, N$). As discussed in Section 24.4, the microstates that describe atomic vibrations in crystalline solids are specified by the numbers of phonons $N_{\boldsymbol{k}s}$ in $3N$ different modes of vibration. In both systems the atoms were treated as distinguishable, and the eigenstates were identified by $3N$ quantum numbers. The wave vectors $\boldsymbol{k}$ for identical bosons, identical fermions, and electromagnetic radiation can have arbitrarily large magnitudes, so that the sums over $\boldsymbol{k}$ contain an infinite number of terms. Although the atoms in monatomic gases may not all be identical, there are groups of identical atoms whose eigenfunctions should in principle satisfy the requirement of exchange symmetry. The effects of exchange symmetry are especially significant for the

Table 26.1 *Quantum mechanical systems*

System	Microstate identifier	Meaning of identifier	Microstate energy
1. Ideal gas of N distinguishable particles	$\alpha = (\boldsymbol{k}_1, \ldots, \boldsymbol{k}_i, \ldots, \boldsymbol{k}_N)$ $\boldsymbol{k}_i$ is an allowed wave vector.	$\boldsymbol{k}_i$ is a single-particle state of particle i.	$E_{N,\alpha} = \sum_i \frac{\hbar^2}{2m}\lvert\boldsymbol{k}_i\rvert^2$ N terms in sum
2. Vibrations of crystalline solids of N atoms	$\alpha = (\ldots, N_{\boldsymbol{k}s}, \ldots)$ $N_{\boldsymbol{k}s} = 0, 1, 2, \ldots, \infty$	$N_{\boldsymbol{k}s}$ is number of phonons in normal mode $\boldsymbol{k}s$	$E_{N,\alpha} = \sum_{\boldsymbol{k}s}\left(N_{\boldsymbol{k}s} + \frac{1}{2}\right)\hbar\omega_{\boldsymbol{k}s}$ $3N$ terms in sum
3. Ideal gas of N identical spin-$\frac{1}{2}$ fermions	$\alpha = (\ldots, N_{\boldsymbol{k}s}, \ldots)$ $N_{\boldsymbol{k}s} = 0, 1$ $\sum_{\boldsymbol{k}s} N_{\boldsymbol{k}s} = N$	$N_{\boldsymbol{k}s}$ is number of fermion in single-particle state $\boldsymbol{k}s$	$E_{N,\alpha} = \sum_{\boldsymbol{k}s} N_{\boldsymbol{k}s}\varepsilon_k$ $\varepsilon_k = \frac{\hbar^2}{2m}k^2$
4. Ideal gas of N identical spin-0 bosons	$\alpha = (\ldots, N_{\boldsymbol{k}}, \ldots)$ $N_{\boldsymbol{k}} = 0, 1, 2, \cdots, N$ $\sum_{\boldsymbol{k}} N_{\boldsymbol{k}} = N$	$N_{\boldsymbol{k}}$ is number of bosons in single-particle state $\boldsymbol{k}$	$E_{N,\alpha} = \sum_{\boldsymbol{k}} N_{\boldsymbol{k}}\varepsilon_k$ $\varepsilon_k = \frac{\hbar^2}{2m}k^2$
5. Electromagnetic radiation in a cavity	$\alpha = (\ldots, N_{\boldsymbol{k}s}, \ldots)$ $N_{\boldsymbol{k}s} = 0, 1, 2, \cdots, \infty$	$N_{\boldsymbol{k}s}$ is number of photons in mode $\boldsymbol{k}s$	$E_\alpha = \sum_{\boldsymbol{k}s} N_{\boldsymbol{k}s}\hbar\omega_k$

electron gases in metals. There is no restriction on the number of photons in the different modes of electromagnetic radiation, while the properties of fermions and bosons are significantly affected by the restriction $\sum_{ks} N_{ks} = N$.

26.3 Fermi–Dirac and Bose–Einstein Statistics

We found in Chapter 21 that particles in systems described by separable Hamiltonians such as (26.1) are statistically independent, but this is not the case when the requirement of exchange symmetry is satisfied. Exchange symmetry introduces an unusual type of interaction that is not reflected in a system's energy. Because of this lack of statistical independence, equilibrium properties are most conveniently obtained by employing the grand canonical ensemble. That this is a matter of convenience not necessity is illustrated in Appendix 26.A, where the canonical ensemble is used to analyze systems of fermions.

Whether fermions or bosons, the energies of the microstates of a system of identical noninteracting particles are

$$E_{N,\alpha} = \sum_j N_j \varepsilon_j = N_0 \varepsilon_0 + N_1 \varepsilon_1 + N_2 \varepsilon_2 + \cdots, \tag{26.30}$$

where N_j is the number of particle in single-particle state j. As there are N particles and every particle is in some state j, it is required that

$$\sum_j N_j = N_0 + N_1 + N_2 + \cdots = N. \tag{26.31}$$

Because of this, the occupation numbers N_j are *not* independent because when the occupation number of one of the states is increased, the occupation number of some other state must decrease.

As the average of a sum equals the sum of the averages, it follows that

$$\langle E_N \rangle = \langle N_0 \varepsilon_0 + N_1 \varepsilon_1 + \cdots \rangle = \langle N_0 \rangle \varepsilon_0 + \langle N_1 \rangle \varepsilon_1 + \cdots \tag{26.32}$$

and

$$\langle N \rangle = \langle N_0 + N_1 + N_2 + \cdots \rangle = \langle N_0 \rangle + \langle N_1 \rangle + \langle N_2 \rangle + \cdots, \tag{26.33}$$

where $\langle N_j \rangle$ is the average occupation number of single-particle states. The internal energy U and the number of particles $\overline{N}$ are given by the ensemble averages

$$\boxed{U = \sum_j \langle N_j \rangle \varepsilon_j \quad \text{and} \quad \overline{N} = \sum_j \langle N_j \rangle.} \tag{26.34}$$

These are valid for both fermions and bosons. The average occupation numbers $\langle N_j \rangle$ can be determined from the grand partition function Q from (25.22), which by using $\beta = 1/k_B T$ becomes

$$Q = \sum_{N=0}^{\infty} e^{\beta \mu N} Z_N, \tag{26.35}$$

where the partition function for systems of N particles is

$$Z_N = \sum_{\alpha} e^{-\beta E_{N,\alpha}}. \tag{26.36}$$

Substituting the energies $E_{N,\alpha}$ from (26.30) into this and summing over all microstates gives

$$Z_N = \underbrace{\sum_{N_0}\sum_{N_1}\cdots\sum_{N_j}}_{N_0+N_1+N_2+\cdots=N}\cdots e^{-\beta(N_0\varepsilon_0+N_1\varepsilon_1+\cdots+N_j\varepsilon_j+\cdots)}, \tag{26.37}$$

where the difference for systems of fermions and bosons is in the possible values of the occupation numbers. For fermions, $N_j = 0, 1$. For bosons, $N_j = 0, 1, \ldots, N$. The restriction $N_0 + N_1 + N_2 + \cdots = N$ written below the summation symbols prevents us from independently evaluating the different sums. This limitation can be avoided by using the grand partition function. Substituting Z_N into the above expression for Q gives

$$Q = \sum_{N=0}^{\infty} e^{\beta\mu N}\underbrace{\sum_{N_0}\sum_{N_1}\cdots\sum_{N_j}}_{N_0+N_1+N_2+\cdots=N}\cdots e^{-\beta(N_0\varepsilon_0+N_1\varepsilon_1+\cdots+N_j\varepsilon_j+\cdots)} \tag{26.38}$$

By taking the factor $e^{\beta\mu N}$ inside the sums and replacing N by $N_0 + N_1 + N_2 + \cdots$, we obtain

$$Q = \sum_{N=0}^{\infty}\underbrace{\sum_{N_0}\sum_{N_1}\cdots\sum_{N_j}}_{N_0+N_1+N_2+\cdots=N}\cdots e^{-\beta[N_0(\varepsilon_0-\mu)+N_1(\varepsilon_1-\mu)+\cdots+N_j(\varepsilon_j-\mu)+\cdots]} \tag{26.39}$$

Equation (26.39) tells us to first add up the Boltzmann factors $e^{-\beta[\cdots]}$ for all combinations of occupation numbers consistent with the requirement that $N_0 + N_1 + N_2 + \cdots = N$ and then add up the results for all values of N. This procedure is equivalent to adding up the Boltzmann factors for *all* combinations of occupation numbers *independent* of the restriction $N_0 + N_1 + N_2 + \cdots$. As a result, the need for the summing over N is effectively eliminated, so that (26.39) simplifies to

$$Q = \sum_{N_0}\sum_{N_1}\cdots\sum_{N_j}\cdots e^{-\beta N_0(\varepsilon_0-\mu)}e^{-\beta N_1(\varepsilon_1-\mu)}\cdots e^{-\beta N_j(\varepsilon_j-\mu)}\cdots, \tag{26.40}$$

where we have written the exponential of a sum as a product of exponentials. Associating each exponential with the corresponding summation gives

$$Q = \sum_{N_0} e^{-\beta N_0(\varepsilon_0-\mu)}\sum_{N_1} e^{-\beta N_1(\varepsilon_1-\mu)}\cdots\sum_{N_j} e^{-\beta N_j(\varepsilon_j-\mu)}\cdots. \tag{26.41}$$

Each of these sums can now be performed independently. Then, finding the logarithm of the grand partition function gives

$$\ln Q = \sum_{j=0}^{\infty}\ln\left(\sum_{N_j} e^{-\beta N_j(\varepsilon_j-\mu)}\right). \tag{26.42}$$

For fermions, N_j has two allowed values, so that

$$\sum_{N_j=0,1} e^{-\beta N_j(\varepsilon_j-\mu)} = 1 + e^{-\beta(\varepsilon_j-\mu)}. \tag{26.43}$$

Substituting this into (26.42) gives

$$\boxed{\ln Q = \sum_{j=0}^{\infty} \ln\left(1 + e^{-\beta(\varepsilon_j-\mu)}\right).} \tag{26.44}$$

For bosons, N_j can have any integer value from zero to infinity, so that

$$\sum_{N_j=0}^{\infty} e^{-\beta N_j(\varepsilon_j-\mu)} = \sum_{N_j=0}^{\infty} \left(e^{-\beta(\varepsilon_j-\mu)}\right)^{N_j}. \tag{26.45}$$

By using the expansion $1 + x + x^2 + x^3 + \cdots = (1-x)^{-1}$, we find that

$$\sum_{N_j=0}^{\infty} e^{-\beta N_j(\varepsilon_j-\mu)} = \frac{1}{1 - e^{-\beta(\varepsilon_j-\mu)}}. \tag{26.46}$$

Substituting into (26.42) gives

$$\boxed{\ln Q = -\sum_{j=0}^{\infty} \ln\left(1 - e^{-\beta(\varepsilon_j-\mu)}\right).} \tag{26.47}$$

Average Occupation Number

The ensemble average of occupation number N_j is

$$\langle N_j \rangle = \sum_{N,\alpha} \rho_{N,\alpha} N_j, \tag{26.48}$$

and the probabilities from (25.19) are

$$\rho_{N,\alpha} = \frac{1}{Q} e^{\beta\mu N} e^{-\beta E_{N,\alpha}}, \tag{26.49}$$

where $\beta = 1/k_B T$. By replacing j with l in the expressions for $E_{N,\alpha}$ and N in (26.30) and (26.31), the probabilities becomes

$$\rho_{N,\alpha} = \frac{1}{Q} e^{-\beta\mu \sum_l N_l} e^{-\beta \sum_l N_l \varepsilon_l} = \frac{1}{Q} e^{-\beta \sum_l N_l(\varepsilon_l - \mu)}. \tag{26.50}$$

As indicated in (26.24) and Table 26.1, index α is an abbreviation for sets of occupation numbers $(N_0, N_1, N_2, \cdots)$ whose values add up to N. By writing out the sums over N and α in (26.48) and then substituting (26.50) into (26.48), we obtain

$$\langle N_j \rangle = \frac{1}{Q} \underbrace{\sum_{N=0}^{\infty} \sum_{N_0} \sum_{N_1} \cdots \sum_{N_l} \cdots}_{N_0+N_1+N_2+\cdots=N} e^{-\beta \Sigma_l N_l(\varepsilon_l-\mu)} N_j, \tag{26.51}$$

which is the same as the equation for Q in (26.39) with the exception of the $1/Q$ and the factor of N_j. We can simplify (26.51) by using the relationship

$$e^{-\beta N_j(\varepsilon_j-\mu)}N_j = -\frac{1}{\beta}\frac{\partial e^{-\beta N_j(\varepsilon_j-\mu)}}{\partial \varepsilon_j}, \tag{26.52}$$

which gives

$$\langle N_j\rangle = -\frac{1}{Q}\frac{1}{\beta}\frac{\partial Q}{\partial \varepsilon_j} = -\frac{1}{\beta}\frac{\partial \ln Q}{\partial \varepsilon_j}. \tag{26.53}$$

By using this and the value of $\ln Q$ for fermions in (26.44), it follows that

$$\langle N_j\rangle = -\frac{1}{\beta}\frac{\partial}{\partial \varepsilon_j}\ln\left(1+e^{-\beta(\varepsilon_j-\mu)}\right) = \frac{e^{-\beta(\varepsilon_j-\mu)}}{1+e^{-\beta(\varepsilon_j-\mu)}}. \tag{26.54}$$

Multiplying numerator and denominator by $e^{+\beta(\varepsilon_j-\mu)}$ and using $\beta = 1/k_BT$ gives

$$\boxed{\langle N_j\rangle = \frac{1}{e^{(\varepsilon_j-\mu)/k_BT}+1}.} \tag{26.55}$$

From the value of $\ln Q$ for bosons in (26.47), we obtain

$$\langle N_j\rangle = \frac{1}{\beta}\frac{\partial}{\partial \varepsilon_j}\ln\left(1-e^{-\beta(\varepsilon_j-\mu)}\right) = \frac{e^{-\beta(\varepsilon_j-\mu)}}{1-e^{-\beta(\varepsilon_j-\mu)}}, \tag{26.56}$$

which becomes

$$\boxed{\langle N_j\rangle = \frac{1}{e^{(\varepsilon_j-\mu)/k_BT}-1}.} \tag{26.57}$$

These expressions for the average occupation numbers of single-particle states are expressions of *Fermi–Dirac* and *Bose–Einstein statistics*, which are also called *Fermi* and *Bose statistics*.

Maxwell–Boltzmann Statistics

Omitting the $+1$ and -1 in the denominators of (26.55) and (26.57) and bringing the exponentials into the numerator gives

$$\langle N_j\rangle = e^{-(\varepsilon_j-\mu)/k_BT}. \tag{26.58}$$

This is an expression of *Maxwell–Boltzmann statistics*. To show that (26.58) describes systems of *distinguishable* particles, we express the eigenvalues of a system of N particles as

$$E_{N,\alpha} = \varepsilon_{j_1} + \varepsilon_{j_2} + \cdots + \varepsilon_{j_N}, \tag{26.59}$$

where j_i is the single-particle state of particle i. The particles are statistically independent, so that the probability that any one of them is in state j is

$$\rho_j = \frac{e^{-\varepsilon_j/k_BT}}{Z_s}, \tag{26.60}$$

where Z_s is the single-particle partition function. The average number of particles in state j is found by multiplying this by the total number of particles in the system

$$\langle N_j \rangle = N\rho_j = N\frac{e^{-\varepsilon_j/k_BT}}{Z_s}. \tag{26.61}$$

Since the value of N is fixed, this result is for a closed system. To introduce the chemical potential, we identify N with $\overline{N}$, which is the average value of N in an open system. We found in (25.48), that

$$\overline{N} = e^{\mu/k_BT}Z_s. \tag{26.62}$$

After replacing $\overline{N}$ by N and substituting the value of N/Z_s implied by (26.62) into (26.61), we obtain

$$\langle N_j \rangle = e^{-\varepsilon_j/k_BT}e^{\mu/k_BT} = e^{-(\varepsilon_j-\mu)/k_BT}, \tag{26.63}$$

which is the expression of Maxwell–Boltzmann statistics given in (26.58).

The distributions specified by Fermi–Dirac, Bose–Einstein, and Maxwell–Boltzmann statistics are different because of the different ways in which fermions, bosons, and distinguishable particles are treated in quantum mechanics. All three distributions were obtained from an assignment of a probability proportional to $e^{-E_{N,\alpha}/k_BT}$. The different distributions result from the different rules for combining single-particle eigenstates to form N-particle eigenstates.

The effect of exchange symmetry on collections of identical particles depends on the type of system. The effect of exchange symmetry is especially significant for electron in metals and for bosons at very low temperatures. However, exchange symmetry is not important in the quantum mechanical analysis of atomic vibrations in solids, where the atoms (which may be identical) are spatially localized by the interactions that bind them together.

Problems

26.1 Find the different eigenfunctions of a system of two particles ($i = 1$ or 2) that can be formed from the four single-particle eigenfunctions, $\psi_1(i)$, $\psi_2(i)$, $\psi_3(i)$, and $\psi_4(i)$. Each eigenfunction corresponds to a different two-particle state.

(a) If the particles are *distinguishable* (*i.e.*, not identical), one possible eigenfunction is $\psi_1(1)\psi_2(2)$. Using this abbreviated notation, write down the corresponding eigenfunctions of the 16 different two-particle states.

(b) If the particles are *identical fermions*, one (un-normalized) two-particle eigenfunction is $\psi_1(1)\psi_2(2) - \psi_1(2)\psi_2(1)$. How many different two-particle states are there? What are the corresponding (un-normalized) eigenfunctions?

(c) If the particles are *identical bosons*, one two-particle eigenfunction is $\psi_1(1)\psi_2(2) + \psi_1(2)\psi_2(1)$. How many different two-particle states are there? What are the corresponding eigenfunctions?

Appendix 26.A

Fermions in the Canonical Ensemble

The equilibrium properties of systems of identical particles were found in Chapter 26 by treating them as open systems and using the grand canonical ensemble. To show that the choice of ensemble is a matter of convenience, not necessity, the average occupation numbers for identical fermions are derived in this appendix by using the canonical ensemble. The derivation presented is due to Baierlein.[1]

The energy of a system of N noninteracting fermions in microstate in α is

$$E_{N,\alpha} = N_0\varepsilon_0 + N_1\varepsilon_1 + N_2\varepsilon_2 + \cdots, \tag{26.A.1}$$

where $N_0 + N_1 + N_2 + \cdots = N$, and the allowed values of N_j for fermions are 0 or 1. As given in (20.9), the probability assigned by the canonical ensemble to microstate α is

$$\rho_\alpha^c = \frac{e^{-\beta E_{N,\alpha}}}{Z_N} = \frac{e^{-\beta(N_0\varepsilon_0+N_1\varepsilon_1+\cdots+N_j\varepsilon_j+\cdots)}}{Z_N}, \tag{26.A.2}$$

where $\beta = 1/k_BT$. The partition function as given in (26.37) is

$$Z_N = \underbrace{\sum_{N_0}\sum_{N_1}\cdots\sum_{N_j}\cdots}_{N_0+N_1+N_2+\cdots=N} e^{-\beta(N_0\varepsilon_0+N_1\varepsilon_1+\cdots+N_j\varepsilon_j+\cdots)} \tag{26.A.3}$$

and the canonical average for the number of particles in single-particle state j is

$$\langle N_j\rangle = \sum_\alpha \rho_\alpha N_j = \underbrace{\sum_{N_0}\sum_{N_1}\cdots\sum_{N_j}\cdots}_{N_0+N_1+N_2+\cdots=N} \frac{e^{-\beta(N_0\varepsilon_0+N_1\varepsilon_1+\cdots+N_j\varepsilon_j+\cdots)}}{Z_N} N_j. \tag{26.A.4}$$

Since the value of the average is determined by the energy of the state, the steps in its derivations are the same for all states. To make the derivation easier to follow, we will find the average for state $j = 1$. After taking Z_N outside the summations and factoring $e^{-\beta N_1\varepsilon_1}$ out of the Boltzmann factor, the canonical average for state $j = 1$ becomes

$$\langle N_1\rangle = \frac{1}{Z_N}\underbrace{\sum_{N_0}\sum_{N_1}\cdots\sum_{N_j}\cdots}_{N_0+N_1+N_2+\cdots=N} e^{-\beta(N_0\varepsilon_0+\cdots+N_j\varepsilon_j+\cdots)} e^{-\beta N_1\varepsilon_1} N_1. \tag{26.A.5}$$

[1] Baierlein R. (1999), *Thermal Physics*, Cambridge University Press, New York, pp.171–172.

As N_1 has two allowed values, 0 and 1, summing over N_1 gives

$$\langle N_1\rangle = \frac{1}{Z_N}\underbrace{\sum_{N_0}\cdots\sum_{N_j}}_{N_0+1+N_2+\cdots=N}\cdots e^{-\beta(N_0\varepsilon_0+\cdots+N_j\varepsilon_j+\cdots)}e^{-\beta\varepsilon_1}, \tag{26.A.6}$$

where $N_1 = 0$ term vanishes. The value of the $N_1 = 1$ term is included in the restriction written under the summations, which can be rewritten as $N_0 + N_2 + \cdots = N - 1$. Also, the factor $e^{-\beta\varepsilon_1}$ can be taken outside of the summations. We now reintroduce a sum over N_1 in a manner that leaves $\langle N_1\rangle$ unchanged. This is done by multiplying by $(1 - N_1)$, including $N_1\varepsilon_1$ in the exponent, and reintroducing N_1 into the restrictions on the summations. The result is

$$\langle N_1\rangle = \frac{e^{-\beta\varepsilon_1}}{Z_N}\underbrace{\sum_{N_0}\sum_{N_1}\cdots\sum_{N_j}}_{N_0+N_1+N_2+\cdots=N-1}\cdots e^{-\beta(N_0\varepsilon_0+N_1\varepsilon_1+\cdots+N_j\varepsilon_j+\cdots)}(1 - N_1). \tag{26.A.7}$$

As N_1 has two values, 0 and 1, the number of terms being summed has in principle been doubled, but half of these terms are zero because $(1 - N_1) = 0$ when $N_1 = 1$. When $N_1 = 0$, we have $(1 - N_1) = 1$ and $N_1\varepsilon_1 = 0$, so that the restriction on the summations becomes $N_0 + N_2 + \cdots = N - 1$. Because of the values for N_1, $N_1\varepsilon_1$, and $(1 - N_1)$, the same set of nonzero terms are summed in (26.A.7) that are summed in (26.A.5).

By replacing N by $N - 1$ in (26.A.3) and (26.A.4), it follows that

$$\underbrace{\sum_{N_0}\sum_{N_1}\cdots\sum_{N_j}}_{N_0+N_1+N_2+\cdots=N-1}\cdots e^{-\beta(N_0\varepsilon_0+N_1\varepsilon_1+\cdots+N_j\varepsilon_j+\cdots)} = Z_{N-1} \tag{26.A.8}$$

and

$$\underbrace{\sum_{N_0}\sum_{N_1}\cdots\sum_{N_j}}_{N_0+N_1+N_2+\cdots=N-1}\cdots\frac{e^{-\beta(N_0\varepsilon_0+N_1\varepsilon_1+\cdots+N_j\varepsilon_j+\cdots)}}{Z_{N-1}}N_j = \langle N_j\rangle_{N-1} \tag{26.A.9}$$

By using these relationships in (26.A.7), we obtain

$$\langle N_1\rangle = \frac{e^{-\beta\varepsilon_1}}{Z_N}(Z_{N-1} - Z_{N-1}\langle N_1\rangle_{N-1}). \tag{26.A.10}$$

This result is exact. We now make a small approximation. Since removing one particle from a system with a macroscopic number of particles causes a negligible change, the difference between the averages for N and $N - 1$ is negligibly small. By approximating the average for $N - 1$ by the average for N, equation (26.A.10) becomes

$$\langle N_1\rangle = e^{-\beta\varepsilon_1}\frac{Z_{N-1}}{Z_N}\left[1 - \langle N_1\rangle\right]. \tag{26.A.11}$$

The ratio Z_{N-1}/Z_N is related to the chemical potential. By using the partition function formula, $F = -k_B T \ln Z$, it follows that

$$\frac{Z_{N-1}}{Z_N} = \frac{e^{-\beta F(N-1)}}{e^{-\beta F(N)}} = e^{+\beta[F(N)-F(N-1)]}. \tag{26.A.12}$$

The dependence Helmholtz free energy on T and V is left implicit. As given in (13.19), the differential of the Helmholtz free energy is

$$dF = -SdT - PdV + \mu dN, \tag{26.A.13}$$

and the chemical potential is

$$\mu = \left(\frac{\partial F}{\partial N}\right)_{TV}. \tag{26.A.14}$$

Although μ is independent of N when expressed as a function of temperature and pressure, it depends on N when it is a function of T and V. By approximating the derivative $\partial F/\partial N$ by $\Delta F/\Delta N$ and setting $\Delta N = 1$, the chemical potential becomes

$$\mu = \frac{F(N) - F(N - \Delta N)}{\Delta N}\Delta N = [F(N) - F(N-1)]. \tag{26.A.15}$$

Substituting this into (26.A.12) gives

$$\frac{Z_{N-1}}{Z_N} = e^{+\beta\mu}. \tag{26.A.16}$$

Equation (26.A.11) then becomes

$$\langle N_1\rangle = e^{-\beta(\varepsilon_1-\mu)}[1 - \langle N_1\rangle]. \tag{26.A.17}$$

Solving this for $\langle N_1\rangle$ gives

$$\langle N_1\rangle = \frac{1}{e^{\beta(\varepsilon_1-\mu)} + 1},$$

where the value of the average is determined by the energy of the single-particle state. Since the steps in derivation would be the same for all other states, we can replace subscript 1 by subscript j. Then, by using $\beta = 1/k_BT$, the average occupation numbers for fermions become

$$\boxed{\langle N_j\rangle = \frac{1}{e^{(\varepsilon_j-\mu)/k_BT} + 1}.} \tag{26.A.18}$$

The only difference between this result and the value of $\langle N_j\rangle$ obtained with grand canonical ensemble is that N is constant in the canonical ensemble, a difference that is not significant because of the small size of the fluctuations in N.

27

Fermi and Bose Gases

In Chapter 26, we determined the effect of exchange symmetry on the occupation numbers of a system's single-particle states. In this chapter, we use those results to investigate the behavior of ideal gases of fermions and bosons. The effects of exchange symmetry are especially significant for the electron gases in metals and in boson gases at very low temperatures where they experience the unusual type of phase transition known as *Bose–Einstein condensation*.

27.1 Ideal Gases

Before we can understand the significance of exchange symmetry, we need more information about the chemical potential. Since an expression for a system's single-particle energies is needed to find its chemical potential μ, we consider ideal gases. As found in Chapter 24, the eigenfunctions and eigenvalues of the particles in an ideal gas are (see (24.15))

$$\psi_{\boldsymbol{k}}(\boldsymbol{r}) = \frac{1}{\sqrt{V}} e^{i\boldsymbol{k}\cdot\boldsymbol{r}} \quad \text{and} \quad \varepsilon_k = \frac{\hbar^2}{2m} k^2, \tag{27.1}$$

where V is volume. The single-particle energies ε_k are determined by the magnitude of the wave vector $\boldsymbol{k}$ of the plane wave described by the exponential $e^{i\boldsymbol{k}\cdot\boldsymbol{r}}$. When the property of spin is included, the single-particle states are identified by $\boldsymbol{k}s$, where s is the spin quantum number. The eigenstates (microstates) of a system of N identical particles are identified by specifying the number of particles $N_{\boldsymbol{k}s}$ in each single-particle state. The energy $E_{N,\alpha}$ of microstate α of an N-particle system is

$$E_{N,\alpha} = \sum_{\boldsymbol{k}s} N_{\boldsymbol{k}s} \varepsilon_k, \quad \text{where} \quad N = \sum_{\boldsymbol{k}s} N_{\boldsymbol{k}s}. \tag{27.2}$$

The ensemble averages of the energy and number of particles are

$$U = \sum_{\boldsymbol{k}s} \langle N_{\boldsymbol{k}s} \rangle \varepsilon_k \quad \text{and} \quad \overline{N} = \sum_{\boldsymbol{k}s} \langle N_{\boldsymbol{k}s} \rangle, \tag{27.3}$$

Thermodynamics and Statistical Mechanics: An Integrated Approach, First Edition.
Robert J Hardy and Christian Binek.

where U is the internal energy. The chemical potential determines the number of particles $\overline{N}$ in the system. The average occupation numbers $\langle N_{ks} \rangle$ for fermions and bosons as given in (26.55) and (26.57) can be combined into a single equation

$$\langle N_{ks} \rangle = \frac{1}{e^{\beta(\varepsilon_k - \mu)} \pm 1}, \tag{27.4}$$

where $\beta = 1/k_B T$. The plus sign (+1) in the denominator is for fermions and the minus sign (−1) is for bosons. Substituting (27.4) into (27.3) gives

$$U = \sum_{ks} \frac{\varepsilon_k}{e^{\beta(\varepsilon_k - \mu)} \pm 1} \quad \text{and} \quad \overline{N} = \sum_{ks} \frac{1}{e^{\beta(\varepsilon_k - \mu)} \pm 1}. \tag{27.5}$$

Finding the chemical potential is facilitated by expressing the sum over $\boldsymbol{ks}$ in the expression for $\overline{N}$ as an integral. Thus,

$$\overline{N} = \sum_s \int dk^3 D_k \frac{1}{e^{\beta(\varepsilon_k - \mu)} \pm 1}, \tag{27.6}$$

where D_k is the number of wave vectors per unit of magnitude k from (24.12). Changing the integral over k-space to an integral over the energy $\varepsilon = \hbar^2 k^2 / 2m$ gives

$$\overline{N} = \sum_s \int_0^\infty \frac{D(\varepsilon)}{e^{\beta(\varepsilon - \mu)} \pm 1} d\varepsilon. \tag{27.7}$$

$D(\varepsilon)$ is the free particle density of states, which is the number of plane wave vectors per unit of energy. Its value from (24.14) is

$$D(\varepsilon) = \frac{V}{4\pi^2} \left(\frac{2m}{\hbar^2} \right)^{3/2} \varepsilon^{1/2}. \tag{27.8}$$

Substituting this into (27.7) gives

$$\overline{N} = \left(\frac{3}{2} \pm \frac{1}{2} \right) \frac{V}{4\pi^2} \left(\frac{2m}{\hbar^2} \right)^{3/2} \int_0^\infty \frac{\varepsilon^{1/2}}{e^{\beta(\varepsilon - \mu)} \pm 1} d\varepsilon, \tag{27.9}$$

where the plus sign (+1) is for fermions and minus sign (−1) is for bosons. The factor of $\left(\frac{3}{2} + \frac{1}{2} \right) = 2$ accounts for the two values of s for spin-$\frac{1}{2}$ fermions, and the factor $\left(\frac{3}{2} - \frac{1}{2} \right) = 1$ accounts for the single value of s for spin-0 bosons. Changing the integration variable from ε to $x = \beta\varepsilon$ and using $\hbar = h/2\pi$ gives

$$\overline{N} = \left(\frac{3}{2} \pm \frac{1}{2} \right) 2\pi V \left(\frac{2m}{h^2 \beta} \right)^{3/2} \int_0^\infty \frac{x^{1/2}}{e^{x - \beta\mu} \pm 1} dx. \tag{27.10}$$

By introducing the thermal de Broglie wavelength $\lambda_{\text{th}} = h/\sqrt{2\pi m k_B T}$ (see (20.71)), and using $\beta = 1/k_B T$, the number of particles in the system becomes

$$\boxed{\overline{N} = \left(\frac{3}{2} \pm \frac{1}{2} \right) \frac{2}{\sqrt{\pi}} \frac{V}{\lambda_{\text{th}}^3} \int_0^\infty \frac{x^{1/2}}{e^{x - \beta\mu} \pm 1} dx.} \tag{27.11}$$

Since the only difference between the expressions for U and $\overline{N}$ in (27.5) is in the factor of ε_k in the expression for U, we can make the same substitutions used to obtain (27.11) to show that

$$U = \left(\frac{3}{2} \pm \frac{1}{2}\right) \frac{2}{\sqrt{\pi}} \frac{V}{\lambda_{\text{th}}^3} k_B T \int_0^\infty \frac{x^{3/2}}{e^{x-\beta\mu} \pm 1} dx, \tag{27.12}$$

where the factor $k_B T$ comes from the extra factor of x that enters when the integration variable is changed from ε to x. The relationship $x = \beta\varepsilon$ implies that $\varepsilon = k_B T x$. These equations can be used to determine the internal energy U and chemical potential μ from the more directly measured quantities, temperature T and number density $\overline{N}/V$.

To find the values for the chemical potential, we rewrite (27.10) as

$$\frac{\lambda_{\text{th}}^3}{V/\overline{N}} = \left(\frac{3}{2} \pm \frac{1}{2}\right) \frac{2}{\sqrt{\pi}} \int_0^\infty \frac{x^{1/2}}{e^{x-\beta\mu} \pm 1} dx, \tag{27.13}$$

where the right-hand side of this equation is a function of $\beta\mu$. Since the left-hand side is the ratio of a volume λ_{th}^3 divided by the volume per particle $V/\overline{N}$, the value of μ is effectively determined by a ratio of volumes, which differs greatly between ordinary gases and the electron gases in metals. For comparison, we find the room temperature values of the chemical potentials for electrons in copper and for gases of helium-3 fermions and helium-4 bosons.

As explained in texts on solid-state physics, the conduction electrons in metals can be modeled as ideal gases with properties that are weakly affected by the periodic arrangement of the atoms. The electron gas in copper is used as an illustration. As copper atoms have one valence electron, the volume per particle $V/\overline{N}$ for the electron gas is the same as the volume per particle for the copper atoms, which can be found by dividing the mass of an atom by the mass density of copper ($8960\,\text{kg m}^{-3}$). Multiplying the atomic weight of copper (63.546) by the atomic mass unit gives $m_{\text{Cu}} = 1.055 \times 10^{-25}\text{kg}$. Thus, the volume per particle for both the atoms and the valence electrons is

$$\frac{V}{\overline{N}} = \frac{1.055 \times 10^{-25}\text{kg}}{8960\,\text{kg m}^{-3}} = 1.18 \times 10^{\ 29}\text{m}^3. \tag{27.14}$$

The volume per particle for an ordinary gas of helium at room temperature ($T = 300\,\text{K}$) and atmospheric pressure ($P = 100\,\text{kPa}$) can be found from the ideal gas equation of state, $PV = \overline{N} k_B T$,

$$\frac{V}{\overline{N}} = \frac{k_B T}{P} = 4.14 \times 10^{-26}\text{m}^{-3}. \tag{27.15}$$

Since the thermal de Broglie wavelength is $\lambda_{\text{th}} = h/\sqrt{2\pi m k_B T}$, the masses of ^{3}He and ^{4}He atoms (3 and 4 amu, respectively) and an electron (9.109×10^{-31} kg) are needed. The values of $V/\overline{N}$, λ_{th}, and μ/k_B for the three systems are given in Table 27.1, where the chemical potentials have been divided by k_B so they can be interpreted as temperatures.

The value of μ/k_B for the electron gas in copper is very different from that for the helium gases. The chemical potentials of the electron gases in metals are large and positive, whereas the chemical potentials of ordinary gases are negative. Since the value of

Table 27.1 *Properties of gases at* $T = 300\,\text{K}$ *and* $P = 100\,\text{kPa}$

	$V/\overline{N}$ (m^3)	λ_{th} (m)	$\lambda_{\text{th}}^3\overline{N}/V$	μ/k_B (K)	ε_F/k_B (K)
Electrons in Cu	1.18×10^{-29}	4.30×10^{-9}	6.77×10^{3}	81 800	81 800
Helium-3 gas	4.14×10^{-26}	5.82×10^{-11}	4.76×10^{6}	−3 880	0.0646
Helium-4 gas	4.14×10^{-26}	5.04×10^{-11}	1.55×10^{6}	−4 010	

μ for helium-3 is negative and $|\mu|$ is large, the exponential $e^{\beta(\varepsilon-\mu)} = e^{\beta(\varepsilon+|\mu|)}$ is very large. Using the data for helium-3 in the Table 27.1 gives $e^{\beta|\mu|} = 4.2\times10^5$. By comparison, the ±1 in the denominators of the expressions for $\langle N_{ks}\rangle$ are negligibly small.

Monatomic Gases

When μ is negative and $|\mu|$ is large, the ±1, in the denominator of the expressions for $\overline{N}$ in (27.9) and in the corresponding expression for U can be neglected, which gives

$$\overline{N} = K_o\int_0^\infty d\varepsilon e^{-(\varepsilon-\mu)/k_BT}\varepsilon^{1/2} \quad \text{and} \quad U = K_o\int_0^\infty d\varepsilon e^{-(\varepsilon-\mu)/k_BT}\varepsilon^{3/2}, \tag{27.16}$$

where $K_o = \left(\frac{3}{2}\pm\frac{1}{2}\right)2\pi V(2m)^{3/2}/h^3$. We can make the integrals in these equations the same by expressing the function $e^{-(\varepsilon-\mu)/k_BT}$ in the equation for U as $-k_BT(\partial e^{-(\varepsilon-\mu)/k_BT}/\partial\varepsilon)$ and integrating by parts. The result is

$$U = \frac{3}{2}k_BTK_o\int_0^\infty d\varepsilon e^{-(\varepsilon-\mu)/k_BT}\varepsilon^{1/2} = \frac{3}{2}k_BT\overline{N}, \tag{27.17}$$

where $U = \frac{3}{2}\overline{N}k_BT$ is the value for a monatomic gas found in Chapter 20. To find the equation of state, we use the logarithms of the grand partition functions for fermions and bosons given in (26.44) and (26.47), which can be expressed as

$$\ln Q = \pm\sum_{ks}\ln\left(1\pm e^{\mu/k_BT}e^{-\varepsilon_k/k_BT}\right). \tag{27.18}$$

When μ is sufficiently negative that e^{μ/k_BT} is very small, we can use the approximation $\ln(1+x) = x$ to show that

$$\ln Q = e^{\mu/k_BT}\sum_{ks}e^{-\varepsilon_k/k_BT} \quad \text{and} \quad \frac{\partial\ln Q}{\partial\mu} = \frac{\ln Q}{k_BT}. \tag{27.19}$$

By combining the relationships $\overline{N} = -(\partial\Phi_{\text{th}}/\partial\mu)_{TV}$ in (25.11) and $-\Phi_{\text{th}} = k_BT\ln Q$ from (25.23), it follows that $\overline{N} = k_BT(\partial\ln Q/\partial\mu)_{TV}$. Combining this with the second equation in (27.19) gives $\overline{N} = \ln Q$. Substituting this into the result, $PV = k_BT\ln Q$ given in (25.23) tells us that the equation of state for monatomic gases is

$$PV = \overline{N}k_BT. \tag{27.20}$$

This is the familiar ideal gas equation of state, which we found in Chapter 20 by treating the particles as distinguishable. Thus, when the chemical potential is negative and large, the requirement of exchange symmetry has negligible effect on the system's properties.

27.2 Fermi Gases

When the effects of exchange symmetry are significant, systems of fermions are referred to as *Femi gases*. The effects are especially significant for electrons in metals. The average occupation numbers for fermions at temperature T are

$$\langle N_{ks}\rangle = \frac{1}{e^{(\varepsilon_k-\mu)/k_BT}+1}, \tag{27.21}$$

which is plotted in **Figure 27.1** as a function of ε_k/μ. The curves in the figure are calculated for $\mu/k_B = 81\,800\,\mathrm{K}$ which is the value for electrons in Cu in Table 27.1. The curves at temperatures T_1, T_2, and T_3 correspond to 300, 600, and 1000 K, respectively. The sharp step at $T = 0$ indicates that one particle occupies every single-particle state with energy ε_k less than μ and that no particles have energies greater than μ. The step becomes slightly rounded as the temperature is increased above absolute zero.

Fermi Energy

The chemical potential of fermions at absolute zero is the *Fermi energy* ε_F.[1] As the states with energies ε_k less than μ are occupied and the states with energies greater than μ are empty, the wave vectors $\boldsymbol{k}$ of the states with energy $\varepsilon_k = \hbar^2k^2/2m$ less than the Fermi

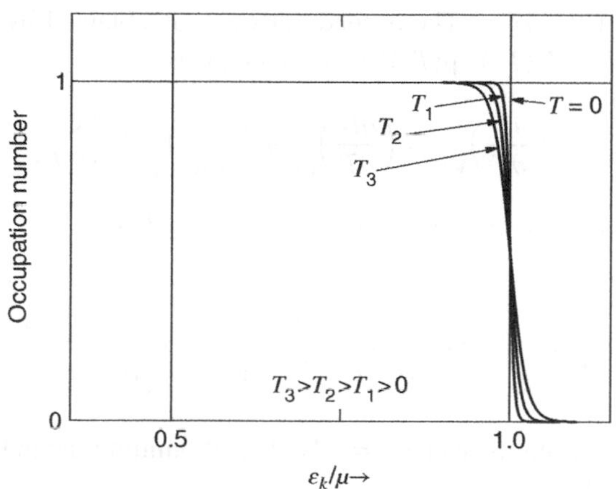

Figure 27.1 *Occupation Numbers $\langle N_{ks}\rangle$ for Fermions*

[1] Some authors call the chemical potential at temperatures other than $T = 0$ the *Fermi energy*, and some authors call the chemical potential at all temperatures the *Fermi level*.

energy are located within a sphere in k-space. By representing the radius of that sphere by k_F, it follows that

$$\varepsilon_F = \frac{\hbar^2 k_F{}^2}{2m} \quad \text{and} \quad k_F = \left(\frac{2m\varepsilon_F}{\hbar^2}\right)^{1/2}. \tag{27.22}$$

Since the number of wave vectors per unit volume of k-space, as given in (24.11), is $D_0 = V/(2\pi)^3$ and since each particle has two spin states, the number of states with energy less than the Fermi energy is

$$\overline{N} = 2D_0 \frac{4\pi}{3} k_F^3 = \frac{V}{3\pi^2}\left(\frac{2m\varepsilon_F}{\hbar^2}\right)^{3/2}, \tag{27.23}$$

where $\frac{4}{3}\pi k_F^3$ is the volume of a sphere of radius k_F. Solving for the Fermi energy gives

$$\varepsilon_F = \frac{\hbar^2}{2m}\left(3\pi^2 \frac{\overline{N}}{V}\right)^{2/3}. \tag{27.24}$$

The Fermi energy for electrons in copper and for a ^{3}He gas are given in Table 27.1. As indicated by the values in the table, the Fermi energies of the free electrons in metals are large and essentially the same as the chemical potentials.

27.3 Low Temperature Heat Capacity

The heat capacity of a gas of free electrons at low temperature has important implications for the theory of metals. The constant-volume heat capacity of a system with a fixed number of particles is $C_V = (\partial U/\partial T)_V$. The needed derivative is obtained by differentiating the composite function $U = U(T, V, \mu(T, V, \overline{N}))$, which gives

$$C_V = \left(\frac{\partial U}{\partial T}\right)_{V\overline{N}} = \left(\frac{\partial U}{\partial T}\right)_{\mu V} + \left(\frac{\partial U}{\partial \mu}\right)_{TV}\left(\frac{\partial \mu}{\partial T}\right)_{\overline{N}V}. \tag{27.25}$$

The final derivative in the above equation can be expressed as

$$\left(\frac{\partial \mu}{\partial T}\right)_{\overline{N}V} = -\left(\frac{\partial \mu}{\partial \overline{N}}\right)_{TV}\left(\frac{\partial \overline{N}}{\partial T}\right)_{\mu V} = -\frac{\left(\partial \overline{N}/\partial T\right)_{\mu V}}{\left(\partial \overline{N}/\partial \mu\right)_{TV}}. \tag{27.26}$$

By combining these equations, using $\beta = 1/k_B T$, and omitting the indication of the variables held fixed, we obtain

$$C_V = -k_B\beta^2\left[\frac{\partial U}{\partial \beta} - \frac{\partial \overline{N}}{\partial \beta}\frac{(\partial U/\partial \mu)}{(\partial \overline{N}/\partial \mu)}\right], \tag{27.27}$$

where $d\beta/dT = -k_B\beta^2$.

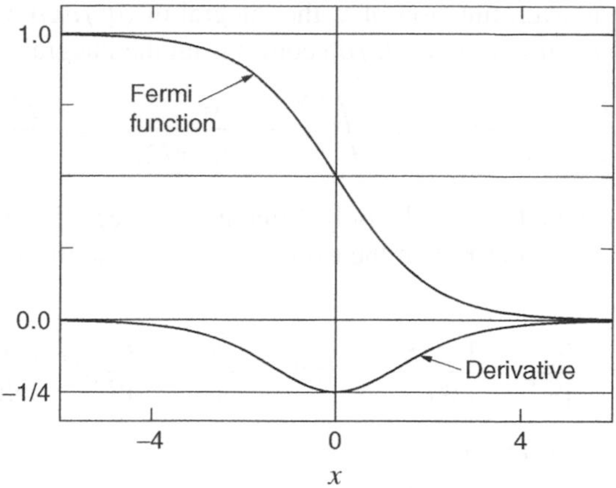

Figure 27.2 *The Fermi Function and Its Derivative*

To find C_V, we need the derivatives of U and $\overline{N}$. The expression for $\overline{N}$ in (27.9) and the corresponding expression for U for spin-$\frac{1}{2}$ fermions are

$$\overline{N} = K_o \int_0^{\infty} \frac{\varepsilon^{1/2}}{e^{\beta(\varepsilon-\mu)}+1} d\varepsilon \quad \text{and} \quad U = K_o \int_0^{\infty} \frac{\varepsilon^{3/2}}{e^{\beta(\varepsilon-\mu)}+1} d\varepsilon, \tag{27.28}$$

where $K_o = \left(V/2\pi^2\right)\left(2m/\hbar^2\right)^{3/2}$. Our calculations are simplified by introducing the *Fermi function* $f(x)$ and its derivative, which are

$$f(x) = \frac{1}{e^x+1} \quad \text{and} \quad \frac{df}{dx} = -\frac{e^x}{(e^x+1)^2} = \frac{-1}{4\cosh^2(x/2)}, \tag{27.29}$$

where $\cosh u = \frac{1}{2}(e^{+u} + e^{-u})$ is the hyperbolic cosine function. The function and its derivative are plotted in **Figure 27.2**.

By using the Fermi function, the derivative of the internal energy with respect to β becomes

$$\frac{\partial U}{\partial \beta} = K_o \int_0^{\infty} \frac{df(\beta(\varepsilon-\mu))}{dx} (\varepsilon-\mu)\varepsilon^{3/2} d\varepsilon. \tag{27.30}$$

Let $x = \beta(\varepsilon - \mu)$, so that $\varepsilon = \mu[1 + x/(\beta\mu)]$ and $d\varepsilon = dx/\beta$. Changing the integration variable from ε to x in $\partial U/\partial\beta$ gives

$$\frac{\partial U}{\partial \beta} = K_o \frac{\mu^{3/2}}{\beta^2} \int_{-\beta\mu}^{+\infty} \frac{df(x)}{dx} x \left(1 + \frac{x}{\beta\mu}\right)^{3/2} dx. \tag{27.31}$$

Expanding the term in parenthesis in powers of $x/(\beta\mu)$ and taking the lower integration limit to $-\infty$ gives

$$\frac{\partial U}{\partial \beta} = \frac{K_o}{\beta^2} \mu^{3/2} \int_{-\infty}^{+\infty} \frac{df(x)}{dx} \left(x + \frac{3}{2}\frac{x^2}{\beta\mu} + \cdots\right) dx. \tag{27.32}$$

Since the (df/dx) is an even function of x, the integral of $(df/dx)x$ vanishes, so that the lowest order nonzero contribution to $\partial U/\partial\beta$ comes from the integral of $(df/dx)x^2$,

$$\int_{-\infty}^{+\infty}\frac{df(x)}{dx}x^2dx = -\int_{-\infty}^{+\infty}\frac{x^2}{4\cosh^2(x/2)}dx = -\frac{\pi^2}{3}. \tag{27.33}$$

By using this result and the fact that the only difference between the expressions for $\overline{N}$ and U in (27.11) and (27.12) is the $x^{3/2}$ in the expression for U and the $x^{1/2}$ in the expression for $\overline{N}$, we find that

$$\frac{\partial U}{\partial\beta} = -\frac{K_o}{\beta^2}\mu^{3/2}\frac{3}{2}\frac{1}{\beta\mu}\frac{\pi^2}{3} \quad \text{and} \quad \frac{\partial\overline{N}}{\partial\beta} = -\frac{K_o}{\beta^2}\mu^{1/2}\frac{1}{2}\frac{1}{\beta\mu}\frac{\pi^2}{3}. \tag{27.34}$$

The derivative U with respect to μ is

$$\frac{\partial U}{\partial\mu} = K_o\int_0^{\infty}\frac{df(\beta(\varepsilon-\mu))}{dx}(-\beta)\varepsilon^{3/2}d\varepsilon. \tag{27.35}$$

After changing the integration variable from ε to x and keeping the lowest order nonvanishing term in the expansion in powers of $x/\beta\mu$, we obtain

$$\frac{\partial U}{\partial\mu} = -K_o\mu^{3/2}\int_{-\infty}^{+\infty}\frac{df(x)}{dx}dx. \tag{27.36}$$

Since $f(+\infty) = 0$ and $f(-\infty) = 1$, it follows that

$$\frac{\partial U}{\partial\mu} = K_o\mu^{3/2} \quad \text{and} \quad \frac{\partial\overline{N}}{\partial\mu} = K_o\mu^{1/2}. \tag{27.37}$$

The accuracy of the approximations that result from neglecting higher order terms in the expansion in powers of $x/\beta\mu$ can be seen by comparing the size $1/\beta\mu$ with the range of values over which (df/dx) is nonzero. As shown in the Figure 27.2, the derivative of the Fermi function (df/dx) is effectively zero beyond $x = \pm 8$. For the data for electrons in Table 27.1, the value of $\beta\mu = \mu/k_BT$ at $T = 100$ K is 818. Hence, the contributions to the derivatives of U and $\overline{N}$ of the higher order terms in the expansion are small at 100 K and become even smaller at lower temperatures.

Substituting (27.34) and (27.37) into the expression for C_V in (27.27) gives

$$C_V = -k_BK_o\left[-\mu^{3/2}\frac{1}{2}\frac{\pi^2}{\beta\mu} + \mu^{1/2}\frac{1}{6}\frac{\pi^2}{\beta\mu}\left(\frac{K_o\mu^{3/2}}{K_o\mu^{1/2}}\right)\right] = k_BK_o\frac{\pi^2}{3}\frac{\mu^{1/2}}{\beta}. \tag{27.38}$$

By combining this with $\beta = 1/k_BT$ and the value of K_o given below (27.28), we find that

$$C_V = \frac{\pi^2}{3}\left[\frac{V}{2\pi^2}\left(\frac{2m}{\hbar^2}\right)^{3/2}\right]\mu^{1/2}k_B{}^2T. \tag{27.39}$$

Since the *density of states for fermions* $D_F(\varepsilon)$ is twice the number of plane wave states per unit of energy given in (27.8), the density of states at the Fermi energy ε_F is

$$D_F(\varepsilon_F) = 2D(\varepsilon_F) = K_o = \frac{V}{2\pi^2}\left(\frac{2m}{\hbar^2}\right)^{3/2}\varepsilon_F^{1/2}. \tag{27.40}$$

By using this and replacing the chemical potential μ by ε_F, the low temperature electronic heat capacity becomes

$$\boxed{C_V = \frac{1}{3}\pi^2 D_F(\varepsilon_F) {k_B}^2 T.} \tag{27.41}$$

The contribution to the heat capacity of the electron gas in metals is small because only the electrons with energies within approximately $k_B T$ of the Fermi energy contribute to it. As shown in Figure 27.1, the occupation numbers for states with energy ε_k slightly less than ε_F are mostly fixed at $N_{ks} = 1$. As a result, the heat capacities of metals are dominated by the contributions of lattice vibrations. However, the electronic contribution can be observed at low temperatures, because the contribution of lattice vibrations is proportional to T^3, whereas the electronic contribution is proportional to T, which implies that $C_V = c_1 T + c_3 T^3$ as absolute zero is approached. Consequently, there is a temperature below which the term proportional to T – the electronic contribution – is larger than the lattice contribution. Values of the electronic contribution have been measured and the agreement with the predictions of (27.41) are quite good.

27.4 Bose Gases

There is no limit on the number of particles that can occupy a single-particle state. Because of this, *Bose gases* behave very differently than Fermi gases. In particular, they experience a phase transition at low temperatures in which a significant number of particles is in the ground state. The average occupation number of spin-0 bosons in single-particle state $\boldsymbol{k}$ is

$$\langle N_{\boldsymbol{k}} \rangle = \frac{1}{e^{(\varepsilon_k - \mu)/k_B T} - 1}, \tag{27.42}$$

and the total number of particles is

$$\overline{N} = \sum_{\boldsymbol{k}} \frac{1}{e^{(\varepsilon_k - \mu)/k_B T} - 1}. \tag{27.43}$$

Since $e^0 = 1$, the denominator in (27.43) would be zero and the value of $\overline{N}$ would be infinite (diverge) if $(\varepsilon_k - \mu)$ were zero in any single-particle state. The energies ε_k of an ideal gas are $\hbar^2 k^2/(2m)$, so that energy is positive and the lowest energy state – the ground state – is essentially zero. Thus, to avoid a divergence, the chemical potential μ must be negative or zero.

The chemical potential μ is determined from the number of particles $\overline{N}$, temperature T, and volume V by equation (27.43). When expressed as an integral (see (27.11)), the value of μ for spin-0 bosons is determined by

$$\overline{N} = \frac{2}{\sqrt{\pi}} V \left(\frac{2\pi m k_B T}{h^2} \right)^{3/2} \int_0^\infty \frac{x^{1/2}}{e^{x - \mu/k_B T} - 1} dx. \tag{27.44}$$

As $-\mu = |\mu|$ when μ is negative or zero, the term $-\mu/k_BT$ in the exponential can be written as $|\mu/k_BT|$. After making this change, the above equation becomes

$$\overline{N} = VT^{3/2}B\left(\frac{\mu}{k_BT}\right), \tag{27.45}$$

where the function $B(u)$ is

$$B(u) = \left(\frac{2\pi mk_B}{h^2}\right)^{3/2}\frac{2}{\sqrt{\pi}}\int_0^\infty \frac{x^{1/2}}{e^{x+|u|}-1}dx. \tag{27.46}$$

Since the integrand $x^{1/2}/(e^{x+|u|}-1)$ increases as $|u|$ decreases, the function $B(u)$ has a maximum value at $u = 0$, which yields to the inequality

$$\boxed{\frac{\overline{N}}{VT^{3/2}} \leq B(0) = B_{\text{max}}.} \tag{27.47}$$

The value of the integral in $B(u)$ at $u = 0$ is

$$\int_0^\infty \frac{x^{1/2}dx}{e^x - 1} = 2.315. \tag{27.48}$$

Hence,

$$B(0) = B_{\text{max}} = 2.612\left(\frac{2\pi mk_B}{h^2}\right)^{3/2}. \tag{27.49}$$

A problem arises with the inequality in (27.47) when $\overline{N}/VT^{3/2}$ is greater than B_{max}, which is a real possibility because $\overline{N}$, T, and V are variables that can be controlled experimentally. The problem is solved by treating the ground state separately from the other states. The occupation number $\langle N_k\rangle$ of the lowest energy state increases as the value of $\overline{N}/VT^{3/2}$ is increased, and it increases more rapidly than the occupation numbers of all other states.

Some numerical estimates help to explain why the ground state can be treated separately. According to (24.6), the components (k_x, k_y, k_z) of the wave vectors $\boldsymbol{k}$ of a particle in a cubic box are $\pi n_x/L$, $\pi n_y/L$, and $\pi n_z/L$, where n_x, n_y, and n_z are positive integers, which implies that

$$k^2 = \boldsymbol{k}\cdot\boldsymbol{k} = \left(\frac{\pi}{L}\right)^2\left(n_x^2 + n_y^2 + n_z^2\right). \tag{27.50}$$

Since ε_k decreases as the value of k^2 becomes smaller, we consider the states with the lowest and next lowest values of k^2, which are

$$k^2 = \frac{\pi^2}{L^2}(1^2+1^2+1^2) = 3\frac{\pi^2}{L^2} \quad \text{and} \quad k^2 = \frac{\pi^2}{L^2}(2^2+1^2+1^2) = 6\frac{\pi^2}{L^2}. \tag{27.51}$$

Thus, the lowest and next lowest energies are

$$\varepsilon = \frac{3}{2m}\frac{\hbar^2\pi^2}{L^2} \quad \text{and} \quad \varepsilon = \frac{6}{2m}\frac{\hbar^2\pi^2}{L^2}. \tag{27.52}$$

Since it is conventional (and convenient) to have the energy of the ground state be zero, the first of these energies is subtracted from $\hbar^2k^2/(2m)$. The resulting energies of the ground state and first exited states are

$$\varepsilon_0 = 0 \qquad \text{and} \qquad \varepsilon_1 = \frac{3}{2m}\frac{\hbar^2\pi^2}{L^2}. \tag{27.53}$$

When both the ground-state energy and chemical potential are zero ($\varepsilon_0 = 0$, $\mu = 0$), the occupation number of the ground state is infinite (diverges). However, this divergence is not apparent in the integral expression for $\overline{N}$ in (27.44) because of the factor of $x^{1/2}$ which entered through the density of states $D(\varepsilon)$ in the derivation of (27.11). This means that the special properties of the ground state were neglected when the sum over states in (27.43) was estimated by the integral in (27.44). Because of this, the ground state should be considered separately. When this is done, the relationship in (27.45) becomes

$$\overline{N} = N_0 + VT^{3/2}B\left(\frac{\mu}{k_B T}\right), \tag{27.54}$$

where N_0 is the number of particles in the ground state.

To interpret the above equation, we define a *critical temperature* T_c by the value of T for which (27.47) is an equality. The resulting value is

$$\boxed{T_c = \left(\frac{\overline{N}}{V}\right)^{2/3}\left(\frac{1}{B_{\text{max}}}\right)^{2/3}.} \tag{27.55}$$

Above T_c, there are nonzero values of the chemical potential that satisfy equation (27.54) without needing to treat the ground state separately, so that the chemical potential is determined by

$$\overline{N} = VT^{3/2}B\left(\frac{\mu}{kT}\right). \tag{27.56}$$

Below T_c, the chemical potential is zero and a nonzero value of N_0 is required. The equation for determining N_0 is then obtained by substituting the value of $B(0)$ into (27.54), which gives

$$\overline{N} = N_0 + VT^{3/2}B_{\text{max}}. \tag{27.57}$$

Bose–Einstein Condensation

The accumulation of a macroscopic fraction of particles in the ground state is called *Bose–Einstein condensation*, and the particles in that state are called the *condensate*. To find the fraction of the particles in the ground state, the definition of T_c is rewritten as

$$B_{\text{max}} = \frac{\overline{N}}{VT_c^{3/2}}, \tag{27.58}$$

which is then substituted into (27.57). After dividing by $\overline{N}$, we obtain

$$\boxed{\frac{N_0}{\overline{N}} = 1 - \left(\frac{T}{T_c}\right)^{3/2}.} \tag{27.59}$$

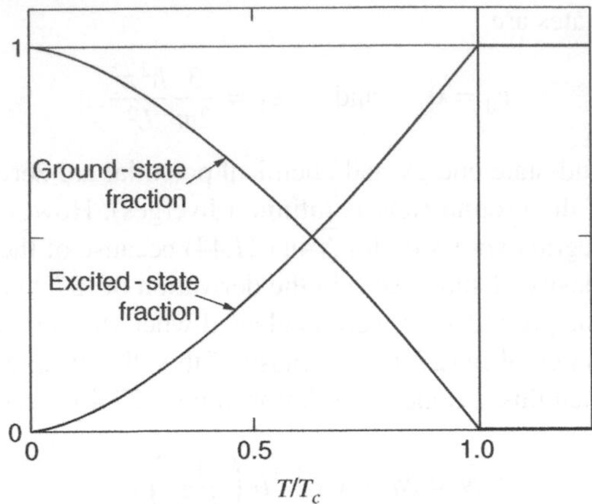

Figure 27.3 *Fractions of Particles in Ground State and Excited States*

The fraction $N_0/\overline{N}$ of particles in the condensate and the fraction in the excited states are plotted in **Figure 27.3**. Above the critical temperature, the fraction in the ground state is too small to show. Below the critical temperature the fraction $N_0/\overline{N}$ increases as the temperature decreases until it becomes 100% at absolute zero.

A gas of ^{4}He atoms at a pressure of 1.0 atm becomes a liquid at 4.2 K and goes through a second-order phase transition at 2.17 K. A new phase of matter – a superfluid – exists below the transition temperature. A superfluid has very unusual properties. Once a circular flow is stated it will persist forever. A thin film of liquid can seemingly defy gravity by flowing up over the lip of a beaker as though it was being siphoned off. It was suggested by Fritz London that the transition at 2.17 K might be an example of Bose–Einstein condensation. The fraction of the system in the ground state is the superfluid, whereas the remainder is a normal fluid, and the two fluids exist without spatial separation throughout the system. Although liquid helium is not an ideal gas, it is generally accepted that the underlying physics of superfluidity is Bose–Einstein condensation. Recently, Bose–Einstein condensation was observed in a dilute gas of alkali atoms. For this work, Eric Cornell, Wolfgang Ketterle, and Carl Wieman were awarded the 2001 Nobel Prize in physics.

Problems

27.1 The valance electrons in sodium can be modeled as an ideal Fermi gas with chemical potential $\mu = 3.24\,\text{eV}$. At absolute zero the average occupation number is $\langle N_{\mathbf{k}s}\rangle = \frac{1}{2}$, and the values of the single-particle energies ϵ_k at which $\langle N_{\mathbf{k}s}\rangle = \frac{1}{2}$ is $\epsilon_k = \mu = \epsilon_F$.

(a) Find the values of ϵ_k at which $\langle N_{\mathbf{k}s}\rangle = 0.99$ at $T = 300\,\text{K}$.

(b) Find the values of ϵ_k at which $\langle N_{\mathbf{k}s}\rangle = 0.1$ at $T = 300\,\text{K}$. Use $1/k_B = 11\,600\,\text{K}\,\text{eV}^{-1}$ to express the energies in electron-volts. (As indicated in Figure 27.1, the answers should be close to $\epsilon_k = \mu$.)

27.2 Bose–Einstein condensation was first observed in a low density gas on June 1995 at the JILA laboratory in Boulder, Colorado. The number of particles, $\overline{N}$ was approximately 2000, the volume was approximately $(20\,\mu\text{m})^3$, and the gas consisted of rubidium atoms with atomic weight 87. Estimate the transition temperature T_c of the experiment. (The temperature is very low!)

28

Interacting Systems

Although interactions are required to bring about the equilibrium distribution of energy, the microscopic origins of many equilibrium properties can be explained by using of independent particle models and employing the canonical energy distribution. Including the effects of the terms in the Hamiltonian that describe the interactions of particles greatly increases the complexity of the mathematics involved. In this chapter, we investigate two examples in which these interactions make essential contributions to the predicted results. We investigate the Ising model which illustrates the critical contribution of interparticle interactions to phase transitions, and derive corrections to the ideal gas equation of state. The use of computer simulations to study interacting systems is the subject of Chapter 29.

28.1 Ising Model

The interactions between the magnetic moments of the system's particles were neglected in our discussion of paramagnetism in Chapter 21. That neglect eliminated the possibility of describing ferromagnetic and other types of magnetic behavior. In this section, we investigate a simple model – the Ising model – in which spin-$\frac{1}{2}$ particles interact with each other through an interaction term proportional to the products of their spin variables. Although the model is highly idealized, it is still very difficult to obtain predictions from it without making significant approximations. Nevertheless, an exact analytic solution does exist for the two-dimensional Ising model. The solution was found by Lars Onsager in a mathematical *tour de force* for which he received the 1968 Nobel Prize in Chemistry. The solution predicts the existence of a second-order phase transition. The existence of this result demonstrates that statistical mechanics, as formulated, is capable of predicting phase transitions without the need for additional assumptions or approximations. An exact solution also exists for the one-dimensional model, but there is no phase transition in one dimension. No exact solution is known for the Ising model in three dimensions.

Thermodynamics and Statistical Mechanics: An Integrated Approach, First Edition.
Robert J Hardy and Christian Binek.

Each spin-$\frac{1}{2}$ particle has a spin variable σ_i with two allowed values, +1 and −1, where $\sigma_i = +1$ indicates spin "up" and $\sigma_i = -1$ indicates spin "down." The energy of the interaction between spin *i* and spin *j* is $-J_{ij}\sigma_i\sigma_j$. The analysis here considers models in which the interactions are limited to nearest neighbors. When all of the nearest neighbor interactions are the same, we can omit the subscripts on J_{ij}, so that the energies of the microstates of the model are

$$E_{N,\alpha} = -J\sum_{\langle i,j\rangle}\sigma_i\sigma_j - B\sum_i \sigma_i, \tag{28.1}$$

where the notation $\langle i,j\rangle$ indicates that each nearest-neighbor pair is included once in the sum over *i* and *j*. Index α is an abbreviation for the numbers $(\sigma_1\sigma_2\ \dots\ \sigma_N)$. We assume that *J* is positive so that the interaction energy of spins *i* and *j* is lowest when $\sigma_i = \sigma_j$, which makes the system ferromagnetic. In antiferromagnetic systems, the constant *J* is negative so that the spins on adjacent sites tend to align in opposite directions. The energy $-B\sigma_i$ describes the interaction of spin *i* with an external magnetic field. As described in Section 21.5, this energy is $-\mu_0\mathcal{H}m_o\sigma_i$, where $\mathcal{H}$ is the *z*-components of the field, m_o is the magnitude of magnetic moment, and μ_0 is permeability of free space, so that $B = \mu_0\mathcal{H}m_o$.

Mean Field Approximation

The energy $E_{N,\alpha}$ is not separable because of the terms $-J\sigma_i\sigma_j$. Because of the difficulty in finding the properties of a system whose energy is not separable, we employ a mean field approximation and use it to predict the magnetic moment of the homogeneous two-dimensional system diagrammed in **Figure 28.1**. The small circles (dots) represent spin-$\frac{1}{2}$ particles, and the lines between them connect nearest neighbors. The lines between the central particle and its six neighbors have been emphasized.

To obtain an approximation that makes the energy separable, we introduce the difference $(\sigma_i - \sigma_{av})$, where σ_{av} is the average value of σ_i. In a homogeneous system, σ_{av} is the same for all spins. Expressing σ_i as $[\sigma_{av} + (\sigma_i - \sigma_{av})]$ gives

$$\sigma_i\sigma_j = \sigma_{av}^2 + \sigma_{av}\left(\sigma_j - \sigma_{av}\right) + \left(\sigma_i - \sigma_{av}\right)\sigma_{av} + \left(\sigma_i - \sigma_{av}\right)\left(\sigma_j - \sigma_{av}\right). \tag{28.2}$$

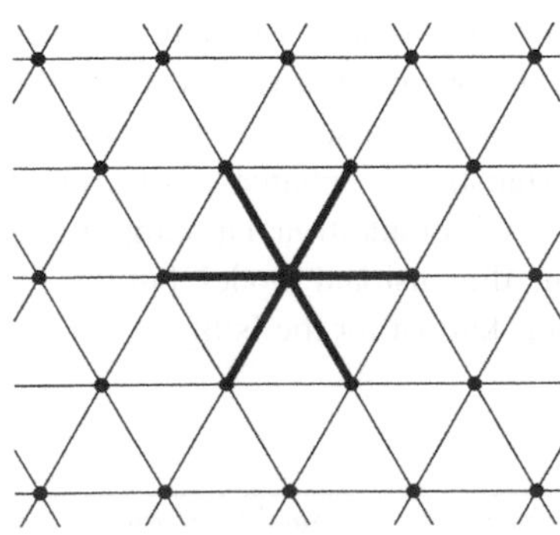

Figure 28.1 Nearest-Neighbor Particles on a Two-Dimensional Lattice

Our approximation neglects the product of differences $(\sigma_i - \sigma_{av})(\sigma_j - \sigma_{av})$. When this is done, the sum of $\sigma_i\sigma_j$ over neighbors becomes

$$J\sum_{\langle i,j\rangle} \sigma_i\sigma_j = J\sum_{\langle i,j\rangle} \sigma_{av}^2 + 2J\sum_{\langle i,j\rangle} \sigma_{av}\left(\sigma_i - \sigma_{av}\right) = 2J\sum_{\langle i,j\rangle} \sigma_{av}\sigma_i - J\sum_{\langle i,j\rangle} \sigma_{av}^2. \tag{28.3}$$

Since $\langle i,j\rangle$ indicates a sum in which each nearest-neighbor pair is included only once, the sums in the final terms of (28.3) can be written as $(z/2)\sum_i$, where z is number of nearest neighbors. The number z is called the *coordination number*. Its value for the lattice in Figure 29.1 is $z = 6$. Its value for a square lattice would be $z = 4$. After making our approximation, the energy $E_{N,\alpha}$ of a system of N spins becomes

$$E_{N,\alpha} = -\left(Jz\sigma_{av} + B\right)\sum_i \sigma_i + \frac{1}{2}NJz\sigma_{av}^2. \tag{28.4}$$

This energy has each spin interacting with an effective field

$$\overline{B} = \left(Jz\sigma_{av} + B\right), \tag{28.5}$$

which combines the external field B with an internal field $Jz\sigma_{av}$. In this *mean field approximation*, the spins interact with the average values of their neighbor's spins instead of with their actual values.

The energy in (28.4) is similar to the energy $-(\mu_0\mathcal{H}m_o)\sum_i\sigma_i$ of the paramagnetic system in (21.72), except that here the coefficient $(\mu_0\mathcal{H}m_o)$ is replaced by $(Jz\sigma_{av} + B)$. There is also the term $\frac{1}{2}NJz\sigma_{av}^2$, which is independent of σ_i. As in the paramagnetic case, the energy is separable and the equilibrium probability of σ_i is proportional to the Boltzmann factor e^{-E/k_BT} for the energy $-\overline{B}\sigma_i$. The normalized probability is

$$\rho_{\sigma_i} = \frac{e^{\beta\overline{B}\sigma_i}}{e^{+\beta\overline{B}\sigma_i} + e^{-\beta\overline{B}\sigma_i}}, \tag{28.6}$$

where $\beta = 1/k_BT$. The average value of σ_i is the same for all spins and is represented here by σ_{av}, instead of $\langle\sigma_i\rangle$, so that

$$\sigma_{av} = \sum_{\sigma_i} \rho_{\sigma_i}\sigma_i = \frac{e^{+\beta\overline{B}}(+1) + e^{-\beta\overline{B}}(-1)}{e^{+\beta\overline{B}} + e^{-\beta\overline{B}}} = \tanh\left(\beta\overline{B}\right), \tag{28.7}$$

which becomes

$$\sigma_{av} = \tanh\left(\beta\left(Jz\sigma_{av} + B\right)\right). \tag{28.8}$$

Zero Field Solution

To see the effect of the spin–spin interactions, we calculate the magnetic moment of the system in the absence of an external magnetic field ($B = 0$), in which case (28.8) becomes

$$\sigma_{av} = \tanh\left(\beta Jz\sigma_{av}\right). \tag{28.9}$$

This is a self-consistent equation for σ_{av} in the sense that σ_{av} appears on both sides of the equal sign. To find the temperature dependence of the moment, we need to solve the

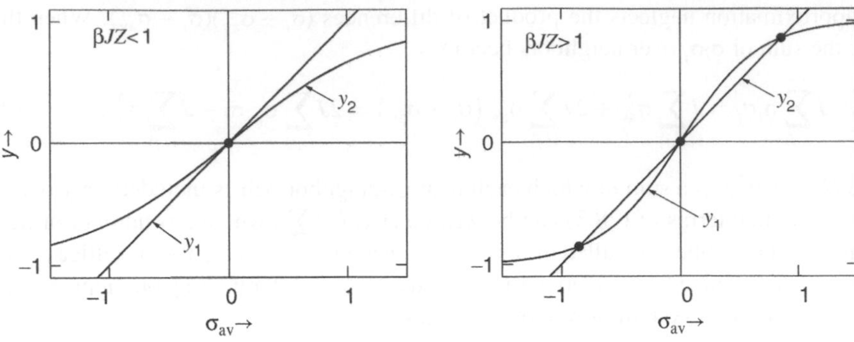

Figure 28.2 *Graphical Solution of Equation (28.9)*

equation for the dependence of σ_{av} on β. This is done by finding where the curve $y_1 = \sigma_{av}$ intersects the curve $y_2 = \tanh(\beta Jz\sigma_{av})$, which can be done graphically or with the help of a computer. Examples are shown in **Figure 28.2**, where curves for $\beta Jz < 1$ and $\beta Jz > 1$ are plotted. The only place where the curves intersect for $\beta Jz < 1$ is at $\sigma_{av} = 0$, while there are three places where $y_1 = y_2$ when $\beta Jz > 1$. The solution with $\sigma_{av} > 0$ indicates the magnetic moment is "up," and $\sigma_{av} < 0$ indicates the moment is "down." The transition from one solution to three solutions occurs when the slope of y_2 becomes greater than the slope of y_1, which happens when $\beta Jz = 1$.

The graphical solution is interpreted by expressing the relationship $\beta Jz = 1$ as $Jz/k_BT = 1$ and defining a critical temperature T_c through the relationship

$$k_BT_c = Jz. \tag{28.10}$$

In terms of T_c, the argument of the hyperbolic tangent in (28.9) is

$$\beta Jz = \frac{T_c}{T}, \tag{28.11}$$

and the inequalities $\beta Jz < 1$ and $\beta Jz > 1$ become $T > T_c$ and $T < T_c$, respectively. The transition from one solution to three occurs at $T = T_c$. As will be shown, the nonzero solutions have a lower free energy than the $\sigma_{av} = 0$ solution.

The equilibrium magnetic moment of a system of N spins is

$$M = Nm_o\sigma_{av}, \tag{28.12}$$

where m_o is the magnitude of the moment of each spin. The two nonzero values of M below the critical temperature represent the spontaneous magnetization characteristic of ferromagnetic behavior, whereas the vanishing value of M above T_c indicates paramagnetic behavior. The temperature dependence of the equilibrium magnetic moment is shown in **Figure 28.3**, where $\sigma_{av} = M/Nm_o$ is plotted as a function of T/T_c. In the absence of small perturbations, the system randomly selects one of the two solutions as the system is cooled through the critical temperature. The phenomenon is known as *spontaneous symmetry breaking*.

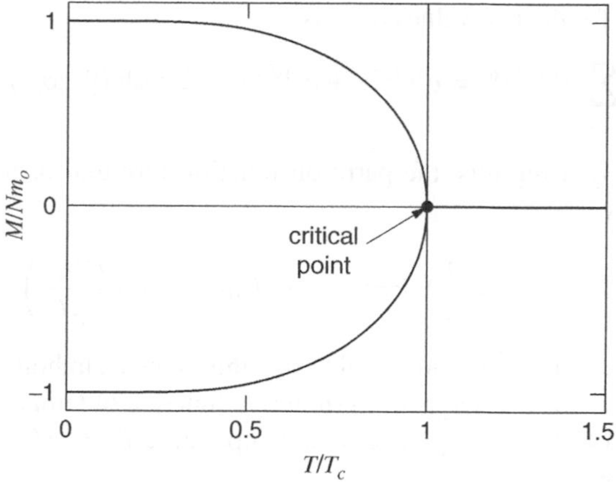

Figure 28.3 *Temperature Dependence of the Magnetic Moment*

Critical Exponent

As T approaches T_c, the magnetic moment drops continuously to zero in a second-phase order. To determine how M behaves as the critical temperature is approached, we expand $\tanh(\beta Jz\sigma_{av})$ in a power series ($\tanh x = x - \frac{1}{3}x^3 + \cdots$). Then, by neglecting higher order terms and using $\beta Jz = T_c/T$, the equation $\sigma_{av} = \tanh(\beta Jz\sigma_{av})$ becomes

$$\sigma_{av} = \sigma_{av}\left(\frac{T_c}{T}\right) - \frac{1}{3}\sigma_{av}^3\left(\frac{T_c}{T}\right)^3, \tag{28.13}$$

which implies that

$$\sigma_{av}^2 = 3\frac{T_c - T}{T_c}\left(\frac{T}{T_c}\right)^2. \tag{28.14}$$

Since $T/T_c = 1 - [(T_c - T)/T_c]$, the lowest order contribution to σ_{av}^2 in powers of $(T_c - T)$ is $\sigma_{av}^2 = 3(T_c - T)/T_c$. Thus, as the critical point is approached, the magnetic moment approaches

$$M = \sqrt{3}Nm_o\left(\frac{T_c - T}{T_c}\right)^{1/2}. \tag{28.15}$$

This indicates that the *critical exponent* in the mean field approximation is $1/2$. As mentioned in Section 14.3, this is the classical value of the exponent. This value is only approximate because of our use of the mean field approximation.

Free Energy

Since the energy in (28.4) becomes $E_{N,\alpha} = \frac{1}{2}NJz\sigma_{av}^2 - Jz\sigma_{av}\sum_i \sigma_i$ when $B = 0$, the partition function in zero field is

$$Z_N = \sum_\alpha e^{-\beta E_{N,\alpha}} = e^{-\frac{1}{2}N\beta Jz\sigma_{av}^2}\sum_{\sigma_1}\cdots\sum_{\sigma_N} e^{\beta Jz\sigma_{av}\Sigma_i\sigma_i}. \tag{28.16}$$

The sum of $e^{\beta Jz\sigma_{av}\sigma_i}$ for the two values of σ_i is

$$\sum_{\sigma_i=\pm 1} e^{\beta Jz\sigma_{av}\sigma_i} = e^{+\beta Jz\sigma_{av}} + e^{-\beta Jz\sigma_{av}} = 2\cosh\left(\beta Jz\sigma_{av}\right), \tag{28.17}$$

where $\beta = 1/k_BT$. By using this, the partition function formula for the free energy $F = -k_BT \ln Z_N$ becomes

$$F\left(T, \sigma_{av}\right) = \frac{1}{2}NJz\sigma_{av}^2 - Nk_BT \ln\left(2\cosh\frac{Jz\sigma_{av}}{k_BT}\right), \tag{28.18}$$

As the magnetic moment M is proportional to σ_{av}, this is the Helmholtz free energy, which expresses the free energy of a magnetic system as a function of temperature and magnetic moment (see Table 14.3). By using the critical temperature $T_c = Jz/k_B$ defined in (28.10), the free energy becomes

$$F\left(T, \sigma_{av}\right) = Nk_BT\left[\frac{T_c}{T}\frac{1}{2}\sigma_{av}^2 - \ln\left(2\cosh\left(\frac{T_c}{T}\sigma_{av}\right)\right)\right]. \tag{28.19}$$

The dependence of the free energy on σ_{av} above and below the critical temperature is shown in **Figure 28.4**, where $F(T, \sigma_{av})/k_BT$ is plotted versus σ_{av}. The equilibrium values of σ_{av} determined by (28.9) are indicated by dots. Above T_c, the solution with the lowest free energy is at $\sigma_{av} = 0$. Below T_c, there are three solutions. The two nonzero solutions are the stable states of the system, because they have the lower free energy. The system randomly selects one of them.

The Ising model in the mean field approximation is a nice illustration of the concepts of the Landau theory. As discussed in Section 14.3, Landau theory introduces an order parameter η and a generalized free energy $F_g(T, \eta)$. In ferromagnetic systems, the order parameter is the magnetic moment M, which in the Ising model is proportional to σ_{av}. Similar to $F_g(T, \eta)$, the free energy $F(T, \sigma_{av})$ is defined for both equilibrium and nonequilibrium values of the order parameter, and the equilibrium value is the value that minimizes the free energy.

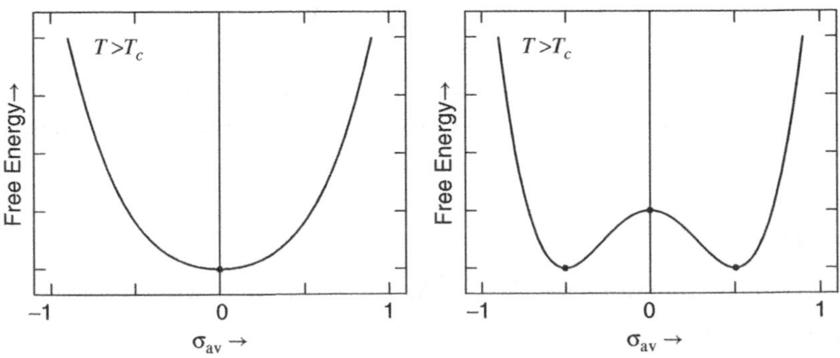

Figure 28.4 Dependence of Free Energy on σ_{av}

28.2 Nonideal Gases

The properties of ideal gases were obtained in Chapter 19 by neglecting the interparticle interaction term $\sum \phi(r_{ij})$ in the Hamiltonian. By neglecting that term, we deduced the ideal gas equation of state. The modifications to the equation of state caused by interparticle interactions are the subject of this section. The Hamiltonian for a monatomic gas with atoms of the same mass is (see (19.13))

$$H_N = \sum_i \frac{1}{2m}|\boldsymbol{p}_i|^2 + \sum_i \phi_{a\text{-}w}(\boldsymbol{r}_i) + \sum_{i<j} \phi(r_{ij}), \tag{28.20}$$

where the atom-wall potentials $\phi_{a\text{-}w}(\boldsymbol{r}_i)$ confine the atoms (particles) within their container. The interaction between atoms i and j is described by the interatomic potentials $\phi(r_{ij})$, where $r_{ij} = |\boldsymbol{r}_i - \boldsymbol{r}_j|$ is the distance between them. By letting indices i and j identify molecules, instead of atoms, the results obtained here are also applicable to gases of multi-atom molecules.

Because described in Section 21.3, the free energy of a classical system can be expressed as the sum of an ideal gas contribution F^{ig} and a configurational contribution,

$$F = F^{ig} + F^{\text{config}}. \tag{28.21}$$

The configurational free energy is

$$F^{\text{config}} = -k_B T \ln Z_N^{\text{config}}, \tag{28.22}$$

where the configurational partition function is

$$Z_N^{\text{config}} = \frac{1}{V^N}\int dr^{3N} e^{-\beta\Phi_N(\boldsymbol{r}_1\cdots\boldsymbol{r}_N)}. \tag{28.23}$$

As the potential energy Φ_N is a sum of potentials, the exponential in the integrand can be expressed as a product of exponentials,

$$e^{-\beta\Phi_N(\boldsymbol{r}_1\cdots\boldsymbol{r}_N)} = \prod_i e^{-\beta\phi_{a\text{-}w}(\boldsymbol{r}_i)} \prod_{i<j} e^{-\beta\phi(r_{ij})}. \tag{28.24}$$

By substituting this into (28.23) and associating the factor $e^{-\beta\phi_{a\text{-}w}(\boldsymbol{r}_i)}$ with the integration over the position variables of the atom, we obtain

$$Z_N^{\text{config}} = \frac{1}{V^N}\int dr_1^3 e^{-\beta\phi_{a\text{-}w}(\boldsymbol{r}_1)} \ldots \int dr_N^3 e^{-\beta\phi_{a\text{-}w}(\boldsymbol{r}_N)} \prod_{i<j} e^{-\beta\phi(r_{ij})}, \tag{28.25}$$

where the atom-wall potential $\phi_{a\text{-}w}(\boldsymbol{r}_i)$ confines the atom within the container. The Boltzmann factor $e^{-\beta\phi_{a\text{-}w}(\boldsymbol{r}_i)}$ is accounted for by limiting the integration over $\boldsymbol{r}_i$ to the volume V of the container, so that

$$Z_N^{\text{config}} = \frac{1}{V^N}\int_V dr_1^3 \cdots \int_V dr_N^3 \prod_{i<j} e^{-\beta\phi(r_{ij})}. \tag{28.26}$$

So far, no approximations have been made.

Since the interatomic potentials $\phi(r)$ depend on the variables of pairs of atoms, the integration in (28.26) cannot be separated into a product of independent integrals for individual atoms. To find the effect of interactions, we need an approximation that yields integrals that are practical to evaluate. We do this by using the fact that the Boltzmann factor $e^{-\beta\phi(r)}$ is essentially equal to 1 for most values of r. The approximation is obtained by expanding $e^{-\beta\phi(r)}$ in a series of powers of the difference

$$f(r) = \left(e^{-\beta\phi(r)} - 1\right), \tag{28.27}$$

so that the Boltzmann factor becomes

$$e^{-\beta\phi(r)} = 1 + f(r). \tag{28.28}$$

The dependence of the interatomic potential $\phi(r)$ and the function $f(r)$ on the distance between atoms is plotted in **Figure 28.5**. The curves shown are for the Lennard–Jones potential defined in (19.7). The difference $f(r)$ drops to zero for separations r greater than about twice the parameter σ that characterizes the range of the potential. The value of σ is typically less than a nanometer.

By using $e^{-\beta\phi(r)} = 1 + f(r)$, the product Boltzmann factors in (28.26) becomes

$$\prod_{i<j} e^{-\beta\phi(r_{ij})} = \left(1 + f_{12}\right)\left(1 + f_{13}\right)\left(1 + f_{14}\right)\cdots, \tag{28.29}$$

where $f_{ij} = f(r_{ij})$. To facilitate the expansion, each difference f_{ij} is multiplied by a bookkeeping parameter α, which is used to define a function

$$F(\alpha) = -k_B T \ln Z_N(\alpha), \tag{28.30}$$

where

$$Z_N(\alpha) = \frac{1}{V^N}\int_V dr_1^3 \cdots \int_V dr_N^3 \prod_{i<j}\left(1 + \alpha f_{ij}\right). \tag{28.31}$$

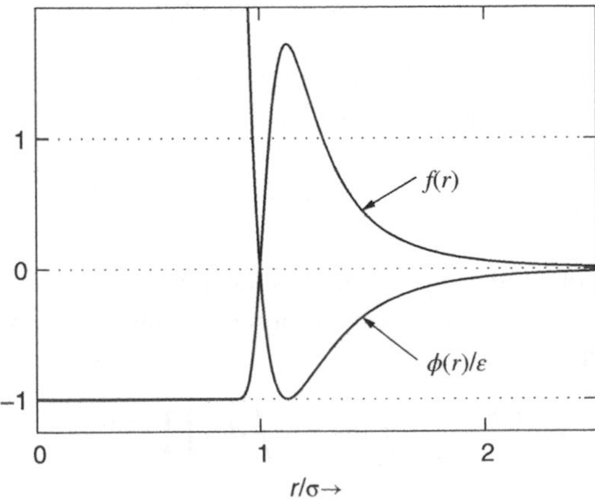

Figure 28.5 Interatomic Potential $\phi(r)$ and Function $f(r)$

Notice that $F(\alpha)$ is equal to the configurational free energy when $\alpha = 1$. The expansion in powers of the differences f_{ij} is obtained by expanding $F(\alpha)$ in a series of powers of α,

$$F^{\text{config}} = F(0) + F_1\alpha + \frac{1}{2}F_2\alpha^2 + \cdots. \tag{28.32}$$

By neglecting terms of order α^2 and higher and setting $\alpha = 1$, we obtain the approximation

$$F^{\text{config}} = F(0) + \left(\frac{\partial F}{\partial \alpha}\right)_0, \tag{28.33}$$

where

$$F(0) = -k_B T \ln Z_N(0) \qquad \text{and} \qquad \left(\frac{\partial F}{\partial \alpha}\right)_0 = -\frac{k_B T}{Z_N}\frac{\partial Z_N}{\partial \alpha}\bigg|_{\alpha=0}. \tag{28.34}$$

The value of $Z_N(0)$ is found by setting $\alpha = 0$ in (28.31), which gives

$$Z_N(0) = \frac{1}{V^N}\int_V dr_1^3 \cdots \int_V dr_N^3. \tag{28.35}$$

Since the integrand is +1, the integration for each particle equals the volume of the container, so that after N integrations we have $Z_N(0) = 1$. Hence,

$$F(0) = -k_B T \ln 1 = 0. \tag{28.36}$$

Taking the derivative with respect to α inside the integrations in (28.31) gives

$$\frac{\partial Z_N}{\partial \alpha} = \frac{1}{V^N}\int_V dr_1^3 \cdots \int_V dr_N^3 \frac{d}{d\alpha}\prod_{i<j}\left(1 + \alpha f_{ij}\right). \tag{28.37}$$

By using the product rule, the derivative of the product of factors $(1 + \alpha f_{ij})$ becomes

$$\begin{aligned}\frac{d}{d\alpha}\prod_{i<j}\left(1 + \alpha f_{ij}\right) &= \frac{d}{d\alpha}\left(1 + \alpha f_{12}\right)\left(1 + \alpha f_{13}\right)(1 + \alpha f_{14})\cdots \\ &= [f_{12}\left(1 + \alpha f_{13}\right)\left(1 + \alpha f_{14}\right)\cdots] + [(1 + \alpha f_{12})f_{13}(1 + \alpha f_{14})\cdots] + \cdots.\end{aligned} \tag{28.38}$$

Evaluating this at $\alpha = 0$ gives

$$\frac{d}{d\alpha}\prod_{i<j}\left(1 + \alpha f_{ij}\right)\bigg|_{\alpha=0} = \sum_{i<j} f_{ij}. \tag{28.39}$$

By substituting this into (28.37) with $\alpha = 0$ and using $f_{ij} = f(r_{ij})$, we obtain

$$\frac{\partial Z_N}{\partial \alpha}\bigg|_{\alpha=0} = \frac{1}{V^N}\int_V dr_1^3 \cdots \int_V dr_N^3 \sum_{i<j} f\left(r_{ij}\right). \tag{28.40}$$

Taking the sum over i and j outside the integrations gives a sum of $\frac{1}{2}N(N-1)$ terms that differ only in the subscripts on r_{ij}, so that each term has the same value. By singling out the first term in the sum, we obtain

$$\frac{\partial Z_N}{\partial \alpha}\bigg|_{\alpha=0} = \frac{1}{2}\frac{N(N-1)}{V^N}\int_V dr_1^3 \cdots \int_V dr_N^3 f_{12}\left(r_{12}\right), \tag{28.41}$$

where $r_{12} = |\boldsymbol{r}_1 - \boldsymbol{r}_2|$. Since only the integrals over $\boldsymbol{r}_1$ and $\boldsymbol{r}_2$ involve the function $f(r_{12})$, each of the other $N-2$ integrals equals volume V, so that

$$\left.\frac{\partial Z_N}{\partial \alpha}\right|_{\alpha=0} = \frac{1}{2}\frac{N(N-1)}{V^2}\int_V dr_1^3 \int_V dr_2^3 f(r_{12}). \tag{28.42}$$

This result can be further simplified by changing the integration variables from $(\boldsymbol{r}_1, \boldsymbol{r}_2)$ to $(\boldsymbol{R}, \boldsymbol{r}_{12})$,

$$\int_V dr_1^3 \int_V dr_2^3 f(r_{12}) = \int dR^3 \int dr_{12}^3 f(r_{12}), \tag{28.43}$$

where $\boldsymbol{R}$ is the center of mass position and $\boldsymbol{r}_{12}$ is the relative displacement. The positions of the atoms are related to $\boldsymbol{R}$ and $\boldsymbol{r}_{12}$ by

$$\boldsymbol{r}_1 = \boldsymbol{R} + \frac{1}{2}\boldsymbol{r}_{12} \quad \text{and} \quad \boldsymbol{r}_2 = \boldsymbol{R} - \frac{1}{2}\boldsymbol{r}_{12}, \tag{28.44}$$

which implies that

$$\boldsymbol{R} = \frac{1}{2}(\boldsymbol{r}_1 + \boldsymbol{r}_2) \quad \text{and} \quad \boldsymbol{r}_{12} = \boldsymbol{r}_1 - \boldsymbol{r}_2. \tag{28.45}$$

Since $f(r_{12})$ is independent of $\boldsymbol{R}$, the integrations over $\boldsymbol{R}$ and $\boldsymbol{r}_{12}$ are independent except when one of the variables $(\boldsymbol{r}_1, \boldsymbol{r}_2)$ is adjacent to a container wall. Because $f(r_{12})$ vanishes for separations r_{12} greater than a few nanometers while the dimensions of the container are measured in centimeters, the interaction with the container walls is negligible. The integration over $\boldsymbol{R}$ yields volume V. Then, by taking the integration over $\boldsymbol{r}_{12}$ cover all space, it follows that

$$\int dr_{12}^3 f(r_{12}) = \int_{-\infty}^{+\infty} dx_{12} \int_{-\infty}^{+\infty} dy_{12} \int_{-\infty}^{+\infty} dz_{12} f(r_{12}). \tag{28.46}$$

where $\boldsymbol{r}_{12} = (x_{12}, y_{12}, z_{12})$. Changing from Cartesian coordinates to spherical coordinates $(r_{12}, \theta, \varphi)$ and integrating over θ and φ gives

$$\int dr_{12}^3 f(r_{12}) = \int_0^\infty 4\pi r^2 f(r) dr. \tag{28.47}$$

By using these results, equation (28.42) becomes

$$\left.\frac{\partial Z_N}{\partial \alpha}\right|_{\alpha=0} = \frac{1}{2}\frac{N^2}{V}\int_0^\infty 4\pi r^2 f(r) dr, \tag{28.48}$$

where $f(r) = e^{-\beta\phi(r)} - 1$. After substituting this into the expression for $(\partial F/\partial\alpha)_0$ in (28.34) and the using the results $F(0) = 0$ and $Z(0) = 1$, the configurational free energy becomes

$$F^{config} = -\frac{1}{2}N k_B T \frac{N}{V}\int_0^\infty 4\pi r^2 (e^{-\beta\phi(r)} - 1) dr. \tag{28.49}$$

Thus, the size of F^{config} is determined by the ratio of the volume over which $\phi(r)$ is non-zero divided by the volume per particle V/N.

Since the Helmholtz free energy equals $F^{ig} + F^{config}$, the pressure of the gas is

$$P = -\left(\frac{\partial F}{\partial V}\right)_T = -\left(\frac{\partial F^{ig}}{\partial V}\right)_T - \left(\frac{\partial F^{config}}{\partial V}\right)_T. \tag{28.50}$$

By using the derivative of F^{config} and the ideal gas equation of state $PV = Nk_BT$, we find that

$$P = Nk_BT\left[\frac{1}{V} - \frac{1}{2}\frac{N}{V^2}\int_0^\infty 4\pi r^2 f(r)\,dr\right]. \tag{28.51}$$

This equation contains the first two terms in the cluster expansion of Ursell and Mayer, which is an expansion in powers of the number density N/V. Higher order terms are obtained by employing the grand canonical ensemble.

Van der Waals model

The correction to the ideal gas equation of state in (28.51) was obtained from the relationship established by the canonical ensemble between thermodynamic properties and the Hamiltonian. It is interesting to compare it with the corrections in the van der Waals model discussed in Chapter 5. As given in (5.9), the van der Waals equation of state is

$$P = \frac{RT}{v-b} - \frac{a}{v^2}, \tag{28.52}$$

where $v = V/n$ is the volume per mole. The parameters a and b were obtained earlier by fitting to experimental data. To facilitate the comparison, we consider low densities for which $1/v$ is small and use the approximation $1/(v-b) \approx (1/v)(1+b/v)$, so that (28.52) becomes

$$P = RT\frac{n}{V}\left[1 + \frac{n}{V}\left(b - \frac{a}{RT}\right)\right]. \tag{28.53}$$

Then, by using $n = N/N_A$ and $R = N_Ak_B$, where N_A is Avogadro's number, the van der Waals equation of state becomes

$$P = k_BT\frac{N}{V}\left[1 + \frac{N}{V}\left(\frac{b}{N_A} - \frac{a}{N_A^2k_BT}\right)\right]. \tag{28.54}$$

The significance of the van der Waals parameters a and b can be understood by introducing a simplified interatomic potential and evaluating the integral in (28.51). The simplified potential employed has a repulsive core at $r \leq \sigma$ and a potential well of depth e_o between $r = \sigma$ and $r = w$,

$$\phi(r) = \begin{cases} \infty, (r < \sigma). \\ -e_o, (\sigma < r < w). \\ 0,\ (r > w). \end{cases} \tag{28.55}$$

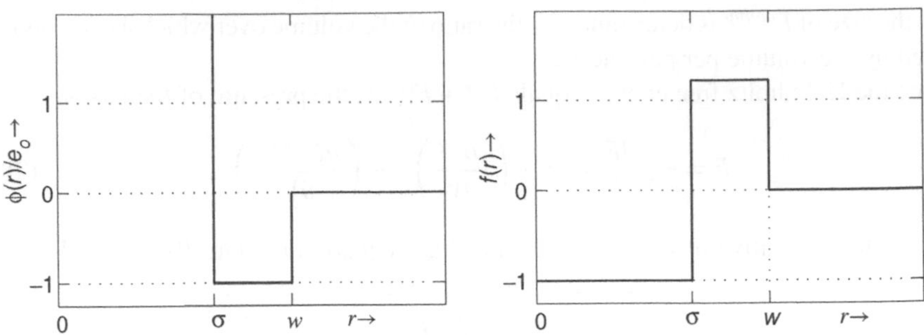

Figure 28.6 *Approximate Potential* $\phi(r)$ *and Function* $f(r)$

The resulting expression for the function $f(r)$ is

$$f(r) = \begin{cases} -1, & (r < \sigma). \\ (e^{\beta e_o} - 1), & (\sigma < r < w). \\ 0, & (r > w). \end{cases} \tag{28.56}$$

The potential $\phi(r)$ and function $f(r)$ are plotted in **Figure 28.6** with $\beta e_o = 1$.

With this simplified potential, the integral in equation (28.51) becomes

$$\int_0^\infty 4\pi r^2 f(r) dr = -\frac{4}{3}\pi\sigma^3 + \frac{4}{3}\pi(e^{\beta e_o} - 1)(w^3 - \sigma^3), \tag{28.57}$$

where $\beta = 1/k_B T$. By substituting this into (28.51) and using the expansion $e^x \approx 1 + x + \cdots$, we obtain

$$P = k_B T \frac{N}{V}\left[1 + \frac{N}{V}\frac{2}{3}\pi\left(\sigma^3 - \frac{e_o}{k_B T}\left(w^3 - \sigma^3\right)\right)\right]. \tag{28.58}$$

Comparing this with (28.54) indicates that

$$b = \frac{2}{3}\pi N_A \sigma^3 \quad \text{and} \quad a = \frac{2}{3}\pi N_A^2 e_o (w^3 - \sigma^3). \tag{28.59}$$

To interpret the significance of these values for a and b, we begin by pointing out the very rapid increase in the potential $\phi(r)$ at $r = \sigma$ seen in Figures 28.5 and 28.6. This rapid increase indicates that σ is the distance between atoms at which they become strongly repelled. Since the distance between the centers of two similar spheres in contact is twice their radius, the parameter σ in the interatomic potential is the effective diameter of an atom. If the atoms were hard spheres of radius σ, the volume per mole in which each atom is free to move would be reduced by the presence of the other atoms by $(N_A - 1)\frac{4}{3}\pi\sigma^3$. Since $N_A >> 1$, the effective volume excluded by the finite size of the atoms $\frac{4}{3}\pi N_A \sigma^3$. The first equation in (28.59) indicates that the parameter b is one-half the value of this excluded volume. Thus, the term containing b in the van der Waals model represents the effects of the repulsive cores of the atoms.

The negative potential energy between $r = \sigma$ and $r = w$ is associated with the forces that attract the particles (atoms) towards each other. Thus, the second equation in (28.59) indicates that the term containing a represents the effects of the attractive forces between atoms. Although only monatomic gases have been considered, our conclusions about the significance of van der Waals parameters also apply to gases containing more complex molecules.

Problems

28.1 Consider the Ising model in zero field ($B = \mu_0 \mathcal{H} m_o = 0$) at temperatures near absolute zero where the average value of σ_i is either +1 or −1. Assume that $\sigma_{av} = +1$. Use the mean field approximation to find the partition function Z_N, the enthalpy $E_{th} = -\partial \ln Z_N / \partial\beta$, and heat capacity $C_{\mathcal{H}} = \partial E_{th} / \partial T$. Compare $C_{\mathcal{H}}$ with its value for spin-$\frac{1}{2}$ systems in Chapter 21.

28.2 The analysis of non-deal gases establishes the relationship $b = \frac{2}{3}\pi N_A \sigma^3$ between the van der Waals parameter b and the parameter σ in the interatomic potential. Use the values of b in Table 5.2 to estimate the values of σ and r_0 for Ne and Ar, where $r_0 = 2^{1/6}\sigma$. Compare with the values of r_0 for the potential in Table 19.1.

29

Computer Simulations

The variety of systems that can be analyzed is greatly extended by the use of computer simulations. The advantage of simulations over analytic studies is the wide variety of interparticle interaction that can be treated. There are two basic approaches. Monte Carlo simulations utilize the probabilistic description of thermal equilibrium given by the canonical ensemble and are intrinsically statistical. The other approach, molecular dynamics, uses the mechanical description of atomic level behavior given by the classical equations of motion. Both approaches estimate thermodynamic properties by calculating averages over collections of microstates. Although the averaging techniques are very different, the two approaches yield the same predictions. The purpose of this chapter is to show the relationship of the algorithms and phase functions used in simulations to the principles of statistical mechanics.

29.1 Averages

Monte Carlo techniques estimate ensemble averages, whereas molecular dynamics evaluates time averages. It is fundamental to statistical mechanics that the two kinds of averages are equivalent. The classical ensemble average of a property described by phase function $X(\boldsymbol{p}_1 \cdots \boldsymbol{r}_N)$ is

$$\langle X \rangle = \int dp^{3N} \int dr^{3N} \rho\left(\boldsymbol{p}_1, \ldots, \boldsymbol{r}_N\right) X\left(\boldsymbol{p}_1, \ldots, \boldsymbol{r}_N\right), \tag{29.1}$$

where $\rho(\boldsymbol{p}_1, \ldots, \boldsymbol{r}_N)$ is the probability density. The *time average* of the property is

$$\langle X \rangle = \frac{1}{\left(t_2 - t_1\right)} \int_{t_1}^{t_2} X\left(\boldsymbol{p}_1(t), \ldots, \boldsymbol{r}_N(t)\right) dt, \tag{29.2}$$

Thermodynamics and Statistical Mechanics: An Integrated Approach, First Edition.
Robert J Hardy and Christian Binek.

where the time dependence of the momentum and position variables is determined by Hamiltonian's equations of motion, or equivalently by Newton's laws. The internal energy U and pressure P are given by the averages of phase functions,

$$\boxed{U = \langle H_N \rangle \quad \text{and} \quad P = \langle P_{\text{ph}} \rangle.} \tag{29.3}$$

The phase function for energy is the Hamiltonian H_N. The phase function for pressure, as found in Chapter 19, is $P_{\text{ph}} = -\partial H_N / \partial V$. An alternate expression for P_{ph} that is more useful in computer simulations is derived in Section 29.2. In addition to properties such as U and P, each of which is given by a single average, other properties can be estimated by evaluating combinations of averages. For example, as shown in Section 22.2, the constant-volume heat capacity is proportional to the variance of the energy,

$$C_V = \frac{1}{k_B T^2} \left[\langle {H_N}^2 \rangle - \langle H_N \rangle^2 \right]. \tag{29.4}$$

29.2 Virial Formula for Pressure

The virial formula for pressure is especially useful in simulations because it eliminates the need to make a detailed description of the interaction of a system with its container. It does this by expressing the interaction with the container in terms of the interactions between the system's constituents. There are two ways to derive the virial formula. One approach is based on the canonical ensemble and the other on the laws of mechanics.

Pressure in the Canonical Ensemble

We are interested in systems with Hamiltonians of the form

$$H_N \left(\boldsymbol{p}_1, \cdots, \boldsymbol{r}_N \right) = K_N \left(\boldsymbol{p}_1, \cdots, \boldsymbol{p}_N \right) + \Phi_N \left(\boldsymbol{r}_1, \cdots, \boldsymbol{r}_N \right), \tag{29.5}$$

where K_N is the kinetic energy of the particles and Φ_N is their potential energy. The advantage of simulations is the wide variety of potential energy functions that can be used. The potential energy was expressed in Chapter 19 as

$$\Phi_N = \Phi_{\text{int}} + \Phi_{a\text{-}w}, \tag{29.6}$$

where Φ_{int} is the energy of the interactions of the system's particles (atoms) with each other and $\Phi_{a\text{-}w}$ is the energy of their interactions with the container walls. The virial formula for pressure is found by reinterpreting the effects of the atom-wall energy $\Phi_{a\text{-}w}$.

Pressure is given by the derivative of the Helmholtz free energy F, which, in the canonical ensemble, is determined by the partition function formula $F = -k_B T \ln Z_N$. When analyzed classically, the free energy can be written as (see (21.38))

$$F = F^{ig} - k_B T \ln Z_N^{\text{config}}, \tag{29.7}$$

where F^{ig} is the free energy of an ideal gas and Z_N^{config} is the *configurational partition function*,

$$Z_N^{\text{config}} = \frac{1}{V^N} \int dr^{3N} e^{-\beta \Phi_N(\boldsymbol{r}_1, \ldots, \boldsymbol{r}_N)} \tag{29.8}$$

and $\beta = 1/k_BT$. Since the pressure of an ideal gas is Nk_BT/V, the pressure of the system is

$$P = -\left(\frac{\partial F}{\partial V}\right)_T = \frac{Nk_BT}{V} + \frac{k_BT}{Z_N^{\text{config}}}\frac{\partial Z_N^{\text{config}}}{\partial V}. \tag{29.9}$$

The atom-wall energy $\Phi_{a\text{-}w}$ is expressed in (19.12) as a sum of single particle contributions $\phi_{a\text{-}w}(\boldsymbol{r}_i)$, which vanishes when particle i is inside the container and is infinite outside of the container. It has the effect of confining the particle within the container by restricting the integration over the position variables $\boldsymbol{r}_i = (x_i, y_i, z_i)$ to the region of space inside the container. Thus,

$$Z_N^{\text{config}} = \frac{1}{V^N}\int_V dr_1^3 \cdots \int_V dr_N^3 e^{-\beta\Phi_{\text{int}}(\boldsymbol{r}_1,\ldots,\boldsymbol{r}_N)}, \tag{29.10}$$

where only the potential energy Φ_{int} of the interactions internal to the system appears in the Boltzmann factor. The derivative of Z_N^{config} with respect to volume can be found by scaling the size of the container. This is done by multiplying the coordinates by the cube-root of volume, so that the position variables of particle i become

$$x_i = \sigma_i^x V^{1/3}, \qquad y_i = \sigma_i^y V^{1/3}, \qquad z_i = \sigma_i^z V^{1/3}, \tag{29.11}$$

where V is the volume of the container. By changing variables from $\boldsymbol{r}_i$ to $\boldsymbol{\sigma}_i$ and using $dr_i^{\,3} = Vd\sigma_i^{\,3}$, the configurational partition function becomes

$$Z_N^{\text{config}} = \int_\Sigma d\sigma_1^3 \cdots \int_\Sigma d\sigma_N^3 \exp\left(-\beta\Phi_{\text{int}}\left(\boldsymbol{\sigma}_1 V^{1/3}, \ldots, \boldsymbol{\sigma}_N V^{1/3}\right)\right), \tag{29.12}$$

where the scaled volume is $\Sigma = 1$. The factor of $1/V^N$ in (29.10) has been cancelled by the factors of V resulting from the change of variables. For example, in a cubic container with walls at $x = 0$, $x = L$, $y = 0$, $y = L$, $z = 0$, and $z = L$, the integral of a function $f(\boldsymbol{r}_i)$ becomes

$$\begin{aligned}\int_V dr_i^3 f(\boldsymbol{r}_i) &= \int_0^L dx_i \int_0^L dy_i \int_0^L dz_i f(x_i, y_i, z_i)\\ &= L^3\int_0^1 d\sigma_i^x \int_0^1 d\sigma_i^y \int_0^1 d\sigma_i^z f(\sigma_i^x L, \sigma_i^y L, \sigma_i^z L) = V\int_\Sigma d\sigma_1^3 f(\boldsymbol{\sigma}_i V^{1/3}).\end{aligned} \tag{29.13}$$

Differentiating the expression for Z_N^{config} in (29.12) gives

$$\frac{\partial}{\partial V} Z_N^{\text{config}} = \int_\Sigma d\sigma_1^3 \cdots \int_\Sigma d\sigma_N^3 e^{-\beta\Phi_{\text{int}}}(-\beta)\frac{\partial}{\partial V}\Phi_{\text{int}}\left(\boldsymbol{\sigma}_1 V^{1/3}, \cdots, \boldsymbol{\sigma}_N V^{1/3}\right), \tag{29.14}$$

where

$$\frac{\partial}{\partial V}\Phi_{\text{int}}\left(\boldsymbol{\sigma}_1 V^{1/3}, \cdots, \boldsymbol{\sigma}_N V^{1/3}\right) = \sum_i \frac{\partial \Phi_{\text{int}}}{\partial\left(\boldsymbol{\sigma}_i V^{1/3}\right)} \cdot \boldsymbol{\sigma}_i \frac{1}{3V^{2/3}}. \tag{29.15}$$

By transforming back to the original variables $\boldsymbol{r}_i$ and using $d\sigma_i^3 = dr_i^3/V$, we obtain

$$\frac{\partial}{\partial V} Z_N^{\text{config}} = \frac{-\beta}{V^N}\int_V dr_1^3 \cdots \int_V dr_N^3 e^{-\beta\Phi_{\text{int}}} \sum_i \frac{\partial \Phi_{\text{int}}}{\partial \boldsymbol{r}_i} \cdot \boldsymbol{r}_i \frac{1}{3V}. \tag{29.16}$$

Since the configurational probability density is

$$\rho^{\text{config}}\left(\boldsymbol{r}_1,\ \ldots\ ,\boldsymbol{r}_N\right) = \frac{e^{-\beta\Phi_{\text{int}}(\boldsymbol{r}_1,\ldots,\boldsymbol{r}_N)}}{\int_V dr_1^3 \cdots \int_V dr_N^3\, e^{-\beta\Phi_{\text{int}}(\boldsymbol{r}_1,\ldots,\boldsymbol{r}_N)}} = \frac{e^{-\beta\Phi_{\text{int}}(\boldsymbol{r}_1,\ldots,\boldsymbol{r}_N)}}{V^N Z_N^{\text{config}}}, \tag{29.17}$$

the configurational contribution to the pressure is

$$k_B T \frac{\partial}{\partial V} \ln Z_N^{\text{config}} = \frac{-1}{3V} \int_V dr_1^3 \cdots \int_V dr_N^3\, \rho^{\text{config}} \sum_i \frac{\partial \Phi_{\text{int}}}{\partial \boldsymbol{r}_i} \cdot \boldsymbol{r}_i. \tag{29.18}$$

Because the quantity averaged is independent of the momentum variables, the configurational average obtained with ρ^{config} is equivalent to a full canonical average, so that (29.18) becomes

$$k_B T \frac{\partial}{\partial V} \ln Z_N^{\text{config}} = \left\langle \frac{-1}{3V} \sum_i \frac{\partial \Phi_{\text{int}}}{\partial \boldsymbol{r}_i} \cdot \boldsymbol{r}_i \right\rangle. \tag{29.19}$$

Combining this with the expression for pressure in (29.9) gives

$$P = \frac{N k_B T}{V} + \left\langle \frac{1}{3V} \sum_i \boldsymbol{F}_{i,\text{int}} \cdot \boldsymbol{r}_i \right\rangle, \tag{29.20}$$

where $\boldsymbol{F}_{i,\text{int}}\left(\boldsymbol{r}_1,\ \ldots\ ,\boldsymbol{r}_N\right) = -\partial\Phi_{\text{int}}\left(\boldsymbol{r}_1,\ \ldots\ ,\boldsymbol{r}_N\right)/\partial\boldsymbol{r}_i$ is the force on particle i exerted by all other particles. The canonical average of the kinetic energy is

$$\langle K \rangle = \left\langle \sum_i \frac{1}{2m_i} |\boldsymbol{p}_i|^2 \right\rangle_c = \frac{3}{2} N k_B T. \tag{29.21}$$

By combining this with (29.20), we obtain

$$P = \frac{1}{3V} \left\langle \sum_i \frac{1}{m_i} |\boldsymbol{p}_i|^2 + \sum_i \boldsymbol{F}_{i,\text{int}} \cdot \boldsymbol{r}_i \right\rangle. \tag{29.22}$$

Thus, the thermodynamic value of pressure is the average of the phase function for pressure, which is

$$\boxed{P_{\text{ph}}\left(\boldsymbol{p}_1,\ \ldots\ ,\boldsymbol{r}_N\right) = \frac{1}{3V} \left(\sum_i \frac{1}{m_i} |\boldsymbol{p}_i|^2 + \sum_i \boldsymbol{F}_{i,\text{int}}\left(\boldsymbol{r}_1,\ \ldots\ ,\boldsymbol{r}_N\right) \cdot \boldsymbol{r}_i \right).} \tag{29.23}$$

This is the *virial formula for pressure*.

Pressure from Newton's Laws

Another way to derive the above phase function is to use Newton's laws to express the effects of the forces on the walls of the container in terms of the forces between the system's particles. To do this, Newton's second law is used to find the time derivative of the following product of the positions $\boldsymbol{r}_i$ and velocities $\dot{\boldsymbol{r}}_i = d\boldsymbol{r}_i/dt$ of the particles:

$$S(t) = \sum_i m_i \boldsymbol{r}_i \cdot \dot{\boldsymbol{r}}_i. \tag{29.24}$$

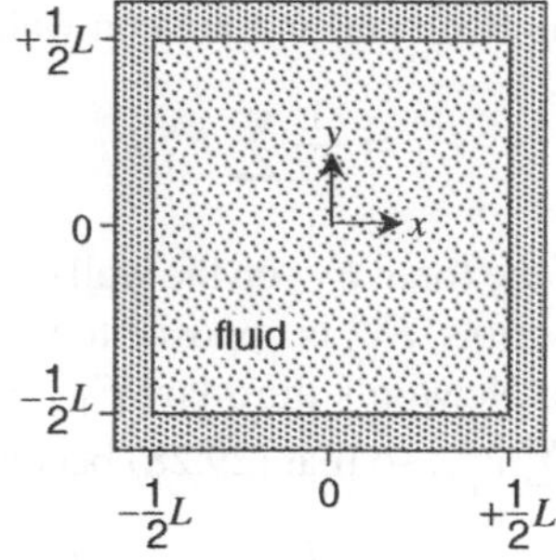

Figure 29.1 *A Fluid Confined in a Cubic Container.*

After distinguishing between the forces exerted by the particles internal to the system and the particles external to it, the time derivative becomes

$$\frac{dS}{dt} = \sum_i m_i \left(\dot{\boldsymbol{r}}_i \cdot \dot{\boldsymbol{r}}_i + \boldsymbol{r}_i \cdot \ddot{\boldsymbol{r}}_i\right) = \sum_i m_i \left|\dot{\boldsymbol{r}}_i\right|^2 + \sum_i \boldsymbol{r}_i \cdot \left(\boldsymbol{F}_{i,\text{int}} + \boldsymbol{F}_{i,\text{ext}}\right). \tag{29.25}$$

$\boldsymbol{F}_{i,\text{int}}$ is the force on particle i exerted by other particles in the system, and $\boldsymbol{F}_{i,\text{ext}}$ is the forces on particle i exerted by the atoms in the container walls. To make the analysis easier to visualize, consider a fluid confined in a cubic container with walls at $x = \pm L/2$, $y = \pm L/2$ and $z = \pm L/2$, as diagrammed in **Figure 29.1**. The six walls of the cube are numbered, so that

$$\boldsymbol{F}_{i,\text{ext}} = \boldsymbol{F}_{i,1} + \boldsymbol{F}_{i,2} + \cdots + \boldsymbol{F}_{i,6}, \tag{29.26}$$

where $\boldsymbol{F}_{i,1}$ is the force on particle i exerted by the wall at $x = +L/2$, $\boldsymbol{F}_{i,2}$ is the force on particle i exerted by the wall at $x = -L/2$, $\boldsymbol{F}_{i,3}$ is the force on particle i exerted by the wall at $y = +L/2$, and so forth.

After combining (29.26) and (29.25), the contribution to dS/dt of wall number 1 becomes

$$\sum_i \boldsymbol{r}_i \cdot \boldsymbol{F}_{i,1} = \sum_i x_i F^x_{i,1} + \sum_i y_i F^y_{i,1} + \sum_i z_i F^z_{i,1}. \tag{29.27}$$

The short range nature of interatomic forces allows us to relate this contribution to the force per unit area (pressure) on that wall. Specifically, the force on particle i exerted by the atoms in the wall vanishes except when the particle is within about a nanometer of it, which implies that the majority system's particles do not contribute to the sums in (29.27). Since the wall is located at $x = +L/2$, the term $\sum_i x_i F^x_{i,1}$ is effectively $(L/2)\sum_i F^x_{i,1}$, so that its contribution to dS/dt is

$$\sum_i \boldsymbol{r}_i \cdot \boldsymbol{F}_{i,1} = \frac{L}{2}\sum_i F^x_{i,1} + 0 + 0. \tag{29.28}$$

The contributions of $\sum_i y_i F^y_{i,1}$ and $\sum_i z_i F^z_{i,1}$ vanish because they involve the components $F^y_{i,1}$ and $F^z_{i,1}$ which are parallel to the wall, so that the components from many particles cancel each other because of the varying directions of the forces.

We now relate $\sum_i \boldsymbol{r}_i \cdot \boldsymbol{F}_{i,1}$ to the force per unit area on wall 1, which is

$$P_1 = \frac{1}{L^2} \sum_i F^x_{1,i}, \tag{29.29}$$

where $F^x_{1,i}$ is the x-component of the force $\boldsymbol{F}_{1,i}$ on the wall exerted by particle i. As indicated by the reversed order of the subscripts, $\boldsymbol{F}_{1,i}$ is opposite to the force $\boldsymbol{F}_{i,1}$ on particle i exerted by the wall, so that according to Newton's third law $\boldsymbol{F}_{1,i} = -\boldsymbol{F}_{i,1}$ and $F^x_{1,i} = -F^x_{i,1}$. Using this in (29.29) gives $P_1 L^2 = -\sum_i F^x_{i,1}$, so that (29.28) becomes

$$\sum_i \boldsymbol{r}_i \cdot \boldsymbol{F}_{i,1} = -\frac{L}{2} P_1 L^2 = -\frac{1}{2} P_1 V, \tag{29.30}$$

where $V = L^3$. The other five walls make similar contributions to dS/dt. When all six contributions are included, the time average of the contribution of $\boldsymbol{F}_{i,\text{ext}}$ to dS/dt becomes

$$\left\langle \sum_i \boldsymbol{r}_i \cdot \boldsymbol{F}_{i,\text{ext}} \right\rangle = -3 \langle P_{\text{av}} \rangle_t V, \tag{29.31}$$

where $P_{\text{av}} = (P_1 + \cdots + P_6)/6$ is the force per unit area averaged over the six faces of the cube.

Although the above result was derived for a cubic container, its validity is not limited to that shape. The surfaces of containers of more general shapes can be idealized as collections of small flat elements of area ΔA_k. Then, because the forces involved are short ranged, the contribution to dS/dt can be expressed as

$$\sum_i \boldsymbol{r}_i \cdot \boldsymbol{F}_{i,\text{ext}} = \sum_k \left(\boldsymbol{r}_k \cdot \boldsymbol{n}_k \Delta A_k \right) \sum_{i \text{ at } k} \frac{\boldsymbol{n}_k \cdot \boldsymbol{F}_{i,\text{ext}}}{\Delta A_k}, \tag{29.32}$$

where the sum over k includes all surface elements and vector $\boldsymbol{r}_k$ is drawn from the center of the enclosed volume to the surface. The unit vector perpendicular (normal) to the element is $\boldsymbol{n}_k$, so that the normal component of the forces on a particle is $\boldsymbol{n}_k \cdot \boldsymbol{F}_{i,\text{ext}}$. The sum labeled "$i$ at k" indicates that the sum includes only those particles that are located next to surface element k. When divided by area ΔA_k, the sum of the components $-\boldsymbol{n}_k \cdot \boldsymbol{F}_{i,\text{ext}}$ is the pressure on element k. The right-hand side of (29.32) has the dimensions of a volume times a pressure, but before we can interpret it as average pressure times volume, we need to find out how much volume is involved. When the time-averaged pressure is uniform, we can factor the average pressure out of the sum over k in (29.32). Then, expressing the sum as an integral gives

$$\left\langle \sum_i \boldsymbol{r}_i \cdot \boldsymbol{F}_{i,\text{ext}} \right\rangle = -\langle P_{\text{av}} \rangle_t \oint \boldsymbol{r} \cdot d\boldsymbol{A}. \tag{29.33}$$

By using the divergence theorem to express this integral over the surface of the container as a volume integral, it follows that $\oint \boldsymbol{r} \cdot d\boldsymbol{A} = \int dr^3 \cdot \nabla \boldsymbol{r}$. Since $\nabla \boldsymbol{r} = 3$, we obtain

$$\left\langle \sum_i \boldsymbol{r}_i \cdot \boldsymbol{F}_{i,\text{ext}} \right\rangle = -3 \langle P_{\text{av}} \rangle V, \tag{29.34}$$

which is the same as (29.31).

By using (29.34) to express the interaction with the container, the time average of dS/dt given in (29.25) becomes

$$\left\langle \frac{dS}{dt} \right\rangle = \left\langle \sum_i m_i |\dot{\boldsymbol{r}}_i|^2 + \sum_i \boldsymbol{r}_i \cdot \boldsymbol{F}_{i,\text{int}} \right\rangle - 3 \langle P_{\text{av}} \rangle V. \tag{29.35}$$

The average of the time derivative is

$$\left\langle \frac{dS}{dt} \right\rangle = \frac{1}{(t_2 - t_1)} \int_{t_1}^{t_2} \frac{dS_x(t)}{dt} dt = \frac{S(t_2) - S(t_1)}{t_2 - t_1}. \tag{29.36}$$

Because $S(t)$ is the sum of terms with random values that tend to cancel, the average $\langle dS/dt \rangle$ becomes negligible when divided by $t_2 - t_1$ and can be neglected. After neglecting $\langle dS/dt \rangle$, equation (29.35) can be rearranged as

$$\langle P_{\text{av}} \rangle = \frac{1}{3V} \left\langle \sum_i \left(m_i |\dot{\boldsymbol{r}}_i|^2 + \boldsymbol{r}_i \cdot \boldsymbol{F}_{i,\text{int}} \right) \right\rangle. \tag{29.37}$$

Since the thermodynamic value of pressure is the average of its phase function, we conclude that

$$\boxed{P_{\text{ph}} (\dot{\boldsymbol{r}}_1, \ldots, \boldsymbol{r}_N) = \frac{1}{3V} \left(\sum_i m_i |\dot{\boldsymbol{r}}_i|^2 + \sum_i \boldsymbol{r}_i \cdot \boldsymbol{F}_{i,\text{int}} (\boldsymbol{r}_1, \ldots, \boldsymbol{r}_N) \right),} \tag{29.38}$$

which is the *virial formula for pressure*. It is the same as (29.23) except that the kinetic energy contribution is expressed in terms of velocities instead of momenta.

Interatomic Potentials

A commonly used method for determining the forces $\boldsymbol{F}_{i,\text{int}}$ that result from interactions between particles is to derive them from the energy Φ_{int} of a sum of *interatomic potentials* $\phi(r)$,

$$\Phi_{\text{int}} (\boldsymbol{r}_1, \ldots, \boldsymbol{r}_N) = \sum_{i<j} \phi(r_{ij}), \tag{29.39}$$

where $r_{ij} = |\boldsymbol{r}_i - \boldsymbol{r}_j|$ is the distance between particles (atoms) i and j. The force on atom i obtained from $\phi(r_{ij})$ is

$$\boldsymbol{F}_{i,j} (\boldsymbol{r}_i, \boldsymbol{r}_j) = -\frac{\partial \phi(r_{ij})}{\partial \boldsymbol{r}_i} = -\phi'(r_{ij}) \frac{\boldsymbol{r}_i - \boldsymbol{r}_j}{r_{ij}}, \tag{29.40}$$

where $\phi'(r) = d\phi(r)/dr$. The force $\boldsymbol{F}_{j,i} = -\partial\phi/\partial\boldsymbol{r}_j$ on atom j is equal to $-\boldsymbol{F}_{i,j}$, as required by Newton's third law. The total force on the atom exerted by all other atoms is

$$\boldsymbol{F}_{i,\text{int}} = \sum_j \boldsymbol{F}_{i,j}, \tag{29.41}$$

where the term with $j = i$ is omitted from the sum. By using $\boldsymbol{F}_{i,j} = -\boldsymbol{F}_{j,i}$ and exchanging indices i and j, the contribution of interatomic forces to $P_{\text{ph}}(\dot{\boldsymbol{r}}_1, \ldots, \boldsymbol{r}_N)$ becomes

$$\sum_i \boldsymbol{r}_i \cdot \boldsymbol{F}_{i,\text{int}} = \sum_{i,j} \boldsymbol{r}_i \cdot \boldsymbol{F}_{i,j} = \sum_{i,j} \frac{1}{2}\left(\boldsymbol{r}_i \cdot \boldsymbol{F}_{i,j} + \boldsymbol{r}_j \cdot \boldsymbol{F}_{j,i}\right)$$
$$= \frac{1}{2}\sum_{i,j}\left(\boldsymbol{r}_i - \boldsymbol{r}_j\right) \cdot \boldsymbol{F}_{i,j} = -\frac{1}{2}\sum_{i,j} r_{ij}\, \phi'\left(r_{ij}\right). \tag{29.42}$$

After substituting this into (29.38), the virial formula for pressure becomes

$$\boxed{P_{\text{ph}}\left(\dot{\boldsymbol{r}}_1, \ldots, \boldsymbol{r}_N\right) = \frac{1}{3V}\left(\sum_i m_i |\dot{\boldsymbol{r}}_i|^2 + \frac{1}{2}\sum_{ij} r_{ij}\, \phi'\left(r_{ij}\right)\right).} \tag{29.43}$$

By replacing $\boldsymbol{p}_i$ by $m_i\dot{\boldsymbol{r}}_i$ in the Hamiltonian, the energy phase function becomes

$$\boxed{H_N\left(\dot{\boldsymbol{r}}_1, \ldots, \boldsymbol{r}_N\right) = \sum_i \frac{1}{2} m_i |\dot{\boldsymbol{r}}_i|^2 + \sum_{i<j} \phi\left(r_{ij}\right).} \tag{29.44}$$

We can calculate the energy and pressure of a system in microstate $(\dot{\boldsymbol{r}}_1, \ldots, \boldsymbol{r}_N)$ from these equations once the functional form of the interatomic potentials $\phi(r)$ has been specified.

29.3 Simulation Algorithms

Molecular Dynamics

To evaluate the phase functions for energy and pressure in a molecular dynamics simulation, we need the positions $\boldsymbol{r}_i$ and velocities $\dot{\boldsymbol{r}}_i$ of the particles as determined by the equations of motion. The numerical method for doing this, which is described below, has a long and productive history and is easily programmed. To find equations for predicting the motion of N point masses (particles), we expand the position $\boldsymbol{r}_i(t)$ of each particle in the Taylor series

$$\boldsymbol{r}_i(t \pm \Delta t) = \boldsymbol{r}_i(t) \pm \dot{\boldsymbol{r}}_i(t)\,\Delta t + \ddot{\boldsymbol{r}}_i(t)\frac{\Delta t^2}{2} \pm \dddot{\boldsymbol{r}}_i(t)\frac{\Delta t^3}{3!} + \cdots, \tag{29.45}$$

where both a future time $(+\Delta t)$ and a past time $(-\Delta t)$ are included. Adding the series for $t - \Delta t$ to the series for $t + \Delta t$ and neglecting higher order terms gives

$$\boldsymbol{r}_i(t + \Delta t) + \boldsymbol{r}_i(t - \Delta t) = 2\boldsymbol{r}_i(t) + \ddot{\boldsymbol{r}}_i(t)\,\Delta t^2. \tag{29.46}$$

Subtracting the series for $t - \Delta t$ from the series for $t + \Delta t$ and dividing by $2\Delta t$ give

$$\dot{\boldsymbol{r}}_i(t) = \frac{\boldsymbol{r}_i(t + \Delta t) - \boldsymbol{r}_i(t - \Delta t)}{2\Delta t}. \tag{29.47}$$

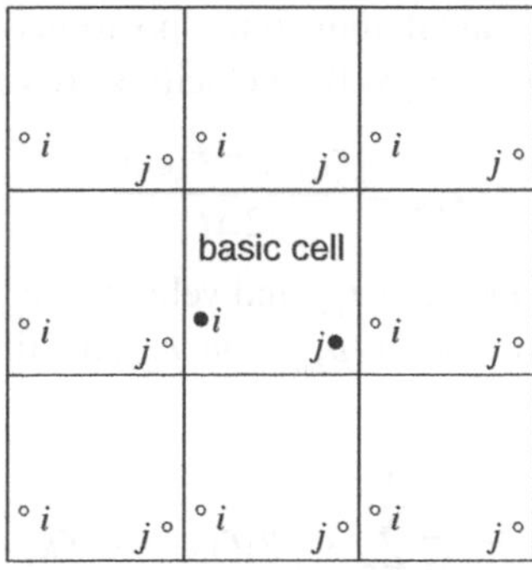

Figure 29.2 Arrangement of Cells for Simulating Periodic Boundaries.

The terms proportional to the third derivative $\dddot{\boldsymbol{r}}_i(t)$ have canceled in the derivation of (29.46) but not in the derivation of (29.47), so that the positions $\boldsymbol{r}_i(t)$ are determined more accurately than the velocities $\dot{\boldsymbol{r}}_i(t)$. By using Newton's second law $m_i\ddot{\boldsymbol{r}}_i(t) = \boldsymbol{F}_i(t)$ to find the acceleration, equation (29.46) becomes

$$\boldsymbol{r}_i(t+\Delta t) = 2\boldsymbol{r}_i(t) - \boldsymbol{r}_i(t-\Delta t) + \boldsymbol{F}_i(t)\frac{\Delta t^2}{m_i}. \tag{29.48}$$

When periodic boundary conditions are used, we only need the force $\boldsymbol{F}_{i,\text{int}}$ on the particle exerted by the other particles in the system. The force on particle i is

$$\boldsymbol{F}_i(t) = -\frac{\partial\Phi_{\text{int}}\left(\boldsymbol{r}_1(t), \dots, \boldsymbol{r}_N(t)\right)}{\partial\boldsymbol{r}_i}, \tag{29.49}$$

where Φ_{int} is the internal potential energy. Periodic boundary conditions are achieved by introducing cells of volume $V = L^3$ as diagramed in **Figure 29.2**, where the system consists of the particles in the basic cell at the center. Each particle has images in the other cells, and each image moves so that it is in the same position relative to its cell as the system's particle is relative to the basic cell. When a particle moves out of the basic cell, it reenters it on the opposite side, and when a particle (black circle) is close enough to interact with one of the images (open circles), the interaction with the image is included in the calculation of the force on the particle. In this way, the environment is uniform throughout the basic cell, and no interaction with a container is required to keep the system within a specific volume.

To find future positions of the particles, we consider evenly spaced times $t_n = n\Delta t$, where $n = 1, 2, 3, \dots,$ and represent the position of particle i at time t_n by $\boldsymbol{r}_{i,n} = \boldsymbol{r}_i(t_n)$. Equation (29.48) then becomes

$$\boldsymbol{r}_{i,n+1} = 2\boldsymbol{r}_{i,n} - \boldsymbol{r}_{i,n-1} + \boldsymbol{F}_{i,n}\frac{\Delta t^2}{m_i}, \tag{29.50}$$

where

$$\boldsymbol{F}_{i,n} = -\frac{\partial\Phi_{\text{int}}\left(\boldsymbol{r}_{1,n}, \dots, \boldsymbol{r}_{N,n}\right)}{\partial\boldsymbol{r}_{i,n}}. \tag{29.51}$$

Notice that these equations for calculating future positions do not involve the velocities. When needed to find the kinetic energy, the velocities are calculated with (29.47), which becomes

$$\dot{\boldsymbol{r}}_{i,n} = \frac{\boldsymbol{r}_{i,n+1} - \boldsymbol{r}_{i,n-1}}{2\Delta t}. \tag{29.52}$$

This procedure for finding the positions $\boldsymbol{r}_{i,n}$ and velocities $\dot{\boldsymbol{r}}_{i,n}$ is known as the *Verlet algorithm*.[1] As the momentum of particle i is $\boldsymbol{p}_{i,n} = m_i \dot{\boldsymbol{r}}_{i,n}$, the time average of the phase function for property X in (29.2) becomes

$$\langle X \rangle = \frac{1}{M}\sum_{n=1}^{M} X\left(m_1\dot{\boldsymbol{r}}_{1,n}, \; \dots \; , \boldsymbol{r}_{N,n}\right), \tag{29.53}$$

where M is the number of time steps calculated.

A molecular dynamics simulation is started by assigning random values to the velocities $\dot{\boldsymbol{r}}_{i,0}$ at the initial time t_0 and choosing a convenient set of initial positions $\boldsymbol{r}_{i,0}$. Since (29.50) uses the positions $\boldsymbol{r}_{i,n}$ and $\boldsymbol{r}_{i,n-1}$ of two previous time steps to predict the position $\boldsymbol{r}_{i,n+1}$ at time t_{n+1}, values for the positions $\boldsymbol{r}_{i,-1}$ at time t_{-1} are needed to start the simulation. They can be obtained by using the series in (29.45) to move back one time step, so that

$$\boldsymbol{r}_{i,-1} = \boldsymbol{r}_{i,0} - \dot{\boldsymbol{r}}_{i,0}\Delta t + \frac{1}{2}\boldsymbol{F}_{i,0}\frac{\Delta t^2}{m_i}. \tag{29.54}$$

Finally, we must select a value for Δt. This is usually done by making several trial runs and choosing a value that yields energy conservation to the desired number of significant figures. The volume V is determined by the periodic boundaries, and the temperature T is estimated from the average value of the kinetic energy, which for a classical systems in three dimensions is

$$\left\langle \frac{1}{2}\sum_{i} m_i |\dot{\boldsymbol{r}}_i|^2 \right\rangle = \frac{3N}{2}k_B T. \tag{29.55}$$

For simulations in two dimensions the factor of $3N$ is replaced by $2N$.

Monte Carlo

In Monte Carlo simulations, we are interested in the positions of particles but not their velocities. Successive configurations $(\boldsymbol{r}_{1,n}, \; \dots \; , \boldsymbol{r}_{N,n})$ are selected so that the unweighted averages of phase functions converge to the averages determined by the canonical ensemble. The index n that we use to label configurations has no significance relative to the system's time evolution. The average of a phase function that depends on the position variables is

$$\langle X \rangle = \frac{1}{M}\sum_{n=1}^{M} X\left(\boldsymbol{r}_{1,n}, \; \dots \; , \boldsymbol{r}_{N,n}\right), \tag{29.56}$$

[1] Verlet L. (1967), *Phys. Rev.* **159**, 98.

where the configurations are distributed with the probability density

$$\rho^{\text{config}}\left(\boldsymbol{r}_1, \dots, \boldsymbol{r}_N\right) = \frac{e^{-\beta\Phi_{\text{int}}(\boldsymbol{r}_1, \dots, \boldsymbol{r}_N)}}{\int dr^{3N} e^{-\beta\Phi_{\text{int}}(\boldsymbol{r}_1, \dots, \boldsymbol{r}_N)}}. \tag{29.57}$$

Averages that depend on the momentum variables are found analytically. As in molecular dynamics, the need to explicitly include the interactions with the container walls is avoided by using periodic boundary conditions.

The procedure for selecting configurations consistent with (29.57) is justified by the theory of Markov processes. An easily implemented procedure for doing this is the *Metropolis method*.[2] In this approach, the configuration in step $n+1$ is calculated from the configuration $(\boldsymbol{r}_{1,n}, \dots, \boldsymbol{r}_{N,n})$ in step n as follows:

1. Select one of the N particles and use index i to identify it. The selection can be done randomly or sequentially.
2. Move the particle from its position $\boldsymbol{r}_{i,n} = (x_{i,n}, y_{i,n}, z_{i,n})$ in step n with uniform probability to a trial position $\boldsymbol{r}'_i = (x'_i, y'_i, z'_i)$ within a small cube centered about $\boldsymbol{r}_{i,n}$, so that

$$x_{i,n} - a < x'_i < x_{i,n} + a, \quad y_{i,n} - a < y'_i < y_{i,n} + a, \quad z_{i,n} - a < z'_i < z_{i,n} + a. \tag{29.58}$$

   The randomness in the trial position is achieved by using a random number generator.
3. Calculate the change in potential energy

$$\Delta E = \Phi_{\text{int}}\left(\dots, \boldsymbol{r}', \dots\right) - \Phi_{\text{int}}\left(\dots, \boldsymbol{r}_{i,n}, \dots\right), \tag{29.59}$$

   where only the position of particle i has been changed.
4. Calculate the Boltzmann factor $e^{-\beta\Delta E}$ and compare it with a random number Rn that is uniformly distributed between 0 and 1.
5. The positions of all of the particles except i remain unchanged, while the position of particle i is selected as follows:

   If $e^{-\beta\Delta E} < Rn$, use the trial position as the new position, so that $\boldsymbol{r}_{i,n+1} = \boldsymbol{r}'_i$.
   If $e^{-\beta\Delta E} > Rn$, leave particle i where it was, so that $\boldsymbol{r}_{i,n+1} = \boldsymbol{r}_{i,n}$.

Next, select a new particle and repeat the procedure until M configurations have been obtained. Although the proof that the Metropolis method yields a canonical average is not given here, we can see in the calculation of the Boltzmann factor $e^{-\beta\Delta E}$ where the canonical ensemble enter the procedure.

Before we can start a simulation, we need to select the parameter a in (29.58) that defines the small cube. It is usually chosen by trial and error so that on the average about half of the particles are not moved to new positions in the selection process described in item 5. Volume V is determined by the periodic boundaries, and temperature T is determined by the inverse temperature $\beta = 1/k_B T$. As in molecular dynamics, the initial set of positions

[2] Metropolis M., Rosenbluth A. W., Rosenbluth M. N., Teller A. N., Teller E. (1953), *J. Chem. Phys.* **21**, 1087.

can be chosen on the basis of convenience. The bias introduced by the choice of the initial configuration is made negligible either by using a large number of steps *M* or by neglecting an initial set of steps before evaluating averages.

Because of its foundation in the canonical ensemble, the Monte Carlo method is confined to the determination of equilibrium properties. Although only equilibrium properties have been considered, molecular dynamics simulations are also useful for studying nonequilibrium behavior. Computer simulations are an active field of research with a wide range of applications. It is recommended that students interested in doing simulations consult one of the many good books on the subject.

A

Mathematical Relations, Constants, and Properties

A.1 Partial Derivatives

Notation: thermodynamic to conventional: $\left(\dfrac{\partial F}{\partial x}\right)_y = \dfrac{\partial F(x,y)}{\partial x}$

The numbers indicate where the formulas occur in Chapter 3.

$$\left(\frac{\partial F}{\partial X}\right)_Z = \left(\frac{\partial F}{\partial X}\right)_Y + \left(\frac{\partial F}{\partial Y}\right)_X \left(\frac{\partial Y}{\partial X}\right)_Z \tag{3.35}$$

$$\left(\frac{\partial F}{\partial Z}\right)_X = \left(\frac{\partial F}{\partial Y}\right)_X \left(\frac{\partial Y}{\partial Z}\right)_X \tag{3.36}$$

$$\left(\frac{\partial Y}{\partial X}\right)_Z \left(\frac{\partial X}{\partial Y}\right)_Z = 1 \tag{3.44}$$

$$\left(\frac{\partial X}{\partial Y}\right)_Z = \frac{1}{(\partial Y/\partial X)_Z} \tag{3.45}$$

$$\left(\frac{\partial Y}{\partial X}\right)_Z \left(\frac{\partial X}{\partial Z}\right)_Y \left(\frac{\partial Z}{\partial Y}\right)_X = -1 \tag{3.51}$$

$$\left(\frac{\partial X}{\partial Z}\right)_Y = -\left(\frac{\partial X}{\partial Y}\right)_Z \left(\frac{\partial Y}{\partial Z}\right)_X \tag{3.52}$$

A.2 Integrals and Series

$$\int_{-\infty}^{+\infty} e^{-ax^2}\,dx = \sqrt{\pi/a}$$

$$\int_{-\infty}^{+\infty} x^2 e^{-ax^2}\,dx = \frac{1}{2}\sqrt{\pi/a^3}$$

Thermodynamics and Statistical Mechanics: An Integrated Approach, First Edition.
Robert J Hardy and Christian Binek.

$$\int_{-\infty}^{+\infty} \frac{x^2 e^x}{(e^x+1)^2} dx = \frac{\pi^2}{3}$$

$$\int_{0}^{+\infty} \frac{x^4 e^x}{(e^x-1)^2} dx = \frac{4\pi^4}{15}$$

$$\int_{0}^{\infty} \frac{x^{\frac{1}{2}}}{e^x-1} dx = 2.315$$

$$\int_a^b f(x)\frac{dg(x)}{dx}dx = f(x)g(x)|_a^b - \int_a^b g(x)\frac{df(x)}{dx}dx \qquad \text{Integration by Parts}$$

$$e^x = \sum_{n=0}^{\infty} \frac{x^n}{n!} = 1 + x + \frac{x^2}{2!} + \frac{x^3}{3!} + \cdots$$

$$\ln(1+x) = x - \frac{1}{2}x^2 + \frac{1}{2}x^3 - \frac{1}{4}x^4 + \cdots$$

$$\frac{1}{1-x} = \sum_{n=0}^{\infty} x^n = 1 + x + x^2 + \cdots$$

$$\frac{1}{1+x} = \sum_{n=0}^{\infty} (-1)^n x^n = 1 - x + x^2 - x^3 + \cdots$$

A.3 Taylor Series

$$f(x+\Delta x) = f(x) + \frac{df(x)}{dx}\Delta x + \frac{d^2 f(x)}{dx^2}\frac{\Delta x^2}{2!} + \frac{d^3 f(x)}{dx^3}\frac{\Delta x^3}{3!} + \cdots$$

$$F(x,y) = F(x_o,y_o) + \frac{\partial F(x_o,y_o)}{\partial x}\Delta x + \frac{\partial F(x_o,y_o)}{\partial y}\Delta y$$
$$+ \frac{\partial^2 F(x_o,y_o)}{\partial x^2}\frac{\Delta x^2}{2} + \frac{\partial^2 F(x_o,y_o)}{\partial x \partial y}\Delta x \Delta y + \frac{\partial^2 F(x_o,y_o)}{\partial y^2}\frac{\Delta y^2}{2} + \cdots$$
$$\Delta x = x - x_o, \Delta y = y - y_o$$

A.4 Hyperbolic Functions

$$\sinh x = \frac{1}{2}(e^{+x} - e^{-x})$$

$$\cosh x = \frac{1}{2}(e^{+x} + e^{-x})$$

$$\tanh x = \frac{e^{+x} - e^{-x}}{e^{+x} + e^{-x}}$$

A.5 Fundamental Constants

Avogadro's number	N_A	$6.022 \times 10^{23}\,\text{mol}^{-1}$
Boltzmann constant	k_B	$1.381 \times 10^{-23}\,\text{J}\,\text{K}^{-1}$
Universal gas constant	$R = k_B N_A$	$8.314\,\text{J}\,\text{mol}^{-1}\,\text{K}^{-1}$
Planck's constant	h	$6.626 \times 10^{-34}\,\text{J}\,\text{s}$
$\hbar$	$\hbar = h/2\pi$	$1.055 \times 10^{-34}\,\text{J}\,\text{s}$
Atomic mass constant	m_u	$1.6605 \times 10^{-27}\,\text{kg}$
Electron mass	m_e	$9.109 \times 10^{-31}\,\text{kg}$
electron charge	e	$1.6020 \times 10^{-19}\,\text{C}$
Speed of light	c	$2.998 \times 10^{8}\,\text{m}\,\text{s}^{-1}$
Vacuum permittivity	ϵ_0	$8.854 \times 10^{-12}\,\text{kg}^{-1}\,\text{m}^{-1}\,\text{s}^2\,\text{C}^2$
Vacuum permeability	μ_0	$4\pi \times 10^{-7}\,\text{kg}\,\text{m}\,\text{C}^{-2}$
Standard atmosphere	1 atm	$101.325 \times 10^{3}\,\text{Pa}$

A.6 Conversion Factors

Force	$1.0\,\text{lb} = 4.448\,\text{N}$
Energy	$4184\,\text{J} = 1\,\text{kcal} = 1000\,\text{cal} = 3.968\,\text{Btu}$
Power	$1.0\,\text{hp} = 746\,\text{W}$
Pressure	$1\,\text{atm} = 101.3\,\text{kPa} = 760\,\text{mmHg} = 760\,\text{torr} = 14.70\,\text{lb}\,\text{in}^{-2}$
Volume	$1\,\text{m}^3 = 10^3\,\text{L} = 10^6\,\text{cm}^3 = 35.31\,\text{ft}^3$

A.7 Useful Formulas

Celsius to Kelvin	$\Theta^{\text{C}} + 273.15 = T^{\text{K}}$
Fahrenheit to Celsius	$\frac{5}{9}(\Theta^{\text{F}} - 32) = \Theta^{\text{C}}$
Volume of a sphere	$V = \frac{4}{3}\pi r^3$
Surface area of a sphere	$A = 4\pi r^2$

A.8 Properties of Water

Triple point	$P_{tr} = 612\,\text{Pa}$	$T_{tr} = 273.16\,\text{K}$	
Critical point	$P_c = 22.12\,\text{MPa}$	$T_c = 647.4\,\text{K}$	$v_c = 3.17 \times 10^{-3}\,\text{m}^3\,\text{kg}^{-1}$
Latent heat of fusion at 1 atm	$L_f = 80\,\text{kcal}\,\text{kg}^{-1}$		
Latent heat of vaporization at 1 atm	$L_v = 539\,\text{kcal}\,\text{kg}^{-1}$		

A.9 Properties of Materials

	Aluminum	Copper	Iron	Lead	Ice (0 °C)	Water
Density ($\text{kg}\,\text{m}^{-3}$)	2 700	8 960	7 880	1 1340	917	1 000
c_P^M ($\text{kcal}\,\text{kg}^{-1}\,\text{K}^{-1}$)	0.216	0.092	0.106	0.0306	0.502	1.000
c_P^M ($\text{kJ}\,\text{kg}^{-1}\,\text{K}^{-1}$)	0.904	0.385	0.444	0.128	2.10	4.184
c_V^M ($\text{kJ}\,\text{kg}^{-1}\,\text{K}^{-1}$)	0.863	0.374	0.436	0.120	2.03	
α_L ($10^{-6}\,\text{K}^{-1}$)	23.2	16.6	11.9	29.0	55	
α_V ($10^{-6}\,\text{K}^{-1}$)	69.6	49.8	35.7	87.0	165	
B_T ($10^9\,\text{Pa}$)	75	135	160	41	7.7	

Answers to Problems

Chapter 1

1. 20.0 °C, 37.0 °C, 293 K, 310 K. **2.** 184 kPa, 26.7 psi.
3. $S(T,P) = n(c_o + R)\ln(T/T_o) - nc_o \ln(P/P_o)$. **4.** 31.3 °C. **5.** 167 J kg^{-1} K^{-1}.

Chapter 2

1. 4320 J, 57 s **2.** 44.2 mol, 50.1 kPa, 76.5 kJ, 0, 76.5 kJ. **3.** 20.75 °C, −83.6 J, −83.6 J, 23.9 J, 83.6 J, 59.7 J. **4.** −0.0027%

Chapter 3

1. 7.94 kJ, **2.** 71.4%, 28.6% **4.** $-nRT/V^2$, nR/P, V/nR, -1, $1/T$, $1/P$, nRT/V.
5a. $p = 2$, $q = 3$. **6.** 53.6 MPa.

Chapter 4

1. 2.06 km (6790 ft). **2.** −1200 J, 2400 J, 0, −1200 J, 1200 J, 0, 3600 J, 0, 2400 J.
3. 20×10^5 Pa, 7.07×10^5 Pa, 0.6 m^3, 800 K, 282 K. **4.** $Q_{a\to b} = 3.5\,nRT_a$, $Q_{b\to c} = -3.105\,nRT_a$, 11.3%. **5.** 736 K, 971 K. **6.** 60%, 70%.

Chapter 5

1. 11.4 °C. **2.** 0.523 kg. **3.** 0.700 min, 5.58 min, 6.97 min, 37.6 min, 0.630 min.
4. 0.9882 atm, 0.9897 atm, 0.9895 atm, 0.9840 atm.

Chapter 6

1. -1.27×10^{-4} K. **2.** 0, 0, 317 K. **3.** 0.473 kg. **4.** −440 J, 0, −440 J, 1.31 g. **5.** 420 K.

Thermodynamics and Statistical Mechanics: An Integrated Approach, First Edition.
Robert J Hardy and Christian Binek.

Chapter 7

1. $2.9 \times 10^{-5}\,°\mathrm{F}$. **3.** 14.5. **4b.** 3.41 W-h. **6.** 29.4 kg, 5.95 kg.

Chapter 8

1. 79.4 K, 0.285 Pa, 1.28×10^{13} Pa. **2a.** 282.9 K, **2c.** −0.284 K. **3c.** $\Theta^{\ln} = -\infty$.

Chapter 9

1. $\frac{1}{2}\mu_0 JR^2\pi$, $\frac{1}{2}\mu_0 JR^2\pi$, 0. **2.** inexact, $-\frac{1}{2}K\ln(x^2+y^2)+c$, aT^2+bTV^3+c. **3.** $p=m-1$, $q=n+1$, $K=mL/(n+1)$. **5a.** $n(c+2bT)dT+(nRT/V)dV$. **6b.** $ncT-n^2bT/V+c$, **6c.** $c-nb/V$. **7a.** $dP=\alpha_V B_T dT-(B_T/V)dV$.

Chapter 10

2. $0.0279\,\mathrm{kcal\,K^{-1}}$. **3.** $2.87\,\mathrm{cal\,K^{-1}}$. **4.** $4.4\times10^{-4}\,\mathrm{cal\,K^{-1}}$. **6.** 0.361 m.

Chapter 11

2. $U=Mc_oT+b_oV[\ln(V/V_o)-1]+c$, $S=Mc_o\ln(T/T_o)+a_o(V-V_o)+c$, $B_S=b_o+TVa_o^2/(Mc_o)$. **3.** $u=c_VT-a/\mathrm{v}+c$, $s=c_V\ln T+R\ln(\mathrm{v}-b)+c$. **4.** $k=0$, $m=1$, $j=1$, $U=nc_oT+b_oT^2V+c$, $S=nc_o\ln T+nR\ln V+2b_oTV+c$, $đQ=(nc_o+2b_oTV)dT+(2b_oT^2+nRT/V)dV$. **5.** 95 °C.

Chapter 12

2. $F=\Phi+b_o\left[\ln\left(V/V_o\right)-1\right]V+Mc_oT\left[1-\ln\left(T/T_o\right)\right]-a_oT(V-V_o)-S_oT$. **3.** $G=n\left[c_PT\left(1-\ln\left(T/T_o\right)\right)+RT\ln(P/P_o)-s_oT+u_o\right]$.

Chapter 13

1b. $S=nc_V\ln T+nR\ln(V-b)+c$, $C_V=nc_V$, $B_T=(nRTV/(V-nb)^2)-2an^2/V^2$.

Chapter 14

1. $\kappa_{\mathrm{el}}=1+[1+b/T]/\varepsilon_o$. **3.** $\mu_0 CV\mathcal{H}^2/T^2$.

Chapter 15

1. 3300. **2.** $28.8\,\mathrm{g\,mol^{-1}}$, $512\,\mathrm{ms^{-1}}$. **3b.** 1.46. **4.** $0.221\,\mathrm{kcal\,kg^{-1}\,K^{-1}}$, $0.0939\,\mathrm{kcal\,kg^{-1}\,K^{-1}}$, $0.0288\,\mathrm{kcal\,kg^{-1}K^{-1}}$.

Chapter 16

1. $\sqrt{2.55\,k_BT/m}$. **2.** $e^{-\alpha}e^{-\beta\,\varepsilon_1}, e^{-\alpha}e^{-\beta\,\varepsilon_2}$. **3a.** 3.4 nm, 120 nm. **3b.** 34 nm, 0.12 mm.

Chapter 17

1. $\Omega = h^{-3N}I_{6N}6N\,E^{6N-1}\delta E$, $S = 6Nk_B \ln E + c$, $U = 6Nk_BT$. **2.** $|E_1 - \widehat{E}_1| = \widehat{E}_1/\sqrt{2cN}$.

Chapter 18

1. $\rho_j = e^{-\beta\varepsilon_j}\left(e^{-\beta\varepsilon_1} + e^{-\beta\varepsilon_2}\right)^{-1}$, $\overline{\varepsilon} = \left(e^{-\beta\varepsilon_1}\varepsilon_1 + e^{-\beta\varepsilon_2}\varepsilon_2\right)(e^{-\beta\varepsilon_1} + e^{-\beta\varepsilon_2})^{-1}$.
2. $U = -\dfrac{1}{2}N\varepsilon_o\dfrac{e^{+\varepsilon_o/2k_BT} - e^{-\varepsilon_o/2k_BT}}{e^{+\varepsilon_o/2k_BT} + e^{-\varepsilon_o/2k_BT}}$, $C = Nk_B(\varepsilon_o/k_BT)^2(e^{+\varepsilon_o/2k_BT} + e^{-\varepsilon_o/2k_BT})^{-2}$.

Chapter 19

1. $-3\,\varepsilon, -6\,\varepsilon$. **2.** $\frac{5}{2}Nk_BT, \frac{5}{2}nR$. **3.** $F = \Phi_{\text{st}} - 3Nk_BT\left[\ln 2\pi mk_BT - \frac{1}{2}\ln mk_o + \frac{1}{3}\ln C_N\right]$, $c_V = 3R$, $P = -\partial\Phi_{\text{st}}/\partial V + \frac{3}{2}Nk_BT\partial \ln k_o/\partial V$.

Chapter 20

1. 1, 9, 36, 9. **2.** $Z_N = \left(e^{+\frac{1}{2}\beta\varepsilon_o} - e^{-\frac{1}{2}\beta\varepsilon_o}\right)^{-3N}$, $U = 3N\left[\left(e^{\beta\varepsilon_o} - 1\right)^{-1} + \frac{1}{2}\right]$, $U = \frac{3}{2}N\hbar\sqrt{k_o/m}$. **3.** $\overline{l} = 1.2 \times 10^{-6}$m, $\lambda_{\text{th}} = 0.75 \times 10^{-10}$ m.

Chapter 21

1. $v_{\text{rms}} = 1.70\,\text{mm}\,\text{s}^{-1}$. **2.** $\rho = e^{-\frac{1}{2}k_o|\mathbf{r}|^2/k_BT}(k_o/2\pi k_BT)^{\frac{3}{2}}$. **3.** $\rho_A = (mg/k_BT)e^{-m_Ag\,z/k_BT}$, $\langle z\rangle_A = k_BT/m_Ag$. **4.** $(e^{-\mu\mathcal{H}/k_BT} + 1 + e^{+\mu\mathcal{H}/k_BT})^N$, $M_{\text{th}} = \dfrac{N\mu}{\mu_0}\dfrac{e^{+\mu\mathcal{H}/k_BT} - e^{-\mu\mathcal{H}/k_BT}}{1 + e^{+\mu\mathcal{H}/k_BT} + e^{-\mu\mathcal{H}/k_BT}}$.

Chapter 22

1. $\langle p_x\rangle = 0$, $\sigma_{p_x} = \sqrt{mk_BT}$. **2.** 3.3×10^5, 2.1×10^{-23} m^3, 34 nm. **3.** $\langle E\rangle = \frac{3}{2}Nk_BT$, $\langle E^2\rangle = \left(\frac{3}{2}N + 1\right)\frac{3}{2}N(k_BT)^2$, $\sigma_E = \sqrt{\frac{3}{2}N}k_BT$, $\sigma_E/\langle E\rangle = \left(\frac{3}{2}N\right)^{-\frac{1}{2}}$.

Chapter 23

1. $Z_N = C_N\left[(2\pi m/\beta)^{3/2}\,\frac{4}{3}\pi R^3\right]^N$, $F = -Nk_BT\ln\left[\left(C_N\right)^{1/N}(2\pi mk_BT)^{3/2}V\right]$.

Chapter 24

1a. $F = +k_B T \sum_j \ln\left(1 - e^{-\beta\hbar\omega_j}\right)$, **1c.** $P = \frac{1}{3}U/V$, **1d.** $P = \frac{2}{3}U/V$.

Chapter 25

1. $S = \left[\left(\mu/k_B T\right) - \frac{5}{2}\right](\Phi_{\text{th}}/T)$, $\overline{N} = -\Phi_{\text{th}}/k_B T$, $U = -\frac{3}{2}\Phi_{\text{th}}$, $U = \frac{3}{2}\overline{N}k_B T$.

Chapter 26

1. 16 distinguishable eigenstates, 6 fermion eigenstates, and 10 boson eigenstates.

Chapter 27

1. $\varepsilon_k = 3.12\,\text{eV}$, $\varepsilon_k = 3.36\,\text{eV}$. **2.** $T_c = 7.3 \times 10^{-9}\,\text{K}$.

Chapter 28

1. $E_{\text{th}} = NJz\left(\frac{1}{2} + \frac{e^{+\beta Jz} - e^{-\beta Jz}}{e^{+\beta Jz} + e^{-\beta Jz}}\right)$, $C_{\mathcal{H}} = 4Nk_B\left(e^{+Jz/k_BT} + e^{-Jz/k_BT}\right)^{-2}\left(\frac{Jz}{k_BT}\right)^2$.
2. $r_0 = 0.270\,\text{nm}$ for Ne; $r_0 = 0.331$ nm and Ar.

Index

Thermodynamics and Statistical Mechanics: An Integrated Approach, First Edition.
Robert J Hardy and Christian Binek.

Printed and bound by CPI Group (UK) Ltd, Croydon, CR0 4YY

27/10/2024

14580312-0002